Chemical Sensors for Volatile Organic Compound Detection

Chemical Sensors for Volatile Organic Compound Detection

Editors

Fanli Meng
Zhenyu Yuan
Dan Meng

Basel • Beijing • Wuhan • Barcelona • Belgrade • Novi Sad • Cluj • Manchester

Editors
Fanli Meng
Northeatern University
Shenyang
China

Zhenyu Yuan
Northeatern University
Shenyang
China

Dan Meng
Shenyang University of
Chemical Technology
Shenyang
China

Editorial Office
MDPI AG
Grosspeteranlage 5
4052 Basel, Switzerland

This is a reprint of articles from the Special Issue published online in the open access journal *Chemosensors* (ISSN 2227-9040) (available at: https://www.mdpi.com/journal/chemosensors/special_issues/VOC_sensors).

For citation purposes, cite each article independently as indicated on the article page online and as indicated below:

Lastname, A.A.; Lastname, B.B. Article Title. *Journal Name* **Year**, *Volume Number*, Page Range.

ISBN 978-3-7258-2245-4 (Hbk)
ISBN 978-3-7258-2246-1 (PDF)
doi.org/10.3390/books978-3-7258-2246-1

Contents

chemosensors

MDPI

Editorial

Chemical Sensors for Volatile Organic Compound Detection

Fanli Meng [1,2,3,4,*], Zhenyu Yuan [1] and Dan Meng [5]

1 College of Information Science and Engineering, Northeastern University, Shenyang 110819, China
2 Hebei Key Laboratory of Micro-Nano Precision Optical Sensing and Measurement Technology, Qinhuangdao 066004, China
3 National Frontiers Science Center for Industrial Intelligence and Systems Optimization, Northeastern University, Shenyang 110819, China
4 Key Laboratory of Data Analytics and Optimization for Smart Industry (Northeastern University), Ministry of Education, Shenyang 110819, China
5 College of Chemical Engineering, Shenyang University of Chemical Technology, Shenyang 110142, China
* Correspondence: mengfanli@ise.neu.edu.cn

The detection of volatile organic compounds (VOCs) is in high demand in various fields, such as environmental pollution monitoring, early disease screening, and food freshness assessment [1–3]. A variety of methods, including spectroscopic analysis [4], mass spectrometry [5], chromatographic analysis [6], electrochemical gas sensors [7], infrared gas sensors [8,9], and semiconductor gas sensors [10–12], have been extensively used for VOC detection. This is especially the case for semiconductor sensors thanks to their high sensitivity, fast response time, and cost-effectiveness [13–15].

Under the unremitting efforts of researchers, the research of semiconductor gas sensors has made a major breakthrough. There have been advanced research results with high sensitivity [16–19], a low minimum detection limit [20–23], room temperature sensing [24–27], and other advanced properties. Li et al. prepared s-Nb_2O_5 @ SnO_2 composite, and its response to 500 ppb acetone at 250 °C was 37 [28]. Xiao et al. prepared TiO_2 NCs-implanted $LaFeO_3$ nanomaterials, and its response to 100 ppm formaldehyde was 221.8 [29]. Hu et al. prepared Au-functionalized MoO3 nanoribbons, which realized the detection of formaldehyde at room temperature [30]. Compared with toluene, ethanol, methanol, and acetone, the sensitivity to formaldehyde is one to two orders of magnitude higher. Li et al. realized the identification of VOC gas at room temperature by using an ultraviolet light regulation method [31].

However, there are still problems to be solved from the perspective of large-scale commercial applications. Firstly, the low temperature response speed of the sensor needs to be improved. Sensors can detect the target gas at room temperature, thus reducing the explosion risk and power consumption. However, their response to the target gas is relatively slow, and the response curve cannot completely recover to the response baseline or is slow after the response. Secondly, the selectivity of semiconductor sensors has not been fundamentally improved. The sensitivity difference between target gas and non-target gas is increased by adjusting the material structure. Thirdly, the practical application ability of the sensor needs further verification. Most research papers only focus on the laboratory stage, and the related gas sensing performance research needs to be carried out with the help of high-purity air and high-precision instruments.

In view of the above difficulties, this Special Issue presents a comprehensive and detailed exploration of the latest achievements in VOC detection based on chemical sensors. Modifying sensitive materials is a practical approach to enhance the gas-sensing performance of semiconductor metal oxide sensors. Li et al. synthesized nickel-doped ZnO-sensitive materials with core–shell structures, doped at ratios of 0.5%, 1%, and 2%, which achieved the rapid detection of toluene [32]. Based on a doping ratio of 1%, the sensors exhibited a response of up to 210 for 100 ppm toluene, with a detection limit

Citation: Meng, F.; Yuan, Z.; Meng, D. Chemical Sensors for Volatile Organic Compound Detection. *Chemosensors* **2023**, *11*, 553. https://doi.org/10.3390/chemosensors11110553

Received: 31 October 2023
Accepted: 1 November 2023
Published: 3 November 2023

as low as 0.5 ppm. Research indicates a significant enhancement in sensor performance through the combined action of a narrower bandgap, higher specific surface area, and ion catalysis. San et al. designed a series of three-dimensional rGO-functionalized flower-shaped In_2O_3 structures, enabling the low-temperature and rapid detection of acetone. The working temperature of 150 °C and a fast response time of 3 s can be attributed to the introduction of an appropriate amount of graphene [33]. Scholars have tried to achieve breakthroughs in detection methods, data analysis, and sensor system design, aiming to further enhance the capability of detecting VOCs in real-world application scenarios. In terms of detection methods, Wang et al. synthesized $Pt/Ti_3C_2T_x$-CNT and employed cataluminescence technology to achieve the non-invasive and rapid detection of the lung cancer biomarker toluene at lower operating temperatures [34]. Regarding the construction of physical systems, Shen et al. developed a wireless luminescent sensor system with excellent luminescent characteristics, featuring high visible light intensity and a high signal-to-noise ratio [35]. Zhang et al. completed a comprehensive review of the detection of triethylamine using chemical sensors [36]. This paper commences with the fundamental characteristics of sensors and typical sensing mechanisms, providing a comprehensive summary of the latest advancements in enhancing the sensing performance of triethylamine sensors from various perspectives. It serves as a detailed entry point for readers interested in exploring VOC detection.

I would like to express my heartfelt thanks to all the authors, reviewers, and editors of this Special Issue, who have enriched its content and contributed to advancing the field of combustible volatile organic compound detection.

Conflicts of Interest: The authors declare no conflict of interest.

References

1. Bari, M.A.; Kindzierski, W.B. Ambient volatile organic compounds (VOCs) in Calgary, Alberta: Sources and Screening Health Risk Assessment. *Sci. Total Environ.* **2018**, *631–632*, 627–640. [CrossRef]
2. Zhao, Q.; Li, Y.; Chai, X.; Xu, L.; Zhang, L.; Ning, P.; Huang, J.; Tian, S. Interaction of Inhalable Volatile Organic Compounds and Pulmonary Surfactant: Potential Hazards of VOCs Exposure to Lung. *J. Hazard Mater.* **2019**, *369*, 512–520. [CrossRef]
3. Gong, Y.; Wei, Y.; Cheng, J.; Jiang, T.; Chen, L.; Xu, B. Health Risk Assessment and Personal Exposure to Volatile Organic Compounds (VOCs) in Metro Carriages-A Case Study in Shanghai. *China. Sci. Total Environ.* **2017**, *574*, 1432–1438. [CrossRef]
4. Wang, C.; Scherrer, S.; Hossain, D. Measurements of Cavity Ring Down Spectroscopy of Acetone in the Ultraviolet and Near-infrared Spectral Regions: Potential for Development of a Breath Analyzer. *Appl. Spectrosc.* **2004**, *58*, 784–791. [CrossRef]
5. Han, B.; Chen, J.; Zheng, L.; Zhou, T.; Li, J.; Wang, X.; Wang, J. Development of an Impurity-profiling Method for Source Identification of Spilled Benzene Series Compounds by Gas Chromatography with Mass Spectrometry: Toluene as a Case Study. *J. Sep. Sci.* **2015**, *38*, 3198–3204. [CrossRef]
6. De Oliveira, D.P.; De Siqueira, M.E.P.B. A Simple and Rapid Method for Urinary Acetone Analysis by Headspace/gas Chromatography. *Quim. Nova* **2007**, *30*, 1362–1364. [CrossRef]
7. Hao, X.; Wu, D.; Wang, Y.; Ouyang, J.; Wang, J.; Liu, T.; Liang, X.; Zhang, C.; Liu, F.; Yan, X.; et al. Gas sniffer (YSZ-based electrochemical gas phase sensor) toward Acetone Detection. *Sens. Actuators B Chem.* **2019**, *278*, 1–7. [CrossRef]
8. Dinh, T.-V.; Choi, I.-Y.; Son, Y.-S.; Kim, J.-C. A review on Non-dispersive Infrared Gas Sensors: Improvement of Sensor Detection Limit and Interference Correction. *Sens. Actuators B Chem.* **2016**, *231*, 529–538. [CrossRef]
9. Xu, Y.; Li, H.; Zhang, X.; Jin, H.; Jin, Q.; Shen, W.; Zou, J.; Deng, S.; Cheung, W.; Kam, W.; et al. Light-Regulated Electrochemical Reaction Assisted Core-Shell Heterostructure for Detecting Specific Volatile Markers with Controllable Sensitivity and Selectivity. *ACS Sens.* **2019**, *4*, 1081–1089. [CrossRef]
10. Fergus, J.W. Perovskite Oxides for Semiconductor-based Gas Sensors. *Sens. Actuators B Chem.* **2007**, *123*, 1169–1179. [CrossRef]
11. Chen, W.Y.; Jiang, X.; Lai, S.N.; Peroulis, D.; Stanciu, L. Nanohybrids of a MXene and Transition Metal Dichalcogenide for Selective Detection of Volatile Organic Compounds. *Nat. Commun.* **2020**, *11*, 1302. [CrossRef]
12. Cho, I.; Sim, Y.C.; Cho, M.; Cho, Y.H.; Park, I. Monolithic Micro Light-Emitting Diode/Metal Oxide Nanowire Gas Sensor with Microwatt-Level Power Consumption. *ACS Sens.* **2020**, *5*, 563–570. [CrossRef]
13. Li, Z.; Yu, J.; Dong, D.; Yao, G.; Wei, G.; He, A.; Wu, H.; Zhu, H.; Huang, Z.; Tang, Z. E-nose based on a high-integrated and low-power metal oxide gas sensor array. *Sens. Actuators B Chem.* **2023**, *380*, 133289. [CrossRef]
14. Ji, H.; Liu, Y.; Zhang, R.; Yuan, Z.; Meng, F. Detection and Recognition of Toluene and Butanone in Mixture based on SnO_2 Sensor via Dynamic Transient and Steady-state Response Analysis in Jump Heating Voltage Mode. *Sens. Actuators B Chem.* **2023**, *376*, 132969. [CrossRef]

15. Ji, H.; Liu, Y.; Zhu, H.; Zhang, H.; Yuan, Z.; Meng, F. Interference Suppression Strategies for Trace Minor Component of Semiconductor Gas Sensor based on Temperature Modulation Mode. *Sens. Actuators B Chem.* **2023**, *388*, 133874. [CrossRef]
16. Liu, Y.; Yao, S.; Yang, Q.; Sun, P.; Gao, Y.; Liang, X.; Liu, F.; Lu, G. Highly Sensitive and Humidity-independent Ethanol Sensors based on In2O3 Nanoflower/SnO_2 Nanoparticle Composites. *RSC Adv.* **2015**, *5*, 52252–52258. [CrossRef]
17. Li, B.; Li, M.; Meng, F.; Liu, J. Highly Sensitive Ethylene Sensors using Pd Nanoparticles and rGO Modified Flower-like Hierarchical Porous α-Fe_2O_3. *Sens. Actuators B Chem.* **2019**, *290*, 396–405. [CrossRef]
18. Zhu, K.M.; Ma, S.Y. Preparations of Bi-doped SnO2 Hierarchical Flower-shaped Nanostructures with Highly Sensitive HCHO Sensing Properties. *Mater. Lett.* **2019**, *236*, 491–494. [CrossRef]
19. Korotcenkov, G.; Cho, B.K. Engineering Approaches for the Improvement of Conductometric Gas Sensor Parameters. *Sens. Actuators B Chem.* **2013**, *188*, 709–728. [CrossRef]
20. Yuan, H.; Aljneibi, S.A.A.A.; Yuan, J.; Wang, Y.; Liu, H.; Fang, J.; Tang, C.; Yan, X.; Cai, H.; Gu, Y.; et al. ZnO Nanosheets Abundant in Oxygen Vacancies Derived from Metal-Organic Frameworks for ppb-Level Gas Sensing. *Adv. Mater.* **2019**, *31*, e1807161. [CrossRef]
21. Suematsu, K.; Harano, W.; Oyama, T.; Shin, Y.; Watanabe, K.; Shimanoe, K. Pulse-Driven Semiconductor Gas Sensors Toward ppt Level Toluene Detection. *Anal. Chem.* **2018**, *90*, 11219–11223. [CrossRef] [PubMed]
22. Jo, Y.K.; Jeong, S.Y.; Moon, Y.K.; Jo, Y.M.; Yoon, J.W.; Lee, J.H. Exclusive and Ultrasensitive Detection of Formaldehyde at Room Temperature Using a Flexible and Monolithic Chemiresistive Sensor. *Nat. Commun.* **2021**, *12*, 4955. [CrossRef] [PubMed]
23. Liu, M.; Ji, J.; Song, P.; Liu, M.; Wang, Q. α-Fe2O3 Nanocubes/Ti3C2Tx MXene Composites for Improvement of Acetone Sensing Performance at Room Temperature. *Sens. Actuators B Chem.* **2021**, *349*, 130782. [CrossRef]
24. Liu, M.; Wang, Z.; Song, P.; Yang, Z.; Wang, Q. Flexible MXene/rGO/CuO Hybrid Aerogels for High Performance Acetone Sensing at Room Temperature. *Sens. Actuators B Chem.* **2021**, *340*, 129946. [CrossRef]
25. Liu, J.; Hu, Z.; Zhang, Y.; Li, H.Y.; Gao, N.; Tian, Z.; Zhou, L.; Zhang, B.; Tang, J.; Zhang, J.; et al. MoS_2 Nanosheets Sensitized with Quantum Dots for Room-Temperature Gas Sensors. *Nano-Micro Lett.* **2020**, *12*, 59. [CrossRef] [PubMed]
26. Li, Z.; Li, H.; Wu, Z.; Wang, M.; Luo, J.; Torun, H.; Hu, P.; Yang, C. Advances in Designs and Mechanisms of Semiconducting Metal Oxide Nanostructures for High-precision Gas Sensors Operated at Room Temperature. *Mater. Horiz.* **2019**, *6*, 470–506. [CrossRef]
27. Zhang, J.; Liu, X.; Neri, G.; Pinna, N. Nanostructured Materials for Room-Temperature Gas Sensors. *Adv. Mater.* **2015**, *28*, 795–831. [CrossRef]
28. Li, C.; Kim, K.; Fuchigami, T.; Asaka, T.; Kakimoto, K.; Masuda, Y. Acetone Gas Sensor Based on Nb_2O_5@SnO_2 hybrid Structure with High Selectivity and Ppt-level Sensitivity. *Sens. Actuators B Chem.* **2023**, *393*, 134144. [CrossRef]
29. Xiao, C.; Tang, Z.; Ma, Z.; Gao, X.; Wang, H.; Jia, L. High Performance Porous LaFeO3 Gas Sensor with Embedded p-n Junctions Enabling Ppb-level Formaldehyde Detection. *Sens. Actuators B Chem.* **2023**, *397*, 134670. [CrossRef]
30. Hu, Z.; Tan, L.; Fu, X.; Fan, Y.; Sun, M.; Zhou, D.; Wang, Z.; Li, Z.; Xu, S.; Gu, H. Au-functionalized MoO_3 Nanoribbons Towards Rapid and Selective Formaldehyde Gas Sensing at Room Temperature. *Results Phys.* **2023**, *49*, 106525. [CrossRef]
31. Li, M.; Chang, J.; Deng, Z.; Mi, L.; Kumar, M.; Wang, S.; He, Y.; Meng, G. Discriminating Gas Molecules at Room Temperature by UV Light Modulation (ULM) of Nonselective Metal Oxide Sensors. *Sens. Actuators B Chem.* **2023**, *378*, 133115. [CrossRef]
32. Li, Z.H.; Li, S.J.; Song, Z.J.; Yang, X.L.; Wang, Z.Y.; Zhang, H.; Guo, L.L.; Sun, C.X.; Liu, H.Y.; Shao, J.K.; et al. Influence of Nickel Doping on Ultrahigh Toluene Sensing Performance of Core-Shell ZnO Microsphere Gas Sensor. *Chemosensors* **2022**, *10*, 327. [CrossRef]
33. San, X.G.; Zhang, Y.; Zhang, L.; Wang, G.S.; Meng, D.; Cui, J.; Jin, Q. One-Step Hydrothermal Synthesis of 3D Interconnected rGO/In2O3 Heterojunction Structures for Enhanced Acetone Detection. *Chemosensors* **2022**, *10*, 270. [CrossRef]
34. Wang, H.Y.; Shi, X.Q.; Liu, F.; Duan, T.M.; Sun, B. Non-Invasive Rapid Detection of Lung Cancer Biomarker Toluene with a Cataluminescence Sensor Based on the Two-Dimensional Nanocomposite Pt/Ti3C2Tx-CNT. *Chemosensors* **2022**, *10*, 333. [CrossRef]
35. Shen, X.Y.; Shi, G.L.; Zhang, Y.X.; Weng, S.Z. Wireless Volatile Organic Compound Detection for Restricted Internet of Things Environments Based on Cataluminescence Sensors. *Chemosensors* **2022**, *10*, 179. [CrossRef]
36. Zhang, H.; Guo, Y.H.; Meng, F.L. Metal Oxide Semiconductor Sensors for Triethylamine Detection: Sensing Performance and Improvements. *Chemosensors* **2022**, *10*, 231. [CrossRef]

chemosensors

MDPI

Article

Porous Pb-Doped ZnO Nanobelts with Enriched Oxygen Vacancies: Preparation and Their Chemiresistive Sensing Performance

Kai-Ge Zheng [1,2], Tian-Yu Yang [1,2] and Zheng Guo [1,2,*]

1 Information Materials and Intelligent Sensing Laboratory of Anhui Province, Institutes of Physical Science and Information Technology, Anhui University, Hefei 230601, China; q19201115@stu.ahu.edu.cn (K.-G.Z.); q21201024@stu.ahu.edu.cn (T.-Y.Y.)

2 Key Laboratory of Structure and Functional Regulation of Hybrid Materials (Anhui University), Ministry of Education, Hefei 230601, China

* Correspondence: zhguo@ahu.edu.cn

Abstract: Among various approaches to improve the sensing performance of metal oxide, the metal-doped method is perceived as effective, and has received great attention and is widely investigated. However, it is still a challenge to construct heterogeneous metal-doped metal oxide with an excellent sensing performance. In the present study, porous Pb-doped ZnO nanobelts were prepared by a simply partial cation exchange method, followed by in situ thermal oxidation. Detailed characterization confirmed that Pb was uniformly distributed on porous nanobelts. Additionally, it occupied the Zn situation, not forming its oxides. The gas-sensing measurements revealed that 0.61 at% Pb-doped ZnO porous nanobelts exhibited a selectively enhanced response with long-term stability toward n-butanol among the investigated VOCs. The relative response to 50 ppm of n-butanol was up to 47.7 at the working temperature of 300 °C. Additionally, the response time was short (about 5 s). These results were mainly ascribed to the porous nanostructure, two-dimensional belt-like morphology, enriched oxygen vacancies and the specific synergistic effect from the Pb dopant. Finally, a possible sensing mechanism of porous Pb-doped ZnO nanobelts is proposed and discussed.

Keywords: zinc oxide; lead doping; partial cation exchange reaction; n-butanol; gas sensing

Citation: Zheng, K.-G.; Yang, T.-Y.; Guo, Z. Porous Pb-Doped ZnO Nanobelts with Enriched Oxygen Vacancies: Preparation and Their Chemiresistive Sensing Performance. *Chemosensors* **2022**, *10*, 96. https://doi.org/10.3390/chemosensors10030096

Academic Editor: Andrea Ponzoni

Received: 21 January 2022
Accepted: 24 February 2022
Published: 1 March 2022

1. Introduction

Due to the outstanding merits of low cost, easy fabrication, high-sensing performance and compatibility with modern electronic devices, resistive-type metal oxide semiconductor (MOS) gas sensors have been widely applied to detect VOCs in various fields, such as human health, environment monitoring, food processing and industrial manufacture. Compared with p-type oxide semiconductors, n-type oxide semiconductors have received more attention since the discovery of oxide semiconductor-based gas sensors in the 1960s, because of their special sensing mechanism and highly relative response [1–4]. Among the n-type MOSs, ZnO as a typical one has been widely applied and investigated in gas sensors because of its high electron mobility, non-toxicity, good stability and simple synthesis route [5]. However, the property of pure ZnO was not satisfied, resulting in the need to improve the intrinsic performance through many ingenious methods, depending on the practical situation of application for further eliminating false-positive or interfering signals and simplifying the design of electronic nose system.

Numerous diverse systematic researches showed that the key sensing parameters of resistive-type MOS gas sensors could be enhanced by the composite with other MOSs, resulting in a controlled chemical affinity toward a specific gas, modified acid-base properties and the tuned catalytic activity [6]. The technology of doping ions in MOSs was also widely applied in the practical field of improving performance, which was responsible for forming oxygen vacancies and changing electronic structures of metal oxides [7].

Oxygen vacancies were one of the intrinsic defects existing in the actual semiconductor samples and provided an effect similar to n-type dopants. Importantly, they increased negatively charged oxygen ions on the surface by their surrounding electron-deficient region. Meanwhile, the doping method primarily affected the shift of the Fermi level and the positions of the energy band, which could remarkably ameliorate the performance of MOSs by altering the energy band [8]. Recently, lead (Pb) doping was used in many application areas of semiconductors, due to its larger ionic radius and specific catalytic activity, which causes the relaxing of tensile strain and an improvement to the sensitization performance [9]. Furthermore, the desirable attributes brought by the doping of Pb ions included a good working stability at a high temperature (due to the high atomic number and stability of Pb), a high resistivity to decrease the baseline current of the sensors (favorable for enhancing the response) and an abundant availability in nature [10]. In addition, the desirable morphologies and structures of sensing materials were also the key points in the improvement of the sensing performance. Among them, the low-dimensional nanomaterial and porous structure were generally applied to practical gas-sensing, because the increased specific surface area always meant a larger resistance variation of sensing materials and more efficient reactions at the surface [11,12].

Herein, Pb-doped ZnO porous nanobelts, owning to their great crystal form and uniformly two-dimensional porous nanostructures, were successfully synthesized via a simply partial cation exchange method with in situ thermal oxidation. Different amounts of Pb-doped ZnO were carefully prepared and, subsequently, the ZnO porous nanobelts with optimal amounts of Pb were systematically explored in relation to their gas-sensing performance to n-butanol. The synthetic strategy presented here could be generally extended to design other semiconductor metal oxide gas sensors with outstanding performances.

2. Experimental Section

2.1. Chemical Reagents

Analytical grade ethanol, ethylene glycol (EG), $N_2H_4 \cdot H_2O$ (mass fraction of N_2H_4, 85%), polyvinylpyrrolidone (PVP, K30), Se powders, $ZnCl_2$ and $Pb(NO_3)_2$ were purchased from Shanghai Chemical Reagent Co., Ltd. (Shanghai, China) without further purification. Milli-Q water with a resistivity of 18.25 MΩ cm was used in the preparation of aqueous solutions.

2.2. Preparation of Pb-Doped ZnO Porous Nanobelts

The whole preparation process is demonstrated in Scheme 1. $ZnSe \cdot 0.5N_2H_4$ precursor nanobelts were firstly synthesized as subsequent templates via a convenient hydrothermal method that was reported in our previous reports [13,14]. In a typical synthetic process, a brown homogeneous solution is formed by the dissolution of 0.4 mmol Se powder into 20 mL of aqueous $N_2H_4 \cdot H_2O$. Then, the $ZnCl_2$ aqueous solution, which was prepared using 0.2 mmol of $ZnCl_2$ and 20 mL of deionized water, was added, drop by drop, to the above brown homogeneous solution under vigorous magnetic stirring. After stirring for about 1 h, the obtained solution was transferred into a 50 mL Teflon-lined stainless steel autoclave and put into an oven at 180 °C. After 12 h, the autoclave was cooled down to room temperature naturally. Finally, the precursor of $ZnSe \cdot 0.5N_2H_4$ nanobelts was obtained by washing them several times with distilled water. Then, the achieved precursor nanobelts were all dispersed into a 20 mL EG solution containing 0.1 g of poly(vinylpyrrolidone) (PVP, K30). Afterwards, a 5 mL EG solution with a certain amount of $Pb(NO_3)_2$ was added, drop by drop, into the homogenous solution. After continuously stirring for 2 h, a dark color uniform solution was formed. Subsequently, the solution was transferred into a 50 mL Teflon-lined stainless steel autoclave and was placed into an oven at 70 °C for 12 h. Finally, the dark green Pb-doped precursor nanobelts were achieved after washing and centrifuging with ethanol and distilled water several times to remove the excess impurities. Through manipulating the added amount of $Pb(NO_3)_2$ (0.001 mmol, 0.002 mmol, 0.004 mmol and 0.006 mmol), the actual doped ratios of 0.44 at%, 0.61 at%, 1.39 at% and 2.01 at% were achieved and identified by XPS. In the following, the as-prepared samples were thermally oxidized in air. Referring to the thermogravimetric analysis of our previous report [15], the samples were firstly heated at a low heating rate of 2 °C/min to 350 °C and maintained for 30 min to totally remove the hydrazine ligand. The temperature was continuously increased to 500 °C at the heating rate of 1 °C/min, for a better crystal structure and morphology in the oxidation process of the doped nanobelts and kept for 30 min. Then, the cooling process was also slow at the rate of 2 °C/min to 100 °C to maintain the characteristic structure of the samples. Finally, after naturally cooling down to room temperature, porous Pb-doped ZnO porous nanobelts were achieved.

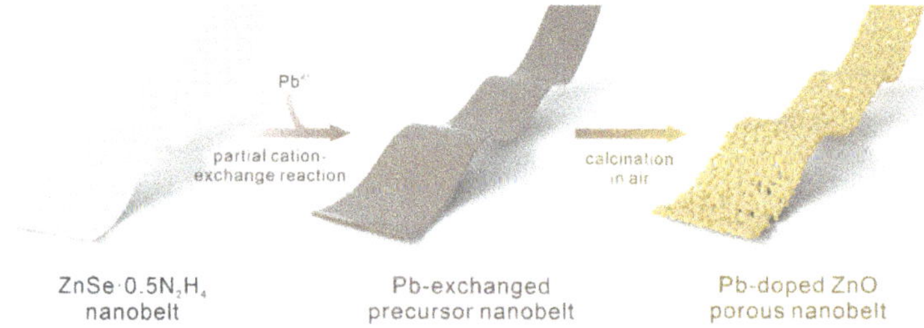

Scheme 1. Synthetic process of porous Pb-doped ZnO nanobelts.

2.3. *Fabrication and Measurements of Gas-Sensing Devices*

An assembly technique combined with in situ calcination was employed to fabricate the gas-sensing devices with a stable and uniform sensing film of Pb-doped ZnO porous nanobelts. As shown in Figure S1a, the square alumina ceramic slices (3 mm × 3 mm × 0.25 mm, length × width × thickness) with a pair of Au electrodes with a 0.3 mm gap on the front side and an Ru-based thick film resistor on the back side were set as the test substrates. The effective area of the sensing film was about 0.42 mm^2. Firstly, a homogeneous solution was formed by dispersing a small amount of the as-prepared Pb-exchanged precursor nanobelts in moderate ethanol and dropped on the front side of the alumina ceramic slice to form a thin film. After this was repeated several times, the front side with the Au electrodes was covered with a dense and uniform Pb-exchanged precursor nanobelt film. Then, they were in situ transformed on the surface of the alumina ceramic slices into a uniform Pb-doped porous ZnO nanobelt sensing film following the above-mentioned thermal oxidation approach.

The sensing measurement of the Pb-doped ZnO porous nanobelts was performed in a homemade gas sensor testing system, as shown in Figure S1b, including a Keithley 6487 picoammeter/voltage sourcemeter, connecting with two Au electrodes to act as both the current reader and voltage source; a heating voltage modulator was employed to offer the working temperature of the sensor by applying a specific voltage on the Ru-based thick film resistor. All gas-sensing measurements were performed in dry air and the certain concentration gas to be tested was injected by a minitype syringe. The relative response of the fabricated gas sensors was defined as $S = R_a/R_g$, where R_a is the resistance of the sensing film in dry air and R_g is the resistance in dry air containing the detected gases. Furthermore, the response time (t_{res}) and recovery time (t_{rec}) were defined as the times required to reach 90% of the final equilibrium value, respectively.

2.4. *Characterization*

X-ray photoelectron spectroscopy (XPS) analyses of the samples was conducted on a Thermo Scientific ESCALAB Xi+ X-ray Photoelectron Spectrometer microprobe (Thermo ESCALAB 250Xi, Waltham, MA, USA). The energy scale was internally calibrated by referencing the binding energy (Eb) of the C 1s peak at 284.60 eV for contaminated carbon. The microstructure and morphology of the as-prepared samples were characterized by transmission electron microscopy (TEM, Tecnai TF-20, 200 kV) and field-emission scanning electron microscopy (FESEM, Quanta 200 FEG, FEI Company, Hillsboro, OR, USA). The crystal components and structures were determined with an X-ray power diffractometer (smart Lab 9KW) using Cu Ka radiation (radiation λ = 1.5418 Å).

3. Results and Discussion

3.1. Preparation and Characterization of the Pb-Doped ZnO Porous Nanobelts

The morphologies of all the achieved nanobelts were first investigated by SEM, as shown in Figure 1. The as-prepared precursors of $ZnSe{\cdot}0.5N_2H_4$ in Figure 1a can be identified as homogeneous nanobelts with a relatively large length–width ratio and smooth surface. After being exchanged with 0.61 at% Pb^{2+} cations, their morphologies, displayed in Figure 1b, are still maintained without any protrusions or other obviously heterogeneous areas. After calcination in air, they are all transformed into a belt-like porous nanostructure, which can be seen in Figure 1c,d, respectively. This morphological evolution is caused by the removal of the N_2H_4 ligand and the oxidation of the Se element following subsequent sublimation at a high temperature. For Pb^{2+}-exchanged precursor nanobelts, the corresponding formed product was a porous Pb-doped ZnO nanobelt. Furthermore, the precursor exchanged with other amounts of Pb^{2+} cations and their corresponding annealed products were also investigated and a similar phenomenon was also observed, as shown in Figure S2.

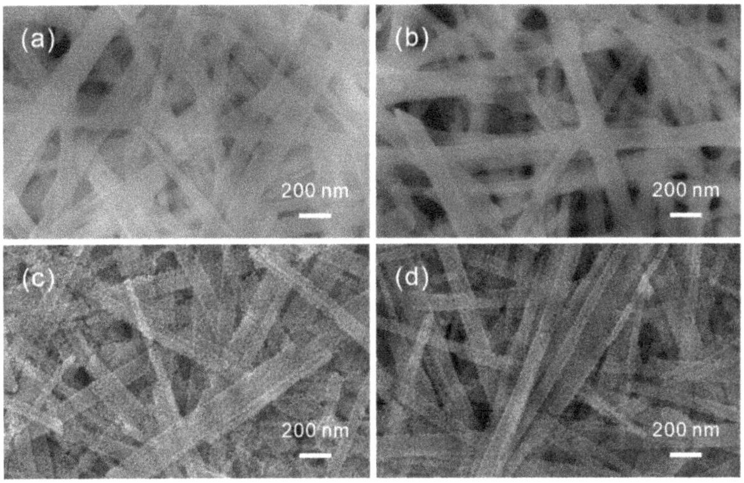

Figure 1. SEM images of $ZnSe{\cdot}0.5N_2H_4$ precursor nanobelts before (**a**) and after (**c**) thermal oxidation, and 0.61 at% Pb-exchanged $ZnSe{\cdot}0.5N_2H_4$ nanobelts before (**b**) and after (**d**) thermal oxidation.

To illuminate the microstructure of the as-prepared Pb^{2+}-exchanged precursor nanobelts before and after calcination in air, the characterization of their TEMs was performed. Similar to the $ZnSe{\cdot}0.5N_2H_4$ precursor nanobelts, the Pb^{2+}-exchanged precursor nanobelts are also not stable and easily decomposed under the high-power electron beam. This phenomenon can be seen from TEM images shown in Figure S3a,b. Notably, after the exchange with Pb^{2+} cations, the achieved nanobelts are uniformly distributed with the Pb element in addition to Zn and Se elements, which can be seen from the elemental mapping shown in Figure S3c–f. From the typical TEM image of a single nanobelt shown in Figure 2a, the as-prepared porous Pb-doped ZnO nanobelts are constructed by numerous interconnected nanocrystal grains with spontaneously generated micropores and nanopores. Evidently, this special nanostructure is favorable for the diffusion of gas molecules, resulting in an enhanced gas-sensing performance. Further crystal information of the Pb-doped ZnO porous nanobelt was confirmed by the high-resolution TEM image in Figure 2b, in which the lattice fringes can clearly be observed with a lattice spacing of 0.247 nm and 0.281 nm, corresponding to the (101) and (100) planes of hexagonal zinc oxide, respectively, with no other crystalline phases. The compositional elemental mapping patterns of Zn, Pb and O in the high-angle annular dark field (HAADF) image of the Pb-doped ZnO porous nanobelt in Figure 2c are displayed in Figure 2d–f, revealing that they are uniformly distributed along the nanobelt. From the EDX spectrum of the obtained Pb-doped ZnO porous nanobelts shown in Figure S4, weak Pb peaks are observed in addition to the strong peaks of Zn and O. The Cu peak most likely originates from the Cu grid.

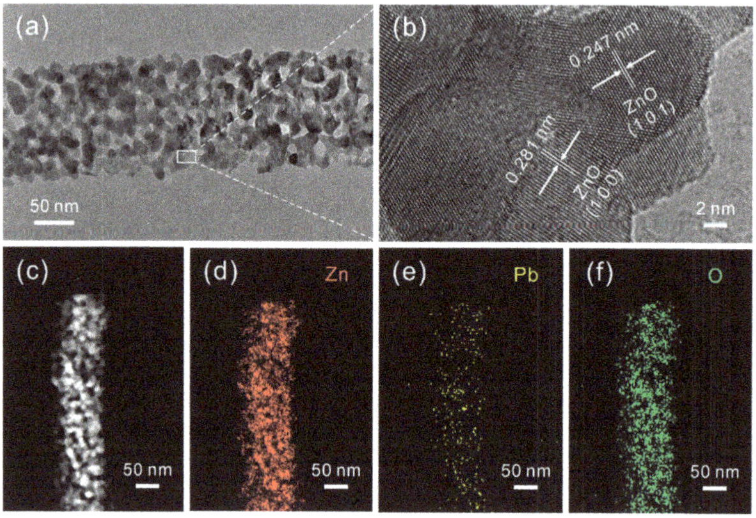

Figure 2. (**a**) TEM image of a single Pb-doped ZnO porous nanobelt, (**b**) its HRTEM image corresponds to the marked area, (**c**) its HAADF image and the corresponding elemental mapping patterns of (**d**) Zn, (**e**) Pb and (**f**) O.

Afterwards, their crystal structures were examined by XRD measurements and all intensities of the peaks were normalized. As shown in Figure 3a, clearly all emerged diffraction peaks of the above-mentioned samples are completely indexed to a hexagonal wurtzite ZnO phase (JCPDS#36-1451) without observing any diffraction peaks related to the Pb compound or other crystalline phases [16]. Additionally, the (100) diffraction peak of ZnO located at the high-resolution XRD patterns shifted with the addition of Pb, as presented in Figure 3b. By increasing the doped amount of Pb, the position of the (100) diffraction peaks shifted to lower angles. This phenomenon confirms that the Pb element, identified as Pb^{2+} by the following XPS analysis, was indeed incorporated into the ZnO lattice and occupied the hexagonal Zn^{2+} cation site [17]. Further lattice refinement of the XRD data displayed in Table S1 revealed the detailed change of their crystal lattices. Calculated by Bragg's law ($n\lambda = 2d\sin\theta$), the interplanar distance (d) values belonging to (100) peaks increase with the raising of Pb concentration, and the lattice parameters (a, c) are similarly expanded. This result is mainly caused by the larger ionic radii of Pb^{2+} (0.119 nm) compared with Zn^{2+} (0.074 nm) [18,19]. It implies the successful substitution of Pb ions for Zn ions, which is consistent with the previous reports [20,21].

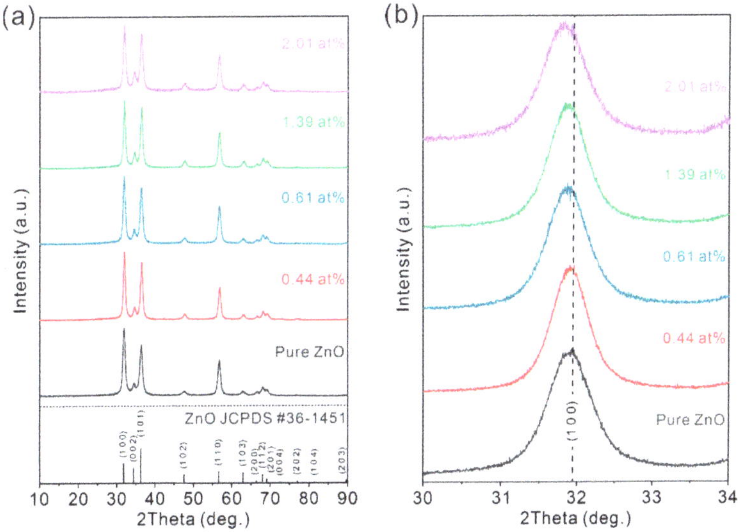

Figure 3. (**a**) XRD patterns of the as-prepared pure ZnO and 0.44 at%, 0.61 at%, 1.39 at% and 2.01 at% Pb-doped ZnO, and (**b**) their corresponding high-resolution XRD patterns at the main diffraction peak of (100).

The composition and chemical states of the as-prepared Pb^{2+}-exchanged precursor nanobelts before and after calcination in air were further investigated by XPS measurements. As the survey spectra shows in Figure 4a, the evident peaks are marked with their corresponding energy level [22]. It identifies the presence of Zn, Se and N elements before calcination and the absence of Se and N elements after calcination, regardless of whether the $ZnSe{\cdot}0.5N_2H_4$ precursor nanobelts were exchanged with Pb or not. This result is ascribed to the total oxidation of the selenide and the entire removal of the hydrazine ligand. However, it is difficult to find the distinct peak of Pb, which is due to the relatively low concentration of Pb^{2+} cation in samples III and IV. Therefore, the high-resolution XPS spectra of Pb 4f in 0.61 at% Pb-doped ZnO porous nanobelts is illustrated in Figure 4b. The two peaks at about 138.71 eV and 143.45 eV, of which the peak area ratio was approximately 4:3, were observed and assigned as Pb $4f_{7/2}$ and Pb $4f_{5/2}$ orbitals, respectively [23]. It exactly matches the previous report, concerning other possible valence states, such as PbO_2 and Pb_3O_4 [24–26]. Meanwhile, the peak located here at about 139.56 eV is recognized as Zn 3s, presenting a high intensity because of the large atomic ratio between the Zn and Pb elements [27]. As shown in Figure 4c, the high-resolution Zn 2p XPS spectra of 0.61 at% Pb-doped ZnO porous nanobelts revealed that two distinct peaks at 1044.12 eV and 1021.1 eV were attributed to Zn $2p_{1/2}$ and Zn $2p_{3/2}$, respectively [28]. Compared with the pure ZnO porous nanobelts displayed in Figure S5, the two characteristic peaks were clearly shifted 0.3 eV to high bond locations after doping with Pb due to the special electronegativity and effective charge movement between the two ions. This illustrates that the band structure of ZnO was affected by the introduction of the dopant of Pb [29]. Figure 4d displays the high-resolution O 1s XPS spectra of 0.61 at% Pb-doped ZnO porous nanobelts. Here, its peak can be deconvoluted into four types of peaks: the peak of lattice O of ZnO (O_{Zn}) at around 530.2 eV, the peak situated at 530.37 eV represents the lattice O of PbO (O_{Pb}), the peak located at around 530.66 eV attributed to the oxygen vacancy (O_{va}), and the peak at a binding energy of 531.68 eV assigned to the adsorbed oxygen (O_{ad}) [30]. The similar analysis procedure of deconvolution and fitting was also applied to the high-resolution O 1s XPS spectra of pure ZnO porous nanobelts in Figure 4e. The ratios of the different kinds of oxygen species previously stated, and their position of binding energy, are listed in Table S2. It should be noted that the corresponding characteristic peaks also have some same directional shifts to a lower binding energy after calcination, due to the different chemical environments of O in different samples. It is well known that the original lattice oxygen in sensing materials is stable and has no influence on the gas sensing properties, and a higher content of O_{ad} can enhance the sensing performance [31,32]. Even though the ratio of Pb to Zn was only 0.61 at% in reality, the increased proportion of O_{va} and O_{ad} was much greater than that. Then, the 0.61 at% Pb-doped ZnO porous nanobelts that obtained a larger ratio of O_{ad}, partially caused by the increased active centers generated from the higher concentration of O_{va}, should possess an enhanced sensing performance.

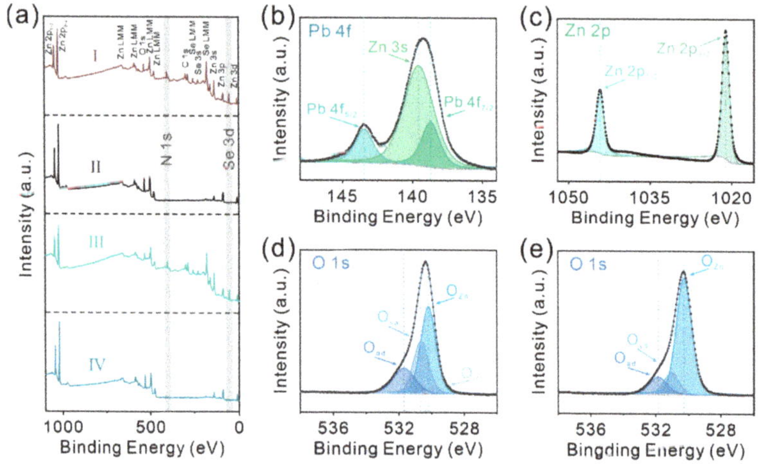

Figure 4. (**a**) The survey XPS spectra of the $ZnSe{\cdot}0.5N_2H_4$ precursor nanobelts before (I) and after (II) thermal oxidation, and precursor nanobelts partially exchanged with 0.61 at% of Pb^{2+} cations before (III) and after (IV) thermal oxidation. The high resolution of XPS spectra in (**b**) Pb 4f, (**c**) Zn 2p and (**d**) O 1s for 0.61 at% Pb-doped ZnO porous nanobelts. (**e**) High-resolution XPS spectrum of O 1s for pure ZnO.

3.2. *The Sensing Performance of Pb-Doped ZnO Porous Nanobelts*

To study the applicable gas-sensing properties of the as-synthesized samples with the dopant of Pb, as well as the effect of oxygen vacancy during the sensing process, the gas-sensing devices based on the porous ZnO nanobelts with different doped amounts of Pb were carefully fabricated following the above-described experimental details and their gas-sensing properties were also provided in a detailed assessment. Figure S6 exhibits the real-time response curves of porous ZnO nanobelts doped with different amounts of Pb toward 50 ppm of n-butanol at different working temperature. Usually, a higher working temperature is beneficial for the reaction of VOCs to the adsorbed oxygen species on the surface of sensing materials, as well as the desorption of them from the surface of the sensors. By increasing the working temperature, the response and recovery times both evidently decreased for the sensor fabricated with porous ZnO nanobelts doped with different amounts of Pb. This could be explained by the fact the that high working temperature accelerated the transport of electrons among the composite and between the sensing material and the tested gas, thus resulting in the shortened response and recovery times [33]. Figure 5a shows the relative response of porous ZnO nanobelts doped with different amounts of Pb toward 50 ppm of n-butanol at different working temperatures. For all samples, the response to the equal concentration of n-butanol was regularly influenced, according to the change in the working temperature. For the samples doped with Pb, the response values of all the samples increased first with the increase in the temperature and reached the maximum at 300 °C. By continuously increasing the working temperature, the response value decreases. This phenomenon was very common in the field of resistance-type gas sensors and could be reasonably interpreted as follows [34]. At a low temperature, the gas molecules did not have enough energy to overcome the energy barrier to react with the oxygen species adsorbed on the materials surface, and the diffusion process was slow. Correspondingly, increasing the working temperature could provide extra energy to boost the reaction between the tested gas molecules and the adsorbed oxygen species. However, more was not always better. Overmuch energy from further increasing the working temperature made the gas molecules so active that they would escape from the surface of the sensing materials before the sensing process; neither were the sensing materials in a stable condition for the reaction. Meanwhile, the sensing response would also be affected by the ion doping as well as the working temperature [35]. With the increased dopants of Pb, more oxygen vacancies were formed, suggesting that more VOC molecules can be easily absorbed and activated on the surface of porous nanobelts. Hence, the sensing performance was enhanced. The 0.61% Pd-doped ZnO exhibited the best response at the working temperature of 300 °C. When its doped amount was large, the excess Pb ions occupied the active sites on the surface of the sensing materials, which is disadvantageous for the gas adsorption and causes the degradation of the gas sensing response. The optimal working temperature was at 300 °C for the Pb-doped ZnO nanobelts, which is lower than 400 °C of the optimal value for pure ZnO porous nanobelts. Hence, the 0.61% doped amount and the operating temperature of 300 °C were selected as the optimal test conditions, where the highest response of 47.7 was obtained toward 50 ppm of n-butanol. Figure 5b illustrates the real-time response curves to 50 ppm n-butanol. The corresponding resistance response curve is shown in Figure S7a. Evidently, the response time is short (about 5 s) due to their porous structures, which is favorable for the diffusion of gas molecules. The recovery time of 108 s was relatively long compared with the response time, due to the special sensing mechanism of the chemical reaction between adsorbed oxygen species and C–H bonds at surfaces of most MOS materials [36]. The more alkyl groups $(–CH_2–)_n$ a VOC molecule owned, the more gaseous products were produced during the sensing process at the surface of the sensing materials. Therefore, the time for their desorption should be longer [37].

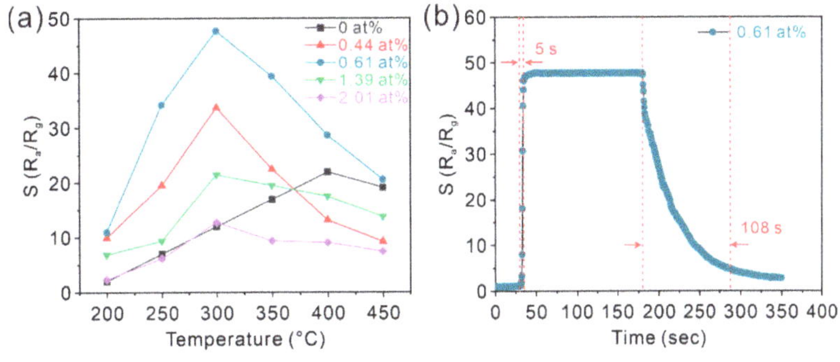

Figure 5. (**a**) The relative response of porous ZnO nanobelts doped with different amounts of Pb toward 50 ppm of n-butanol at different working temperatures. (**b**) The real-time response curve of 0.61 at% Pb-doped ZnO porous nanobelts toward 50 ppm of n-butanol at 300 °C.

Furthermore, the 0.61 at% Pb-doped ZnO and ZnO porous nanobelts were investigated in a wide concentration range of n-butanol, ranging from 1 ppm to 50 ppm at the optimal working temperature of 300 °C, and their real-time response curves are presented in Figure 6a, and its resistance curve is shown in Figure S7b. The results show a fast response of the sensors and a stable value of response in the saturation region, which has a regular relationship with the concentration of n-butanol. The relative response increased with increasing the concentration of n-butanol. A good reversibility could also be demonstrated from the fact that the resistance of the sensors was able to recover to the baseline value. Moreover, 0.61 at% Pb-doped ZnO porous nanobelts demonstrated a distinctly improved response at every point of concentration compared with pure ZnO nanobelts, which was over 4-fold at 50 ppm. The above results are linearly fitted in Figure 6b. It can be seen that the gas sensors fabricated with 0.61 at% Pb-doped ZnO porous nanobelts displayed a higher relative response with a wide detection range than that of pure ZnO porous nanobelts. Furthermore, as seen in Table S3 [38–51], compared with previously reported n-butanol sensors based on various semiconductor nanomaterials, the fabricated sensor showed a relatively better sensing performance.

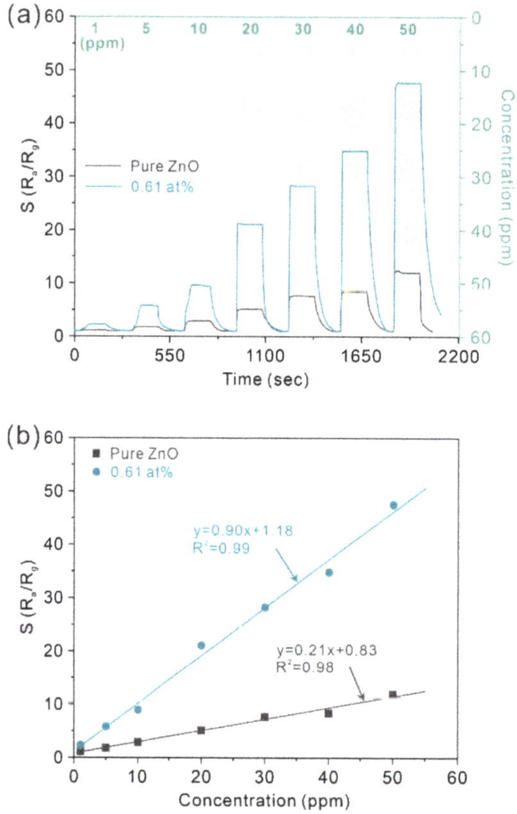

Figure 6. (**a**) The real-time response curves of pure ZnO porous nanobelts and 0.61 at% Pb-doped ZnO porous nanobelts toward different concentrations of n-butanol at 300 °C, and (**b**) their linear fitting curves between the relative responses and concentrations of n-butanol.

To investigate the relative selectivity of the sensor fabricated with 0.61 at% Pb-doped ZnO porous nanobelts, nine kinds of typical VOCs (m-xylene, ethanol, formaldehyde, n-butanol, acetone, benzene, isopropanol, methanol and diethyl ether) were explored at the same concentration of 50 ppm. As shown in Figure 7a, the results reveal that the 0.61 at% Pb-doped ZnO porous nanobelts exhibited a high response of 47.7 to n-butanol, a noticeable response of 34.3 to isopropanol, but significantly reduced responses toward other tested VOCs. Compared with pristine ZnO porous nanobelts, 0.61 at% Pb-doped ZnO porous nanobelts showed better response values toward all the investigated gases. Especially for n-butanol, methanol and isopropanol, they are more than four, two and three times of those of the pristine ZnO porous nanobelts, respectively. This suggests that the sensing performance of ZnO porous nanobelts is greatly improved by the doping of Pb. This result manifested the high sensing selectivity of the 0.61 at% Pb-doped ZnO porous nanobelts toward n-butanol, which could be ascribed to the particular active interfaces from the doped Pb ions at the surfaces, and also the synergistic effect between the molecular polarity of n-butanol and the polar surface provided by Pb. [41,52]. As shown in Figure 7b, the long-term stability of the 0.61 at% Pb-doped ZnO porous nanobelts to 50 ppm n-butanol was examined during 100 days at 300 °C. For each measurement per 10 days, the relative response can be well-maintained without noticeable changes. The average of the relative response was 47.53 with a comparatively small variance of 0.324, which demonstrated a distinguished repeatability. This excellent stability is probably attributed to the fact that the sensing film is uniform, following the above in situ calcination process, and difficult to fall off from the alumina ceramic slice electrode. Additionally, the morphology of porous nanobelts, the key to obtain the excellent performance, was fundamentally preserved under the practical long-term use, due to the relative low working temperature of 300 °C, compared with the calcination temperature of 500 °C. Therefore, the n-butanol sensor based on the as-synthesized 0.61 at% Pb-doped ZnO porous nanobelts could be considered as a remarkable candidate for the actual application.

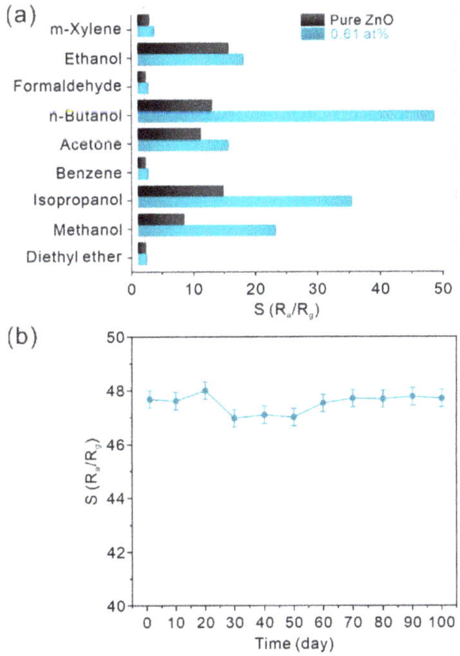

Figure 7. (**a**) The relative responses of pure ZnO porous nanobelts and 0.61 at% Pb-doped ZnO porous nanobelts toward 50 ppm of different VOCs at 300 °C, (**b**) the stability of 0.61 at% Pb-doped ZnO porous nanobelts to 50 ppm n-butanol during 100 days with variance.

3.3. Sensing Mechanism

The main principle of the sensing mechanism of MOS sensors is a phenomenon related to the process of chemical adsorption and desorption of gas molecules and the reaction between the gas molecules at the surfaces of sensing materials, which could be recorded and analyzed by measuring the change of the electrical resistance at the external circuit of the test system [53]. The gas sensor based on our as-synthesized 0.61 at% Pb-doped ZnO porous nanobelts should also abide by this principle, as shown by the sketch in Figure 8. When the sensor was exposed to the air atmosphere, the free oxygen molecules (O_2) in air were adsorbed on the sensing material and then successively converted into oxygen ions (O_2^-) by the capture of electrons from the sensing material. After the initial physical absorption, the oxygen ions continually ionized to the states of O^- and O^{2-} with the working temperature increasing, which established the electron depletion layer at the surface of the sensing materials [54]. Additionally, the energy barriers formed at the contact area among the nanobelts also enlarged the value of the resistance. Therefore, the movement of the carriers in the sensor were greatly hindered, resulting in a relatively large resistivity. In addition, the resistivity of 0.61 at% Pb-doped ZnO porous nanobelts was much greater than that of the pristine ZnO porous nanobelts, due to the increased lattice defects and impurity ions caused by doped Pb, and was maintained regardless of the raising of the working temperature, as displayed in Figure S8. The electrical conductivity (σ) was defined as $\sigma = n\mu e$, where n was the concentration of the carries, μ was the carrier mobility and e was the electron charge. The concentration of electrons was increased following the increment of oxygen vacancies by Pb doping, but, inversely, the resistance increased when the tendency of the decreasing carrier mobility was faster than that of increasing the electronic concentration. For the reason of the increased effective mass (m*) near the Fermi level, the ionic impurity scattering, especially the partly incorporated Pb in the lattice and the microstructural complexity, the carrier mobility was dramatically decreased with the doping of Pb [19,21]. In short, the lattice scattering played a vital role here. Furthermore, the grain boundary, the principal factor of the lattice scattering, was enriched due to the high porosity and small grain sizes, which resulted in an enhanced influence of lattice scattering. Thus, the resistance increased even if the concentration of the electrons increased. The increased resistivity caused by the previously stated reasons lowered the baseline, ultimately, and enhanced the response of the as-prepared gas sensor [55].

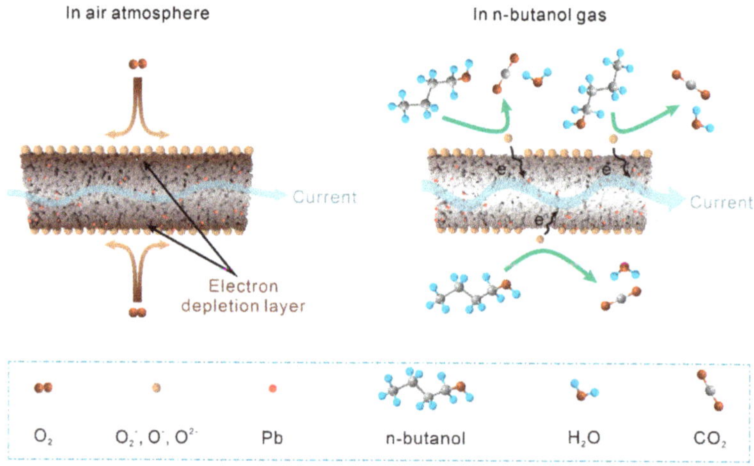

Figure 8. Possible sensing mechanism of porous Pb-doped ZnO nanobelts toward n-butanol.

When the testing chamber contained a certain amount of n-butanol by injection, the chemisorbed oxygen ions and their derivatives on the surface of the sensing materials would react with the gas molecules of the n-butanol, and then caused the backdonation of the trapped electrons to the depletion layers, resulting in the decrease in the sensor resistance due to the reduction in the range of depletion region and the height of the barriers between the nanobelts [56,57]. Meanwhile, the H_2O on the surface of the sensing film, one of the byproducts in the reaction described above, would further decrease the resistivity of the sensor [58]. Regarding the enhanced gas-sensing property of 0.61 at% Pb doped ZnO porous nanobelts, it is worth noting that the gas-sensing process of pristine ZnO porous nanobelts was short of the sufficient adsorbed oxygen molecules and oxygen vacancies generated from the dopant of Pb. However, the effective sensitized region was diminished by the accumulation of the dopant, which contributed to the necessity to maintain a certain amount of the doped Pb.

4. Conclusions

In summary, porous Pb-doped ZnO nanobelts were successfully fabricated via a simply partial cation exchange method, followed by in situ thermal oxidation. The doped

Pb were uniformly distributed on the ZnO nanobelts and their doped amounts can be tuned in the process of the cation exchange reaction. The high porosity and uniformity of the doped Pb with enriched oxygen vacancies can result in an excellent gas-sensing performance with a long-term stability to n-butanol. At the optimal working temperature of 300 °C, the response time of 0.61 at% Pb-doped ZnO nanobelts toward 50 ppm n-butanol was as short as 5 s. Additionally, its relative response was up to 47.7, which was four times higher than that of pristine ZnO porous nanobelts. This work provides a novel approach to synthesize other uniform metal-doped ZnO porous nanobelts and is also believed to be further extended to prepare other doped metal oxide semiconductor nanostructures.

Supplementary Materials: The following supporting information can be downloaded at: https://www.mdpi.com/article/10.3390/chemosensors10030096/s1, Figure S1: (a) Structure of the as-fabricated sensing device, (b) Schematic setup of gas-sensing measurement; Figure S2: SEM images of 0.44 at% (a), 1.39 at% (b) and 2.01 at% (c) Pb^{2+}-exchanged ZnSe precursor nanobelts, and 0.44 at% (d), 1.39 at% (e) and 2.01 at% (f) Pb-doped ZnO porous nanobelts; Figure S3: (a) TEM image of single Pb-doped precursor nanobelt, (b) its high magnified TEM, (c) its HAADF image and the corresponding elemental mapping patterns of (d) Zn, (e) Pb and (f) Se; Figure S4: EDX spectrum of the obtained Pb-doped ZnO porous nanobelts; Figure S5: High-resolution Zn 2p XPS spectra of pure ZnO porous nanobelts and 0.61 at% Pb-doped ZnO porous nanobelts; Figure S6: The real-time response curves of porous ZnO nanobelts doped with (a) 0 at%, (b) 0.44 at%, (c) 0.61 at%, (d) 1.39 at% and (e) 2.01 at% of Pb toward 50 ppm of n-butanol at different working temperature; Figure S7: (a) The real-time resistance response curve of 0.61 at% Pb-doped ZnO porous nanobelts toward 50 ppm of n-butanol at 300 °C, (b)The real-time resistance curve of pure ZnO porous nanobelts and 0.61 at% Pb-doped ZnO porous nanobelts toward different concentrations of n-butanol at 300 °C; Figure S8: The relationship between resistance and temperature of the sensing film fabricated with pure ZnO porous nanobelts and 0.61 at% Pb-doped ZnO porous nanobelts; Table S1: Structural parameters of as-synthesized samples analyzed at (100) from XRD data; Table S2: Fitting results of O 1s XPS spectrum of pure ZnO porous nanobelts and 0.61 at% Pb-doped ZnO porous nanobelts; Table S3: Relative response of various metal oxide nanostructures toward n-butanol, A reported in the literatures and the present work.

Author Contributions: Conceptualization, K.-G.Z. and Z.G.; formal analysis, K.-G.Z., Z.G. and T.-Y.Y.; investigation, K.-G.Z.; data curation, K.-G.Z. and T.-Y.Y.; writing—original draft preparation, K.-G.Z.; writing—review and editing, Z.G.; supervision, Z.G.; project administration, K.-G.Z.; funding acquisition, Z.G. All authors have read and agreed to the published version of the manuscript.

Funding: This work was supported by the National Natural Science Foundation of China (61774159 and 22001001), Anhui Provincial Natural Science Foundation (2008085QF320).

Institutional Review Board Statement: Not applicable.

Informed Consent Statement: Not applicable.

Data Availability Statement: Not applicable.

Conflicts of Interest: The authors declare no conflict of interest.

References

1. Zhao, Y.; Zhang, W.; Yang, B.; Liu, J.; Chen, X.; Wang, X.; Yang, C. Gas-sensing enhancement methods for hydrothermal synthesized SnO_2-based sensors. *Nanotechnology* **2017**, *28*, 452002. [CrossRef]
2. Yuan, Z.; Han, E.; Meng, F.; Zuo, K. Detection and Identification of Volatile Organic Compounds Based on Temperature-Modulated ZnO Sensors. *IEEE Trans. Instrum. Meas.* **2019**, *69*, 4533–4544. [CrossRef]
3. Zhao, Z.; Tian, J.; Sang, Y.; Cabot, A.; Liu, H. Structure, synthesis, and applications of TiO_2 nanobelts. *Adv. Mater.* **2015**, *27*, 2557–2582. [CrossRef] [PubMed]
4. Jeong, S.Y.; Kim, J.S.; Lee, J.H. Rational Design of Semiconductor-Based Chemiresistors and their Libraries for Next-Generation Artificial Olfaction. *Adv. Mater.* **2020**, *32*, 2002075. [CrossRef]
5. Bai, S.; Tian, K.; Han, N.; Guo, J.; Luo, R.; Li, D.; Chen, A. A novel rGO-decorated $ZnO/BiVO_4$ heterojunction for the enhancement of NO_2 sensing properties. *Inorg. Chem. Front.* **2020**, *7*, 1026–1033. [CrossRef]
6. An, D.; Liu, N.; Zhang, H.; Sun, Q.; Li, C.; Li, Y.; Zhang, Q.; Lu, Y. Enhanced n-butanol sensing performance of SnO_2-based gas sensors by doping In_2O_3 via co-precipitation method. *Sens. Actuators B Chem.* **2021**, *340*, 129944. [CrossRef]

7. Nakayama, R.; Maesato, M.; Lim, G.; Arita, M.; Kitagawa, H. Heavy Hydrogen Doping into ZnO and the H/D Isotope Effect. *J. Am. Chem. Soc.* **2021**, *143*, 6616–6621. [CrossRef]
8. Zunger, A.; Malyi, O.I. Understanding Doping of Quantum Materials. *Chem. Rev.* **2021**, *121*, 3031–3060. [CrossRef]
9. Montazeri, A.; Jamali-Sheini, F. Enhanced ethanol gas-sensing performance of Pb-doped In_2O_3 nanostructures prepared by sonochemical method. *Sens. Actuators B Chem.* **2017**, *242*, 778–791. [CrossRef]
10. Yousefi, R.; Zak, A.K.; Jamali-Sheini, F.; Huang, N.M.; Basirun, W.J.; Sookhakian, M. Synthesis and characterization of single crystal PbO nanoparticles in a gelatin medium. *Ceram. Int.* **2014**, *40*, 11699–11703. [CrossRef]
11. Rao, A.; Long, H.; Harley-Trochimczyk, A.; Pham, T.; Zettl, A.; Carraro, C.; Maboudian, R. In Situ Localized Growth of Ordered Metal Oxide Hollow Sphere Array on Microheater Platform for Sensitive, Ultra-Fast Gas Sensing. *ACS Appl. Mater. Interfaces* **2017**, *9*, 2634–2641. [CrossRef] [PubMed]
12. Gu, J.; Zhang, B.; Li, Y.; Xu, X.; Sun, G.; Cao, J.; Wang, Y. Synthesis of spindle-like Co-doped $LaFeO_3$ porous microstructure for high performance n-butanol sensor. *Sens. Actuators B Chem.* **2021**, *343*, 130125. [CrossRef]
13. Guo, Z.; Li, M.-Q.; Liu, J.-H.; Huang, X.-J. Cation Exchange Synthesis and Unusual Resistive Switching Behaviors of Ag_2Se Nanobelts. *Small* **2015**, *11*, 6285–6294. [CrossRef] [PubMed]
14. Kang, P.; Zheng, K.-G.; Wang, Z.; Chen, L.; Guo, Z. Cation-exchange synthesis of PbSe/ZnSe hetero-nanobelts with enhanced near-infrared photoelectronic performance. *Nanotechnology* **2021**, *32*, 335504. [CrossRef] [PubMed]
15. Jin, X.-B.; Li, Y.-X.; Su, Y.; Guo, Z.; Gu, C.-P.; Huang, J.-R.; Meng, F.-L.; Huang, X.-J.; Li, M.-Q.; Liu, J.-H. Porous and single-crystalline ZnO nanobelts: Fabrication with annealing precursor nanobelts, and gas-sensing and optoelectronic performance. *Nanotechnology* **2016**, *27*, 355702. [CrossRef] [PubMed]
16. Tian, H.; Fan, H.; Li, M.; Ma, L. Zeolitic Imidazolate Framework Coated ZnO Nanorods as Molecular Sieving to Improve Selectivity of Formaldehyde Gas Sensor. *ACS Sens.* **2015**, *1*, 243–250. [CrossRef]
17. Pavithra, M.; Raj, M.J. Influence of ultrasonication time on solar light irradiated photocatalytic dye degradability and antibacterial activity of Pb doped ZnO nanocomposites. *Ceram. Int.* **2021**, *47*, 32324–32331. [CrossRef]
18. Yang, D.; Su, X.; Yan, Y.; Hu, T.; Xie, H.; He, J.; Uher, C.; Kanatzidis, M.G.; Tang, X. Manipulating the Combustion Wave during Self-Propagating Synthesis for High Thermoelectric Performance of Layered Oxychalcogenide $Bi_{1-x}Pb_xCuSeO$. *Chem. Mater.* **2016**, *28*, 4628–4640. [CrossRef]
19. Li, X.B.; Ma, S.Y.; Li, F.M.; Yang, F.C.; Liu, J.; Zhang, X.L.; Wang, X. Blue-green and red luminescence from non-polar ZnO:Pb films. *Appl. Surf. Sci.* **2013**, *270*, 467–472. [CrossRef]
20. Aboud, A.A.; Shaban, M.; Revaprasadu, N. Effect of Cu, Ni and Pb doping on the photo-electrochemical activity of ZnO thin films. *RSC Adv.* **2019**, *9*, 7729–7736. [CrossRef]
21. Yin, Z.; Liu, Z.; Yu, Y.; Zhang, C.; Chen, P.; Zhao, J.; He, P.; Guo, X. Synergistically Optimized Electron and Phonon Transport of Polycrystalline BiCuSeO via Pb and Yb Co-Doping. *ACS Appl. Mater. Inter.* **2021**, *13*, 57638–57645. [CrossRef] [PubMed]
22. Liu, W.; Gu, D.; Li, X. Ultrasensitive NO_2 Detection Utilizing Mesoporous ZnSe/ZnO Heterojunction-Based Chemiresistive-Type Sensors. *ACS Appl. Mater. Interf.* **2019**, *11*, 29029–29040. [CrossRef] [PubMed]
23. Xing, C.; Chen, X.; Huang, W.; Song, Y.; Li, J.; Chen, S.; Zhou, Y.; Dong, B.; Fan, D.; Zhu, X.; et al. Two-Dimensional Lead Monoxide: Facile Liquid Phase Exfoliation, Excellent Photoresponse Performance, and Theoretical Investigation. *ACS Photon.* **2018**, *5*, 5055–5067. [CrossRef]
24. Kumar, P.; Liu, J.; Ranjan, P.; Hu, Y.; Yamijala, S.S.; Pati, S.K.; Cheng, G.J. Alpha Lead Oxide (alpha-PbO): A New 2D Material with Visible Light Sensitivity. *Small* **2018**, *14*, 1703346. [CrossRef]
25. Liu, X.; Min, L.; Yu, X.; Zhou, Z.; Sha, L.; Zhang, S. Changes of photoelectrocatalytic, electrocatalytic and pollutant degradation properties during the growth of β-PbO_2 into black titanium oxide nanoarrays. *Chem. Eng. J.* **2021**, *417*, 127996. [CrossRef]
26. Jiang, W.; Wang, S.; Liu, J.; Zheng, H.; Gu, Y.; Li, W.; Shi, H.; Li, S.; Zhong, X.; Wang, J. Lattice oxygen of PbO_2 induces crystal facet dependent electrochemical ozone production. *J. Mater. Chem. A* **2021**, *9*, 9010–9017. [CrossRef]
27. Zhang, D.; Yang, Z.; Wu, Z.; Dong, G. Metal-organic frameworks-derived hollow zinc oxide/cobalt oxide nanoheterostructure for highly sensitive acetone sensing. *Sens. Actuators B Chem.* **2018**, *283*, 42–51. [CrossRef]
28. Jiang, W.; Low, J.; Mao, K.; Duan, D.; Chen, S.; Liu, W.; Pao, C.-W.; Ma, J.; Sang, S.; Shu, C. Pd-Modified ZnO-Au Enabling Alkoxy Intermediates Formation and Dehydrogenation for Photocatalytic Conversion of Methane to Ethylene. *J. Am. Chem. Soc.* **2021**, *143*, 269–278. [CrossRef]
29. Wang, Z.; Gao, S.; Fei, T.; Liu, S.; Zhang, T. Construction of ZnO/SnO2 Heterostructure on Reduced Graphene Oxide for Enhanced Nitrogen Dioxide Sensitive Performances at Room Temperature. *ACS Sens.* **2019**, *4*, 2048–2057. [CrossRef]
30. Yan, W.; Xu, H.; Ling, M.; Zhou, S.; Qiu, T.; Deng, Y.; Zhao, Z.; Zhang, E. MOF-Derived Porous Hollow Co3O4@ZnO Cages for High-Performance MEMS Trimethylamine Sensors. *ACS Sens.* **2021**, *6*, 2613–2621. [CrossRef]
31. Huang, B.; Zhang, Z.; Zhao, C.; Cairang, L.; Bai, J.; Zhang, Y.; Xie, E. Enhanced gas-sensing performance of ZnO@In_2O_3 core@shell nanofibers prepared by coaxial electrospinning. *Sens. Actuators B Chem.* **2018**, *255*, 2248–2257. [CrossRef]
32. Liu, W.; Xu, L.; Sheng, K.; Chen, C.; Zhou, X.; Dong, B.; Bai, X.; Zhang, S.; Lu, G.; Song, H. APTES-functionalized thin-walled porous WO3 nanotubes for highly selective sensing of NO2 in a polluted environment. *J. Mater. Chem. A* **2018**, *6*, 10976–10989. [CrossRef]
33. Li, L.; He, S.; Liu, M.; Zhang, C.; Chen, W. Three-Dimensional Mesoporous Graphene Aerogel-Supported SnO2 Nanocrystals for High-Performance NO2 Gas Sensing at Low Temperature. *Anal. Chem.* **2015**, *87*, 1638–1645. [CrossRef] [PubMed]

34. Chen, D.; Zhang, K.; Zhou, H.; Fan, G.; Wang, Y.; Li, G.; Hu, R. A wireless-electrodeless quartz crystal microbalance with dissipation DMMP sensor. *Sens. Actuators B Chem.* **2018**, *261*, 408–417. [CrossRef]
35. Koo, W.-T.; Choi, S.-J.; Kim, S.-J.; Jang, J.-S.; Tuller, H.L.; Kim, I.-D. Heterogeneous Sensitization of Metal–Organic Framework Driven Metal@Metal Oxide Complex Catalysts on an Oxide Nanofiber Scaffold Toward Superior Gas Sensors. *J. Am. Chem. Soc.* **2016**, *138*, 13431–13437. [CrossRef]
36. Swaminathan, N.; Henning, A.; Jurca, T.; Hayon, J.; Shalev, G.; Rosenwaks, Y. Effect of varying chain length of n-alcohols and n-alkanes detected with electrostatically-formed nanowire sensor. *Sens. Actuators B Chem.* **2017**, *248*, 240–246. [CrossRef]
37. Chen, M.; Zhang, Y.; Zhang, J.; Li, K.; Lv, T.; Shen, K.; Zhu, Z.; Liu, Q. Facile lotus-leaf-templated synthesis and enhanced xylene gas sensing properties of Ag-$LaFeO_3$ nanoparticles. *J. Mater. Chem. C* **2018**, *6*, 6138–6145. [CrossRef]
38. Wang, Y.; Zeng, Y.; Wang, L.; Lou, Z.; Qiao, L.; Tian, H.; Zheng, W. Ultrathin nanorod-assembled SnO2 hollow cubes for high sensitive n-butanol detection. *Sens. Actuators B Chem.* **2018**, *283*, 693–704. [CrossRef]
39. Liu, F.; Huang, G.; Wang, X.; Xie, X.; Xu, G.; Lu, G.; He, X.; Tian, J.; Cui, H. High response and selectivity of single crystalline ZnO nanorods modified by In_2O_3 nanoparticles for n-butanol gas sensing. *Sens. Actuators B Chem.* **2019**, *277*, 144–151. [CrossRef]
40. Yang, W.; Xiao, X.; Fang, B.; Deng, H. Nanorods-assembled ZnO microflower as a powerful channel for n-butanol sensing. *J. Alloys Compd.* **2021**, *860*, 158410. [CrossRef]
41. Hoppe, M.; Ababii, N.; Postica, V.; Lupan, O.; Polonskyi, O.; Schütt, F.; Kaps, S.; Sukhodub, L.F.; Sontea, V.; Strunskus, T.; et al. (CuO-Cu2O)/ZnO:Al heterojunctions for volatile organic compound detection. *Sens. Actuators B Chem.* **2018**, *255*, 1362–1375. [CrossRef]
42. Zhao, R.; Wei, Q.; Ran, Y.; Kong, Y.; Ma, D.; Su, L.; Yao, L.; Wang, Y. One-dimensional In2O3 nanorods as sensing material for ppb-level n-butanol detection. *Nanotechnology* **2021**, *32*, 375501. [CrossRef] [PubMed]
43. Li, W.; Ren, Y.; Guo, Y. ZrO2/ZnO nanocomposite materials for chemiresistive butanol sensors. *Sens. Actuators B Chem.* **2020**, *308*, 127658. [CrossRef]
44. Wang, S.; Gao, X.; Yang, J.; Zhu, Z.; Zhang, H.; Wang, Y. Synthesis and gas sensor application of $ZnFe_2O_4$–ZnO composite hollow microspheres. *RSC Adv.* **2014**, *4*, 57967–57974. [CrossRef]
45. Wang, M.; Shen, Z.; Zhao, X.; Duanmu, F.; Yu, H.; Ji, H. Rational shape control of porous Co3O4 assemblies derived from MOF and their structural effects on n-butanol sensing. *J. Hazard. Mater.* **2019**, *371*, 352–361. [CrossRef]
46. Bai, J.; Li, Y.; Liu, Y.; Wang, H.; Liu, F.; Liu, F.; Sun, P.; Yan, X.; Lu, G. $Au_{39}Rh_{61}$ Alloy Nanocrystal-Decorated $W_{18}O_{49}$ for Enhanced Detection of *n*-Butanol. *ACS Sens.* **2019**, *4*, 2662–2670. [CrossRef]
47. Xu, Y.; Zheng, L.; Yang, C.; Zheng, W.; Liu, X.; Zhang, J. Chemiresistive sensors based on core-shell ZnO@TiO_2 nanorods designed by atomic layer deposition for n-butanol detection. *Sens. Actuators B Chem.* **2020**, *310*, 127846. [CrossRef]
48. Lv, L.; Cheng, P.; Wang, Y.; Xu, L.; Zhang, B.; Lv, C.; Ma, J.; Zhang, Y. Sb-doped three-dimensional $ZnFe_2O_4$ macroporous spheres for N-butanol chemiresistive gas sensors. *Sens. Actuators B Chem.* **2020**, *320*, 128384. [CrossRef]
49. Wang, Y.; Zhang, B.; Liu, J.; Yang, Q.; Cui, X.; Gao, Y.; Chuai, X.; Liu, F.; Sun, P.; Liang, X.; et al. Au-loaded mesoporous WO_3: Preparation and n-butanol sensing performances. *Sens. Actuators B Chem.* **2016**, *236*, 67–76. [CrossRef]
50. He, P.; Fu, H.; Yang, X.; Xiong, S.; Han, D.; An, X. Variable gas sensing performance towards different volatile organic compounds caused by integration types of ZnS on In_2O_3 hollow spheres. *Sens. Actuators B Chem.* **2021**, *345*, 130316. [CrossRef]
51. Dang, F.; Wang, Y.; Gao, J.; Xu, L.; Cheng, P.; Lv, L.; Zhang, B.; Li, X.; Wang, C. Hierarchical flower-like NiCo2O4 applied in n-butanol detection at low temperature. *Sens. Actuators B Chem.* **2020**, *320*, 128577. [CrossRef]
52. Cho, S.Y.; Yoo, H.W.; Kim, J.Y.; Jung, W.B.; Jin, M.L.; Kim, J.S.; Jeon, H.-J.; Jung, H.-T. High-Resolution p-Type Metal Oxide Semiconductor Nanowire Array as an Ultrasensitive Sensor for Volatile Organic Compounds. *Nano Lett.* **2016**, *16*, 4508–4515. [CrossRef] [PubMed]
53. Shehada, N.; Brönstrup, G.; Funka, K.; Christiansen, S.; Leja, M.; Haick, H. Ultrasensitive Silicon Nanowire for Real-World Gas Sensing: Noninvasive Diagnosis of Cancer from Breath Volatolome. *Nano Lett.* **2014**, *15*, 1288–1295. [CrossRef]
54. Alali, K.T.; Liu, J.; Liu, Q.; Li, R.; Zhang, H.; Aljebawi, K.; Liu, P.; Wang, J. Enhanced acetone gas sensing response of ZnO/$ZnCo_2O_4$ nanotubes synthesized by single capillary electrospinning technology. *Sens. Actuators B Chem.* **2017**, *252*, 511–522. [CrossRef]
55. Wei, D.; Huang, Z.; Wang, L.; Chuai, X.; Zhang, S.; Lu, G. Hydrothermal synthesis of Ce-doped hierarchical flower-like In_2O_3 microspheres and their excellent gas-sensing properties. *Sens. Actuators B Chem.* **2018**, *255*, 1211–1219. [CrossRef]
56. Spagnoli, E.; Krik, S.; Fabbri, B.; Valt, M.; Ardit, M.; Gaiardo, A.; Vanzetti, L.; Della Ciana, M.; Cristino, V.; Vola, G.; et al. Development and characterization of WO_3 nanoflakes for selective ethanol sensing. *Sens. Actuators B Chem.* **2021**, *347*, 130593. [CrossRef]
57. Masoumi, S.; Shokrani, M.; Aghili, S.; Hossein-Babaei, F. Zinc oxide-based direct thermoelectric gas sensor for the detection of volatile organic compounds in air. *Sens. Actuators B Chem.* **2019**, *294*, 245–252. [CrossRef]
58. Postica, V.; Lupan, O.; Gapeeva, A.; Hansen, L.; Khaledialidusti, R.; Mishra, A.K.; Drewes, J.; Kersten, H.; Faupel, F.; Adelung, R.; et al. Improved Long-Term Stability and Reduced Humidity Effect in Gas Sensing: SiO_2 Ultra-Thin Layered ZnO Columnar Films. *Adv. Mater. Technol.* **2021**, *6*, 2001137. [CrossRef]

chemosensors

Article

Wireless Volatile Organic Compound Detection for Restricted Internet of Things Environments Based on Cataluminescence Sensors

Xinyi Shen [1], Guolong Shi [1,*], Yongxing Zhang [2,*] and Shizhuang Weng [3]

1 School of Information and Computer, Anhui Agricultural University, Hefei 230036, China; sxy@stu.ahau.edu.cn
2 Anhui Province Key Laboratory of Pollutant Sensitive Materials and Environmental Remediation, Department of Materials Science and Engineering, Huaibei Normal University, Huaibei 235000, China
3 National Engineering Research Center for Agro-Ecological Big Data Analysis & Application, Anhui University, 111 Jiulong Road, Hefei 230601, China; weng1989@mail.ustc.edu.cn
* Correspondence: shigl@ahau.edu.cn (G.S.); zyx07157@mail.ustc.edu.cn (Y.Z.)

Abstract: Cataluminescence-based sensors do not require external light sources and complex circuitry, which enables them to avoid light scattering with high sensitivity, selectivity, and widely linear range. In this study, a wireless sensor system based on hierarchical CuO microspheres assembled from nano-sheets was constructed for Volatile Organic Compound (VOC) online detection. Through sensor characteristics and data process analysis, the results showed that the luminous sensor system has good luminous characteristics, including the intensity of visible light, high signal/noise (S/N) values, and very short response and recovery times. Different VOC concentration values can be detected on multiple wavelength channels and different Cataluminescence signal spectra separations can process multiple sets of Cataluminescence data combinations concurrently. This study also briefly studied the mechanism action of the Cataluminescence sensor, which can specifically be used for VOC detecting.

Keywords: Cataluminescence; internet of things; wireless sensor system; gas dynamic monitoring

Citation: Shen, X.; Shi, G.; Zhang, Y.; Weng, S. Wireless Volatile Organic Compound Detection for Restricted Internet of Things Environments Based on Cataluminescence Sensors. *Chemosensors* **2022**, *10*, 179. https://doi.org/10.3390/chemosensors10050179

Academic Editor: Bilge Saruhan-Brings

Received: 12 April 2022
Accepted: 3 May 2022
Published: 9 May 2022

1. Introduction

As the types of decorative materials in the human living environment increase, the pollutants in the air in trace amounts are gradually changing. The colorless irritating gas is difficult for ordinary people to detect and may be inhaled for a long time, which destroys the body's hematopoietic function and appears to have central nervous system anesthetic effects, with severe cases even inducing lung cancer, leukemia, and other diseases. The United States Environmental Agency defines Volatile Organic Compounds (VOCs) as carbon-containing compounds other than carbon dioxide, carbonic acid, carbonates, metal carbides, and carbonaceous compounds that are involved in photocatalytic reactions in the atmosphere, including TVOC (Total Volatile Organic Compounds), aldehydes and benzene, toluene, xylene, etc. [1]. At present, although domestic and foreign researchers have established corresponding limit standards for aldehydes and benzene-based organic compounds, VOC substances detection with high sensitivity, low-cost, low-energy, fast, and reliable sensing methods has received continuous attention. The colorimetric method can only determine the total amount of organic volatile gases; gas chromatography has high separation efficiency, but the limitation of the boiling point of derivatives leads to low determination results; liquid chromatography is difficult for comparing the large number of interferences in VOC with the completely separated components; and the liquid chromatography-mass spectrometry instrument has a complex structure and high cost, making it difficult to use widely. Thus, it is important to study the VOC detection method with low cost, long life, and strong portability.

In the environment of an oxygen atom, the gas molecules can be used for heterogeneous catalytic oxidation on the surface of a solid catalyst. Therefore, catalytic light-emitting sensors have become one of the most effective methods for gas detection. Cataluminescence refers to the reaction of substances on the surface of catalytic materials to produce luminescence. Catalytic light-emitting sensors have received widespread attention due to their visible sensitivity, specific selectivity, and rapid response [2–6]. Cataluminescence sensors consume only oxygen and sample molecules in the air during the luminescence process, while the fixed solid catalyst is nearly not consumed. Therefore, the gas sensor using nanomaterial as a sensitive material overcomes the shortcomings of traditional sensor reagent consumption and shedding and is expected to develop into a new class of practical chemiluminescent gas sensors [7–9]. In the past few years, people have developed CTL sensors with excellent performance, excellent selectivity, and fast response and recovery [10–13]. Zhang Runkun and coworkers developed a gas sensor system for the detection of luminescence (CTL) emissions of hexafluoride groups by using a Zn-doped SnO_2 composite. The results show that the linear detection range of the sensor system is lower than the standard detection concentration and the pattern recognition method can be used to detect the identifiability of the sensor [14]. Based on the Cataluminescence (CTL) emission of MgO/In_2O_3 composite (with a mass ratio of 1:2), Liu developed a gas sensor for the determination of dimethyl ether and isopropanol. The sensor has high sensitivity and selectivity to analyte at 245 °C [15].

High sensitivity, good selectivity, and fast response contribute to the interest in Cataluminescence and also provide significant challenges to achieving significant enhancements in Cataluminescence-based sensor systems. However, in the process of routine analysis and application of chemiluminescence sensors, there are still some deficiencies found. These are mainly due to the short life of the sensor and the signal drift caused by the consumption of luminescent reactants, which limits the application of the chemiluminescence sensor in practical analysis [16]. For this reason, it is urgent to manufacture a chemiluminescent sensor with high sensitivity and stability, a simple preparation process, and a long service life. The continuous development of nanotechnology has provided new opportunities for the research of chemiluminescence sensors [17]. Some gases can produce strong chemiluminescence on the surface of specific nano-materials, so nano-materials can be used to design different types of Cataluminescence sensors for sensitive materials [18]. Moreover, with the development of wireless communication technologies, a lot of sensor systems have been developed for use in remote monitoring and control applications consisting of environmental, wildlife monitoring, road tunnel monitoring, building structural health monitoring, and gas leakage and detection. Wireless communications have many advantages, including remote monitoring, low cost, fast response, self-organizing, and flexible layout, while reports of their application in analytical smart CTL sensors that detect and identify multiple hazardous vapors in real time are rare. As mentioned above, a wireless CTL sensor system is needed for harmful gas monitoring which has a wireless communication technology function. For the determination of benzene and toluene gas, Wang Jian and his colleagues have raised a sensitive CTL sensor. However, while the method is successful to steam detection, it also has short board in remote control and the transfer of data [19].

Some nano-oxides and metal or noble-doped nanomaterials have better chemiluminescence signals and can be used to construct Cataluminescence sensing materials. However, most of them require noble metal or metal doping which is costly and complicated to synthesize. CuO is a p-type metal oxide with good characteristics like no toxicity, availability, and a smaller band gap (1.3–2.4 eV). The combination of CuO and other metal oxides will produce unique p-n hybrid materials which show the quality of high-efficiency photocatalytic materials [20]. In this study, nano-oxide CuO with low cost and abundant sources was selected as the catalytic material, the preparation method was simple, and the cost was low.

Specifically, the technical contributions of this paper are the following: In this study, the Cataluminescence sensor system was used as a practical tool for detecting harmful gases. A theory of dynamic and static characteristics of gas sensors was introduced and a wireless sensor system based on hierarchical CuO microspheres assembled from nano-sheets was constructed in Section 2. The third section introduces the theoretical derivation of the system and discusses the experimental results and analysis in detail. Finally, Section 4 summarizes the whole paper.

2. Experiments and Methods

2.1. Experiment Reagents

In this study, we used a reagent of a high-purity (98.5%), analytical-grade sample, and did not further purify it for experimental use.

2.2. Preparation of the Sensor Material

A simple chemical solution method was used to prepare hierarchical CuO microspheres assembled from nano-sheets, that is, 2.4 mmol of $Cu(CH_3COO)_2 \cdot H_2O$ was completely dissolved in 60 mL of deionized aqueous solution. Then 4 mL of ammonia (25–28%) was added to the above mixture and stirred slowly until a clear solution was formed. The solution was sealed in a conical flask with a cover and heated at 82 °C for 150 min. Then it was naturally cooled to room temperature to obtain a synthetic black powder which was washed several times with anhydrous alcohol and deionized water before finally being dried in the oven for more than 12 h [21]. We prepared hierarchical CuO microspheres assembled from nano-sheets by this simple chemical solution method.

2.3. Preparation of Test Gas

The method for preparing a certain concentration of the gas sample was as follows: a micro-sampler drew a certain amount of the sample solution to be tested (analytically pure liquid reagent) into a fixed volume flask and sealed the flask with a rubber stopper. The relationship between the mass of the reagent added to the flask and the volume of the flask after the liquid sample was completely volatilized. The sample to be tested was converted into the corresponding gas density (ppm) at normal temperature and pressure and the gas concentration was calculated as follows:

$$C = \frac{22.4V_a \times d}{V_f \times M} \tag{1}$$

where 24.5 is the molar volume of the gas (L/mol), V_a represents the volume (μL) of the liquid injected into the flask, V_f represents the volume of the flask (m^3), d represents the density of the liquid (mg/μL); M represents the molecular weight of the liquid substance [22].

2.4. Experiment Instruments

The product was characterized by an X-ray diffractometer (XRD), field emission scanning electron microscope (FESEM, quanta200feg), and transmission electron microscope (TEM, JEOL JEM-2100). Micromeritics ASAP 2020 M + C Brunauer-Emmet-Teller (BET) equipment was used to test the specific surface area of the sample (4.84 m^2/g) and N_2 gas was used as the carrier gas.

The system diagram of the wireless luminous sensor is shown in Figure 1. Dispersed material was placed on a CTL catalyst with an inner diameter of approximately 10 mm. The dispersed material was a thickness of 0.5 mm. The air pump could control the gas flow and quickly disperse the test gas. The sensor is equipped with an optical filter and the working temperature can be adapted by adjusting the temperature controller. Catalysis occurred on the surface of the catalytic material as it flowed through the CTL catalyst part. The wireless ultra-weak chemiluminescence sensor system collected the luminous intensity of the photomultiplier tube (PMT) and we chose to use the optical filter in the range of 400–650 nm to test the wavelength.

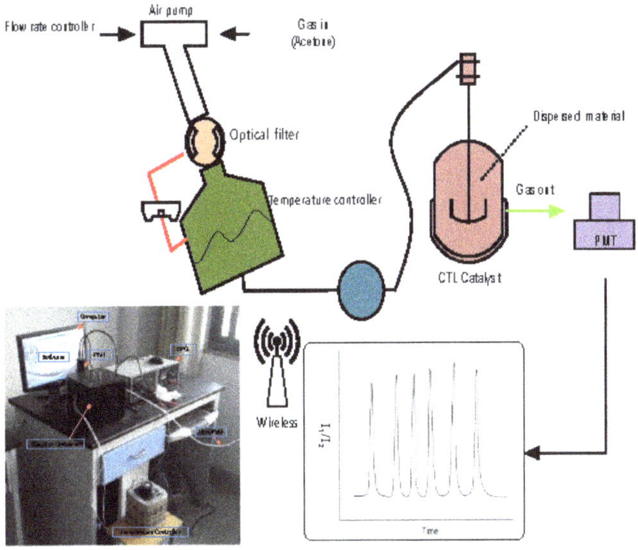

Figure 1. Diagram of the online wireless Cataluminescence sensor system.

The wireless harmful gas online detection system is an interactive system integrating Internet of Things sensor technology. It operates based on a data acquisition module, a data transmission module, and a remote monitoring platform. The data acquisition module consisted of phase modulation, analog modulation, amplitude modulation, and frequency modulation, where the cloud service platform was divided into wireless transmission module data collection and telecommunication module. A general schematic diagram of wireless CTL harmful gas online-detection system is shown in Figure 2.

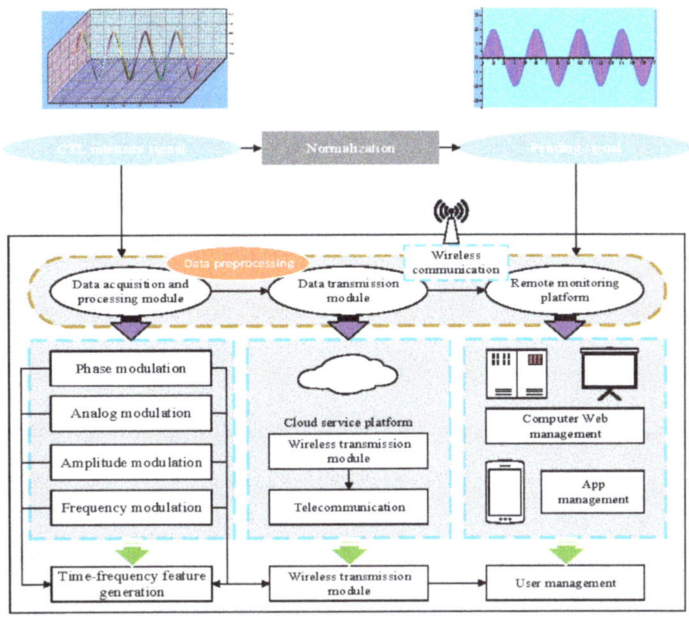

Figure 2. General schematic diagram of the wireless CTL harmful gas online detection system.

2.5. Theory of Sensor Characteristics and Data Process

A layer of catalytic material was coated onto the ceramic heating core and its thickness was controlled to be about 0.4 mm. After being naturally dried, it was placed in a quartz tube with a gas inlet and an outlet at the wall (Φ = 14 mm). When an experimental device was connecting, the temperature of the heating core was controlled through a voltage regulator and the temperature of the catalyst on the heating core was measured by a thermocouple. The gas flow meter was used to adjust the air flow rate through the tube made of quartz. The gas to be tested was carried by pure air and flowed through the quartz tube and a catalytic oxidation reaction occurred on the catalyst to release the photons. The photons passed through the filter below and were amplified by a photomultiplier tube and detected by the luminescence system [23].

In this study, the calculation process of relative Cataluminescence intensity was as follows:

$$S = S_{\max} \times N \tag{2}$$

where $S_{\max}$ was the peak response of Cataluminescence, N was the background average value of the measurements. The detection limit of the sensor was calculated as follows:

$$LOD = 3\sigma/k \tag{3}$$

where σ was the blank standard deviation and k was the slope of the working curve. The static characteristic Equation (4) was used to describe and express the relationship between the output of the detection system and input of which

$$y(x) = a_0 + a_1 x + a_2 x_2 + \ldots + a_i x_i + \ldots + a_n x_n \tag{4}$$

where x was the input quantity; $y(x)$ was the output; and $a_0, a_1, a_2, \ldots, a_i, \ldots, a_n$ was a constant coefficient term. For the linear time invariant CTL detection system, the constant coefficient linear differential equation characterizing its dynamic characteristics was:

$$\begin{aligned} & a_n \frac{d^n Y(t)}{dt^n} + a_{n-1} \frac{d^{n-1} Y(t)}{dt^{n-1}} + \ldots + a_1 \frac{dY(t)}{dt} + a_0 Y(t) \\ & = b_m \frac{d^m X(t)}{dt^m} + b_{m-1} \frac{d^{m-1} X(t)}{dt^{m-1}} + \ldots + b_1 \frac{dX(t)}{dt} + b_0 X(t) \end{aligned} \tag{5}$$

when initial t = 0, the initial conditions that output $y(t)$ = 0 and input $X(t)$ = 0, and the initial values of their derivatives to time were all zero, then the transfer Function (6) of CTL measurement system was:

$$H(s) = \frac{Y(s)}{X(s)} = \frac{b_m s^m + b_{m-1} s^{m-1} + \ldots + b_1 s + b_0}{a_n s^n + a_{n-1} s^{n-1} + \ldots + a_1 s + a_0} \tag{6}$$

Expression of amplitude and the phase frequency characteristic:

$$A(\omega) = |H(j\omega)| = \frac{K}{\sqrt{\left[1 - \left(\frac{\omega}{\omega_0}\right)^2\right]^2 + \left[2\zeta\left(\frac{\omega}{\omega_0}\right)\right]^2}} \tag{7}$$

$$\varphi(\omega) = -\arctan \frac{?\zeta \frac{\omega}{\omega_0}}{1 - \left(\frac{\omega}{\omega_0}\right)^2} \tag{8}$$

2.6. Characterization of Sensor Material

Figure 3a shows the product's X-ray diffraction (XRD) pattern. All the sharp diffraction peaks of the product can be labeled as the monoclinic CuO (JCPDS No.: 48-1548). SEM (Scanning Electron Microscopy, Figure 3b) and TEM (Transmission Electron Microscopy, Figure 3c) images showed that the structures of the hierarchical CuO microspheres were constituted with nano sheets. The O^{2-} ions were successfully doped into the CuO lattice and compared to pure Cu nano-materials. For CuO, the use of reflectance spectra in UV showed an extension of the absorption edge to the visible region, which can be attributed to the Cu^{2+} dopants.

Figure 3. (**a**) X-ray diffractometer pattern, (**b**) SEM, and (**c**) TEM images of the hierarchical CuO microspheres assembled from nano-sheets.

The Cu2p XPS spectrum of CuO microspheres assembled with nano-sheets is shown in Figure 4. Figure 4 shows the Cu2p XPS spectrum of CuO microspheres assembled with nano-sheets. For the Cu2p spectrum, the peaks corresponded to the 2p3/2 and 2p1/2 transitions of copper at 932.3 and 952.2 eV, respectively. Meanwhile, satellites on the higher binding energy sides were also visible (indicated by *). These values were comparable to the values reported for the Cu2p levels (Cu^{2+}) species in CuO [24,25]. In this study, the hierarchical CuO microspheres were synthesized and the strength of CTL effects on VOCs was investigated.

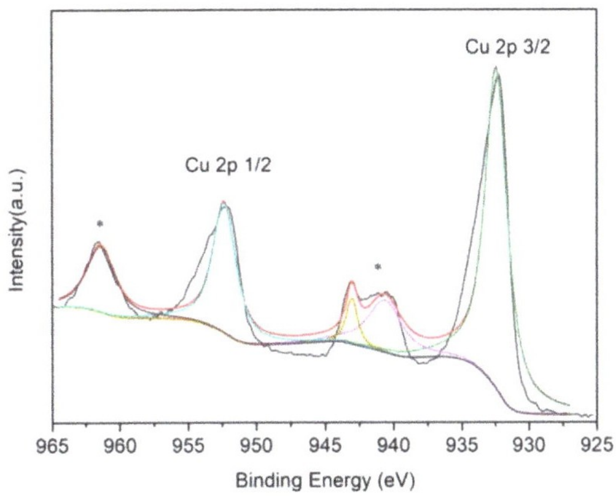

Figure 4. The Cu2p XPS spectrum of CuO microspheres assembled with nano-sheets.

3. Luminescence Detection Analysis

3.1. Optimization of Air Flow Rate

The experimental temperature was adjusted to 220 °C and 200 ppm acetone vapor was injected at a wavelength of 425 nm. The relationship between the flow rate of the carrier gas and the acetone luminescence intensity was measured. As shown in Figure 5, when the flow rate of the carrier gas was between 50 and 310 mL/min, the CTL intensity increased with the increase in the carrier gas flow rate. When the flow rate of the carrier gas was greater than 310 mL/min, the CTL intensity decreased with the increase in the carrier gas flow rate. This indicated that at lower flow rates, the diffusion rate of oxygen molecules may be lower than that of acetone, thus the luminescence intensity increased with the flow rate. The condition of the constant reaction rate was that the flow rate of the carrier gas was more than 310 mL/min. The flow rate was too large and the acetone molecules were not sufficiently reacted on the sensitive material to be lost, so the luminescence intensity decreased as the flow rate increased at a higher flow rate. Therefore, this study determined 310 mL/min to be the optimum carrier gas flow rate.

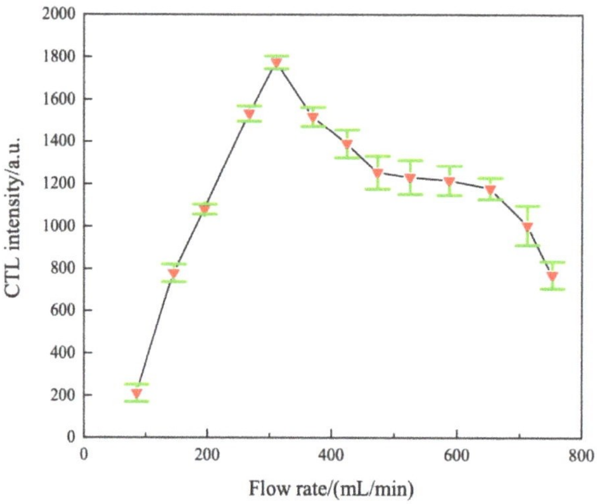

Figure 5. Selection of the optimal carrier gas flow rate (temperature: 220 °C, wavelength: 425 nm, concentration: 200 ppm).

3.2. *Optimal Wavelength and Temperature Selection*

The catalytic emission wavelength of acetone on the surface of the test CuO material was first tested by using an optical filter from 400 to 630 nm (400, 420, 480, 530, 560, 590, 630). Figure 6 shows the change in the Cataluminescence intensity of acetone at different wavelengths with a temperature of 220 °C and a flow rate of 310 mL/min. Since the incandescent lamp from the ceramic heater emits longer wavelength radiation, the noise signal (S/N) will increase. Thus, the S/N ratio was used to display the actual illumination intensity. The calculation method of the signal-to-noise ratio is 10 lg (P_S/P_N), where P_S and P_N represent the effective power of the signal and noise, respectively, and can also be converted into the ratio relationship of voltage amplitude: 10 lg (V_S^2/V_N^2), where V_S and V_N represent the effective value of the signal and noise voltage, respectively. As shown in Figure 6, acetone has a maximum emission wavelength of 590 nm and a maximum signal-to-noise ratio (S/N). As a consequence, the best test wavelength of the CuO luminescence sensor is 590 nm. At the same time, it can be seen from Figure 6 that the Cataluminescence signal is the strongest at 590 nm and the curve shows its characteristic spectral shape. This may be due to the energy difference of the high energy state luminescent intermediates produced in the Cataluminescence reaction. This difference varies from substance to material and was therefore of great importance for substance identification and selective detection.

This study used filters to separate each group of test CTL signal spectra. Different CTL signal spectra separations can process multiple sets of CTL data combinations concurrently (Figure 7). The CTL signal spectra change in the channels with the center wavelengths of *w*1, *w*2, *w*3, and *w*4, and will not interact with each other. During our experiment, center wavelengths were selected from 400 to 630 nm (400, 420, 480, 530, 560, 590, 630). If the sensing wavelengths are sufficiently separated, wavelength separation can be used in the frequency domain or gating in the time domain to discern the evolution of each catalytic luminescence feature in the central channel wavelength. For example, in the vicinity of the central wavelengths $w1$, $w2$, $w3$ and $w4$, these four groups of wavelengths are independent and do not interfere with each other, so different relative catalytic luminescence values can be detected on the different channels. Different relative acetone concentration values can be detected on multiple wavelength channels. Based on the principle of anti-collision technology in spectral separation, multiple pieces of CTL spectral information can be processed at one time. The filters are independent and will not interfere with each other, so different relative CTL values can be detected in multiple channels.

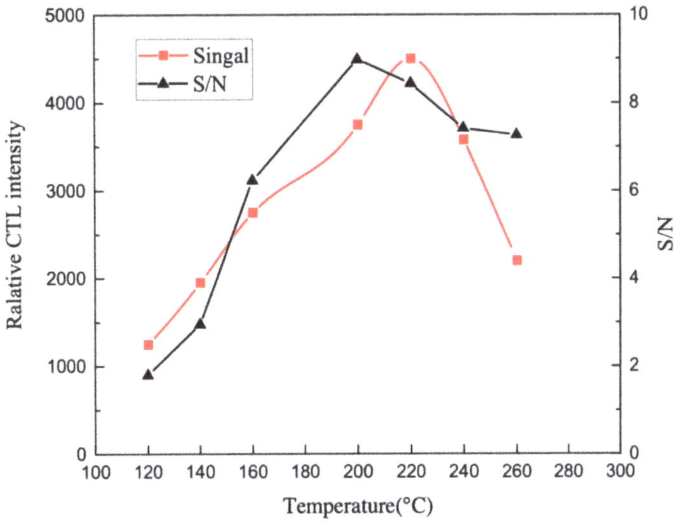

Figure 6. The luminescence intensity of acetone changed with different wavelengths (temperature: 220 °C, flow rate: 310 mL/min, concentration: 500 ppm).

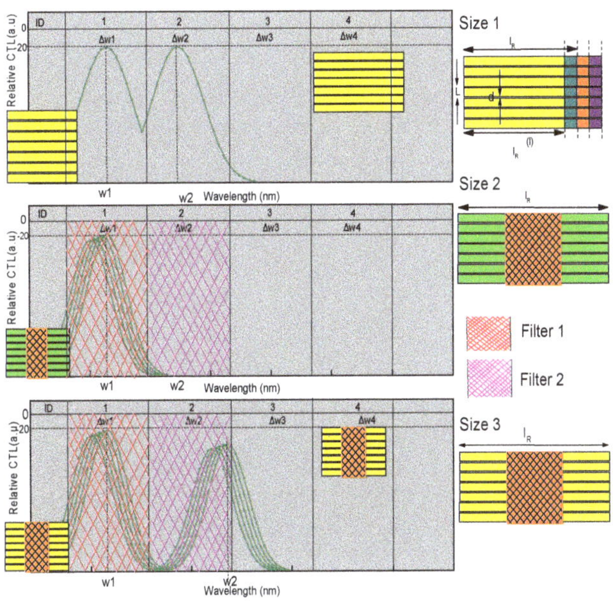

Figure 7. Different CTL signal spectra separated by multiple wavelength channels.

The impact of operating temperature on luminescence sensor systems has remained a serious issue. In this research, the influence of reaction temperature on luminescence intensity was studied. Figure 8 shown that, as the test temperature increased, the luminescence intensity of acetone also increased. This was due to the fact that at high temperatures, O atoms were doped into the layered CuO microsphere composite assembled from nanosheets and the conversion of acetone was higher. However, due to the increase in noise, the signal-to-noise ratio decreases significantly at about 220 °C. Given to these results, we chose 220 °C as the test temperature for further study.

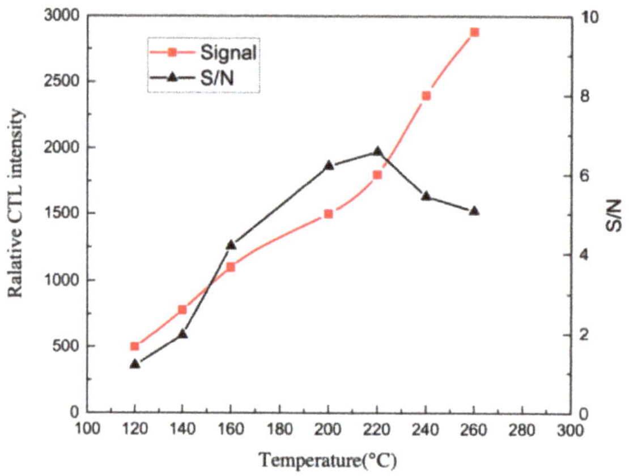

Figure 8. The luminescence intensity of acetone changed with different temperatures (wavelength: 590 nm, flow rate: 310 mL/min, concentration: 200 ppm).

3.3. Linearity Characteristics

In conclusion, under the best experimental conditions, this study investigated the relationship between the concentration platform of acetone and luminescence intensity. The results showed that there was a good linear relationship with luminescence intensity in the range of 15.0~2200.0 ppm, the correlation coefficient was $r = 0.9991$, and the detection limit was 1.65 ppm (Figure 9). Referring to the detection limit of 1.65 ppm, the sensitivity of the sensor can meet the detection requirements of acetone. The relative standard deviation of the six-time luminescence intensity measured in parallel for 100 ppm acetone gas was 1.9%. The detection performance of the method was comparable to that reported in the literature [26] and has a wider linear range. In addition, the detection method did not require pre-processing and the response speed was fast.

3.4. Linearity Characteristics

For the same catalyst, when the reactants are different, the catalytic effect it exhibits may vary widely. In order to verify the selectivity of nano-catalysts for the catalytic oxidation of acetone, this study investigated the responses of ethanol, dimethyl ether, n-butanol, propanol, and other gases that may interfere with the detection of acetone by this sensor. Under the optimal detection conditions of 590 nm, 210 °C, and 310 mL/min, different concentrations of ethanol, dimethyl ether, n-butanol, and propanol vapor were loaded into the reaction chamber through the air. The output signal represented the production of a response signal index, as the heating rod itself had the cataluminescence response for the test gas, which contained the output signal of the heating rod polished with catalytic material and the output signal of the clean heating rod. The output signal, which produced a response signal with specific combinations of catalytic materials at different concentrations of ethanol (a) and acetone (b) vapors, is shown in Figure 10.

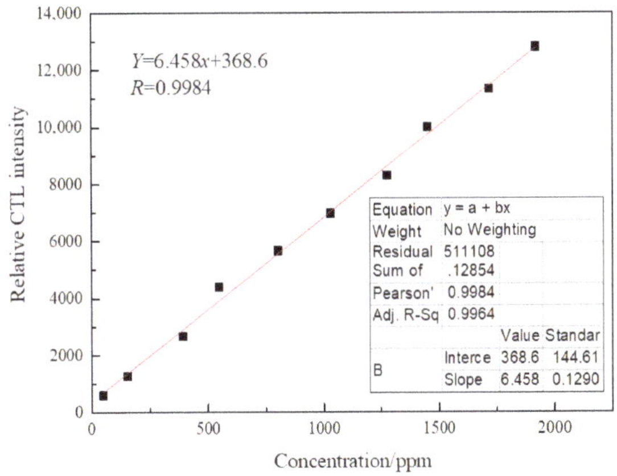

Figure 9. The calibration curve for acetone (temperature: 210 °C, flow rate: 310 mL/min, wavelength: 590 nm).

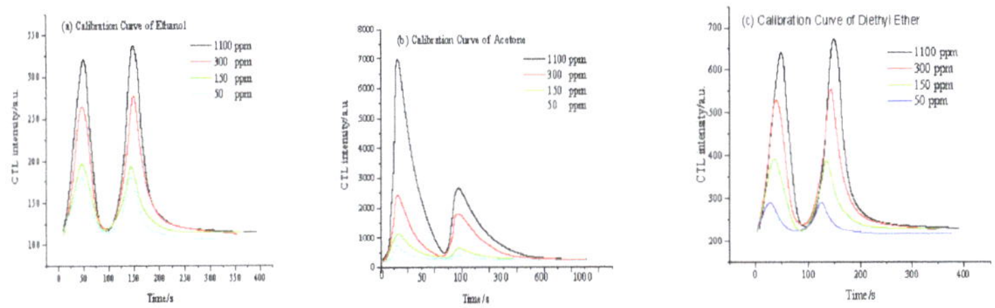

Figure 10. CTL response profiles of ethanol with (**a**) acetone and (**b**) diethyl ether and (**c**) signal values (flow rate: 310 mL/min, temperature: 220 °C, wavelength: 590 nm).

It can be concluded that the first-order CTL phenomenon due to the different alcohol molecules is derived from the catalytic oxidation of gas molecules to produce a first-order product. The generation of secondary CTL may result from the catalytic oxidation of the primary product to the secondary product. The remaining sample molecules are catalytically oxidized. Due to the different molecular structures of the different alcohols, the products produced by the stepwise CTL reaction are different. The adsorption performance, reactivity, and reaction rates of these products on the catalytic materials are different. Therefore, the luminescence intensity produced by the CTL reactions at different levels was also different, resulting in different I_1/I_2 values of different alcohol vapors on the same catalytic material combination. Since the luminous efficiency, reaction rate, and conversion rate of the same alcohol molecule on different catalytic materials are different, the I_1/I_2 values of the same alcohol vapors in different catalytic material combinations are also different. Therefore, based on the I_1/I_2 eigenvalues, it is possible to identify alcohol vapors with similar properties [27]. It is worth noting that the various alcohol vapors were in the range of experimentally selected concentrations (50 to 1000 ppm) and the I_1/I_2 values remained unchanged, which is very advantageous for practical applications of the CTL sensor. The interference of ethyl ether and ethanol on acetone was 7.5% and 4.2%; other gases do not interfere with the determination.

3.5. Selectivity, Repeatability and Stability of Sensor System

Since bad selectivity may lead to false positive results, selectivity is an important performance index for chemical sensors. Some common organic solvents and indoor air pollutants may interfere with the experimental results. Under optimized operating conditions, acetone was tested by introducing 1200 ppm of each compound at the same concentration into the sensor. As shown in Figure 11, there are emissions of ethanol and acetone. The luminous intensity of ether is only 7.5% of that of acetone, and that of ethanol is about 4.2% of that of acetone. This indicated that the sensor had good selectivity for acetone. The selectivity of the sensor can be determined by measuring the response generated by the interference gas of a specific concentration, that is, the sensor response generated by the target gas of a specific concentration. Cross-sensitivity reflects whether the measurement results are accurate, so the sensitivity and selectivity of the ideal sensor should be as high as possible. Clearly, the CuO-based luminescence sensor had better selectivity for acetone.

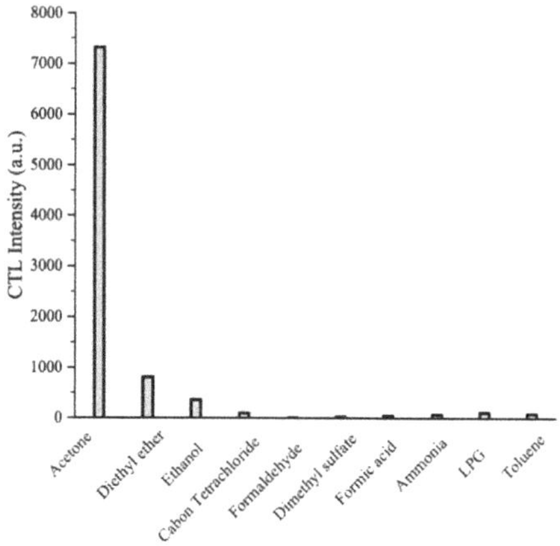

Figure 11. The luminescence responses of different gases (flow rate: 310 mL/min, temperature: 220 °C, wavelength: 590 nm, concentration: 1200 ppm).

For verification of the reproducibility and stability of this luminescence system, the study was conducted by performing six replicate measurements during 100 s. Response-recovery time: Response time refers to the time it takes for the gas sensor signal to rise from 0 to 90% of the equilibrium signal value; recovery time refers to the time it takes for the sensor's ventilation balance to return to 10% of the signal. In our study, response-recovery time was 'Response time'+' recovery time'. As shown in Figure 12a, the sensor displayed a fast response speed and recovery speed of within 4 s and within 20 s, respectively. The relative standard deviation (RSD, n = 6) of the six replicates was 4.1%, which indicated that the sensor had good reproducibility. To study the sensor's stability, the luminous intensity of 900 ppm acetone can be measured every 24 h under optimized conditions. As shown in Figure 12b, when the sensor worked continuously for 168 h, there was no change in luminous intensity. The RSD of the repeated measurements of the seven experiments was 3.7%. Because the sensor shows good response and recovery speeds and long-term stability, the sensor can monitor acetone quickly and continuously for a long time.

3.6. Mechanism Discussion

According to the molecular information obtained by the wireless catalytic luminescence sensor, a possible mechanism is proposed. The catalytic reaction which described the transformation of an organic compound on the CuO surface included two processes: (1) There were a large number of active centers on CuO surface which can adsorb acetone. (2) The surface of CuO can catalyze and oxidize the adsorbed acetone. (3) The high luminescence is due to the formation of active intermediates [16,28].

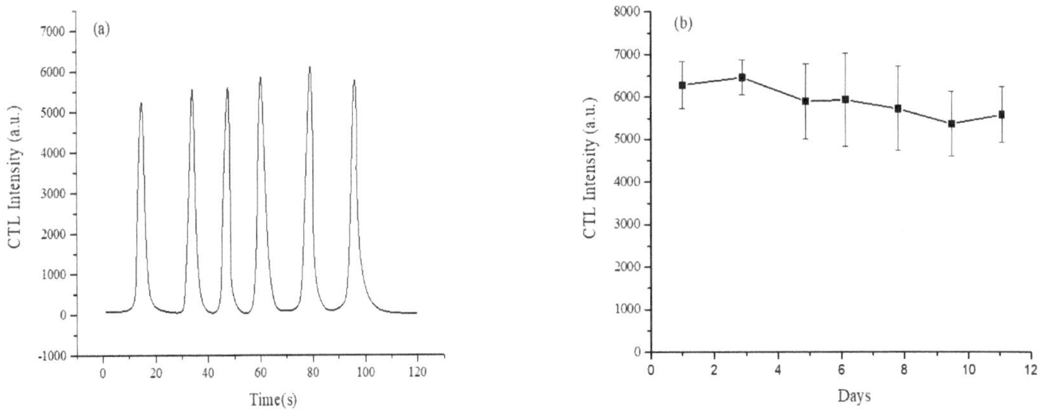

Figure 12. (**a**) Typical results obtained from six replicate determinations of acetone within 100 s. (**b**) The change trend of the luminescence intensity of acetone within one week (168 h). Flow rate: 310 mL/min, working temperature: 220 °C, wavelength: 590 nm, concentration: 900 ppm.

When acetone vapor passed through the CuO surface and was adsorbed, it might be oxidized by O_2 to active intermediates (CH_3CO^* or CO_2^*) under the catalysis of CuO. The catalytically oxidized of acetone on CuO surface for luminescence production may be as follows [29,30]:

$$CH_3COCH_3 + O_2 \rightarrow CH_3CO^* \tag{9}$$

$$CH_3CO^* \rightarrow CH_3CH_2COOH + hv \tag{10}$$

or

$$CH_3COCH_3 + O_2 \rightarrow CO_2^* \tag{11}$$

$$CO_2^* \rightarrow CO_2 + hv \tag{12}$$

The active intermediates of CH_3CO^* or CO_2^* were unstable and released energy (hv) in the form of luminescence. The results made acetone have a high CTL signal.

3.7. Sample Analysis

For the sake of testing the feasibility of the sensor for determining acetone in a mixed gas, the composition of the air was simulated (20 times), and three test samples containing a known concentration of acetone were measured in a linear range to evaluate the practical application of the CTL sensor. Sample 1 was a mixture of 500 ppm acetone and methanol with the same concentration, Samples 2 were prepared by adding 500 ppm acetone to 500 ppm diethyl ether, and Samples 3 were prepared by adding 500 ppm acetone to 500 ppm ethanol, respectively. The three samples were analyzed by CTL sensor and the concentration of acetone was calculated by a linear regression equation based on CTL intensity (Table 1). Acetone in the three artificial samples was well-quantified and the recovery was satisfactory [31]. The results indicate that the sensor was used for routine monitoring of the effects of acetone in the air and control the quality of the acetone, methanol, and diethyl ether mixture.

Table 1. Determination Result of Acetone in Artificial Samples.

Sample No.	Composition	Spiked Values (ppm)	Acetone Measured Values (ppm)	Average CTL Intensity	Recovery Time (s)
1	Acetone (500 ppm) Methanol (500 ppm)	498.4 482.9	481.1 ± 0.6	4023	97% × $T_{acetone}$
2	Acetone (500 ppm) Diethyl ether (500 ppm)	488.9 473.3	476.9 ± 1.1	4269	101% × $T_{acetone}$
3	Acetone (500 ppm) Ethanol (500 ppm)	479.1 485.2	486.4 ± 0.3	4108	94% × $T_{acetone}$

4. Conclusions

At this stage, some theoretical and applied studies of Cataluminescence are still not mature enough and further research is still needed. For a sensor, the main indicators to evaluate performance are sensitivity, selectivity, stability, and response speed. In this study, a catalytic light-emitting sensor was constructed by using alkaline earth metal oxide CuO nanomaterials with abundant sources, simple preparation, and no noble metal doping as catalysts, and the rapid detection of acetone was realized. This method has low energy consumption, simple device construction, low cost, fast response, no sample pretreatment, and simple and easy operation, which improve the applicability of the Cataluminescence detection method. At the same time, this study also broadened the scope of Cataluminescence sensors application and has potential application value in the fields of industrial testing, food safety, and agricultural production.

Author Contributions: Conceptualization, X.S.; methodology, G.S.; software, Y.Z.; validation, G.S. and X.S.; formal analysis, G.S.; investigation, X.S.; data curation, X.S.; writing—original draft preparation, G.S.; writing—review and editing, S.W.; visualization, Y.Z.; project administration, Y.Z.; funding acquisition, G.S. and S.W. All authors have read and agreed to the published version of the manuscript.

Funding: This work was Funded by China Postdoctoral Science Foundation (No. 2021M692473), Natural Science Foundation of Anhui Province (2108085QF260), and The Open Research Fund of National Engineering Research Center for Agro-Ecological Big Data Analysis & Application, Anhui University (No. AE202103), Anhui Provincial Department of Education Research Project (KJ2021A0179), the Key Natural Science Research Project for Colleges and Universities of Anhui Province (KJ2021ZD0056).

Institutional Review Board Statement: Not applicable.

Informed Consent Statement: Not applicable.

Data Availability Statement: The data presented in this study are available in the article.

Conflicts of Interest: The authors declare no conflict of interest.

References

1. Meng, F.; Lu, Z.; Zhang, R.; Li, G. Cataluminescence sensor for highly sensitive and selective detection of iso-butanol. *Talanta* **2019**, *1630*, 910–918. [CrossRef] [PubMed]
2. Yu, S.; Zhang, D.; Zhang, Y.; Pan, W.; Meteku, B.E.; Zhang, F.; Zeng, J. Green light-driven enhanced ammonia sensing at room temperature based on seed-mediated growth of gold-ferrosoferric oxide dumbbell-like heteronanostructures. *Nanoscale* **2020**, *36*, 18815–18825. [CrossRef] [PubMed]
3. Zhang, L.; Liu, T.; Ren, R.; Zhang, J.; He, D.; Zhao, C.; Suo, H. In situ synthesis of hierarchical platinum nanosheets-polyaniline array on carbon cloth for electrochemical detection of ammonia. *J. Hazard. Mater.* **2020**, *392*, 122342. [CrossRef] [PubMed]
4. Wu, Y.; Peng, S.; Xie, Q.; Xu, P. Nonlinear least squares with local polynomial interpolation for quantitative analysis of IR spectra. *Spectrochim. Acta Part A Mol. Biomol. Spectrosc.* **2019**, *206*, 147–153. [CrossRef]
5. Pfeifer, J.; Simon, M.; Heinritzi, M.; Piel, F.M.; Weitz, L.; Wang, D.S.; Granzin, M.; Müller, T.; Bräkling, S.; Kirkby, J.; et al. Measurement of ammonia, amines and iodine compounds using protonated water cluster chemical ionization mass spectrometry. *Atmos. Meas. Tech.* **2020**, *13*, 2501–2522. [CrossRef]

6. Perraud, V.; Li, X.; Smith, J.N.; Finlayson-Pitts, B.J. Novel ionization reagent for the measurement of gas-phase ammonia and amines using a stand-alone atmospheric pressure gas chromatography (APGC) source. *Rapid Commun. Mass Spectrom.* **2019**, *34*, e8561. [CrossRef]
7. Gu, G.; Zhu, H. A portable embedded explosion gas detection and identification device based on intelligent electronic nose system. *Sens. Rev.* **2016**, *36*, 57–63. [CrossRef]
8. Shi, G.; He, Y.; Li, B.; Zuo, L.; Yin, B.; Zeng, W.; Ali, F. Analysis and modeling of wireless channel characteristics for Internet of Things scene based on geometric features. *Future Gener. Comp. Syst.* **2019**, *101*, 492–501. [CrossRef]
9. Tang, Y.; Su, Y.; Yang, N.; Zhang, L.; Lv, Y. Carbon nitride quantum dots: A novel chemiluminescence system for selective detection of free chlorine in water. *Anal. Chem.* **2014**, *9*, 4528–4535. [CrossRef]
10. Shi, G.; He, Y.; Luo, Q.; Li, B.; Zhang, C. Portable device for acetone detection based on cataluminescence sensor utilizing wireless communication technique. *Sens. Actuators B Chem.* **2018**, *257*, 451–459. [CrossRef]
11. Koo, W.; Choi, S.; Kim, S.; Jang, J.; Tuller, H.L.; Kim, I. Heterogeneous sensitization of metal–organic framework driven metal@metal oxide complex catalysts on an oxide nanofiber scaffold toward superior gas sensors. *J. Am. Chem. Soc.* **2016**, *138*, 13431–13437. [CrossRef] [PubMed]
12. Chu, Y.; Zhang, Q.; Li, Y.; Xu, Z.; Long, W. A cataluminescence sensor for propionaldehyde based on the use of nanosized zirconium dioxide. *Microchim. Acta* **2014**, *181*, 1125–1132. [CrossRef]
13. He, H.; Zhang, D.; Guo, F.; Sun, F. A versatile microporous zinc (II) metal–organic framework for selective gas adsorption, cooperative catalysis, and luminescent sensing. *Inorg. Chem.* **2018**, *57*, 7314–7320. [CrossRef] [PubMed]
14. Zhang, R.; Cao, X.; Liu, Y.; Chang, X. Development of a simple cataluminescence sensor system for detecting and discriminating volatile organic compounds at different concentrations. *Anal. Chem.* **2013**, *85*, 3802–3806. [CrossRef]
15. Liu, W.; Wang, Y.; Bai, Z.; Li, Y.; Wang, Y.; Chen, L.; Xu, L.; Diwu, J.; Chai, Z.; Wang, S. Hydrolytically stable luminescent cationic metal organic framework for highly sensitive and selective sensing of chromate anions in natural water systems. *ACS Appl. Mater. Int.* **2017**, *19*, 16448–16457. [CrossRef]
16. Shi, G.; He, Y.; Zhang, Y.; Yin, B.; Ali, F. Detection and determination of harmful gases in confined spaces for the Internet of Things based on cataluminescence sensor. *Sens. Actuators B Chem.* **2019**, *296*, 126686. [CrossRef]
17. Cheng, H.; Zhou, Z.; Liu, T. Electro-spinning fabrication of nitrogen, phosphorus co-doped porous carbon nanofiber as an electro-chemiluminescent sensor for the determination of cyproheptadine. *RSC Adv.* **2020**, *39*, 23091–23096. [CrossRef]
18. Shen, C.; Lou, Q.; Liu, K.; Dong, L.; Shan, C. Chemiluminescent carbon dots: Synthesis, properties, and applications. *Nano Today* **2020**, *35*, 100954. [CrossRef]
19. Wang, J.; Jiang, M.; Yan, L.; Peng, R.; Huangfu, M.; Guo, X.; Li, Y.; Wu, P. Multifunctional Luminescent Eu (III)-Based Metal–Organic Framework for Sensing Methanol and Detection and Adsorption of Fe (III) Ions in Aqueous Solution. *Inorg. Chem.* **2016**, *55*, 12660–12668. [CrossRef]
20. Tantubay, K.; Das, P.; Sen, M.B. Ternary reduced graphene oxide–CuO/ZnO nanocomposite as a recyclable catalyst with enhanced reducing capability. *J. Environ. Chem. Eng.* **2020**, *8*, 103818. [CrossRef]
21. Sharma, G.P.; Bansal, A.; Singh, R. Thermal coefficients of Earth fuller reinforced with nano-oxide particles. *Nano Express* **2021**, *2*, 010024. [CrossRef]
22. Zhang, Y.; Zhang, J.; Jiang, Y.; Duan, Z.; Liu, B.; Zhao, Q.; Wang, S.; Yuan, Z.; Tai, H. Ultrasensitive flexible NH3 gas sensor based on polyaniline/SrGe4O9 nanocomposite with ppt-level detection ability at room temperature. *Sens. Actuators B Chem.* **2020**, *319*, 128293. [CrossRef]
23. Chow, D.M.; Sinefeld, D.; Kolkman, K.E.; Ouzounov, D.G.; Akbari, N.; Tatarsky, R.; Bass, A.; Xu, C.; Fetcho, J.R. Deep three-photon imaging of the brain in intact adult zebrafish. *Nat. Methods* **2020**, *17*, 605–608. [CrossRef] [PubMed]
24. Gao, D.; Yang, G.; Li, J.; Zhang, J.; Zhang, J.; Xue, D. Room-temperature ferromagnetism of flowerlike CuO nanostructures. *J. Phys. Chem. C* **2010**, *114*, 18347–18351. [CrossRef]
25. Wang, X.; Xi, G.; Xiong, S.; Liu, Y.; Xi, B.; Yu, W.; Qian, Y. Solution-phase synthesis of single-crystal CuO nanoribbons and nanorings. *Cryst. Growth Des.* **2007**, *7*, 930–934. [CrossRef]
26. Huang, X.; Huang, Z.; Zhang, L.; Liu, R.; Lv, Y. Highly efficient cataluminescence gas sensor for acetone vapor based on UIO-66 metal-organic frameworks as preconcentrator. *Sens. Actuators B Chem.* **2020**, *312*, 127952. [CrossRef]
27. Khnykov, A.Y.; Vdovichenko, A.Y.; Morozov, P.V.; Zavyalov, S.A.; Shevchenko, V.G.; Chvalun, S.N. Gas-sensing properties of poly (p-xylylene)-Titanium thin film nanocomposite, prepared by vapor deposition polymerization. *Sens. Actuators B Chem.* **2020**, *320*, 128367. [CrossRef]
28. Li, L.; Hu, Y.; Deng, D.; Song, H.; Lv, Y. Highly sensitive cataluminescence gas sensors for 2-butanone based on g-C3N4 sheets decorated with CuO nanoparticles. *Anal. Bioanal. Chem.* **2016**, *408*, 8831–8841. [CrossRef]
29. Zeng, N.; Long, Z.; Wang, Y.; Sun, J.; Ouyang, J.; Na, N. An acetone sensor based on plasma-assisted cataluminescence and mechanism studies by online ionizations. *Anal. Chem.* **2019**, *91*, 15763–15768. [CrossRef]
30. Wang, S.; Shi, W.; Lu, C. Chemisorbed oxygen on the surface of catalyst-improved cataluminescence selectivity. *Anal. Chem.* **2016**, *88*, 4987–4994. [CrossRef]
31. Watanabe, E.; Seike, N. Detection of herbicide clopyralid at nanogram per gram level in agricultural products using easy-to-use micro liquid–liquid extraction followed by analysis with ultraperformance liquid chromatography–tandem mass spectrometry. *J. Chromatogr. A* **2020**, *1630*, 461578. [CrossRef] [PubMed]

chemosensors

MDPI

Article

A New Kind of Chemical Nanosensors for Discrimination of Espresso Coffee

Giuseppe Greco [1], Estefanía Núñez Carmona [2,*], Giorgio Sberveglieri [1,3], Dario Genzardi [1] and Veronica Sberveglieri [1,2]

1 Nano Sensor Systems S.r.l. (NASYS), Spin-Off University of Brescia, Brescia, Via Camillo Brozzoni, 9, 25125 Brescia, Italy; giuseppe.greco@nasys.it (G.G.); giorgio.sberveglieri@nasys.it (G.S.); dario.genzardi@nasys.it (D.G.); veronica.sberveglieri@ibbr.cnr.it (V.S.)

2 National Research Council, Institute of Bioscience and Bioresources (CNR-IBBR), Via J.F. Kennedy, 17/i, 42124 Reggio Emilia, Italy

3 Department of Information Engineering, University of Brescia, 25123 Brescia, Italy

* Correspondence: estefania.nunezcarmona@ibbr.cnr.it

Abstract: There are different methods to extract and brew coffee, therefore, coffee processing is an important factor and should be studied in detail. Herein, coffee was brewed by means of a new espresso professional coffee machine, using coffee powder or portioned coffee (capsule). Four different kinds of coffees (Biologico, Dolce, Deciso, Guatemala) were investigated with and without capsules and the goal was to classify the volatiloma of each one by Small Sensor System (S3). The response of the semiconductor metal oxide sensors (MOX) of S3 where recorded, for all 288 replicates and after normalization ΔR/R0 was extracted as a feature. PCA analysis was used to compare and differentiate the same kind of coffee sample with and without a capsule. It could be concluded that the coffee capsules affect the quality, changing on the flavor profile of espresso coffee when extracted different methods confirming the use of s3 device as a rapid and user-friendly tool in the food quality control chain.

Keywords: gas sensors; coffee capsule; Volatilome; Small Sensor Systems

Citation: Greco, G.; Carmona, E.N.; Sberveglieri, G.; Genzardi, D.; Sberveglieri, V. A New Kind of Chemical Nanosensors for Discrimination of Espresso Coffee. *Chemosensors* **2022**, *10*, 186. https://doi.org/10.3390/chemosensors10050186

Academic Editors: Fanli Meng, Zhenyu Yuan and Dan Meng

Received: 11 April 2022
Accepted: 10 May 2022
Published: 16 May 2022

1. Introduction

Coffee is one of the most consumed beverages in the world. Moreover, it is the second one after water and its consumption constantly increases [1]. Coffee brewing methods can be changed depending on the geographic, cultural and social environment as well as individual preferences [2]. Espresso coffee (EC) is the traditional Italian extraction method that requires more control of many parameters, such as temperature, pressure, grind size and knowledge. Bartender training is fundamental. He must adopt a series of measures in the extraction phase and perform checks and interventions of ordinary management and periodic maintenance of the equipment. The quality of coffee is globally appreciated. However, it is difficult to manage. Big companies have developed portioned coffee segments to overcome the aforementioned issues. The application of this new strategy to extract EC is related to coffee capsules. The coffee capsule is a single-dose coffee product, which is roasted, ground and perfectly portioned in a small aluminum or plastic container (generally in cylindrical form). Furthermore, coffee capsules exhibit an extended shelf life of the product compared to the not portioned ones. This is possible owing to the capsule packaging mainly using polypropylene. The coffee aroma is composed of over 800 volatile substances that characterize the mixture and its intensity can be changed depending on the extraction method [3]. The well-known and the most abundant classes of chemical compounds are aldehydes, phenols, ketones, pyridines and furans. These compounds can be easily extracted by several coffee extraction methods combining the pressure and the temperature of processes. Coffee supply chain have a lot of steps, until reaching the final one of extraction. Coffee brewing is a solid-liquid extraction process and the extraction kinetics is heavily influenced by the aforementioned parameters regarding the different chemical compounds present in roasted coffee that will develop subsequently the volatile fraction. The common consumer is careful to the most important coffee characteristics. One of these is the coffee because it is considered to be one of the fundamental parameters in order to categorize its overall quality [2]. The right processing of the raw material (green coffee beans) is crucial to obtain a high-quality product. There are a lot of factors that could influence coffee final properties. Specifically, the main ones that surely have an influence on coffee extraction and its sensor profile are six [4]. These are: plant varieties, growing region/conditions, processing methods (from coffee cherries to green coffee bean), roasting levels, grinding size and brewing methods [5]. The conditions and the parameters of the roasting process are so important in order to let the precursors developed in the desired way. The green beans contain them, and their influence can change regarding the changeable aforementioned conditions [6]. Although the aroma and flavors are characterized by the origin of green coffee, the roasting process affects the production of volatile compounds resulting in differences in the complexity of coffee aroma. In general, roasting ruptures the cell structure of green coffee beans exposing it to heat that drives out the moisture and releases the aromatic compounds that have been chemically bound in the beans [4].

Several classes of molecules are volatilized from the bean such as: carbon dioxide, aldehydes, ketones, ethers, acetic acid, methanol, oils and glycerol. The volatilization of different volatile compounds can change with regard to different temperatures, and as pyrolysis continues, the volatile fraction of the product continues to develop until the process ends [7]. Hence, the extraction process radically changes the aroma fingerprint. Regarding to this, new types of chemical gas sensors have recently been studied in order to classify food samples through their volatiloma, and, consequently, to have a better knowledge of the volatile fraction.

The necessity of developing accurate analytical methods has prompted the present study. The other techniques, such as Gas-Chromatography and Mass Spectrometry (GC-MS), are useful but they need intense training, they are time consuming and generally have higher costs. On the contrary, the use of an array of metal oxide (MOX) gas sensors allows classification of different kinds of samples with high sensitivity, quick responses and low costs. MOX sensors are conductometer sensors, so called thanks to their ability to transduce a chemical signal in an electrical resistance signal [8]. Interactions happen between volatile compounds and sensing material. MOX gas sensors are non-specific sensors; hence, they can be suited for different classes of compounds with different sensibility.

Furthermore, this is the main reason that sensor arrays are used in most applications. Overall, the aims of this work are:

- Discrimination of four different typologies of "Molinari" coffee capsules (Biologico, Deciso, Dolce, Guatemala) using a new innovative technology based on semiconductor metal oxide sensors (MOX).
- To underline differences between coffee extracted with and without capsule packaging.

2. Materials and Methods

2.1. Sample Preparation

This study is based on the analysis of four types of coffee capsules: "Biologico", "Deciso", "Dolce" and "Guatemala". A professional machine "Spaziale S2" (LA SPAZIALE S.p.A., Casalecchio di Reno, BO, Italia) was used for the coffee extraction. The coffee extraction was performed in 18 L containers of "Rocca Galgana" (Citerna di Fornovo di Taro, Parma, Italy) mineral water to overcome the inhomogeneities due to the water [1,7].

This study was carried out considering two approaches: (i) coffee was prepared by entering the coffee capsule in the correspondent extraction tool; (ii) coffee was pulled out from each capsule and extracted without capsule packaging. The quantity of coffee extracted was 62.5 mL (espresso shot coffee [9]) and it was collected using a beaker. Then, 20 mL chromatographic vials were filled with 1.2 mL of coffee. 36 vials for each sample were made obtaining 144 vials (36 Biologico, 36 Deciso, 36 Dolce, 36 Guatemala) of coffee extracted with capsule and 144 vials (36 biologico, 36 deciso, 36 dolce, 36 guatemala) of coffee extracted without capsule (Table 1).

Table 1. Samples Analyzed.

Sample	With Capsule	Without Capsule
Biologico	36	36
Deciso	36	36
Dolce	36	36
Guatemala	36	36
Total	**144**	**144**

The vials containing the sample were sealed with an aluminum cap and a of polytetrafluoroethylene (PTFE) and silicone septa. In order to heat samples and achieve the equilibrium of the volatile compounds between the headspace [10] and the liquid phase and to reduce variables, the vials were initially incubated at 30 °C for 10 min. Afterwards, the extraction phase was performed, where the autosampler syringe is exposed in the head-space of the vial for 1 min to allow the absorption of volatile compounds.

2.2. *Small Sensor System (S3)*

2.2.1. Analysis Conditions

The S3 device, acronym for Small Sensor Systems, is constructed by Nano Sensor Systems S.r.l. (Reggio Emilia, Italy, www.nasys.it (accessed on 18 February 2022)) This device has been already used with remarkable success in other previous studies in the field of quality control and food technology [7]. S3 is connected to the autosampler HT2010H (HTA s.r.l., Brescia, Italy) equipped with a carousel with 42 positions to accommodate the vials for sampling. The system allows to collect and analyze the data acquired in the cloud making S3 an IoT device for the management and control of signals [8]. S3 is composed of three essential parts [7]:

1. The sensor steel chamber which contains the six MOX sensors. This allows the sensor to be separated from the environment, except for an inlet and an outlet path for the passage of volatile compounds. Other types of sensors are placed in order to control several parameters during the analysis. These are the temperature, humidity, and flow in the chamber.
2. The fluid dynamic circuit is composed of a pump (Knf, model: NMP05B), polyurethane pipes, a solenoid valve and a metal cylinder, which contains activated carbon. The activated carbon is used for filtering any type of odors present in the environment that can alter the final response. The solenoid valve, positioned at the chamber inlet to control the pump flow with a maximum of 250 sccm.
3. The electronic board elaborates the sensor responses through the detection of electrical resistance. In addition, it controls the operating temperature of sensors, which is an important parameter for the detection of volatile compounds. Finally, the system is able to send the data to the Web App dedicated to the S3 device through an internet connection.

The sensor response is based on the change of its resistance over time caused by interaction with different kinds of volatile compounds or surrounding environment. The reactions between the oxygen species adsorbed on the surface of the sensor and the target molecules lead to the variation in the concentration of charge carriers in the sensing material affecting its electrical conductance [9].

2.2.2. S3 Data Processing

Processing of the S3 sensor signals was performed using MATLAB® R2019b software (MathWorks, Natick, MA, USA) in order to extract the features of the sensor response. Sensors' responses in terms of resistance (Ω) were normalized to the first value of the acquisition (R0). For all of the sensors, the difference between the first value and the minimum value during the analysis time was calculated. Hence, the ΔR/R0 parameter was calculated and has been use as a feature for all the sensors response to the 36 replicates of each sample. The standard deviation of the ΔR/R0 parameter was calculated for each group of sample measurement prior to proceed with PCA analysis revealing a maximum uncertainty of the 10%. Once the data matrix was calculated, principal component analysis (PCA) was applied to the data in order to verify the variation of the VOCs in the different samples [10].

This technique consists of clustering the sample variables through linear combinations that describe the link between one sample to the others. This results in the main components (PC), which are far fewer than the original variables. These new variables are structured in such a way as to be orthogonal to each other (not correlated). Moreover, most of the variability of the samples is present in the first main components. As a consequence PC1 shows the largest variation. Next, PC2, which represents the second largest variation. This can continue until all the variables are explained. These conditions allow the detection of any groupings [11]. They are also known as clusters that represent samples united by characteristics. PCA is not a classification technique, but a technique that may provide the distribution of the samples within the main components considered in the hyperplane.

3. Results and Discussion

3.1. Sensor Response Regarding the Discrimination of Molinari Coffee Capsules

The response curve of a sensor is represented in Figure 1, where on the x-axis is indicated the time (s), while in the y-axis the normalized resistance is reported. The use of normalized parameters allows us to obtain data without dimensions, hence they are a-dimensional variables, where the variability is equal to 1 and the average is 0. The choice to use normalized variables is advantageous in the analysis of samples that have different units of measurement or size, which would prevent accurate analysis of samples.

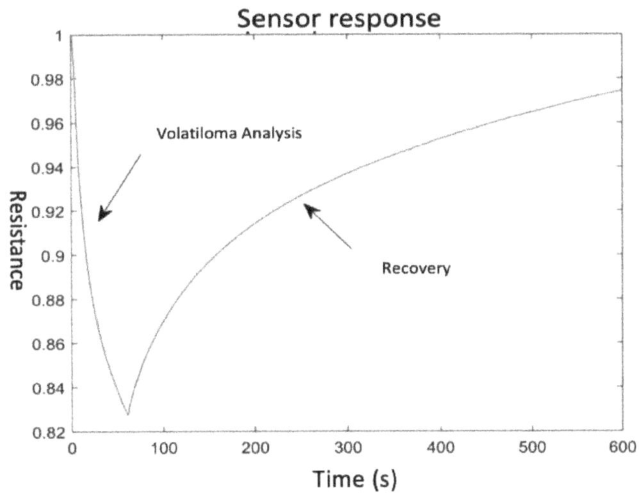

Figure 1. Schemes of the typical sensor response.

As can be seen in Figure 1, the signal starts from a resistance value of 1, which is reduced. The studying of samples using the sensing system is based on two phases: the first one is the analysis of the coffee volatiloma, where the volatile compounds come into contact with the sensors. This interaction leads to a decrease in the electrical resistance of sensing material. The second stage is called recovery where the base line is restored.

In this case, filtered ambient air is introduced to the chamber, the adsorbed gaseous compounds are desorbed from the surface of sensors and their resistances are recovered to their steady stage values. The first phase lasted 1 min followed by the recovery to the base-line resistance, which required 9 min. Hence the total analysis time is 10 min. Small Sensor System (S3) analysis was able to discriminate the samples, and describe the volatile fingerprint. Figure 2 illustrates a complete separation of the Biologico coffee capsule from the other ones on the hyperplane. This means that it is the only one that doesn't have any match, while Dolce Deciso and Guatemala coffee capsules are on the same side. In particular, Dolce and Deciso are mainly overlapped. Guatemala's are well separated among PC1 and PC2. Therefore, the sensors must have perceived some differences, although not as much as in Biologico's case.

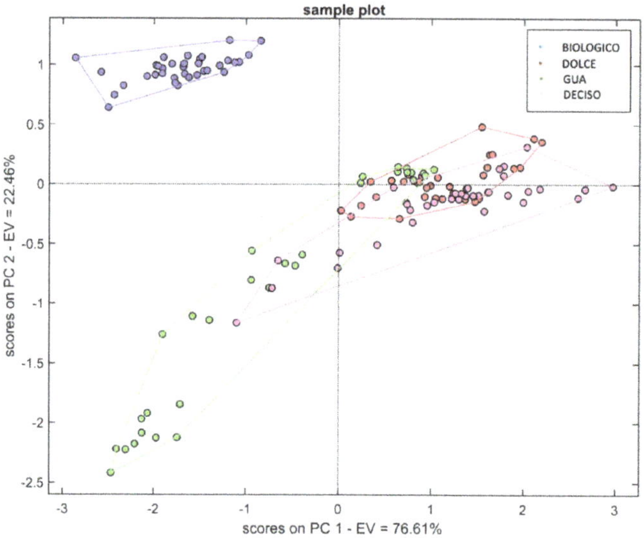

Figure 2. PCA Molinari coffee capsules.

3.2. *Coffee with and without Capsule Packaging Comparison*

The most important parameter about coffee is its aroma. It is the first factor which take consumer attention and influence his appreciation and selection; hence, it is considered an important quality trait. Espresso quality depends on several parameters which must be gauged in order to obtain the proper ratio of desirable components for the most desirable product [12].

In particular the ratio of water to coffee, time and grind size impact the brewing process. In addition, Espresso Coffee is a product characterized by an intense viscosity caused by high concentration of Total Dissolved Solids (TDS) [13].

The latter is usually a highly desired quality of the espresso brew because it gives an appreciated fullness to the final product. Beyond that, TDS is a non-specific analysis technique generally used to do a quantitative analysis regarding the extraction performance, also called extraction yield. Moreover, the most common methods to assess aroma are based on expensive equipment or human senses through sensory evaluation, which is time-consuming and requires highly trained assessors to avoid subjectivity [14].

On the contrary, the S3 analysis could easily find volatiloma differences between coffee extracted with and without capsule packaging. Figures 3–6 reports Principal Component Analysis (PCA) performed to process the obtained data. It is possible to say overall that PCA showed the EV never under 78%. This represents an optimum result since at least 78% of the total variability of the samples was enclosed between the hyperplane (enclosed between the first two principal components).

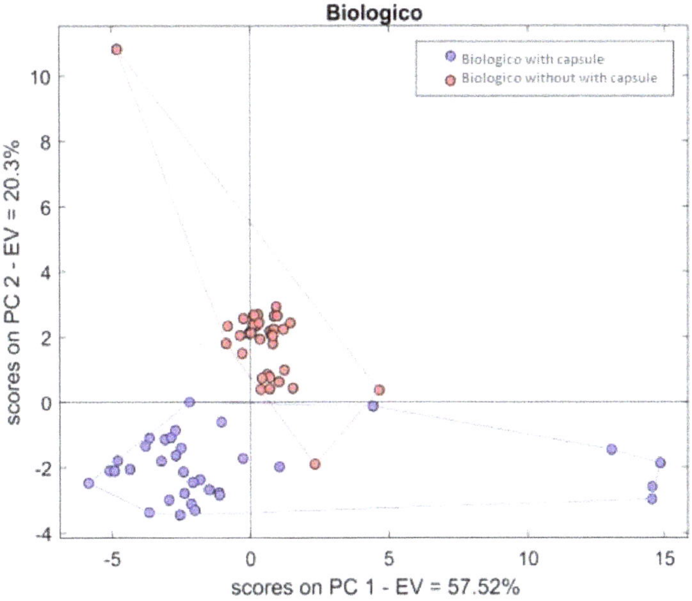

Figure 3. S3 comparison Biologico with and without capsule.

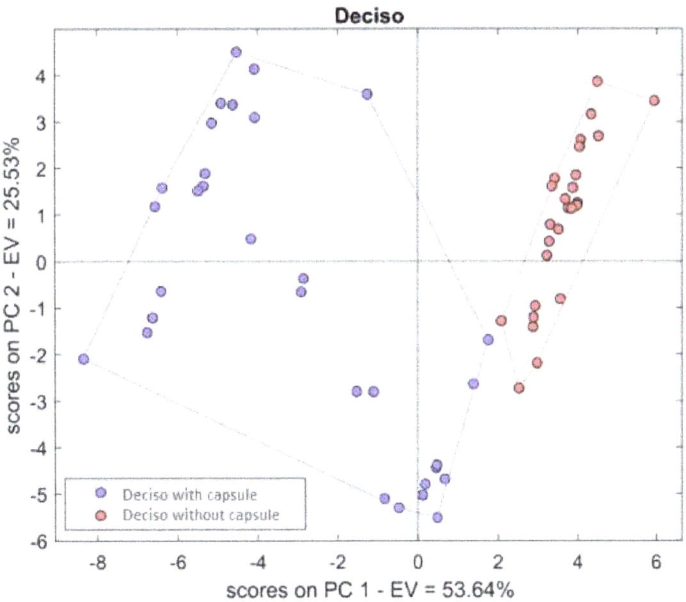

Figure 4. S3 comparison Deciso with and without capsule.

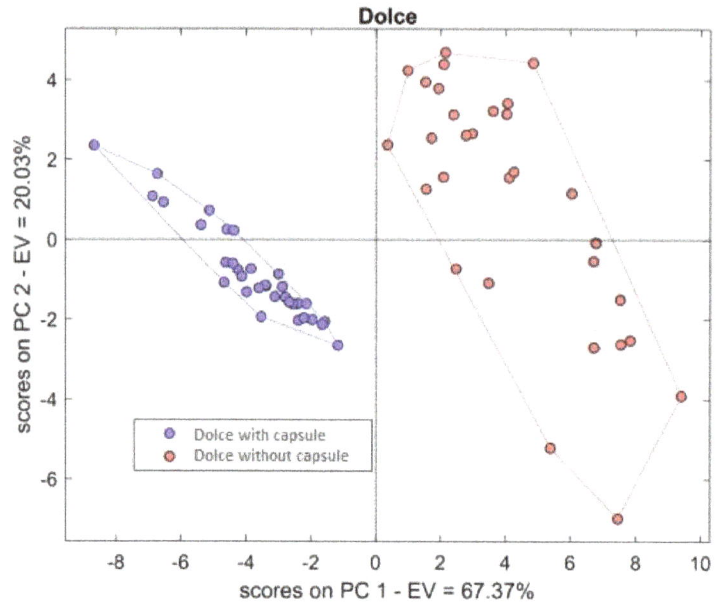

Figure 5. S3 comparison Dolce with and without capsule.

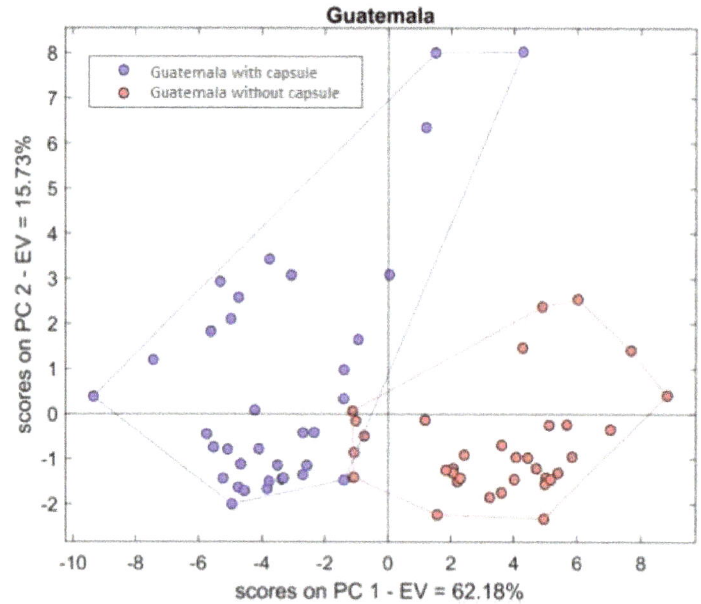

Figure 6. S3 comparison Guatemala with and without capsule.

PC1 is always the component with a larger load, reaching the maximum value of 67% in the case of the Deciso samples. The six single sensors perfectly discriminated the samples (one for each type) where with and without capsule packaging coffees were compared. The six single sensors perfectly discriminated the samples (one for each type) where with and without capsule packaging coffees were compared. As a matter of fact, Ferini et al. [15] made some investigations about the differences of coffee product obtained with different techniques. One of them was to check possible differences between espresso and coffee capsules through a human sensorial test. At this point the consumers, even if coffee capsules are intellectually judged as similar to espresso coffee, noticed differences when tasting the beverages.

This is probably due to the compression of the coffee capsule, which improves the extraction of the volatile fraction. Thus, the coffee volatile fraction is far from simple, considering the huge number of different chemical compounds that must be considered. As a matter of fact, over a thousand different chemical entities have been detected in coffee beans such as aldehydes, phenols, ketones, pyridines and furans. A significant number of these compounds will be extracted during brewing. It's a comparatively small subset of chemicals that affect the aroma. Among all the variables, there are two main factors which must be considered to investigate the effect of compounds' aroma: the concentration of the compound and the compound's odor threshold, or the minimum concentration at which it is possible to detect its smell [16]. The result of these two factors gives the compounds 'odor activity value' (OAV) [17], responsible parameter of the overall aroma.

Several chemicals are contributors and responsible about the evolution of the coffee volatile fraction. The most significant chemicals which enrich the volatiloma coffee product are the following:

- Aldehyde compounds, which generally add a fruity green aroma
- Furans, which contribute with caramel-like odors
- Pyrazines, which have an earthy scent.
- Guaiacol and related phenolic compounds offer smoky, spicy tones, and pyrroles.
- Alkaloids, as caffeine or trigonelline. Caffeine content is highly variable in a range between 0.64 mg/mL and 4.89 mg/mL [18] considering different parameters (pressure and temperature) values.
- Chlorogenic acids (CGAs), the most important are 3-, 4- and 5-CQA [19].

Over the last 40 years, the term "electronic nose" (EN) has defined a device equipped with an array of gas sensors capable of providing a response as a function of a stimulus provided by volatile chemical compounds (VOCs). Numerous studies have started from this idea, which have led to significant improvements and advantages, especially useful for providing a device capable of monitoring situations and applications in real-time., especially in the food and beverages industry [20], having regard to the gas sensors capabilities of providing a response as a function of a stimulus provided by volatile chemical compounds. Known as also the aroma sensor, mechanical nose, multi-sensor array, artificial nose, odor sensing system, or electronic olfactometry, the initial goal of EN technology was to simulate the mammalian nose to obtain a fast response regarding the characteristics of the analyte, high sensitivity for odours and high discrimination between them [21]. As a matter of fact, EN sensing is mainly used for the classification and characterization of the overall aroma pattern of EC samples, so it is generally coupled with other analytical techniques such as GC [22], which can qualitatively and quantitatively characterize EC odor-active compounds [23].

4. Conclusions

This study is based on the analysis of 4 different types of coffee samples (Biologico, Deciso, Dolce, Guatemala). The analyses were carried out by the Small Sensor System (S3) device, where the sensors were able to identify the differences of the biologico capsule from the other ones and to certify the differences between the samples with and without capsule packaging.

The obtained results (PCA Figures 3–6) indicate that the coffee aroma profile is incredibly influenced by the compression of the coffee pod, which improves the extraction of the volatile fraction. Indeed, the results show how the coffee obtained using the capsules has a different volatile fraction. Moreover, these ones would also provide standardized organoleptic characteristics of each coffee without the need of trained operators. This can be a huge advantage for household use but also for professional use allowing anyone easily to prepare a high quality final product. Overall, two main advantages can be highlighted:

the first one is that the coffee aroma intensity can be maintained over the time and avoid the deterioration that would occur for the coffee without capsule. The second one is that the volatile fraction of the final product of the capsule is higher than the one without the capsule packaging.

Author Contributions: Conceptualization, G.G., E.N.C., D.G. and V.S.; methodology, G.G., E.N.C., D.G. and V.S.; software, G.G.; validation, G.G., E.N.C., D.G. and V.S.; formal analysis, G.G., E.N.C., D.G. and V.S.; investigation, G.G., E.N.C., D.G. and V.S.; resources, G.G., E.N.C., D.G. and V.S.; data curation, G.G., E.N.C., D.G. and V.S.; writing—original draft preparation, G.G., E.N.C., D.G. and V.S.; writing—review and editing, G.G., E.N.C., D.G., and V.S.; visualization, G.G., E.N.C., D.G. and G.S.; supervision, G.G., E.N.C., D.G., G.S. and V.S.; project administration, G.G., E.N.C., D.G. and V.S.; funding acquisition, E.N.C. and V.S. All authors have read and agreed to the published version of the manuscript.

Funding: This research received no external funding.

Institutional Review Board Statement: Not applicable.

Informed Consent Statement: Not applicable.

Data Availability Statement: Not applicable.

Acknowledgments: Authors want to thank "Caffè Molinari" (Via Francia, Modena, Italy) for the gently sample supply.

Conflicts of Interest: The authors declare no conflict of interest.

References

1. Greco, G.; Núñez-Carmona, E.; Abbatangelo, M.; Fava, P.; Sberveglieri, V. How Coffee Capsules Affect the Volatilome in Espresso Coffee. *Separations* **2021**, *8*, 248. [CrossRef]
2. Nancy, C.; Fernandez-Alduenda, M.; Moreno, F.; Ruiz, Y. Coffee extraction: A review of parameters and their influence on the physicochemical characteristics and flavour of coffee brews. *Trends Food Sci. Technol.* **2020**, *96*, 45–60.
3. Herron, R.; Lipphardt, A.; Euler, L.; Nolvachai, Y.; Marriott, P.L. Person-portable gas chromatography-toroidal ion trap mass spectrometry analysis of coffee bean volatile organic compounds. *Int. J. Mass Spectrom.* **2022**, *473*, 116797. [CrossRef]
4. Illy, A.; Viani, R. (Eds.) *Espresso Coffee The Science of Quality*; Elsevier Academic Press: San Diego, CA, USA, 2004.
5. Natnicha, B.; Adhikari, K.; Chambers, E. Evolution of sensory aroma attributes from coffee beans to brewed coffee. *LWT-Food Sci. Technol.* **2011**, *44*, 2185–2192.
6. Bottazzi, D.; Farina, S.; Milani, M.; Montorsi, L. A numerical approach for the analysis of the coffee roasting process. *J. Food Eng.* **2012**, *112*, 243–252. [CrossRef]
7. Núñez-Carmona, E.; Abbatangelo, M.; Sberveglieri, V. Internet of Food (IoF), Tailor-Made Metal Oxide Gas Sensors to Support Tea Supply Chain. *Sensors* **2021**, *21*, 4266. [CrossRef] [PubMed]
8. Navarini, L.; Rivetti, D. Water quality for Espresso coffee. *Food Chem.* **2010**, *122*, 424–428. [CrossRef]
9. Abbatangelo, M.; Núñez-Carmona, E.; Duina, G.; Sberveglieri, V. Multidisciplinary Approach to Characterizing the Fingerprint of Italian EVOO. *Molecules* **2019**, *24*, 1457. [CrossRef] [PubMed]
10. Núñez-Carmona, E.; Abbatangelo, M.; Zappa, D.; Comini, E.; Sberveglieri, G.; Sberveglieri, V. Nanostructured MOS Sensor for the Detection, Follow up, and Threshold Pursuing of Campylobacter Jejuni Development in Milk Samples. *Sensors* **2020**, *20*, 2009. [CrossRef] [PubMed]
11. Gonzalez Viejo, C.; Tongson, E.; Fuentes, S. Integrating a Low-Cost Electronic Nose and Machine Learning Modelling to Assess Coffee Aroma Profile and Intensity. *Sensors* **2021**, *21*, 2016. [CrossRef] [PubMed]
12. Eiermann, A.; Smrke, S.; Guélat, L.M.; Wellinger, M.; Rahn, A.; Yeretzian, C. Extraction of single serve coffee capsules: Linking properties of ground coffee to extraction dynamics and cup quality. *Sci. Rep.* **2020**, *10*, 17079. [CrossRef] [PubMed]
13. Petracco, M. The Cup. In *Espresso Coffee: The Chemistry of Quality*; Illy, A., Viani, R., Eds.; Academic Press: Cambridge, MA, USA, 1995; p. 187.
14. Kim, N.S.; Lee, J.H.; Han, K.M.; Kim, J.W.; Cho, S.; Kim, J. Discrimination of commercial cheeses from fatty acid profiles and phytosterol contents obtained by GC and PCA. *Food Chem.* **2014**, *143*, 40–47. [CrossRef] [PubMed]
15. Ferini, J.L.; Morales, M.V.; da Silva, T.A.; Pedreira, J.R.M.; de Godoy, N.T.; de Oliveira Garcia, A.; Tfouni, S.A.V. Consumers' perception of different brewed coffee extractions using the sorting technique. *J. Sens. Stud.* **2021**, *36*, e12633. [CrossRef]
16. The Chemical Compounds behind the Aroma of Coffee. 17 February 2015. Available online: https://www.compoundchem.com/2015/02/17/coffee-aroma/ (accessed on 18 February 2022).
17. Wu, C.; Liu, C.; Yan, L.; Chen, H.; Shao, H.; Meng, T. Assessment of odor activity value coefficient and odor contribution based on binary interaction effects in waste disposal plant. *Atmos. Environ.* **2014**, *103*, 231–237. [CrossRef]

18. Gloess, A.; Vietri, A.; Wieland, F.; Smrke, S.; Schönbächler, B.; Sánchez López, J.A.; Petrozzi, S.; Bongers, S.; Koziorowski, T.; Yeretzian, C. Evidence of different flavour formation dynamics by roasting coffee from different origins: On-line analysis with PTR-ToF-MS. *Int. J. Mass Spectrom.* **2014**, *365–366*, 324–337. [CrossRef]
19. Angelino, D.; Tassotti, M.; Brighenti, F.; Del Rio, D.; Mena, P. Niacin, alkaloids and (poly)phenolic compounds in the most widespread Italian capsule-brewed coffees. *Sci. Rep.* **2018**, *8*, 17874. [CrossRef] [PubMed]
20. Crozier, T.W.M.; Stalmach, A.; Lean, M.E.J.; Crozier, A. Espresso coffees, caffeine and chlorogenic acid intake: Potential health implications. *Food Funct.* **2012**, 3, 30–33. [CrossRef] [PubMed]
21. Angeloni, S.; Mustafa, A.M.; Abouelenein, D.; Alessandroni, L.; Acquaticci, L.; Nzekoue, F.K.; Petrelli, R.; Sagratini, G.; Vittori, S.; Torregiani, E.; et al. Characterization of the Aroma Profile and Main Key Odorants of Espresso Coffee. *Molecules* **2021**, *26*, 3856. [CrossRef] [PubMed]
22. Lolli, V.; Acharjee, A.; Angelino, D.; Tassotti, M.; Del Rio, D.; Mena, P.; Caligiani, A. Chemical Characterization of Capsule-Brewed Espresso Coffee Aroma from the Most Widespread Italian Brands by HS-SPME/GC-MS. *Molecules* **2020**, *25*, 1166. [CrossRef] [PubMed]
23. Karakaya, D.; Ulucan, O.; Turkan, M. Electronic Noseand Its Applications: A Survey. *Int. J. Autom. Comput.* **2020**, *17*, 179–209. [CrossRef]

chemosensors

Article

Dynamic Measurement of VOCs with Multiple Characteristic Peaks Based on Temperature Modulation of ZnO Gas Sensor

Xue Shi [1,†], Hua Zhang [1,†], Hanyang Ji [1] and Fanli Meng [1,2,3,*]

1 College of Information Science and Engineering, Northeastern University, Shenyang 110819, China; 1970698@stu.neu.edu.cn (X.S.); zhanghua@ise.neu.edu.cn (H.Z.); jihy1996@sina.cn (H.J.)
2 Hebei Key Laboratory of Micro-Nano Precision Optical Sensing and Measurement Technology, Qinhuangdao 066004, China
3 Key Laboratory of Data Analytics and Optimization for Smart Industry, Northeastern University, Ministry of Education, Shenyang 110819, China
* Correspondence: mengfanli@ise.neu.edu.cn
† These authors contributed equally to this work and should be regarded as co-first authors.

Abstract: Volatile organic compounds (VOC) harm human health seriously in the air. Therefore, it is essential to recognize VOC gases qualitatively and quantitatively. The dynamic measurement method can improve the selectivity of metal oxide semiconductor (MOS) gas sensors to VOC, but there is a problem of the insufficient number of characteristic peaks. From the experimental point of view, the primary judgment basis for the correct qualitative and quantitative recognition of VOC gases by the dynamic measurement method is the characteristic peak of the dynamic response signal. However, the traditional dynamic measurement method generally only has two characteristic peaks. In this experiment, the voltage was changed at the time of the second characteristic peak by controlling the constant dynamic response period. Taking ethyl alcohol as an example, the experimental results show that the characteristic peak of the dynamic response signal does not increase when the voltage is constant. However, a new characteristic peak will appear based on a continuously rising heating voltage. The characteristic peaks of the dynamic response of n-propyl alcohol, isopropyl alcohol, and n-butyl alcohol were also increased based on the rising heating voltage waveform. Based on the K-Nearest-Neighbors algorithm, the qualitative and quantitative recognition rate of the four alcohol homologue gases reached 100%.

Keywords: dynamic response signal; characteristic peaks; ZnO gas sensor; periodic cycle heating waveform; alcohol homologue gases

Citation: Shi, X.; Zhang, H.; Ji, H.; Meng, F. Dynamic Measurement of VOCs with Multiple Characteristic Peaks Based on Temperature Modulation of ZnO Gas Sensor. *Chemosensors* **2022**, *10*, 226. https://doi.org/10.3390/chemosensors10060226

Academic Editors: Simonetta Capone and Andrea Ponzoni

Received: 10 May 2022
Accepted: 13 June 2022
Published: 15 June 2022

1. Introduction

MOS gas sensors have been widely used in detecting VOCs gases. MOS gas sensors have some advantages, such as small size [1,2], high sensitivity [3–6], and simple fabrication [7]. At present, the detection method of VOCs gases by MOS gas sensor includes static measurement method [8–11] and dynamic measurement method. The types and concentrations of VOCs gases in the atmosphere are not single [12,13]. The heating voltage and power are unchanged in the static measurement method with time. The response signals are obtained at a specific heating temperature. The static response signals have little information and cross-sensitivity. The same current response value can be obtained with the different types and concentrations of VOC gases. So that it is difficult to recognize complex gas components correctly [14–18] under the static measurement method. The dynamic measurement method is to change the heating voltage and get the response signals at different temperatures. Iwata et al. indicate the heater waveform can obtain enhanced information on gas species [19]. The change in temperature is realized by changing the shape and voltage of the input cyclic heating waveforms [20]. Therefore, the dynamic measurement method not only needs to accurately measure the output dynamic response signal, but also needs to measure and record the cyclic heating waveforms. The dynamic measurement method expands the characteristic information in the time dimension, and the same sensor can be regarded as several different gas sensors at different temperatures, which solves the cross-sensitivity problem in the static measurement method. Hossein-Babaei et al. revealed that the dynamic response signal will present different characteristic peaks because the sensitive material will react with different types and concentrations of VOCs gases [21]. Bouricha et al. also detect methanol, toluene, acetone, isopropanol, and ethanol using a dynamic measurement method, which shows different responses [22]. As a result, different gases will have different output current curves under the same heating temperature curve because of the different sensitive responses of different gases at different temperatures, and the influence of the adsorption and desorption process, from which we can distinguish the types of gases. The amplitude of the output current curve has a concentration difference, so we can judge its concentration information by the amplitude. The dynamic response signals for different types and concentrations of VOCs gases are labeled as a sample. Through signal preprocessing and classification algorithms, the VOCs can be quantitatively and qualitatively recognized efficiently, rapidly, and accurately. The dynamic measurement method is significant to improve the selectivity of MOS gas sensor to various VOCs gases.

Based on the dynamic measurement method to improve gas selectivity mainly from two aspects. One is to optimize the late qualitative and quantitative recognition algorithm. Gaggiotti et al. recognize aldehydes (1-hexanal, 1-nonanal, trans-2-nonenal) from the other VOCs by the PCA [23]. Ji et al. adopt PCA weak separation with dynamic measurement to recognize VOCs gases [24]. The identification for different VOCs gases only reached 82% by GRNN [25]. Pan et al. proposed a lightweight network called multiscale convolutional neural network with attention (MCNA). The MCNA identifies ambient gases through signals of semiconductor gas sensor [26]. But the complexity of the algorithm is very high. The other is to increase the characteristic peaks of the dynamic response signal. The position and amplitude of characteristic peaks are the primary basis for qualitative and quantitative recognition of VOCs gases [22,25]. At present, the dynamic measurement method increases the number of characteristic peaks through a lot of exploratory experiments. The response reveals time is long, and the periodic cycle heating waveform is randomly selected. So, it is very challenging and accidental to optimize the characteristic peak of the dynamic response signals base on the traditional dynamic measurement method.

The unsaturated phenomenon was solved by a dynamic interval temperature modulation pattern in the paper [27]. The result revealed that the number of dynamic response signal characteristic peaks is only two under the experiment. As a result, the dynamic response signals of the four alcohol homologues are remarkably similar. The recognition rate of qualitative and quantitative is only 97.62%. Therefore, it is necessary to optimize the characteristic peak of dynamic response signals to improve the selectivity of which ZnO gas sensor to the VOCs gases.

In this paper, the experiments are carried out based on improving the conventional cyclic heating triangular wave to explore the optimization of characteristic peaks. Taking ethanol as an example, the same period of the cyclic heating triangular is applied. Firstly, the constant heating voltage is applied when the second characteristic peak of the response signal appears. After that, the heating voltage is continuously increased when the second characteristic peak appears. Then the experimental method was used to detect the other three alcohol homologue gases. The characteristic peaks of the dynamic response signal all are optimized. This paper provides an experimental idea and method to increase the number of characteristic peaks of the dynamic response signal.

2. Experiment

2.1. Synthesis of ZnO Nanomaterials

The hydrothermal method is one of the mainstream synthesis methods for metal oxide nanomaterial. Li et al. synthesized SnO_2 microspheres by hydrothermal method which had high sensitivity for formaldehyde [28]. The α-MoO_3@SnS_2 nanosheets by hydrothermal method display a high response of 114.9 for 100 ppm TEA [29]. The nanomaterials are closely related to the experiment condition. Chu et al. investigated heating duration and the heating temperature on the gas-sensing properties of the sensors. It is found that the sensor based on MoO_3 nanomaterial (200 °C, 4 h) exhibits the best performance [30]. The sensor-based on $LaFeO_3$ (800 °C, 2 h) can detect trimethylamine [31]. So, the experiment needs to be accurately adjusted.

The synthesis of ZnO nanomaterials is shown in Figure 1. $Zn(AC)_2$, absolute ethyl alcohol, urea, and other AR are purchased from Shanghai Aladdin Biochemical Technology Company Limited. The distilled water is generated by the ultrapure water machine provided by the Hefei intelligent Institute of the Chinese Academy of Sciences. All drugs are weighed by a precision analytical balance (ME155DU). The preparation of ZnO nanomaterials needs nine steps. Step A is to prepare solution. A total of 1.1 g $Zn(AC)_2$ was dissolved in 30 mL absolute ethyl alcohol and placed in a 60 mL beaker A. A total of 0.6 g urea was dissolved in 30 mL distilled water and placed in a 60 mL beaker B. Step B is ultrasound. Both standard solutions were ultrasonic for 15 min to clarify the solution. The purpose of ultrasound is to allow sufficient dissolution of $Zn(AC)_2$ in absolute ethyl alcohol and urea in distilled water. The ultrasonic cleaning machine (DT) from Hubei Dingtai Hengsheng Technology Equipment Company Limited. Step C is stirring. The breaker B was placed at room temperature, and the breaker A was poured into the breaker B. The mixed solution was stirred for 2 h under the magnetic stirrers (C-MAG HS 7, made in IKA). Step D is hydrothermal reaction. The 60 mL mixed solution is evenly placed in four 30 mL autoclaves (ax-fO25, made in Hangzhou Tongniu Technology Company Limited). The hydrothermal reaction was at 180 °C for 6 h. Step E is centrifugation. The purpose of centrifugation is to alternately clean the precipitate with absolute ethyl alcohol and distilled water. The speed of centrifuge (TGL-16B, made in Shanghai Anting Scientific Instrument Factory) is set at 6000 r/min. Finally, the distilled water on the upper layer is poured out to obtain white precipitation. Step F is drying. The white precipitation was placed in an oven at 80 °C (LDO, made in Shanghai Longyue Instrument Equipment Company Limited) for 6 h to form $Zn(OH)_2$ white crystals. Step G is grinding. The obtained white crystals are ground into a white powder in an agate mortar. Step H is calcination at a high temperature. The white powder was calcined in a muffle furnace (OTF-12000X, made in Hefei Material Crystal Material Technology Company Limited). The calcination temperature is 800 °C and the calcination time is 3 h. Finally, the ZnO is placed in a transparent plastic tube.

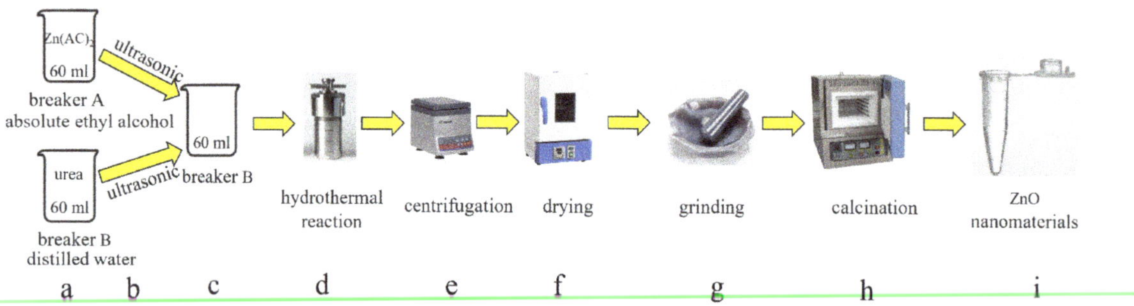

Figure 1. Synthesis of ZnO nanomaterials.

2.2. Characterization Details

Energy Dispersive Spectrometer (EDS) is an instrument used to analyze the element type and content of the materials. (Al5%, Mg1%) co-doped ZnO by EDS can get the O, Zn, Mg, and Al elements [32]. The X-axis of EDS represents the energy value, and the unit is keV. The Y-axis represents the counts of x-rays emitted by different elements. X-Ray Diffraction (XRD) can obtain the crystal peak of the nanomaterial. Shen et al. characterize CuO by

XRD and all the sharp diffraction peaks of the product can be labeled as the monoclinic CuO (JCPDS No.: 48-1548) [33]. The crystal can be compared with a standard PDF card to determine the type of nanomaterial. The purity of the nanomaterial can be determined by EDS and XRD characterization.

2.3. Experimental Measurement Process

The experimental apparatus diagram is shown in Figure 2. The experimental apparatus is mainly composed of a programmable DC power supply, precision source measure unit, computer, gas chamber, and gas cylinder. The programmable DC power supply is designed to provide different cycle heating waveforms. The precision source measure unit is used to collect dynamic response signals under different cycle heating waveforms. The computer is used to display dynamic response signals. The reaction between ZnO nanomaterials and the VOCs gas is carried out in the gas chamber. The gas cylinder is used to fill the gas chamber with synthetic air. The synthetic air is passed into the air chamber to simulate the natural atmospheric environment before passing into the VOCs gas. Then the programmable DC power supply input the heating waveform. The precision source measure unit also shows a periodic change when the VOCs gases through the syringe into the gas chamber. Finally, the dynamic response signal (usually five cycles) is collected.

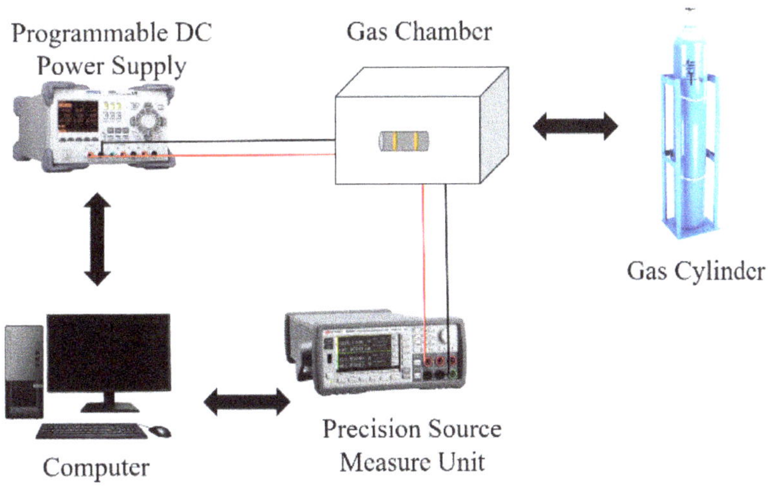

Figure 2. The diagram of the experimental apparatus.

According to a previous study [27], the cycle heating triangular waveform of the saturated dynamic response signal for dynamic measurement of alcohol homolog gases has a period of 35 s and a symmetry of 57% (TW-1). Taking ethyl alcohol as an example, the second characteristic peak appeared at about 22 s. After 22 s, the voltage continued to fall, but no new characteristic peak appeared in the dynamic response signal. In this experiment, the periodic heating cycle was maintained unchanged. Then, 100 ppm ethyl alcohol was measured by applying constant heating voltage (TW-2) and increasing heating voltage (TW-3). Figure 3 shows three different heating triangular waveforms. After that, 100 ppm n-propyl alcohol, isopropyl alcohol, and n-butyl alcohol were detected based on the TW-3 waveform. The dynamic response signal curves of the three alcohol homolog gases were obtained. Finally, the concentration gradients of the four alcohol homolog gases were measured at 100–500 ppm under the TW-3 waveform.

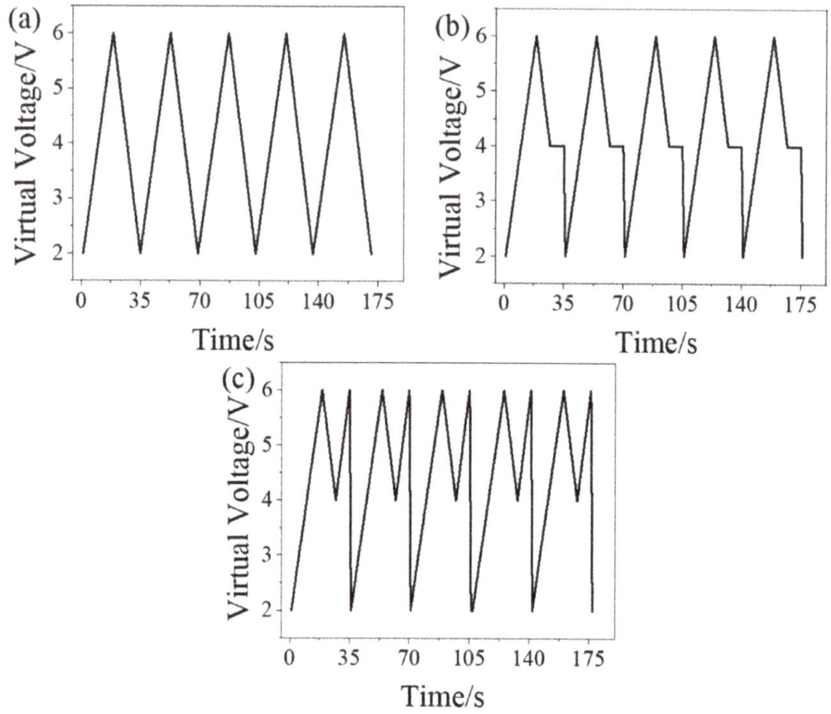

Figure 3. The heating waveforms (**a**) TW-1 (**b**) TW-2 (**c**) TW-3.

3. Experimental Results and Discussion

3.1. Characterization of the ZnO Nanomaterials

The XRD and EDS of ZnO nanomaterial prepared by the hydrothermal method is shown in Figure 4.

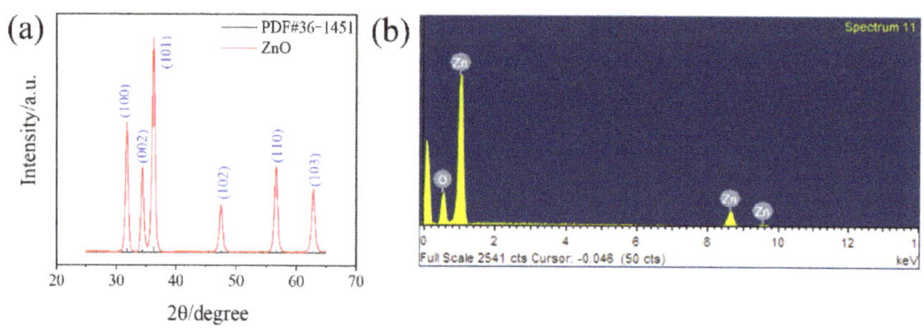

Figure 4. The characterizations of ZnO nanoparticles (**a**) XRD (**b**) EDS.

The XRD is in the 25° to 65° at a scanning rate of 5°/min. The six peaks appear at 31.77°, 34.42°, 36.25°, 47.54°, 56.60°, and 62.86° which correspond to the (100), (002), (101), (102), (110), and (105) crystal surfaces of the ZnO PDF standard card (36-1451). There are no peaks of other crystals. This indicates that ZnO has a high purity. According to Figure 4b, Zn and O are found in the ZnO nanomaterials without other elements. The purity of ZnO nanomaterial synthesized by the hydrothermal method is high. The ZnO nanomaterial with higher purity produces stable dynamic response signals to various VOCs gases, which is suitable for dynamic measurement. Figure 5a shows the distribution of O elements and Zn elements at three different positions under Figure 5b. Each position is about 78% Zn elements and 22% O elements. It can be seen from the table that the distribution of Zn element and O element at different positions are relatively uniform.

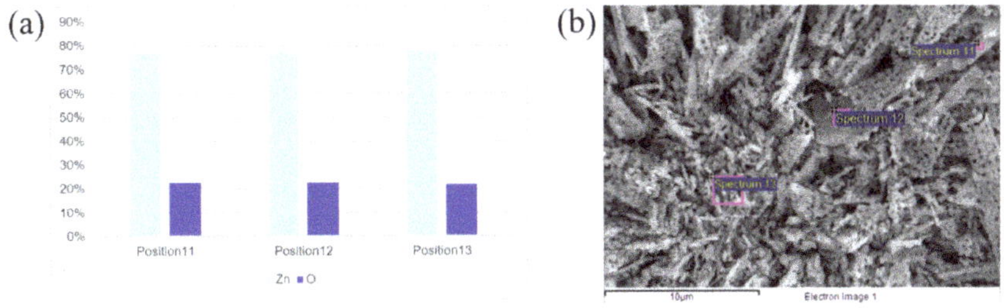

Figure 5. (**a**) The plots of Zn and O elements at different positions (**b**) three different positions of electron image.

3.2. *Dynamic Response Signal of Ethyl Alcohol*

Figure 6a shows the concentration gradient of the characteristic peak of the four alcohol homologous gases at 100–400 ppm under the TW-1. It can be seen that the response signal current of different VOCs gases at different concentrations are generally different. As the concentration of the measured gases increases, the output current also increases. However, the characteristic peak current of 200 ppm ethyl alcohol and 250 ppm n-propyl alcohol are the same, resulting in an error in recognizing the two gases from the characteristic peak current. As the concentration of the measured gases increases, the current value of the highest peak of the ethyl alcohol increases more slowly than that of the isopropyl alcohol, which results in the coincidence of the current value of the highest peak of ethyl alcohol and isopropyl alcohol at a later same concentration. Therefore, a characteristic peak current of the dynamic response signal cannot be used as a basis for qualitative and quantitative recognition of the alcohol homologue gases. Figure 6b is the current concentration gradient diagram of the two characteristic peaks of the four alcohol homologs gases at 100–400 ppm under TW-1. It can be seen that the response signal current values of the tested VOCs gases at different concentrations are generally different from the first peak and the secondary peak. However, the response ratios of ethyl alcohol at 100–400 ppm and n-propyl alcohol at 150–350 ppm will overlap, which makes the recognition of these two gases appear to be wrong. So, when the peak number of the dynamic response signal is 2, the qualitative and quantitative recognition of the alcohol homologue gases will be wrong.

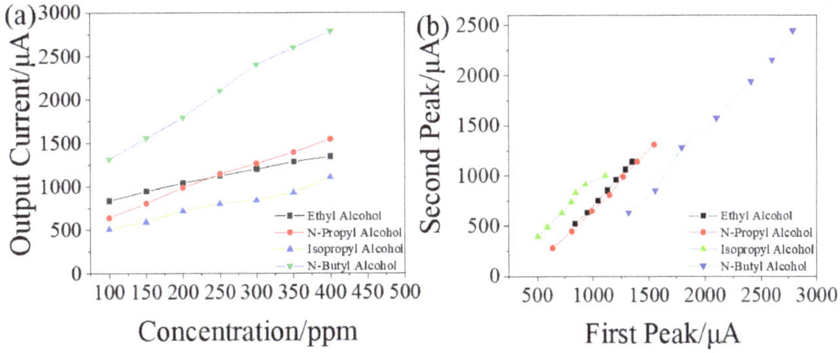

Figure 6. (**a**) The concentration gradient diagram of peak gas concentrations of four alcohol homologues under TW-1 (**b**) The first peak and secondary peak concentration gradients of four alcohol homologues under TW-1.

Under the TW-2, the dynamic response signal curve of 100 ppm ethyl alcohol was obtained, as shown in Figure 7a. It can be seen that the number of characteristic peaks of the ethyl alcohol did not increase when the heating voltage was kept constant at the second characteristic peak. In the 20 s of the cycle, the second characteristic peak appears in the dynamic response signal. At 20 s–28 s of the cycle, the current signal has a sharp decline trend. After 28 s, the current signal begins to decline slowly. Under the TW-1, the second characteristic peak of the ethyl alcohol appeared at the 22 s of the cycle. Under the TW-2, the second characteristic peak of the ethyl alcohol appeared 2 s earlier. The essential influencing factor of dynamic measurement is the temperature, and the change of circulating heating voltage is the change of the temperature. The temperature will not change much if the voltage is maintained at 4 V. At the same time, the type of oxygen ions adsorbed on the ZnO nanomaterials surface remains unchanged, and no new chemical or physical reactions occur between oxygen ions and hydroxyl groups of ethyl alcohol. Therefore, the dynamic response signal of the 100 ppm ethyl alcohol under TW-2 does not appear a new characteristic peak.

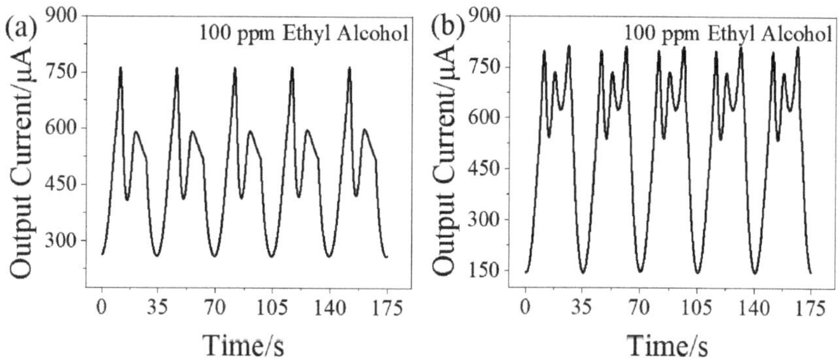

Figure 7. The dynamic response signal curves of ethyl alcohol-based on (**a**) TW-2 and (**b**) TW-3.

The dynamic response signal curve of 100 ppm ethyl alcohol under TW-3 is shown in Figure 7b. By comparing Figure 7a,b, it can be seen that the number of characteristic peaks of dynamic response signals can indeed be increased by increasing the heating voltage when the second characteristic peak appears. The second characteristic peak appears in the 17 s. The third characteristic peak appears at the 26 s. The second characteristic peak of 100 ppm ethyl alcohol under TW-1 appeared at the 22 s. The second characteristic peak appeared at 20 s under TW-2. The second characteristic peak appeared at 17 s under TW-3, and the third characteristic peak appeared at 26 s. In other words, with the increase of heating voltage, the second characteristic peak will appear earlier. Under the TW-3, the heating voltage reaches 4.25 V at 26 s, and the second characteristic peak of the TW-1 also appears at about 4 V. Under the TW-3, there is a characteristic peak when the heating voltage is 4 V. It is possible that new oxygen ions appear on the surface of ZnO nanomaterials, and a new physical or chemical reaction occurs with the hydroxyl group, which results in a new characteristic peak.

3.3. Dynamic Response Signal of Other Alcohol Homologue Gases

Based on the TW-2, the 100 ppm n-propyl alcohol, isopropyl alcohol, and n-butyl alcohol were detected. The dynamic response signal was shown in Figure 8. It can be seen that it is similar to ethyl alcohol. There is no new characteristic peak appeared in the three alcohol homologue gases under the TW-2. The number of characteristic peaks was two. For 100 ppm ethyl alcohol, n-propyl alcohol, isopropyl alcohol, and n-butyl alcohol, the improved cyclic heating triangle wave did not increase the number of characteristic peaks, but the position of the characteristic peak was different. The second characteristic peak of the n-propyl alcohol appeared at 28 s. The second characteristic peak of the isopropyl alcohol appeared at 25 s. The second characteristic peak of the n-butyl alcohol appeared at 29 s. For 100 ppm n-propyl alcohol, the output current decreases slowly after 28 s of the cycle. For 100 ppm isopropyl alcohol, its dynamic response signal curve is similar to the ethyl alcohol. There is a sharp current decline trend in the 25–30 s, and the output current shows a slow decline trend after the 30 s. For 100 ppm n-butyl alcohol, the dynamic response curve is similar to the n-propyl alcohol. After the appearance of the second characteristic peak, the output current always presents a trend of gentle decline. This may be because the number of hydroxyl groups in n-propyl alcohol and n-butyl alcohol is the same, so they have similar dynamic response signals when they react with the surface of ZnO nanomaterials for oxygen adsorption.

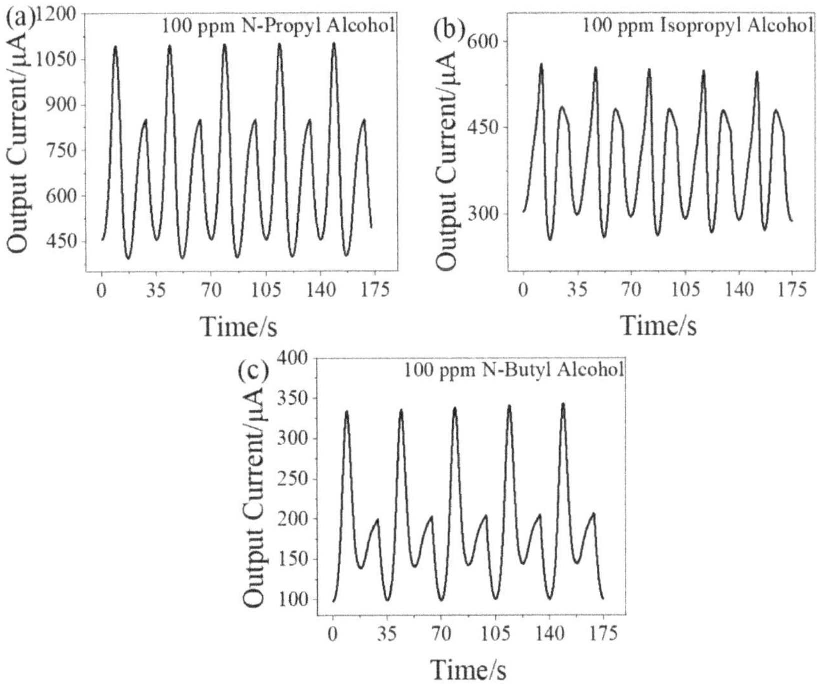

Figure 8. The dynamic response signal curves based on TW-2 (**a**) n-propyl alcohol (**b**) isopropyl alcohol (**c**) n-butyl alcohol.

The 100 ppm n-propyl alcohol, isopropyl alcohol, and n-butyl alcohol were detected based on the TW-3. As shown in Figure 9, when the second characteristic peak appears, the three alcohol homologues all appear as the new characteristic peaks with the increase of heating voltage. The second characteristic peak of the n-propyl alcohol appeared in the 23 s, and the third characteristic peak appeared in the 30 s of the cycle. The second characteristic peak of the isopropyl alcohol appeared in the 26 s of the cycle, and the third characteristic peak appeared in the 30 s of the cycle. The number of characteristic peaks of n-butyl alcohol reached 4. The second, third, and fourth characteristic peaks appeared in the 18 s, 25 s, and 28 s respectively. Compared with the time of the second characteristic peak under the TW-3, the time of the characteristic peak will advance with the increase of heating voltage. This may be because the overall temperature will be higher with the increase of the heating voltage. The reaction between oxygen ions adsorbed on the ZnO surface and hydroxyl group will occur physical or chemical reaction at an earlier time. Thus, the characteristic peaks appear.

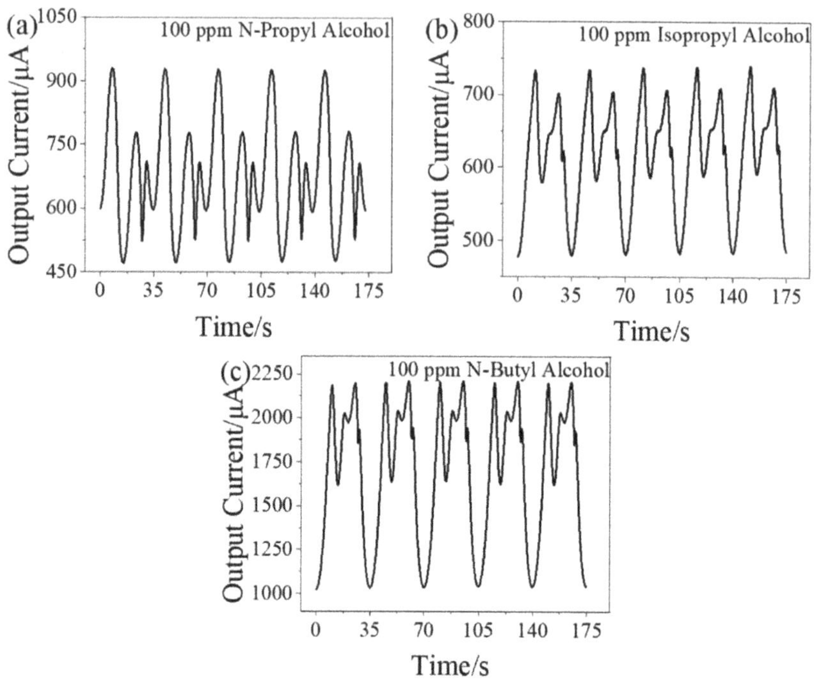

Figure 9. The dynamic response signal curves based on TW-3 (**a**) n-propyl alcohol (**b**) isopropyl alcohol (**c**) n-butyl alcohol.

3.4. Dynamic Response Mechanism

There are hydroxyl groups in the alcohol homolog gas, located on H and O. The electronegativity of O is higher than that of H, and the ability to attract electrons is very strong. The electrons in the O-H bond are heavily biased toward O, which results in the polarity of the O-H bond. It is easily broken to form O ions and H ions. Along the way, other products will emerge. Literature shows that ethyl alcohol, part of it can react with adsorbed oxygen to produce CO_2 and H_2O directly. Most lose H ions as they are captured by the adsorbent oxygen. Three intermediates $CH_3CH_2O^-$, CH_3CO^-, and CH_3COO^- are also generated when CO_2 is generated. The reaction between alcohol homologue gases and oxygen anions is completed in multiple steps. The hydroxyl groups have a slow equilibrium and competitive effect on the surface of sensitive materials. The types of oxygen ions adsorbed on the surface of sensitive materials will change with the temperature change, and the reaction with hydroxyl groups will change. Under the TW-1, the temperature continues to decrease, and the reaction between oxygen ions and hydroxyl groups may be incomplete. When the heating voltage is kept constant at 4 V, the temperature will decrease slowly. The continuous high temperature makes oxygen ions entirely react with hydroxyl groups, which results in the significant increase of dynamic response current signals of the four alcohol homologues gases. Under the TW-2, the reaction temperature of oxygen ions and hydroxyl groups increases, but it is possible that the reaction temperature at this time does not change the type of oxygen ions on the surface of ZnO nanomaterials, and the electronic activity does not change. There is no new chemical reaction between oxygen ions on the surface of ZnO nanomaterials and alcohol homologue gases. Therefore, the characteristic peak is not optimized. The essence of the increase in the TW-3 waveform heating voltage is the increase in temperature, which may be the emergence of new oxygen ion states with the increase in temperature. There is a new transformation between oxygen ions with the increase of electronic activity. The reaction between hydroxyl and oxygen negative ions also accelerates, resulting in the emergence of the third characteristic peak.

3.5. Qualitative and Quantitative Analysis of Alcohol Homologue Gases

Based on TW-3, nine different concentrations of four alcohol homologue gases are dynamically detected, and the concentrations are 100 ppm, 150 ppm, 200 ppm, 250 ppm, 300 ppm, 350 ppm, 400 ppm, 450 ppm, and 500 ppm, respectively. Figure 10 shows the relationship between the three characteristic peak output currents and concentrations of four alcohol homologue gases under TW-3. In particular, n-butyl alcohol obtained the second, third and fourth characteristic peaks. The characteristic peak current increases with the increasing concentration. The characteristic peak currents of ethyl alcohol, n-propyl alcohol, isopropyl alcohol, and n-butyl alcohol are almost parallel to each other as the concentration increases. However, the three characteristic peak currents obtained under the TW-3 do not have the cross-sensitivity problem of four alcohol homologue gases. Compared with Figure 6, the increase in the number of characteristic peaks can indeed increase the recognition rate of measured VOCs gases. The concentration gradient curve is shown in Figure 11. It can be seen that the positions of characteristic peaks of different gases are different. With the increase of gas concentration, the dynamic response current value of each gas increases, but the positions of characteristic peaks do not change. The location and number of characteristic peaks of the same gas are the same, but the current of the characteristic peak increases with the increase of gas concentration. The three characteristic peaks of ethyl alcohol occur at 10 s, 17 s, and 22 s of the cycle respectively; The three characteristic peaks of N-Propyl alcohol appear in the 8 s, 23 s, and 30 s of the cycle respectively; The three characteristic peaks of isopropyl alcohol appear at the 12 s, 26 s and 30 s of the cycle respectively; The four characteristic peaks of n-butyl alcohol appeared at the 10 s, 18 s, 25 s and 28 s of the cycle respectively. It may be that different chemical constituents are involved in the reaction of the different gases. As a result, the measured gases with oxygen atoms need different times to reach equilibrium and the characteristic peaks appear at different times. Different concentrations of the same gas have the same chemical composition and react with the oxygen atoms. The number of chemicals increases, so the current signal increases as the concentration increases.

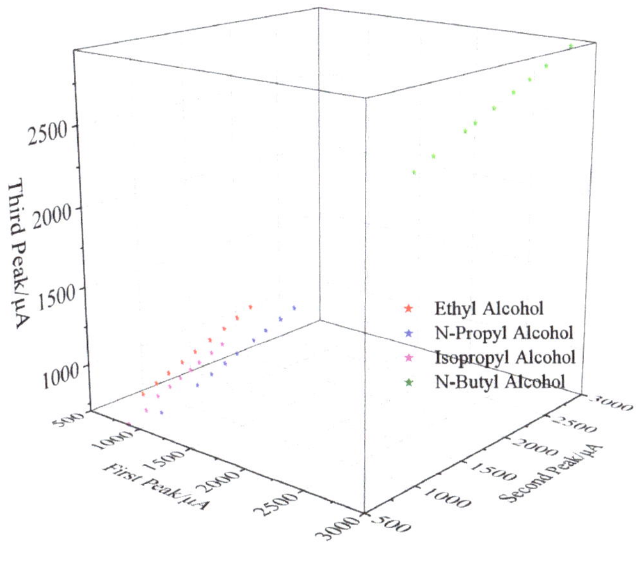

Figure 10. The diagram between characteristic peak output currents vs ppm of four alcohol homologue gases under TW-3.

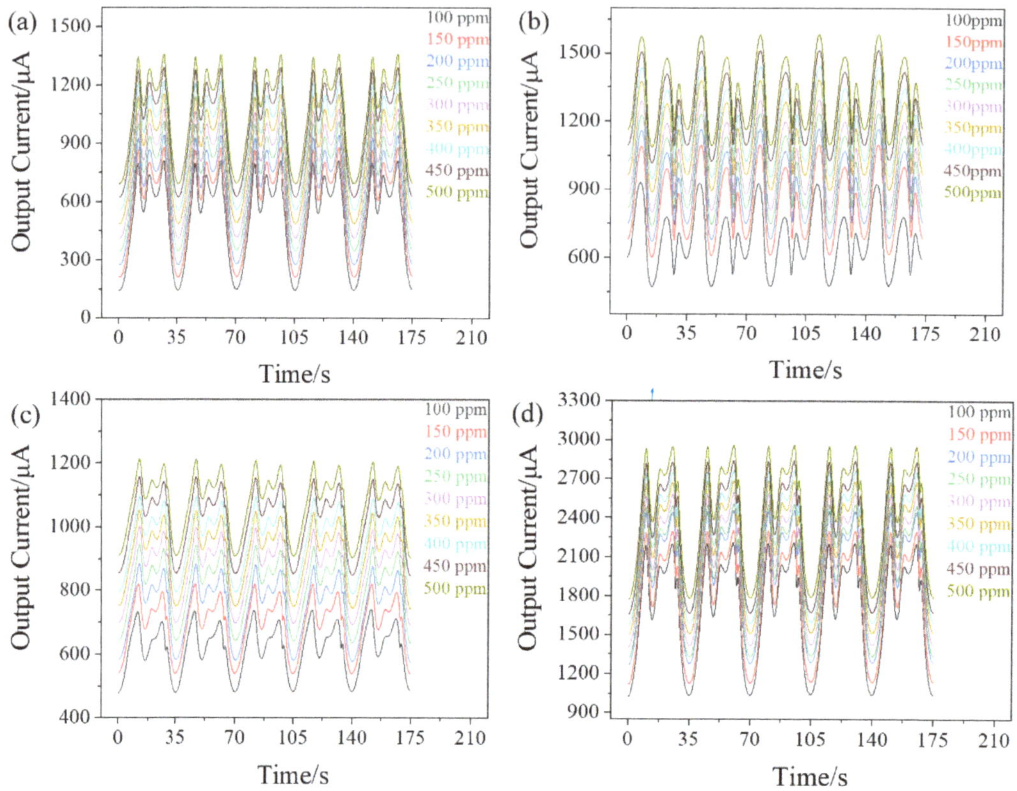

Figure 11. The 100–500 ppm concentration gradient curve under TW-3 of four alcohol homologues gases. (**a**) ethyl alcohol (**b**) n-propyl alcohol (**c**) isopropyl alcohol (**d**) n-butyl alcohol.

K-nearest Neighbor (KNN) is a point corresponding to the feature space for each sample of the same dimension. The output is the result labels. Feature points with similar distances will be classified into the same result label. KNN needs to calculate the distance from the test sample point to each other sample point. When the dimension increases, the number of different sample points from a sample point will decrease within the unit distance, which will cause us to look for a longer length to find the adjacent value. In this experiment, each tag is a sample containing 350-dimension data. Too large of a dimension is easy to cause dimension disaster. Therefore, Principal Component Analysis (PCA) is used to reduce the dimension of the data set. The proportion curve of dimension reduction information is shown in Figure 12. Its principle is to recombine many original indexes with certain correlations into a new group of linear and mutually independent comprehensive indexes. Dimensionality reduction needs to reduce information, but also ensure that important information is not lost. In the experiment, nine concentration information of 4 sample gases were collected, with a total of 36 result labels. Each result label is a sample containing 350-dimensional data. Each sample is repeatedly collected five times, a total of 180 samples. It can be known from the proportion of the information curve after the dimension reduction, the proportion of information in three-dimensional data is 99.7%. After that, the three-dimensional data are recognized by the KNN algorithm, and the accuracy of dynamic reorganization of four alcohol homologue gases by the ZnO Gas sensor is 100%.

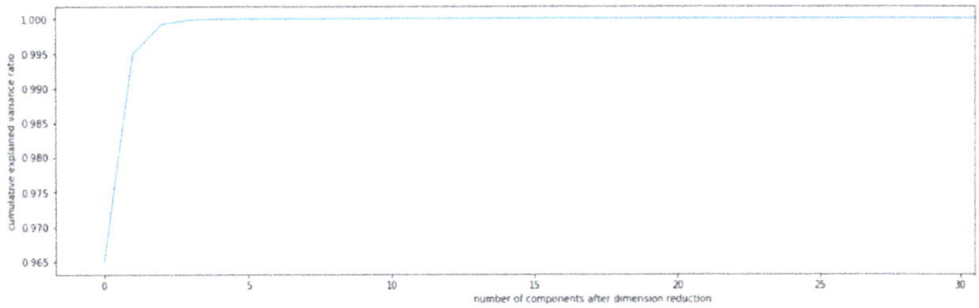

Figure 12. The proportion of dimensions and information.

4. Conclusions

In this paper, the characteristic peak of the dynamic response signal is increased by optimizing the periodic cyclic heating triangular waveform. The results show that the number of characteristic peaks can increase if increasing the heating voltage at the last characteristic peaks of the dynamic response signal. There are two characteristic peaks of ethyl alcohol in the TW-1 and TW-2 waveforms, and three characteristic peaks in the TW-3 waveform. Based on the TW-3 waveform, 100 ppm n-propyl alcohol, isopropyl alcohol, and n-butyl alcohol were tested. The number of characteristic peaks of dynamic response signals of the three alcohol homologue gases also is optimized. Under the TW-3 waveform, the concentration gradients of the four alcohol homologue gases of 100 ppm, 150 ppm, 200 ppm, 250 ppm, 300 ppm, 350 ppm, 400 ppm, 450 ppm, and 500 ppm were tested respectively. A total of 180 samples and 36 labels were obtained. After the dimensionality reduction of the data through PCA, the KNN algorithm has a recognition rate of 100% for the qualitative and quantitative analysis of four alcohol homologue gases.

Author Contributions: Conceptualization, F.M.; methodology, H.J.; software, H.Z.; validation, F.M.; formal analysis, H.Z.; investigation, H.J.; resources, X.S.; data curation, X.S.; writing—original draft preparation, X.S.; writing—review and editing, H.Z.; visualization, H.J.; supervision, F.M.; project administration, F.M.; funding acquisition, F.M. All authors have read and agreed to the published version of the manuscript.

Funding: This work was supported by the National Natural Science Foundation of China (62071112, 61833006, 62033002, and 61973058), the Fundamental Research Funds for the Central Universities in China (N2201008), the 111 Project (B16009), and Hebei Natural Science Foundation (F2020501040).

Institutional Review Board Statement: Not applicable.

Informed Consent Statement: Not applicable.

Data Availability Statement: Not applicable.

Conflicts of Interest: The authors declare no conflict of interest.

References

1. Zhao, S.; Shen, Y.; Yan, X.; Zhou, P.; Yin, Y.; Lu, R.; Han, C.; Cui, B.; Wei, D. Complex-surfactant-assisted hydrothermal synthesis of one-dimensional ZnO nanorods for high-performance ethanol gas sensor. *Sens. Actuators B Chem.* **2019**, *286*, 501–511. [CrossRef]
2. Zhao, S.; Shen, Y.; Zhou, P.; Zhong, X.; Han, C.; Zhao, Q.; Wei, D. Design of Au@WO_3 core∓shell structured nanospheres for ppb-level NO_2 sensing. *Sens. Actuators B Chem.* **2019**, *282*, 917–926. [CrossRef]
3. Kumarage, G.; Comini, E. Low-Dimensional Nanostructures Based on Cobalt Oxide (Co_3O_4) in Chemical-Gas Sensing. *Chemosensors* **2021**, *9*, 197. [CrossRef]
4. Acharyya, S.; Nag, S.; Kimbahune, S.; Ghose, A.; Pal, A.; Guha, P.K. Selective Discrimination of VOCs Applying Gas Sensing Kinetic Analysis over a Metal Oxide-Based Chemiresistive Gas Sensor. *ACS Sens.* **2021**, *6*, 2218–2224. [CrossRef] [PubMed]
5. Lin, T.; Lv, X.; Hu, Z.; Xu, A.; Feng, C. Semiconductor Metal Oxides as Chemoresistive Sensors for Detecting Volatile Organic Compounds. *Sensors* **2019**, *19*, 233. [CrossRef]
6. Acharyya, S.; Jana, B.; Nag, S.; Saha, G.; Guha, P.K. Single resistive sensor for selective detection of multiple VOCs employing SnO_2 hollowspheres and machine learning algorithm: A proof of concept. *Sens. Actuators B Chem.* **2020**, *321*, 128484. [CrossRef]
7. Zhang, J.; Yu, Y.; Fang, P.; Liu, L.; Yue, H.; Qu, J.; Han, A. Anodization of aluminum in a sealed container. *Electrochem. Commun.* **2021**, *129*, 107086. [CrossRef]

8. Subin David, S.P.; Veeralakshmi, S.; Sandhya, J.; Nehru, S.; Kalaiselvam, S. Room temperature operatable high sensitive toluene gas sensor using chemiresistive Ag/Bi_2O_3 nanocomposite. *Sens. Actuators B Chem.* **2020**, *320*, 128410. [CrossRef]
9. Ghazi, M.; Janfaza, S.; Tahmooressi, H.; Tasnim, N.; Hoorfar, M. Selective detection of VOCs using microfluidic gas sensor with embedded cylindrical microfeatures coated with graphene oxide. *J. Hazard. Mater.* **2022**, *424*, 127566. [CrossRef]
10. Zhu, L.; Zeng, W.; Ye, H.; Li, Y. Volatile organic compound sensing based on coral rock-like ZnO. *Mater. Res. Bull.* **2018**, *100*, 259–264. [CrossRef]
11. Liu, B.; Zhu, Q.; Pan, Y.; Huang, F.; Tang, L.; Liu, C.; Cheng, Z.; Wang, P.; Ma, J.; Ding, M. Single-Atom Tailoring of Two-Dimensional Atomic Crystals Enables Highly Efficient Detection and Pattern Recognition of Chemical Vapors. *ACS Sens.* **2022**, *7*, 1533–1543. [CrossRef] [PubMed]
12. Jiménez-Garzaa, O.; Guob, L.; Byunc, H.; Carrierid, M.; Bartoluccid, G.; Barrón-Vivancoe, B.; Baccarelli, A. Aberrant promoter methylation in genes related to hematopoietic malignancy in workers exposed to a VOC mixture. *Toxicol. Appl. Pharm.* **2018**, *339*, 65–72. [CrossRef] [PubMed]
13. Kim, D.; Shin, W.; Hong, S.; Jeong, Y.; Jung, G.; Park, J.; Lee, J. Effects of Electrode Structure on H_2S Sensing and Low-Frequency Noise Characteristics in In_2O_3-Based Resistor-Type Gas Sensors. *IEEE Sens. J.* **2022**, *22*, 6311–6320. [CrossRef]
14. Huang, J.; Wu, Y.; Gu, C.; Zhai, M.; Sun, Y.; Liu, J. Fabrication and gas-sensing properties of hierarchically porous ZnO ar-chitectures. *Sens. Actuators B Chem.* **2011**, *155*, 126–133. [CrossRef]
15. Zheng, K.; Yang, T.; Guo, Z. Porous Pb-Doped ZnO Nanobelts with Enriched Oxygen Vacancies: Preparation and Their Chemoreceptive Sensing Performance. *Chemosensors* **2022**, *10*, 96. [CrossRef]
16. Hemmati, S.; Anaraki Firooz, A.; Khodadadi, A.; Mortazavi, Y. Nanostructured SnO_2-ZnO sensors: Highly sensitive and selective to ethanol. *Sens. Actuators B Chem.* **2011**, *160*, 1298–1303. [CrossRef]
17. Gu, C.; Shanshan, L.; Huang, J.; Shi, C.; Liu, J. Preferential growth of long ZnO nanowires and its application in gas sensor. *Sens. Actuators B Chem.* **2013**, *177*, 453–459. [CrossRef]
18. Bai, S.; Guo, J.; Shua, X.; Xiang, X.; Luo, R.; Li, D.; Chen, A.; Liu, C. Surface functionalization of Co_3O_4 hollow spheres with ZnO nanoparticles for modulating sensing properties of formaldehyde. *Sens. Actuators B Chem.* **2017**, *245*, 359–368. [CrossRef]
19. Iwata, T.; Saeki, M.; Okura, Y.; Yoshikawa, T. Gas discrimination based on enhanced gas-species related information obtained by a single gas sensor with novel temperature modulation. *Sens. Actuators B Chem.* **2022**, *354*, 131225. [CrossRef]
20. Meng, F.; Ji, H.; Yuan, Z.; Chen, Y.; Zhang, H.; Qin, W.; Gao, H. Dynamic Measurement and Recognition Methods of SnO_2 Sensor to VOCs Under Zigzag-Rectangular Wave Temperature Modulation. *IEEE Sens. J.* **2021**, *21*, 10915–10922. [CrossRef]
21. Hossein-Babaei, F.; Amini, A. A breakthrough in gas diagnosis with a temperature-modulated generic metal oxide gas sensor. *Sens. Actuators B Chem.* **2012**, *166–167*, 419–425. [CrossRef]
22. Bouricha, B.; Sekrafi, T.; Labidi, A.; Dridi, C. VOCs Identification Method Based on One Single ZnTTP Sensor. *IEEE Sens. J.* **2022**, *22*, 671–677. [CrossRef]
23. Gaggiotti, S.; Hurotb, C.; Weerakkodyb, J.; Matheyb, R.; Buhotb, A.; Mascinia, M.; Houb, Y.; Compagnone, D. Development of an optoelectronic nose based on surface plasmon resonance imaging with peptide and hairpin DNA for sensing volatile organic compounds. *Sens. Actuators B Chem.* **2020**, *303*, 127188. [CrossRef]
24. Ji, H.; Qin, W.; Yuan, Z.; Meng, F. Qualitative and quantitative recognition method of drug-producing chemicals based on SnO_2 gas sensor with dynamic measurement and PCA weak separation. *Sens. Actuators B Chem.* **2021**, *348*, 130698. [CrossRef]
25. Yuan, Z.; Han, E.; Meng, F.; Zuo, K. Detection and Identification of Volatile Organic Compounds Based on Temperature Modulated ZnO Sensors. *IEEE Trans. Instrum. Meas.* **2020**, *69*, 4533–4544. [CrossRef]
26. Pan, J.; Yang, A.; Wang, D.; Chu, J.; Lei, F.; Wang, X.; Rong, M. Lightweight Neural Network for Gas Identification Based on Semiconductor Sensor. *IEEE Trans. Instrum. Meas.* **2022**, *71*, 1–8. [CrossRef]
27. Meng, F.; Shi, X.; Yuan, Z.; Ji, H.; Qin, W.; Shen, Y.; Xing, Z. Detection of four alcohol homologue gases by ZnO gas sensor in dynamic interval temperature modulation mode. *Sens. Actuators B Chem.* **2022**, *350*, 130867. [CrossRef]
28. Chu, X.; Zhu, X.; Dong, Y.; Zhang, W.; Bai, L. Formaldehyde Sensing Properties of SnO–Graphene Composites Prepared via Hydrothermal Method. *J. Mater. Sci. Technol.* **2015**, *31*, 913–917. [CrossRef]
29. Dong, X.; Han, Q.; Kang, Y.; Li, H.; Huang, X.; Fang, Z.; Yuan, H.; Chi, Z.; Wu, G.; Xie, W. Rational construction and triethylamine sensing performance of foam shaped α-MoO_3@SnS_2 nanosheets. *Chin. Chem. Lett.* **2022**, *33*, 567–572. [CrossRef]
30. Chu, X.; Liang, S.; Sun, W.; Zhang, W.; Chen, T.; Zhang, Q. Trimethylamine sensing properties of sensors based on MoO_3 microrods. *Sens. Actuators B Chem.* **2010**, *148*, 399–403. [CrossRef]
31. Chu, X.; Zhou, S.; Zhang, W.; Shui, H. Trimethylamine sensing properties of nano-$LaFeO_3$ prepared using solid-state reaction in the presence of PEG400. *Mater. Sci. Eng.* **2009**, *164*, 65–69. [CrossRef]
32. Jaballah, S.; Alaskar, Y.; Alshunaifi, I.; Ghiloufi, I.; Neri, G.; Bouzidi, C.; Dahman, H.; El Mir, L. Effect of Al and Mg Doping on Reducing Gases Detection of ZnO Nanoparticles. *Chemosensors* **2021**, *9*, 300. [CrossRef]
33. Shen, X.; Shi, G.; Zhang, Y.; Weng, S. Wireless Volatile Organic Compound Detection for Restricted Internet of Things Environments Based on Cataluminescence Sensors. *Chemosensors* **2022**, *10*, 179. [CrossRef]

chemosensors

MDPI

Review

Metal Oxide Semiconductor Sensors for Triethylamine Detection: Sensing Performance and Improvements

Hua Zhang, Yinghao Guo and Fanli Meng *

College of Information Science and Engineering, Northeastern University, Shenyang 110819, China; zhanghua@ise.neu.edu.cn (H.Z.); 2100745@stu.neu.edu.cn (Y.G.)
* Correspondence: mengfanli@ise.neu.edu.cn

Abstract: Triethylamine (TEA) is an organic compound that is commonly used in industries, but its volatile, inflammable, corrosive, and toxic nature leads to explosions and tissue damage. A sensitive, accurate, and in situ monitoring of TEA is of great significance to production safety and human health. Metal oxide semiconductors (MOSs) are widely used as gas sensors for volatile organic compounds due to their high bandgap and unique microstructure. This review aims to provide insights into the further development of MOSs by generalizing existing MOSs for TEA detection and measures to improve their sensing performance. This review starts by proposing the basic gas-sensing characteristics of the sensor and two typical TEA sensing mechanisms. Then, recent developments to improve the sensing performance of TEA sensors are summarized from different aspects, such as the optimization of material morphology, the incorporation of other materials (metal elements, conducting polymers, etc.), the development of new materials (graphene, TMDs, etc.), the application of advanced fabrication devices, and the introduction of external stimulation. Finally, this review concludes with prospects for using the aforementioned methods in the fabrication of high-performance TEA gas sensors, as well as highlighting the significance and research challenges in this emerging field.

Keywords: triethylamine; metal oxide semiconductor; sensing performance; morphology; new materials

Citation: Zhang, H.; Guo, Y.; Meng, F. Metal Oxide Semiconductor Sensors for Triethylamine Detection: Sensing Performance and Improvements. *Chemosensors* **2022**, *10*, 231. https://doi.org/10.3390/chemosensors10060231

Academic Editor: Boris Lakard

Received: 25 May 2022
Accepted: 15 June 2022
Published: 17 June 2022

1. Introduction

Emissions of noisome gases are ubiquitous in daily life and industrial production. Outdoor fuel combustion and transportation and indoor emissions from building and decorative materials, furniture, household appliances, cleaning agents, and the human body itself all produce noxious fumes. Volatile organic compounds (VOCs) are a common class of hazardous gases in the air that can have a huge negative impact on human health. As a member of VOCs, triethylamine is a colorless oily substance that is slightly soluble in water and easily dissolves in organic solvents such as ethanol; it is toxic and flammable and has a strong ammonia odor [1–3].

In many chemical synthesis processes, TEA is considered a multifunctional and efficient organocatalyst and solvent [4]. Because of its relative safety, commercial availability, and low price, it is often used in industrial production as a synthetic dye and preservative, and because of its excellent physical and chemical properties, it is also used in large quantities in chemical experiments [5]. However, when the TEA concentration is too high, it endangers our physical health by causing injuries such as skin burns and headaches as well as pulmonary edema and poisoning by accidental swallowing; its vapor can also strongly irritate the eyelids and mucous membranes [6,7]. It also has the risk of rapid burning and explosion when exposed to open fire, high temperature, and strong oxidizing agents [8,9]. Both the European Commission and the American Conference of Governmental Industrial Hygienists have recommended that the threshold concentration of TEA exposed to air be 1 ppm [10,11]. Therefore, TEA gas sensors with low detection limits that can detect quickly need to be developed.

The gas-detection methods developed so far include quartz crystal microbalance gas sensor [12], visual colorimetric detection [13], headspace gas chromatography [14], electrochemical sensors [15–17], and chemiresistive semiconductor gas sensors. The detection methods mentioned above can all detect a certain concentration of TEA, but several of them have a long detection time, high detection cost, and complicated detection operation. Therefore, the chemiresistive semiconductor sensors, which can be fast, accurate, and highly sensitive; have low detection limits; and can be manufactured in batches, have received widespread attention from scientific researchers around the world. To date, semiconductor gas sensors are mostly made of metal oxide semiconductors (MOSs), which have excellent physicochemical properties such as wide bandgap, unique microstructure [18–20], higher sensitivity to gases, and fast response time; most importantly, lower fabrication costs make them circulate in the market in large quantities. The most commonly used n-type semiconductor metal oxide materials are ZnO, SnO_2, Fe_2O_3, and MoO_3 [21], and p-type semiconductor metal materials are Co_3O_4, CuO, and NiO [22]. It is found that they all respond to TEA gas, but all have the shortcomings of low detection limit, poor stability, and high operating temperature to be solved. To improve the gas-sensing performance of TEA sensors, experimenters have been studying the uninterrupted optimization of material morphology and the compositions of different materials (MXenes, TMDs, and graphene materials) in terms of all sorts of sensing characteristics.

In the body of this review, the working principle and gas-sensing characteristics of gas sensors are introduced. More importantly, several known methods for improving the sensing performance of TEA-sensing materials are described in turn in the following sections, such as the optimization of the morphology and surface structures, combinations with other materials like metal elements and conducting polymers, the development of new materials, the application of advanced fabrication devices, and external stimulation. The conclusion and outlook are summarized at the end.

2. Gas-Sensing Characteristics

It is usually necessary to use some specific indicators to evaluate the gas-detection ability of a sensor. At present, the characteristic parameters commonly used to measure gas-sensing performance include optimal working temperature, sensitivity, selectivity, stability, repeatability, response and recovery time, and the lowest detection limits.

2.1. Optimal Working Temperature

The sensing characteristics of MOSs depend on the carrier concentration, which is relevant to the working temperature. Only at the optimum operating temperature can the sensitive materials fully stimulate the chemical activity to push the gas sensor to its maximum response. Several response curves are shown in Figure 1 [23], which often show an increase–maximum–decrease trend. When the temperature is too low, TEA molecules are inert and cannot overcome the activation energy barrier to react with the adsorbed oxygen [24]; then, as the temperature increases, the whole reaction will accelerate. However, at higher temperatures, the gas molecules get enough energy to rapidly escape from the material surface without affecting the conductivity of the sensor, resulting in the decline response [25].

At the present time, the optimal working temperature of most semiconductor sensors is high, often in hundreds of degrees Celsius, which causes huge power consumption. Developing a special gas sensor with a low operating temperature is also one of the current challenges. According to research, metals have surface redox reaction or catalytic properties, so noble doped or loaded can reduce the demand for energy input [26]. Additionally, the construction of heterojunctions can regulate carrier transport, or layered or core–shell gas-sensing materials will be more conducive to gas adsorption and desorption because of the porous structure and large specific surface area [27–29]. The above three methods are used to reduce the operating temperature of gas sensors.

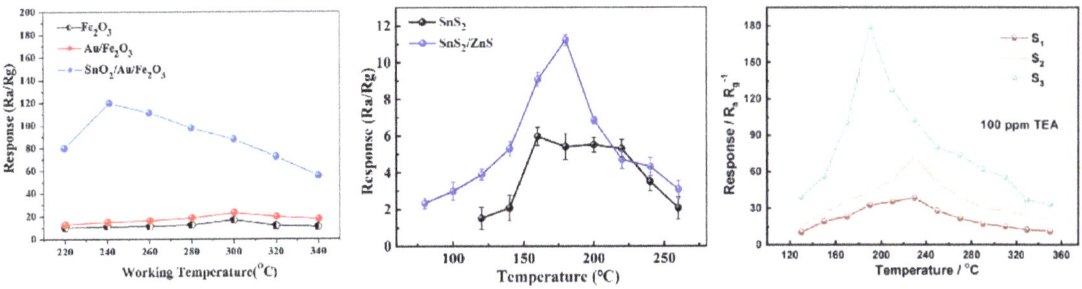

(a) (b) (c)

Figure 1. (**a**) The response curves of SnS_2 and SnS_2/ZnS to 50 ppm TEA under various working temperatures [23]; (**b**) The response curves of Fe_2O_3, Au/Fe_2O_3, and $SnO_2/Au/Fe_2O_3$ to 100 ppm TEA at different operating temperatures [30]; (**c**) The response curves of sample-1, -2, -3 to 100 ppm TEA at various operating temperatures [31].

2.2. Sensitivity

The sensitivity K of a gas sensor is an indicator of the responsiveness of a gas sensor to the target gas. It represents the compliance relationship between the electrical parameters of the gas sensor and the target gas concentration. There is no doubt that the greater the K value, the better the performance of the gas-sensing materials. The sensitivity K is usually expressed as K = Ra/Rg (n-type semiconductor) or K = Rg/Ra (p-type semiconductor) [32,33]. Ra and Rg represent the resistance of the sensor in air and in the target gas, respectively.

2.3. Selectivity

The selectivity of a gas sensor refers to its ability to recognize and measure a gas without interference from non-target gases in multigas environments [34]. In short, selectivity is the ability of a gas sensor to identify the measured gas more accurately in mixed gas. Only when the sensitivity of the target gas is several times or even tens of times higher than that of other interfering gases can a sensor be said to have good selectivity.

2.4. Stability

Stability is an important index for evaluating the property of a sensor. It refers to whether the sensor can work for a long time and still maintain or approach the initial performance within the predetermined working range [35]. Considering the practical application of gas sensors, the sensor's ability to maintain long-term stability is very necessary. Normally, the response of the prepared sensor over one or several months will be measured to ensure its stability.

2.5. Repeatability

Repeatability is defined as the ability of a semiconductor resistance sensor to restore its resistance to its original value and maintain its high sensing performance after target gas measurement. If the sensor cannot recover the resistance value in the normal gas environment, it maybe that the target gas has an irreversible impact on the sensor that makes it no longer operational [36].

2.6. Response Time (τ_{res}) and Recovery Time (τ_{rec})

Response time (τ_{res}) is defined as the time required for the resistance value of a sensor in the air to reach 90% of the resistance value in the measured gas. Similarly, recovery time (τ_{rec}) is specified as the time required to recover to 90% of the resistance value in the air after removing the target gas [37]. The preparation of sensors that can quickly detect TEA gas is also one of the directions that people are vigorously studying.

2.7. The Lowest Detection Limits

The lowest detection limit is one of the major indexes pursued by many TEA sensors. It refers to the minimum gas concentration that can make the sensor respond under certain conditions, that is, the minimum detection concentration. High-performance sensors with low detection limits can often detect parts from per million (ppm) or even lower to parts per billion (ppb) [38]. This makes it possible to capture mixed harmful gases in the air much earlier when gases leaks.

A comparison of the performance of several TEA sensors is given in Table 1.

Table 1. The sensing characteristics of several typical triethylamine sensors.

Nanomaterial Shapes	τ_{res}/τ_{rec} (s)	T (°C)	Conc. (ppm)	Lim. (ppm)	Res.	Ref.
SnS_2/ZnS microspheres	2/8	180	50	-	11.21	[23]
ZIF-67/PBA arrays	5/182	180	100	-	11.7	[24]
$ZnFe_2O_4$–ZnO mesoporous	0.9/23	240	50	-	21.23	[25]
Au−PdO Modified Cu-Doped $K_2W_4O_{13}$ Nanowires	17/27	120	10	1	282	[26]
mesoporous ZnO/Co_3O_4 nanosheets	17/25	240	50	0.087	67.8	[27]
COFs@SnO_2@carbon nanospheres	7/5	RT	2	0.2	95.1	[28]
ZnO/SnO_2 micro-camellia	27/12	100	100	1	780	[29]
yolk-shell $SnO_2/Au/Fe_2O_3$ nanoboxes	7/10	240	100	0.05	126.84	[30]
$Zn_2SnO_4/ZnSnO_3$	19/37	190	100	0.5	179.7	[31]
ZnO/Co_3O_4 nanomeshes	30/55	100	5	-	3.2	[32]

Note: τ_{res} and τ_{rec} represent response times and recovery times, respectively. T and Conc. indicate the optimal working temperature and detection concentration, respectively. Lim. is the abbreviation of the lowest detection limit. And Res. and Ref. represent response and reference, respectively. RT represents room temperature.

3. Triethylamine Gas Sensing Mechanism

To date, scientists have used various ways to explain the gas-sensing mechanism of semiconductor sensors. Several widely used theories are electron depletion layer theory, hole accumulation layer theory, bulk resistance control mechanism, and gas diffusion control mechanism.

3.1. Electron Depletion Layer (EDL) Theory

For n-type semiconductors, the gas-sensing mechanism is more often explained using electron depletion layer theory. As an n-type semiconductor, the main carriers are negatively charged electrons [39]. When the sensor is in an air environment, oxygen molecules are adsorbed on the material surface, and the material traps its free electrons in the conduction band, forming the surface adsorption of O^{2-}, O^- and $O_2{}^-$,described by Equations (1)–(4) [40]. The result is the formation of an electron depletion layer on the sensor surface, which leads to an increase in resistance. At this point, TEA gas is passed and will undergo a redox reaction with the adsorbed oxygen on the material surface, while the captured electrons will be released back into the conduction band, the process of which is described by Equation (5) [40]. This process will significantly reduce the Schottky barrier height and decrease the depletion layer thickness, leading to lower resistance. An typical image of the mechanism is shown in Figure 2 [41].

$$O_2(gas) \to O_2(ads) \tag{1}$$

$$O_2(ads) + e^- \to O_2{}^-(ads) \quad (T < 146.85°C) \tag{2}$$

$$O_2^-(ads) + e^- \to 2O^-(ads) \quad (146.85°C < T < 396.85°C) \tag{3}$$

$$O^-(ads) + e^- \to O^{2-}(ads) \quad (T > 396.85°C) \tag{4}$$

$$(C_2H_5)_3N + O^- \to CO_2 + H_2O + NO_2 + e^- \tag{5}$$

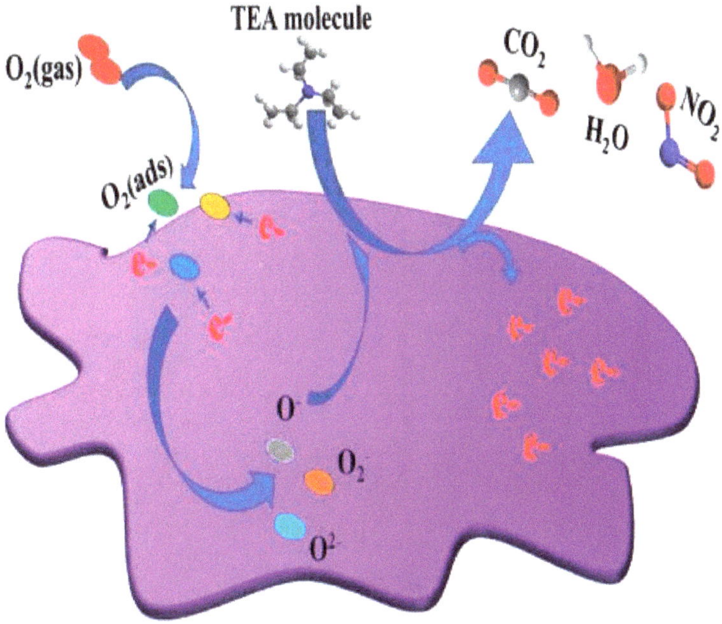

Figure 2. One typical n-type gas-sensing mechanism for an $SnWO_4$ sensor exposed to air and TEA gas [41].

3.2. *Hole Accumulation Layer (HAL) Theory*

For p-type semiconductors, the gas-sensing mechanism is more often explained using hole accumulation layer theory. As a p-type semiconductor, the main carriers are positively charged holes. In the air, oxygen molecules come in contact with the material and draw electrons out of the conduction band; they are then adsorbed on the surface to form chemisorbed oxygen (O^{2-}, O^-) [42]. The loss of electrons results in the formation of HAL on the surface and a concomitant decrease in the resistance of the sensor. The process is represented by Equations (6) and (7) [43]. When the sensor is exposed to TEA gas, the gas molecules adsorbed on the material surface undergo a chemical reaction and decompose into ammonia and ethylene, and the chemisorbed oxygen oxidizes the ammonia and ethylene to nitrogen dioxide, carbon dioxide, and water. Then, the free electrons return to the material surface, the thickness of HAL decreases, and the resistance of the sensor gradually returns to the original value. The process is represented by Equations (8)–(11) [43]. An example image of the mechanism is shown in Figure 3.

$$O_2(gas) + e^- \rightarrow O_2(ads) \quad (6)$$

$$O_2(ads) + e^- \rightarrow 2O^-(ads) \quad (7)$$

$$(C_2H_5)_3N \rightarrow NH_3 + 3C_2H_4 \quad (8)$$

$$NH_3 + O^-(ads) \rightarrow NO_2 + H_2O + e^- \quad (9)$$

$$C_2H_4 + O^-(ads) \rightarrow CO_2 + H_2O + e^- \quad (10)$$

$$(C_2H_5)_3 + O^-(ads) \rightarrow CO_2 + NO_2 + H_2O + e^- \quad (11)$$

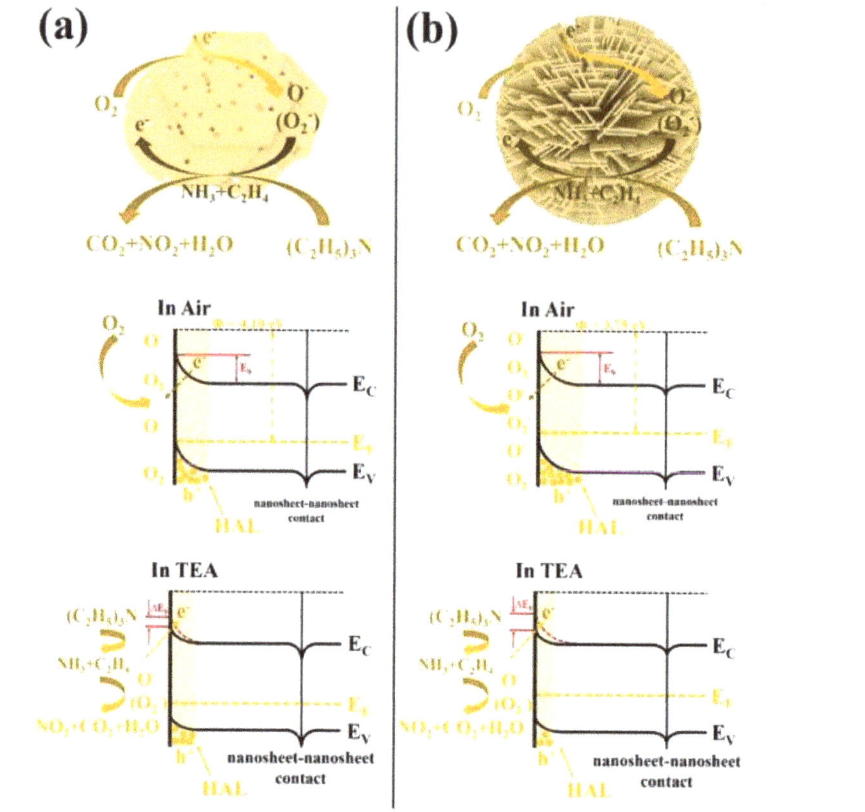

Figure 3. Typical p-type gas-sensing mechanisms for (**a**) Co_3O_4 sheet-300 and (**b**) Co_3O_4 sphere-300 [43].

4. Optimizing the Material Shape and Surface Structure

As is known to all, the gas-sensing performance of nanomaterials is partly determined by their ability to convert oxygen in the air into lattice oxygen. The stronger the conversion ability, the better the gas-sensing performance will be [44]. Obviously, the conversion ability is affected by the morphological structure of nanomaterials because the larger the specific surface area and porosity of the nanomaterials, the more active sites will adsorb oxygen, and the stronger their ability to adsorb oxygen. Researchers have prepared such nanomaterials as zero-dimensional (nanocrystals [45], nanoparticles [46], quantum dots), one-dimensional (nanorods [47,48], nanowires [49], nanofibers), two-dimensional (nanosheets [50], nanofilms), three-dimensional (nanofoam, nanospheres [51], nanoflowers [52]), and other nanomaterials by hydrothermal method, solvothermal method, and sol-gel method. In addition, nanocrystals also have complex surface structures, with clear shapes and exposed crystal surfaces [53]. Semiconductor nanocrystals usually have different surface structures on the atomic scale and exhibit different physical and chemical properties. It can be inferred that adjusting the different exposed crystal planes of nanomaterials also plays a key role in the sensing performance of TEA.

4.1. Zero-Dimensional (0D) Nanomaterials

Low-dimensional nanomaterials such as 0D nanoparticles and nanocrystals are the earlier nanomaterials prepared for gas sensors [54,55]. Because of their small specific surface area and low porosity compared with other dimensional nanomaterials, they often need some ways to improve their gas-sensing characteristics [56], although some 0D materials do have good sensing performance.

Yu et al. [57] synthesized hierarchical hollow $GaFeO_3$ by a kind of solvothermal method and a subsequent annealing strategy. They obtained Fe_4 $[Fe(CN)_6]_3$ MOF precursors through a facile hydrothermal process and then used Ga^{3+} as the modifier to modify the precursors. The hollow porous nanocube microstructure was vividly depicted by TEM, SEM, and XRD in Figure 4. They found that the microsized morphologies and hollow interior structures of $GaFeO_3$ microcubes can be feasibly modulated by controlling the thermolysis temperatures. The ultrasmall $GaFeO_3$ nanostructure exhibited rapid response times (9 s), good selectivity, and excellent stability.

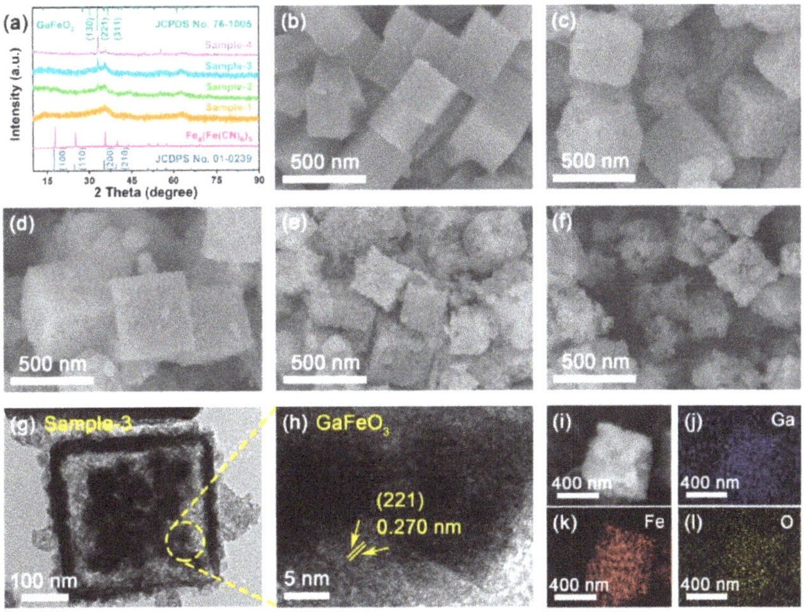

Figure 4. (**a**) From bottom to top, the simulated and experimental XRD patterns of Fe_4 $[Fe(CN)_6]_3$ MOFs (JCPDS No. 01–0239), the experimental XRD patterns of Samples-1, -2, -3, -4, and the simulated XRD pattern of $GaFeO_3$ phase (JCPDS No. 76–1005). The FE-SEM images of (**b**) Fe_4 $[Fe(CN)_6]_3$ MOF precursors and (**c**–**f**) porous $GaFeO_3$ microcubes of Samples-1, -2, -3 and -4 obtained at 400, 450, 500, and 550 °C, respectively. (**g**) TEM and (**h**) HRTEM images of Sample-3 obtained at 500 °C. (**i**) The selected SEM image of Sample-3 and the corresponding elemental mapping images of (**j**) Ga, (**k**) Fe, and (**l**) O, respectively [57].

Meng et al. [58] successfully synthesized SnO_2 nanoparticles with abundant oxygen vacancies (OVs) by a combined hydrothermal route and ice-water bath stirring method. The sensors based on this material reveal excellent selectivity, ppb level detection limit, and long-term stability. The outstanding sensing characteristics of the sensor could be attributed to the abundant OVs, improving the O_2 adsorptivity and enhancing electron transfer. Du et al. [59] synthesized SnO_2 quantum dots with controllable size by changing the amounts of hydrazine in the hydrothermal process. The experimental results show that the smaller the SnO_2 quantum dot material, the better the sensing performance for VOC gases. More critically, it has a faster response time and a lower detection limit for TEA than other VOC gases.

4.2. One-Dimensional (1D) Nanomaterials

Hierarchical materials with 1D structures can be used to design adjustable surface-active regions to obtain more reactive sites, boost electron transmission efficiency, and optimize electron transmission channels [60–62]. Moreover, 1D hierarchical structures can inhibit aggregation, resulting in significantly improved gas dispersion and transfer [63].

Lv et al. [64] successfully fabricated well-crystallized ZnO nanorods using a simple solution route with dodecyl benzene sulfonic acid sodium salt as a modifying agent. The experiment results reveal that this material has a low working temperature (150 °C), excellent selectivity, and superior sensitivity to ppm level TEA. In another study, Liu et al. [65] fabricated ultralong NiO nanowires assembled with NiO nanocrystals by adjusting the hydrothermal reaction temperature and time to enhance the gas-sensing properties. The TEM images of the structural representations of the material are shown in Figure 5, and typical transient response curves of several powders of NiO are exhibited in Figure 6. The curves indicate the improvement in gas sensing of ultralong NiO nanowire compared with the other two nanomaterials.

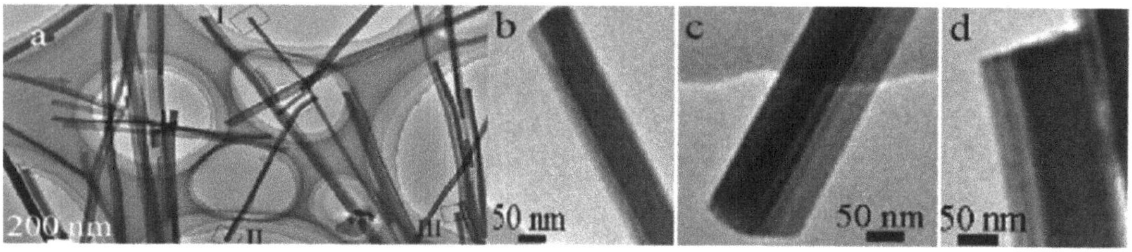

Figure 5. (**a**) The typical TEM images of the $NiC_2O_4 \cdot 2H_2O$ precursors prepared at 220 °C for 12 h. (**b–d**) TEM images at high magnification from rectangles I–III in [65].

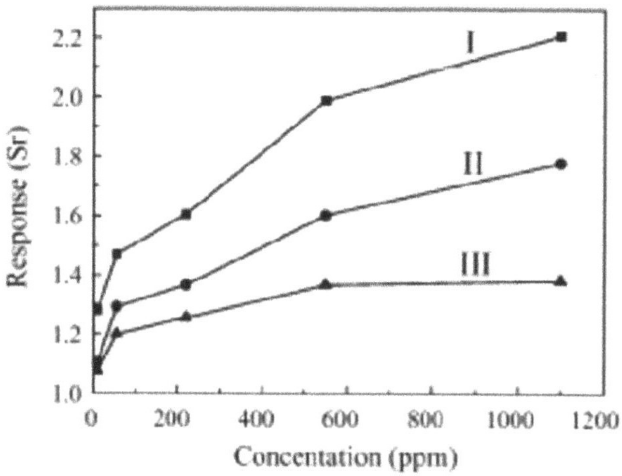

Figure 6. The response of the sensors is based on (I) the ultra-long nanowires, (II) nanowires, and (III) powders of NiO to triethylamine of different concentrations [65].

Zou et al. [66] reported a hollow SnO_2 microfiber that was prepared via a unique sustainable biomass conversion strategy for highly efficient TEA detection. The authors established that the unique structure efficiently immobilized Sn^{2+} cations during the subsequent calcination process to synthesize the hollow SnO_2 microfiber. The excellent TEA gas property in turn contributed to the synergism of the 1D carbon morphology and the porous hollow structure.

4.3. Two-Dimensional (2D) Nanomaterials

Mesoporous or layered 2D nanomaterials composed of nanoparticles often have large specific surface area and porosity, which also gives them good sensitivity, long-term cycle stability, and excellent selectivity, which is more conducive to the detection of TEA [67]. Researchers today are still focusing on research and development related to different 2D nanostructures to improve gas-sensing performance.

Zhang et al. [68] used polystyrene-polyacrylic acid ZIF-8 as the collaborative template by calcination precursor for 2 h to obtain mesoporous ZnO nanosheets. The team completely characterized the morphology as well as the microstructure and specifically tested the gas properties for TEA. The test results show that the high response to 50 ppm TEA can reach 43.771 at 268 °C. Meanwhile, the low detection limit (1 ppm) and rapid response/recovery are also advantages of the material. In one recent report, Liu et al. [69] designed 2D ultrathin SnO_2 nanofilms for manufacturing low-temperature TEA gas sensors. They studied the effect of calcination temperature on crystal crystallization and found that with the increase in temperature, the crystal size decreased, and the uniformity and density of the surface particles gradually increases. Furthermore, the decrease in crystallinity combined with the increase in interfacial defects led to the suppression of grain boundary migration and an increase in the energy barrier for grain growth.

4.4. Three-Dimensional (3D) Nanomaterials

There are now a variety of 3D porous core-shell, layered, microsphere, and other nanostructures assembled by low-dimensional nanomaterials with greatly increased specific surface areas and active sites that can promote the adsorption and transmission of gas molecules [70]. The increased specific surface areas and pore diameters of 3D porous hollow structures also avoid agglomeration [71,72].

Wang et al. [73] prepared a 3D porous ZnO foam structure through a simple solvothermal method by dissolving $Zn(NO_3)_2{\cdot}6H_2O$ solid in ethylene glycol solution and calcining the precipitate at 350 and 450. The sample obtained at 350 °C showed an excellent response to low-concentration TEA. One type of WO_3 hollow microsphere for fast TEA gas sensing was successfully prepared by Zhai et al. [74]. They used a simple low-temperature solvothermal strategy followed by an annealing process in atmospheres. The fabricated WO_3 hollow microspheres material shows excellent selectivity and fast response time to 50 ppm TEA at 220 °C. The response is almost greater than that for all of the previously reported TEA sensors. In the SEM images shown in Figure 7, we can clearly see the microsphere structure of the prepared material. Its large specific surface area and hollow structure play a key role in improving its response speed. Sui et al. [75] constructed three-dimensional novel flower-like α-MoO_3 with hierarchical structure via a facile solvothermal route without any surfactant or template and performed subsequent calcination at 400 °C in the air for 2 h. The experimental results present that at the working temperature of 250 °C, this material to 100 ppm TEA not only attains a high response (416) but a low detection limit (0.5 ppm). However, the long recovery time is also a big problem to be solved.

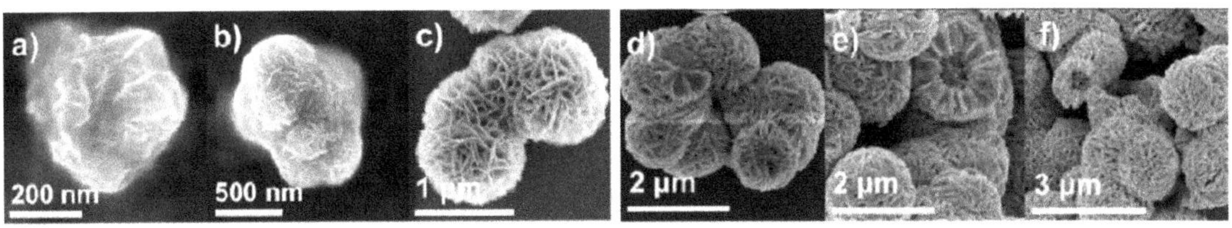

Figure 7. (**a–e**) SEM images of the hollow microsphere samples prepared after 10 min; 15 min; 30 min; 2 h; and 6 h hydrothermal reaction time. (**f**) SEM image of the WO_3 hollow microspheres after annealing [74].

4.5. Exposed Crystal Plane

Lattice is the regular arrangement of atoms in the crystal. In the process of spontaneous growth, crystals can develop polyhedral shapes composed of planes with different orientations [76]. The planes in these polyhedral shapes are called crystal faces. Crystal will produce two different situations: anisotropic and isotropic. The anisotropy of crystals means that the periodicity and density of the arrangement of protons are different along with different directions of the lattice, resulting in different physical and chemical properties of crystals in different directions. The opposite is isotropy, that is, the physical and chemical properties of crystals in different directions are the same. The crystals often show anisotropy, so they often show different gas-sensing characteristics when exposed to different crystal planes [77].

There are few reports on the effect of exposed crystal on TEA gas sensing; however, Xu et al. [78] prepared SnO_2 nanorods with {200} crystal faces and successfully made normal SnO_2 nanorods with outstanding TEA selectivity. Xiang et al. [79] make a profound study of the TEA-sensing characteristics of nonpolar (11–20) and polar (0001) GaN thin films. The results show that Mn atoms at (111) surface show remarkably improved sensing properties, which allows the active unsaturated Mn atoms to adsorb oxygen, creating electrons and catalyzing the gas-sensing reaction.

In Table 2, the reaction of nanomaterials with different morphology and exposed crystal surface to TEA is sorted out.

Table 2. TEA-sensing properties of different gas-sensing material shapes.

	Nanomaterial Shapes	τ_{res}/τ_{rec} (s)	T (°C)	Conc. (ppm)	Lim. (ppm)	Res.	Ref.
0D	hollow $GaFeO_3$ microcubes	9/49	200	200	-	7.4	[57]
	SnO_2 nanoparticles	163/163	260	100	0.001348	430.65	[58]
	SnO_2 quantum dots	1/47	240	100	1	153	[59]
1D	ZnO nanorods	15/15	150	1	0.1	39	[64]
	NiO nanowires	-	350	9	2	3.5	[65]
	Hollow SnO_2 Microfiber	14/12	270	100	2	49.5	[66]
2D	ZnO nanosheet	7/21	268	50	1	43.771	[68]
	SnO_2 nanofilms	-	150	100	-	19.2	[69]
3D	porous ZnO foam	1/1	350	100	<5	79.5	[73]
	WO_3 hollow microspheres	1.5/22	220	50	-	16	[74]
	flower-like α-MoO_3	3 s/1283	250	100	0.5	416	[75]
crystal face	SnO_2 nanorods {200} crystal faced	6 s/465	120	50	-	64	[78]
	polar (0001) GaN thin films	7.9 s/20.7	480	200	0.2	5.23	[79]

5. Combinations of Different Materials

Although the metal oxide of a single metal has a certain response to VOCs, most of the materials have a high intrinsic value band, leading to weak gas adsorption capacity and resulting in low sensitivity, poor selectivity, and insufficient response speed. In the process of continuously improving the performance of the sensor, researchers found that combinations of different materials can often increase the adsorption sites and reduce the reaction energy consumption [80]. Several material combination methods for improving the sensing ability will be introduced below.

5.1. Effect of Metal Elements and Non-Metallic Elements

Among the sea of methods for enhancing the gas-sensing characteristics of gas-sensing materials, the use of metal or nonmetallic elements to dope and modify the target materials is the most convenient and popular [81]. Both the catalysis of metal materials and the electronic properties of nonmetallic materials can reduce the adsorption energy of gas [82], and the formation of defects on the surface and modification of the electronic structure can also play a role in improving sensing properties [83]. The action of metal elements can be simply divided into two categories, doped and surface decoration, and the action of nonmetallic elements can be roughly divided into doped and conductive polymer composite modification.

5.1.1. Role of Transition Metals, Noble Metals, and Rare Earth Elements

Through a large number of studies, it is found that noble metals and transition metals have strong catalytic properties and can replace the metal ion sites in the original materials to form more oxygen vacancies and surface adsorption sites, so as to improve the adsorption performance of gas. Therefore, transition metals such as Fe, Cr, Ni, Co, Ce, Cu, and noble metals such as Ag, Au, Pt, and Pd have been utilized as dopants into nanostructured [84,85]. In addition, rare earth elements such as Y, Sc, La, Ce, Ho, and Gd with unique electronic shells as dopants can cause lattice deformation and defects, increasing the number of oxygen vacancies and improving the gas adsorption capacity [86].

Liu et al. [87] loaded Pt nanoparticles on InO_3 novel hierarchical ZnO microspheres via a two-step hydrothermal method. They not only compared the TEA gas-sensing performance of the prepared Pt-ZnO nanospheres with that of commercial ZnO but also discussed the role of Pt decoration in enhancing the gas-sensing properties. Figure 8 depicts that the responses of the self-made ZnO nanospheres mixed with Pt at 200 °C are much higher than those of commercial ZnO and the self-made nanospheres without Pt. Zhu et al. [88] dropped different holmium ions concentrations of 0 at.%, 0.11 at.%, 0.45 at.%, and 0.53 at.% in SnO_2 nanoparticles (NPs) via gas–liquid phase chemical deposition following annealing. Figure 9 the presents the TEM images of the process, showing that Ho particles are already dropped in SnO_2. Meanwhile, Figure 10 reveals the brief enhancement of its TEA gas-sensing properties.

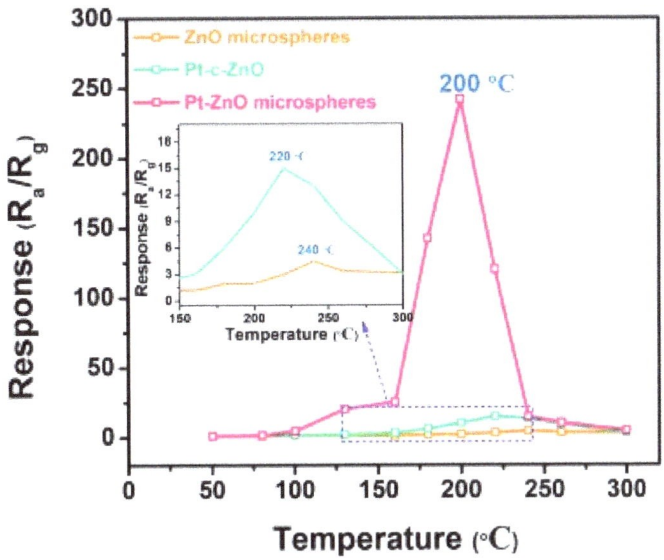

Figure 8. The response curves of the ZnO, Pt-c-ZnO, and Pt-ZnO microspheres to 100 ppm TEA at various temperatures [87].

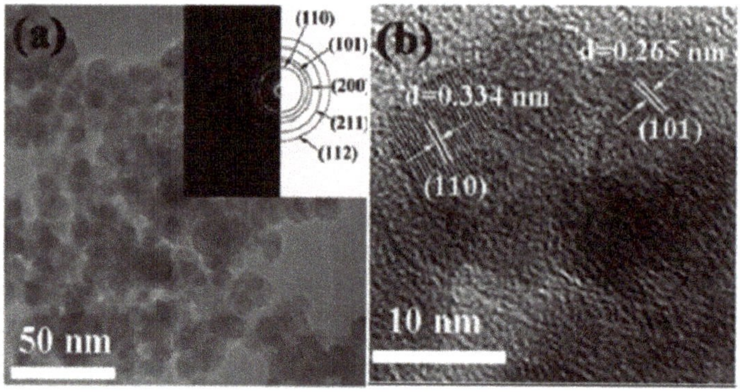

Figure 9. (**a**) TEM images of 0.45 at% SnO_2: Ho^{3+} nanoparticles. The inset is the SAED image of 0.45 at% SnO_2: Ho^{3+} nanoparticles. (**b**) The HRTEM image of 0.45 at% SnO_2: Ho^{3+} nanoparticles [88].

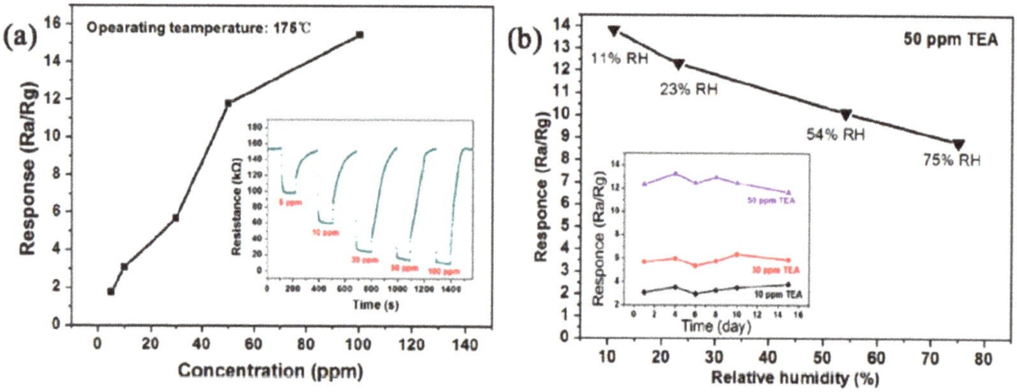

Figure 10. (**a**) A line chart of the relationship between the sensor response and the TEA concentration; (**b**) nanoparticle-based sensor responses to 50 ppm TEA under different RHs [88].

Zhang et al. [89] prepared Cr-doped SnO_2 microrods using a facile hydrothermal method to detect TEA. The gas sensitivity test found excellent selectivity and sensitivity to TEA; what was more surprising was its marvelous response time (1 s) as well as its humidity resistance. The latter above all makes it one good commercial sensor. In one recent study, Bi et al. [90] synthesized Rh-SnO_2 nanosheets by facile hydrothermal synthesis and subsequent surface impregnation precipitation and heat treatment. The sensor had better stability and gas selectivity, and its temperature response was nearly 15 times higher, than that of the nondoped sensor at the best working temperature (325 °C).

5.1.2. The Roles of Nonmetallic Elements

Adding nonmetallic elements to the material can reduce the Fermi level and enhance the electron transfer between the material and gas molecules, so as to improve the selectivity and sensitivity of the material to the target gas [91]. At present, some researchers have added carbon and other nonmetallic elements to gas-sensing materials to improve their gas-sensing properties [92].

To overcome the problems of high concentrations of chemical hydrothermal solutions and complex chemical reactions, Peng et al. [93] applied the magnetron sputtering process to prepare the core-shell structure and then combined it with reduced graphene oxide to prepare boron-doped reduced graphene oxide- (BRGO) coated Au@SnO_2. Figure 11 is the structure of the prepared materials. The enhancement of its gas-sensing properties may be largely attributed to the wrapping with BRGO. In another report, Zhang et al. [94] prepared carbonized polymer dots doped with hierarchical tungsten to solve the problems of the high working temperature and high detection limit of TEA sensors. They not only developed the preparation method of the mixed material but also studied its gas-sensing performance to TEA and the effect of the carbon compound point on the performance improvement, which laid a foundation for the further study of carbon doping in the future.

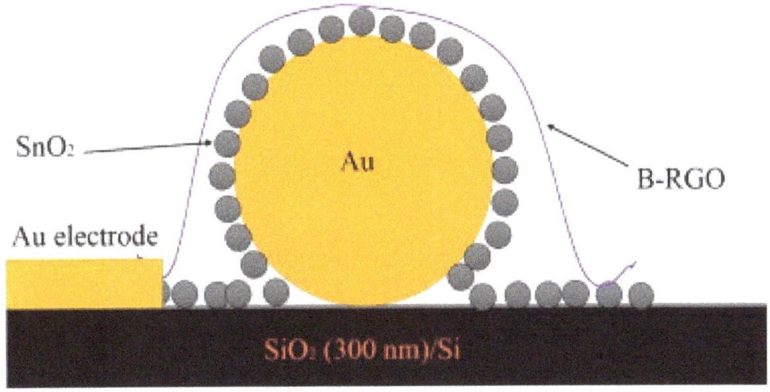

Figure 11. The schematic diagram of the B-RGO/SnO_2@Au heterostructure-based sensor [93].

5.2. Noble Metal Oxides and Transition Metal Oxides

Similar to the idea of using metals to enhance gas-sensing properties, the combination of different metal oxides can also improve gas-sensing performance [95]. In semiconductor physics, semiconductors with electrons as the main charge carriers are called n-type semiconductors, semiconductors with holes as the main charge carriers are called p-type semiconductors, and heterojunctions will be produced when different semiconductors contact [96]. Generally, the semiconductor oxides with different properties have different gas-sensing mechanisms and effects. According to investigations, common types of heterojunction are p-p, p-n, n-p, and hybrid junctions [97]. Because of the existence of heterojunctions, the material resistance is often very high, and in order to balance the Fermi level, it will produce a greater response when exposed to the target gas [98].

5.2.1. N-N Heterojunction

The n-type semiconductor metal compounds tend to have better gas-sensing performance and are the most used class of semiconductor metal compounds. Investigators have combined a large number of various n-type semiconductor compounds in an attempt to enhance their performance with TEA detection. The main principle of n-n heterojunction for enhancing the gas-sensing performance is as follows. The main carrier of n-n heterojunction is electrons, where the carriers will flow from the high side of the conduction band to the low side of the conduction band; an EDL is formed on the high side of the conduction band, and a charge accumulation layer is formed on the low side of the conduction band. This change will result in a larger response of the sensing material before and after exposure to the target gas with a larger range of electron changes [99].

Among the many examples, Xu et al. [100] prepared n-n heterojunctions via assembling SnO_2 nanosheets and TiO_2 nanoparticles employing the PLD method. They characterized the shape of the material by SEM and TEM, and they analyzed the elemental composition of the material by EDS, XRD, and XPS. The gas sensors were completely tested for several gas-sensing indexes such as repeatability, selectivity, and operating temperature. The test reports show that the sensor has not only excellent TEA selectivity but fast response time and recovery time. Xue et al. [101] prepared CeO_2/SnO_2 nanoflowers using the one-step hydrothermal method. The result shows that the response of 5 wt.% CeO_2 content composite excellently improved. The sensing mechanism diagrams of pure SnO_2 and CeO_2/SnO_2 are shown in Figure 12.

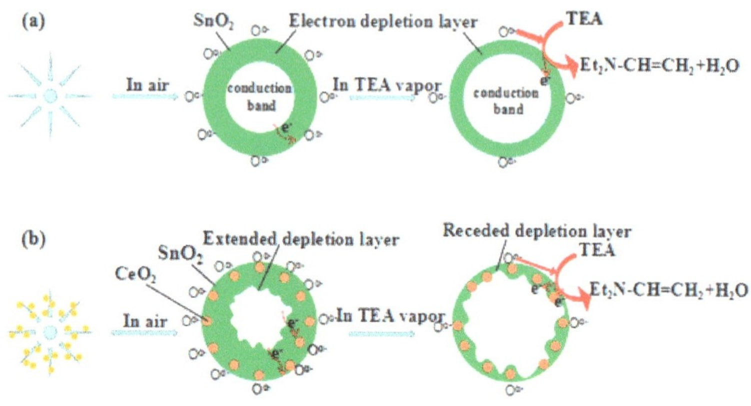

Figure 12. Sensing mechanism diagrams of (**a**) pure SnO_2 and (**b**) CeO_2/SnO_2 nanostructures [101].

5.2.2. P-P Heterojunction

Based on previous experimental experience, the gas-sensing performance of p-type semiconductor metal compounds is poor compared with that of n-type semiconductor metal compounds, but the gas sensing performance of p-type semiconductors can be relatively enhanced by forming p-p heterojunctions. The enhancement mechanism of the gas-sensing performance through p-p heterojunctions is mainly as follows. P-P heterojunctions have holes as the main carriers, in which the carriers will flow from the side with a higher valence band to the side with a lower valence band, forming a hole depletion layer on the side with a higher valence band and a hole accumulation layer on the side with a lower valence band [102]. The change in resistance due to this change is beneficial for the increase in gas-sensing performance [103].

Wang et al. [104] prepared $NiWO_4$@NiO p-p heterostructure via a self-sacrificing template method. The response of the heterojunction material prepared by the template was enhanced from 2.5 to 65 compared with pure NiO at an optimum operating temperature (240 °C) with 50 ppm of TEA gas. The team found that the p-p heterojunction needs to keep the balance of the Fermi energy level; electrons will be transferred from $NiWO_4$ to nickel monoxide, and the electron-hole pair recombination leads to higher resistance. The mechanism diagram is shown in Figure 13.

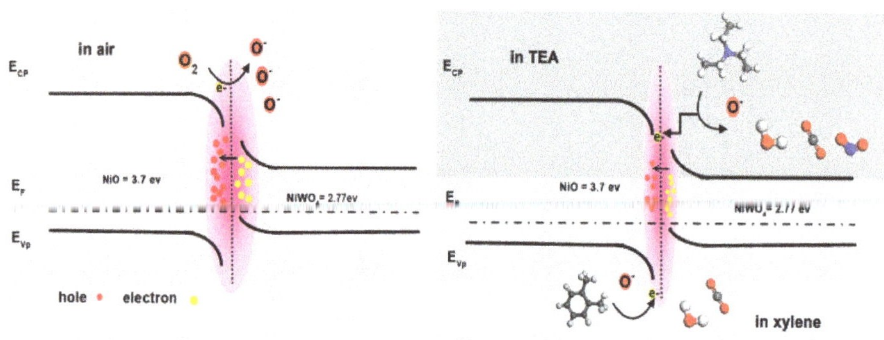

Figure 13. A schematic diagram of the improved gas-sensing mechanism of p-p $NiWO_4$ decorated NiO [104].

5.2.3. P-N Heterojunction

P-N heterojunctions, the most representative heterojunction, applied in a variety of different fields to enhance the physicochemical properties of semiconductors, are also one of the most important means of enhancing the TEA-sensing performance. P-N heterojunctions to improve gas-sensing performance are briefly described below. For p-n junction MOS sensing materials, the electron will migrate from n-type material to p-type material, and a large number of electrons are lost in the p-type material, forming a thick EDL, while holes are transferred to the p-type material, leading to band bending [105]. This process accelerates the carrier transfer efficiency in the accelerated material, so that the p-n heterojunction material obtains a higher resistance in the post-in and target gas reactor process to obtain the reaction-generated free electrons, lowering the resistance to obtain a large response [106].

For instance, Zeng et al. [107] fabricated a p-n heterojunction of Co_3O_4/WO_3 by a ZIF template. In this study, the sensor shows superb selectivity, excellent long-term stability, and a linear response for TEA. The authors explained the high performance of the sensor according to the large specific surface area, porosity, and most importantly, the depletion principle. In another study, Yu et al. [108] fabricated mesoporous and hierarchical hollow-structured In_2O_3-NiO composites via solvothermal reaction and subsequent cation exchange. The result in Figure 14 shows the obvious descent of the working temperature. Figure 15 depicts the gas-sensing mechanisms as well as the energy band states of NiO, In_2O_3, and In_2O_3-NiO.

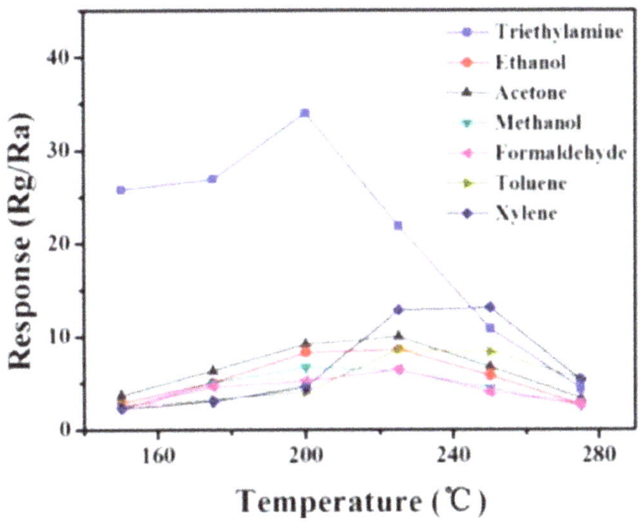

Figure 14. The selectivity of In_2O_3-NiO composites to seven different VOCs at various temperatures [108].

5.3. Conducting Polymer

Although MOS materials are widely used as sensors, their high operating temperatures have always affected their practical applications. To solve the temperature problem, experimenters have begun to gradually apply conductive polymers in semiconductor gas sensors. Among them, the most widely used were the well-known polyaniline (PANI), polypyrrole (PPy), polythiophene (PT), and poly (3,4-ethylene dioxythiophene) (PEDOT). In 2016, Bai et al. [109] apply polyaniline to fabricate a TEA detection sensor. The fabricated polyaniline@SnO_2 shows excellent selectivity and ultra-low detection temperature, which are the important performance indicators of TEA detection. This material opens the way to combining conducting polymer with semiconductor mental oxide for TEA detection.

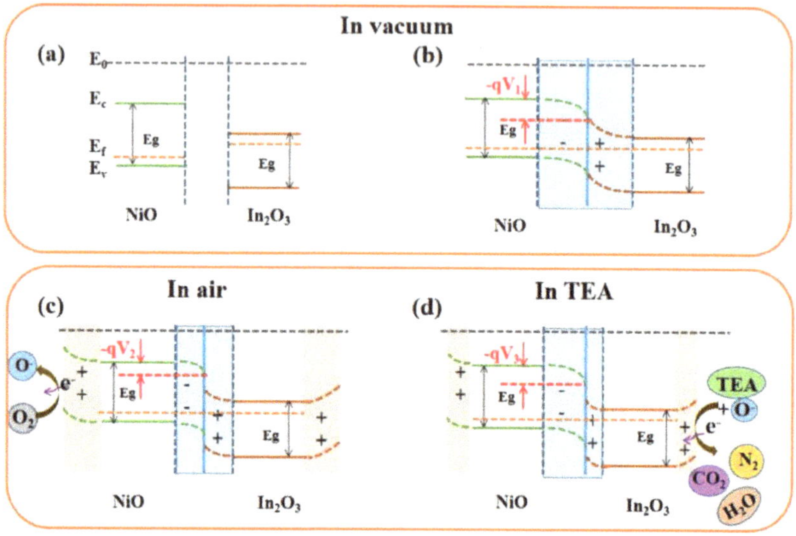

Figure 15. Schematic diagrams of the energy band states of NiO, In_2O_3, and In_2O_3-NiO, (**a**,**b**) in a vacuum, (**c**) in air, and (**d**) in TEA [108].

5.4. Quantum Dots

Quantum dots are one species of material used in chemiresistive sensors. Since their first introduction in 1983, scientists found that the bandgap properties of quantum dot materials can be changed by adjusting the quantum dot size. The materials have been applied in various fields including gradually in the sensing field [110]. In a report on TEA detection, Liu et al. [111] combined CdS quantum dots with ZnO to form a kind of n-n heterojunction structure, improving the response time (2 s) and decreasing the working temperature (200 °C). The following year, Liu et al. [112] again used quantum dots technology to combine $CsPbBr_3$ quantum dots with ZnO. This material has a higher response and lower detection limit compared with the previous experiment. The images of response versus temperature are exhibited in Figure 16. The two experiments by Liu et al. show that quantum dot materials have great potential for TEA detection, and more quantum dot materials may be used for TEA detection in the future.

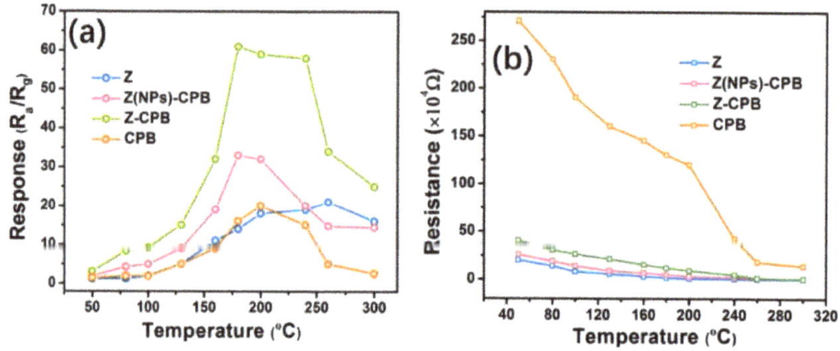

Figure 16. (**a**) Responses of as-obtained sensors versus different temperatures to 100 ppm of TEA. (**b**) The plots of Ra change for all sensors at different working temperatures to 100 ppm of TEA [112].

6. New Materials Application

Although conventional MOSs such as ZnO and SnO_2 have certain gas-sensing abilities, they are still difficult to work with at room temperature and have high detection limits, which hinders their practical application [113,114]. Therefore, it is imperative to find new materials to solve their existing problems. In recent years, graphene and its derivatives, carbon nanotubes, 2D transition metal carbides and nitrides (MXene), 2D transition metal dichalcogenide (TMDs), and spinel and perovskite materials have been gradually developed by scholars to enhance gas detection [115–117].

6.1. Graphene and Its Derivatives-Based

Graphene is a single-layer carbon sheet with a hexagonal filled lattice structure, which has many outstanding physical and chemical characteristics, such as good low temperature, high carrier mobility, excellent optical properties, large specific surface area, and excellent thermal conductivity [118]. The performance of graphene derivatives, graphene oxide (GO) and reduced graphene oxide (RGO) tablets, is further improved for their reactive oxygen groups [113,119]. Graphene and its derivatives combined with various functional materials are widely used in lithium-ion batteries, metal oxide supercapacitors, and other applications [120]. Because of its ability to adsorb gas at room temperature, large specific surface area, and high conductivity, graphene is also gradually being used in gas sensors [121].

Bai et al. [122] prepared a hybrid of pine dendritic $BiVO_4$/RGO via a one-step hydrothermal method enhancing the response of 10 ppm TEA at a working temperature of 180 °C. Figure 17 shows SEM images that indicate sufficient contact between the two materials and the full interaction of both. Figure 18 displays the nitrogen adsorption/desorption isotherm of $BiVO_4$/13.0 wt.% RGO hybrid; the figure reveals its large specific surface area, which is a benefit for TEA detection.

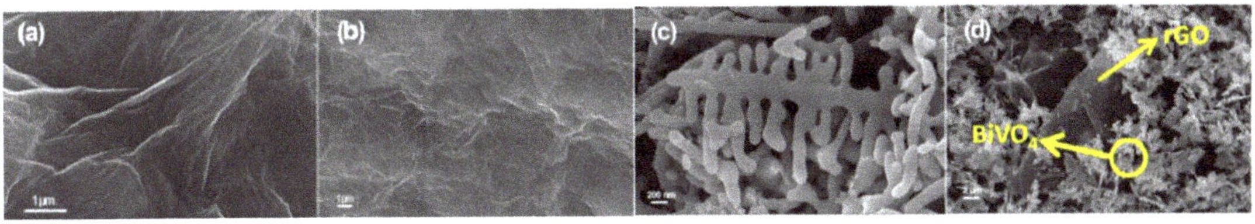

Figure 17. SEM images of (**a**) GO; (**b**) RGO; (**c**) pure pine dendritic $BiVO_4$; (**d**) $BiVO_4$/RGO hybrid [122].

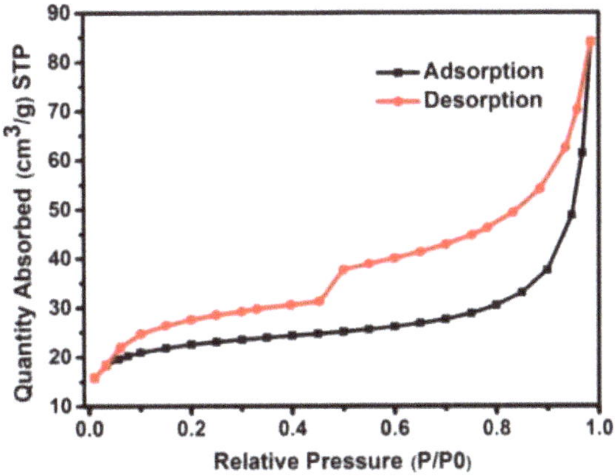

Figure 18. The nitrogen adsorption/desorption isotherm of $BiVO_4$/13.0 wt.% RGO hybrid [122].

Wei et al. [123] successfully prepared a MOF-derived α-Fe_2O_3 porous spindle combined with RGO using Fe-MIL-88 as a precursor through a simple solvothermal method to improve TEA sensing performance. Figure 19 presents dynamic response–recovery curves. It clearly shows that the new material decorated with RGO not only greatly increased response but accelerated response time. Responses of RGO/α-Fe_2O_3 nanocomposites display better linearity and superior sensing performance.

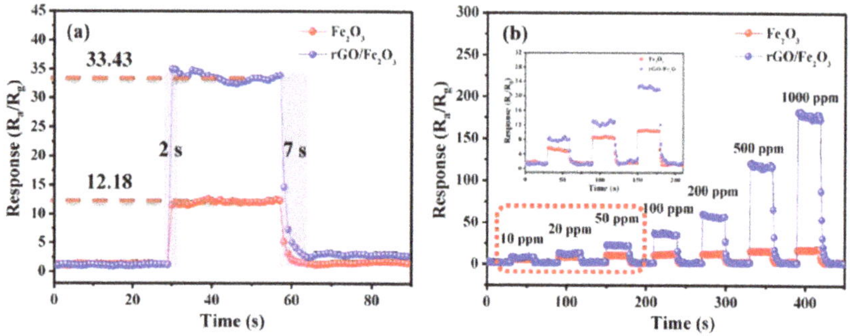

Figure 19. (**a**) Dynamic response–recovery curves (100 ppm TEA) of pure α-Fe_2O_3 spindles and RGO/α-Fe_2O_3 nanocomposites at 280 °C. (**b**) Concentration-dependent response curves of pure α-Fe_2O_3 spindles and RGO/α-Fe_2O_3 nanocomposites to 10–1000 ppm TEA [123].

Yu et al. [124] modified Co_3O_4 nanoparticles by wrapping RGO via a simple hydrothermal method. In their experiments, they found that Co_3O_4 with {100} crystal planes and {112} crystal planes could be prepared by changing the amount of sodium hydroxide. The experimental results show that the nanostructures with {112} surface contain more adsorbed oxygen, which can enhance the adsorption of TEA. Figure 20 shows the 3D electron density difference between the Co_3O_4 (100)/graphene interface and the Co_3O_4 (112)/graphene.

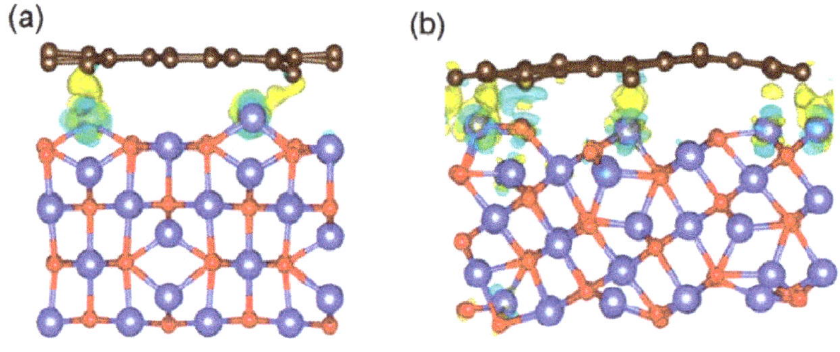

Figure 20. 3D electron density differences on the Co_3O_4 (100)/graphene interface (**a**) and the Co_3O_4 (112)/graphene interface (**b**). The cyan and yellow contours represent electron density depressions and accumulations, respectively [124].

6.2. MXenes-Based

Since the first MXene report in 2011, MXene materials have received extensive attention [125,126]. Various MXene materials such as $Ti_3C_2T_x$, Mo_2CT_x, and V_2CT_x have been successively prepared and gradually used in the research on gas sensors [127]. MXenes are early-transition metal carbides/carbonitrides and nitrides with a general formula $M_{n+1}X_nT_x$, where M represents transition metals (like Sc, Ti, V, Cr, Zr, Nb, Mo, and Ta), X is C or N, and T represents surface terminal groups (like O, F, and OH) [128].

MXene is a multifunctional material that is mostly used in catalysis, ion batteries, and gas storage [129]. Due to its special geometric and electronic structure, good conductivity, large specific surface area, excellent gas adsorption capacity, rich active sites, and excellent stability, it has been the focus of much attention in the field of gas sensing in recent years [130]. In one recent study, Liang et al. [131] applied MXene material for the first time to a TEA gas sensor and achieved a good response. They synthesized 2D/2D SnO_2 nanosheets/$Ti_3C_2T_x$ MXene nanocomposites by a simple hydrothermal method and achieved fast response/recovery time, relatively high sensitivity, high stability, and excellent selectivity to the synergistic effect of SnO_2 and MXene in SNTM composite and its highly interconnected porous structure determine. Figure 21 is its XPS spectrum.

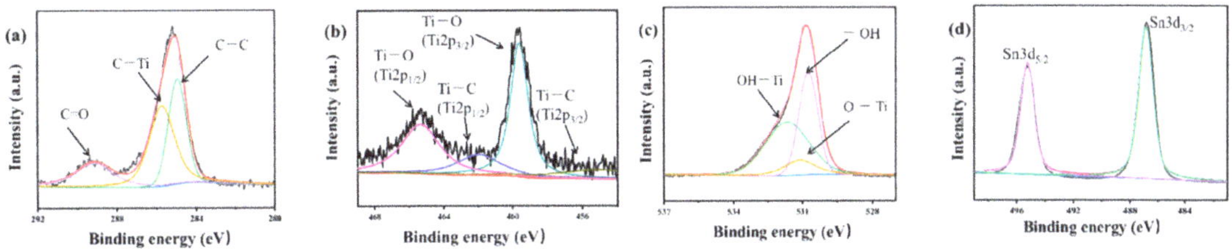

Figure 21. The high-resolution XPS spectra of (**a**) C 1s; (**b**) Ti 2p; (**c**) O 1s; (**d**) Sn 3d, respectively [131].

6.3. TMD-Based Materials

In addition to graphene and its derivatives as well as MXenes, 2D transition metal dichalcogenide (TMD) nanosheets are also applied in gas sensing. TMDs are a kind of layered material [132]. The basic chemical formula can be written as MX2, where M represents transition metal elements, including Ti, V, Ta, Mo, and W, and X represents chalcogenide atoms (S, Se, Te, etc.); MoS_2, WS_2, $MoSe_2$, and WSe_2 are a few of the typical materials [133]. Because of its unique layered structure, large specific surface area, and excellent physicochemical as well as electronic properties, TMD has great application potential in the field of gas sensors [134].

However, there are few cases of its application in TEA detection. Xu et al. [135] proposed a novel n-n heterojunction material based on MoS_2/ZnO, and its size and microstructure were designed by controlling the annealing rate. They determined that the material remaining in a weak acid environment CTAB can gradually be adsorbed on MoS_2 nanosheets to form the n-n heterojunction. Then, they carefully researched its TEA sensing mechanism and enhanced the sensitivity by the unique microbridge structure. Yang et al. [136] studied the gas selectivity of 3D MoS_2/GO hybrid nanostructures under TEA and other VOCs. They found that the material had better selectivity for TEA gas at working temperature (260 °C). It may be that the synergistic effect of the layered TMDs and GO produce strange sensing characteristics. Figure 22 is the response curves.

6.4. Perovskite Structure and Spinel Structure

The general formula of perovskite structure is ABX_3, and the structure commonly used in gas sensors is ABO_3 [137,138]. Perovskite-type ternary compounds have a wide bandgap, excellent thermal stability, and chemical compositions including the partial substitution of A and/or B positions with aliovalent elements of variable sizes and values [139,140]. However, chemical instability, low porosity, and poor low-temperature stability hinder its application in gas sensors [141,142]. There are only a few perovskite materials applied to TEA sensing. Zheng et al. [143] obtained the $CoSnO_3$ nanoboxes via the calcination of $CoSn(OH)_6$ precursors derived from a solution–precipitation method. After a series of experiments, they came to the conclusion that the material is a kind of p-type semiconductor that can detect TEA under a low temperature with a low detection limit.

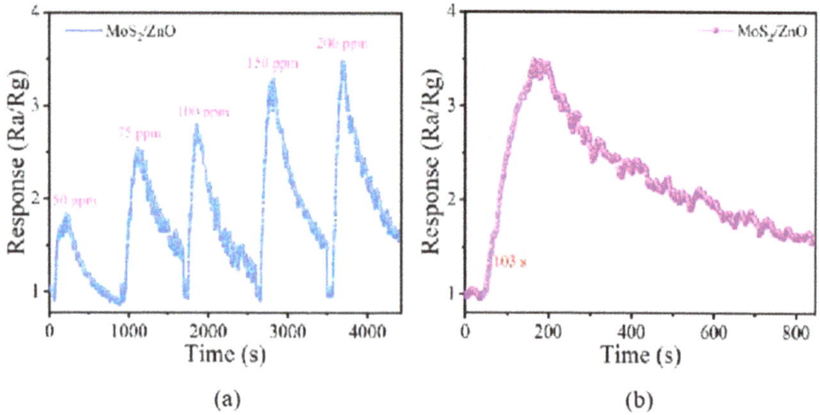

Figure 22. (**a**) The response curves of MoS_2/ZnO to different concentrations of TEA at room temperature. (**b**) The response–recovery curve of MoS_2/ZnO to 200 ppm TEA at room temperature [135].

Hao et al. [144] modified porous $LaFeO_3$ microspheres with RGO following a simple electrostatic self-assembly strategy. Figure 23a shows the sensing mechanism that the tight interface contact between $LaFeO_3$ microspheres and RGO tablets. RGO is wrapped on the surface of the microspheres to form p-p heterojunctions, causing the adsorbed oxygen to trap free electrons from the conductive band, in turn making the electrical resistance decrease. Figure 23b reveals the excellent selectivity of $LaFeO_3$, $RGO/LaFeO_3$ nanomaterials to 100 ppm of five different VOCs at 240 °C.

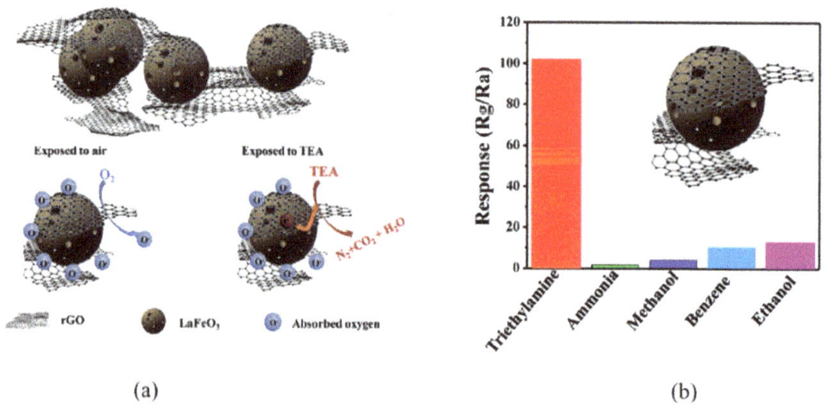

Figure 23. (**a**) The nanostructure and the possible gas-sensing mechanism of the $RGO/LaFeO_3$. (**b**) The response of $LaFeO_3$, $RGO/LaFeO_3$ to 100 ppm of five different VOCs at 240 °C [144].

In recent years, spinel structures have also been used in TEA gas detection because of their good thermal stability and sensitivity to VOCs [145,146]. The common formula for normal spinel structures is $A^{tetra}(B_2)^{octa}O_4$ where A is the distribution of positive second-order metal cations in tetrahedral voids, and B is the distribution of positive trivalent metal cations in octahedral voids [147]. The common formula for inverse spinel structures is $B^{tetra}(AB)^{octa}O_4$. The distribution of A and B is just the opposite of that formal spinel structure [148]. The different cation spices and charge states of two sites in the polyhedral crystalline structure will significantly affect the physicochemical properties of nanomaterials [149]. Ma et al. [150] prepared hierarchical spinel-type corn-like MGa_2O_4 (M = Ni, Co) architectures using a facile hydrothermal method and subsequent calcination. Subsequently, they completely and scientifically investigated the phase compositions and microstructures, which may be the essential influence factor of excellent gas-sensing properties. Yang et al. [151] investigated the impact of the calcination temperature on the morphology and TEA-sensing properties of $NiCo_2O_4$ microsphere nanomaterial. The experiment results reveal that the material shows good selectivity to TEA but that its high working temperature and slow response time need to be improved. The synthesis process and calcination results of $NiCo_2O_4$ micromaterials are displayed in Figure 24.

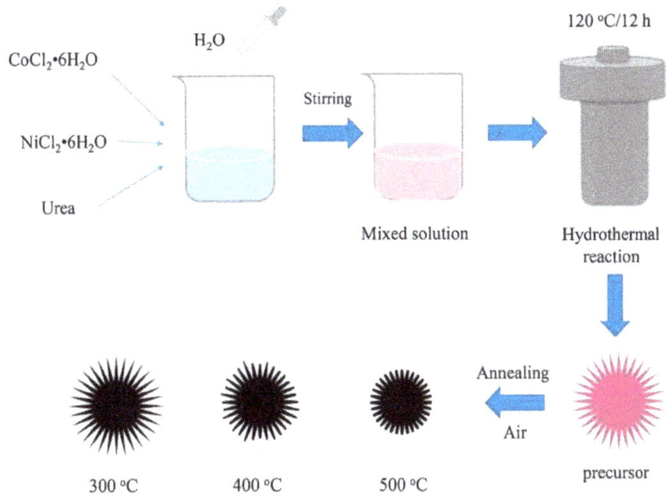

Figure 24. The synthesis process of $NiCo_2O_4$ nanospheres annealed at various temperatures [151].

The comparison of gas-sensing characteristics of nanomaterials listed in this paper will be given in Table 3 at the end of this section.

Table 3. The gas-sensing properties of various materials.

Nanomaterials	τ_{res}/τ_{rec} (s)	T (°C)	Conc. (ppm)	Res.	Lim. (ppm)	Ref.
Pine dendritic $BiVO_4$/RGO	5.9/11.4	180	10	5.9	2	[122]
α-Fe_2O_3 porous spindle/RGO	2/7	280	50	24	-	[123]
Co_3O_4/RGO	-	25	100	10	-	[124]
2D/2D SnO_2 nanosheets/$Ti_3C_2T_x$ MXene	1/1	140	50	33.4	5	[131]
MoS_2/ZnO bridge-like	35/142	200	100	31.08	0.097	[135]
MoS_2/GO hybrid nanostructures	7/11	260	1	2.8	1	[136]
$CoSnO_3$ nanoboxes	-	100	5	2.7	0.134	[143]
RGO-wrapped porous $LaFeO_3$ microspheres	3/4	240	50	103.5	1	[144]
corn-like MGa_2O_4 (M = Ni, Co)	136/41	270	100	7.6	-	[150]
Hierarchical $NiCo_2O_4$ microspheres	49/54	300	50	-	0.145	[151]

7. The Application of Advanced Instruments to Make Sensing Materials

The different fabrication methods of gas-sensing materials usually create different morphologies of the materials with different gas-sensing properties. Therefore, using some advanced equipment to prepare or modify sensing materials is also one typical way to improve gas-sensing properties. Generally, because of the simple synthesis operation, and low production costs, one-step hydrothermal [152] or solvothermal [153], sol-gel [154], and coprecipitation methods are popular. However, micromaterials prepared by atomic layer deposition (ALD) [155], utilizing DC-sputtering technology, pulsed laser deposition (PLD) [156], or electrospinning [157] can achieve remarkable responses.

Take the ultrathin ZnO films (20 nm) as an example, Li et al. [158] prepared them by depositing ZnO on SiO_2 wafers by ALD. The preparation process for ALD deposition is exhibited in Figure 25a, and Figure 25b depicts the scheme of the Ar plasma process. The improvement of the TEA gas-sensing properties is attributed to oxygen vacancies acting as electron donors and multiple existing active sites. PLD is a technology that uses a laser to bombard a target material and deposit the bombarded plasma on the substrate for thin-film growth. The easy agreement with the target material composition is the biggest advantage of PLD, which is the main mark to distinguish it from other technologies. Moreover, the advantages of high deposition rate, short test cycle, high orientation, and high film resolution have allowed experimenters grow thin films using PLD for almost all materials [159]. Song et al. [160] successfully fabricated a kind of n-n heterojunction that combined ZnO nanorods with α-Fe_2O_3 nanoparticles by the PLD method. Superior response, lower detection concentration, and shorter response time of the material were caused by its larger specific surface area, which adsorbs more oxygen ions.

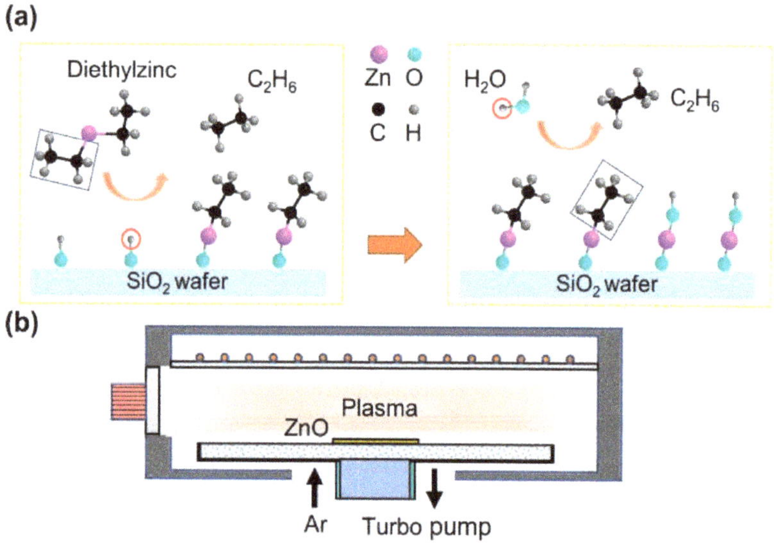

Figure 25. (**a**) ALD deposition process; (**b**) Ar plasma treatment [158].

Electrospinning is also a common method of preparing the nanofibers that are often used to obtain 1D nanoscale continuous fibers such as hollow fibers [161]. Because of the unique reaction process that mixes different materials into the electrospinning solution, the process can often yield 1D materials with good pore structure, large specific surface area, and excellent electron transport properties for a variety of applications [162]. A single needle electrospinning setup is drawn in Figure 26. Although electrospinning is efficient and convenient, there are few application examples in the manufacture of TEA gas sensors. One of the cases is Ma et al. [163], who fabricated several In_2O_3 hierarchical structure materials by electrospinning controlling calcination time and temperature. And the materials could be used in room temperature TEA detection, preserving superior sensitivity. A comprehensive analysis of the gas-sensing performance of the material was accomplished, and the team attributed the improved sensing performance to the unique material structure, multiple active sites, and special electron transmissions obtained by electrospinning.

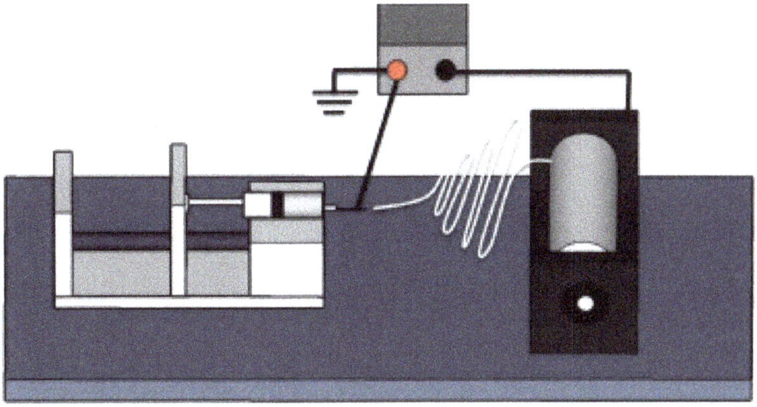

Figure 26. Schematic diagram of the electrospinning device [164].

A summary of the performance of the sensors manufactured with the various devices in this section is given in Table 4 at the end of this section.

Table 4. The gas-sensing properties of various materials fabricated by different devices.

Nanomaterials	τ_{res}/τ_{rec} (s)	T(°C)	Conc. (ppm)	Res.	Lim. (ppm)	Ref.
ultrathin ZnO films	531/46	180	10	1.8	-	[158]
α-Fe_2O_3 nanoparticle/ZnO nanorod	4/86	300	50	63	1	[160]
InO_2	184/-	40	50	87.8	5	[163]
SnO_2	26/13	235	50	54.9	1	[165]

8. External Stimuli on Sensing Performance

In the practical application and production process of these sensors, we should not only consider the hard indexes such as the sensitivity, selectivity, and stability of the gas sensors to the target gas but factors such as the power consumption, repeatability, and manufacturing cost of the sensors. If the working temperature of a sensor can reach room temperature, the power consumption of the sensor will be greatly reduced, which can decrease both the fabricating and use costs. Decades ago, people began to look for auxiliary methods such as voltage biasing, UV light, and visible light excitons to enhance gas-sensing performance without relying on high temperature [166,167].

Liu et al. [168] investigated the enhancement of the TEA gas-sensing mechanism after the light irradiation of $ZnO/ZnFe_2O_4$ composites and developed a synthesis strategy by calcinating Zn_2Fe-LDH at different temperatures. The preparation process is shown in Figure 27a. The changing curves of sensor resistance after illumination are exhibited in Figure 27b. Clearly, light irradiation improves TEA gas sensing. Shanmugam et al. [169] reported an experiment with UV light to improve the gas adsorption properties of CeO_2 nanomaterials. The test result shows materials with higher TEA adsorption capacity compared with other VOCs attributable to Ov in the prepared CeO_2 NPs and photo-generated electron-hole pairs of the material. Yang et al. [170] investigated the impacts of ultraviolet light irradiation on ZnO-SnO_2 heterojunction nanobelts for TEA sensing. The conclusion is that under UV laser illumination, response/recovery times (1.8 s/18 s) significantly accelerate.

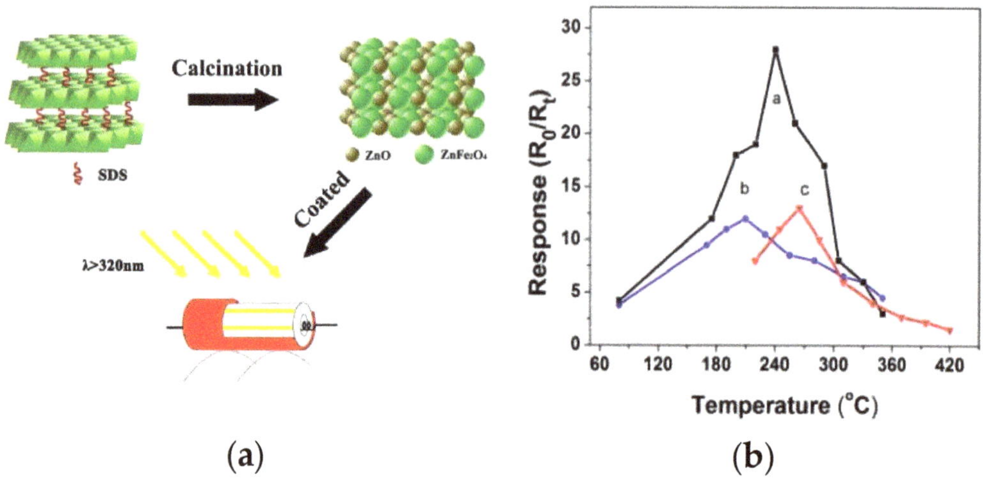

Figure 27. (**a**) The preparation and illumination of $ZnO/ZnFe_2O_4$. (**b**) The responses of $ZnO/ZnFe_2O_4$-600-a, $ZnO/ZnFe_2O_4$-800-b, and $ZnO/ZnFe_2O_4$-1000-c at various temperatures [168].

9. Conclusions and Outlook

Nanomaterial MOS sensors have gradually become an important development trend in gas detection. In this review paper, we briefly introduce the sensing characteristics of these gas sensors and two gas-sensing mechanisms for TEA detection. The major section of this review elaborates on several strategies to enhance the gas detection performance, such as the optimization of material structure, combining with other materials, using new materials, preparing materials with more advanced instruments, and stimulating the materials with external excitation.

In general, the alteration of the morphology and surface structure of materials can be either a dimensional increase to enlarge the specific surface area, porosity, and active adsorption sites for improved gas absorption or a change in the exposed crystal plane to improve gas selectivity and sensitivity. Various metal and nonmetal single elements can be doped into materials to improve gas-sensing performance due to their catalytic properties or electronic sensitization mechanisms. Other chemical compounds can also be incorporated to form p-p, p-n, and n-n heterojunctions to improve response. Recent advances also indicated the potential of quantum dot materials and conducting polymer materials in the optimization of sensitive performance. Furthermore, new materials such as graphene and its derivatives, MXenes ($Ti_3C_2T_x$, Mo_2CT_x, V_2CT_x), TMDs (MoS_2, WS_2, $MoSe_2$, and WSe_2), perovskite, and spinel materials are also used in TEA detection and possess unique characteristics such as reducing the working temperature. Advanced technology methods such as ALD and PLD also allow for preparing materials with better microstructures. Finally, external simulation can be an additional strategy for improving gas-sensing properties; for example, UV stimulation may allow gas-sensing materials to play an unexpected role.

Technology advances give rise to great numbers of materials with potential TEA-sensing capability, but further improvements are still in need. The previous summary illustrates that current TEA detection materials remain highly dependent on the traditional MOSs such as ZnO and SnO_2. However, there is still much room for improvement in its gas-sensing properties. First, the current poor response and selectivity of TEA gas materials should improve because many of the specific surface areas and porosities of these materials cannot adsorb enough target gas. Currently, polymetallic composites are becoming one of the main trends in TEA sensors, and the improvement of composite materials often increases the specific surface area and porosity of the material. Second, the operating temperature of most TEA sensors is still high, most above 100 °C; graphene material is usually one of the choices to solve the high working temperature because of the unique properties of graphene composite materials. Third, many TEA-sensing materials are still in the laboratory stage, and practical TEA sensors that can be used in industrial production need to be developed and produced. This review provides a summary of existing techniques and materials and aims to improve the sensitivity performance of the TEA sensors, which may pave the way for the development of more advanced sensors in the future.

Author Contributions: Writing—review and editing, H.Z.; funding acquisition, F.M.; supervision, H.Z. and F.M.; investigation, Y.G.; resources, Y.G.; writing—original draft preparation, Y.G.; project administration, H.Z.; validation, H.Z. All authors have read and agreed to the published version of the manuscript.

Funding: This work was supported by the National Natural Science Foundation of China (62071112, 61833006 and 62033002) and the Fundamental Research Funds for the Central Universities in China (N2201008).

Institutional Review Board Statement: Not applicable.

Informed Consent Statement: Not applicable.

Data Availability Statement: Not applicable.

Conflicts of Interest: The authors declare no conflict of interest.

References

1. Yuan, Z.; Zhao, J.; Meng, F.; Qin, W.; Chen, Y.; Yang, M.; Ibrahim, M.; Zhao, Y. Sandwich-like composites of double-layer Co_3O_4 and reduced graphene oxide and their sensing properties to volatile organic compounds. *J. Alloys Compd.* **2019**, *793*, 24–30. [CrossRef]
2. Yun, P.; Ma, S.; Xu, X.; Wang, S.; Liu, W.; Wang, L.; Alhadi, A. Bi_2WO_6 nanoparticles-decorated ZnO nanosheets and their enhanced gas sensing properties. *Vacuum* **2021**, *194*, 110627. [CrossRef]
3. Yun, P.D.; Ma, S.Y.; Xu, X.L.; Wang, S.Y.; Han, T.; Sheng, H.; Pei, S.T.; Yang, T.T. Excellent triethylamine sensor with ultra-fast response and recovery time based on bulk Bi_2WO_6 material. *Mater. Lett.* **2021**, *285*, 129162. [CrossRef]
4. Li, C.B.; Xiao, F.; Xu, W.; Chu, Y.; Wang, Q.; Jiang, H.; Li, K.; Gao, X.W. Efficient self-photo-degradation of cationic textile dyes involved triethylamine and degradation pathway. *Chemosphere* **2021**, *266*, 129209. [CrossRef] [PubMed]
5. Wang, Z.; Lin, R.; Fang, W.; Li, G.; Guo, Y.; Qin, Z. Triethylamine as an initiator for cracking of heptane. *Energy* **2006**, *31*, 2773–2790. [CrossRef]
6. Yin, M.; Zhang, L.; Qiu, T.; Chen, Y.; Qi, S.; Wei, X.; Tian, X.; Ge, K.; Qiu, J.; Xu, D. Double-layer capsule of mesoporous ZnO@SnO_2 for sensitive detection of triethylamine. *Analyst* **2021**, *146*, 6193–6201. [CrossRef]
7. Brahmachari, G.; Nayek, N.; Nurjamal, K.; Karmakar, I.; Begam, S. Triethylamine—A Versatile Organocatalyst in Organic Transformations: A Decade Update. *Synthesis* **2018**, *50*, 4145–4164. [CrossRef]
8. Abbas, H.; Marappan, G.; Chidabaram, D.; Govindasamy, S.; Jayaraman Surya, V.; Sivalingam, Y. Graphene Oxide based Gas Sensor for Triethylamine Detection at Room Temperature. *IOP Conf. Ser. Mater. Sci. Eng.* **2022**, *1219*, 012031. [CrossRef]
9. Cai, T.; Chen, L.; Ren, Q.; Cai, S.; Zhang, J. The biodegradation pathway of triethylamine and its biodegradation by immobilized Arthrobacter protophormiae cells. *J. Hazard. Mater.* **2011**, *186*, 59–66. [CrossRef]
10. Xu, Y.; Ma, T.; Zheng, L.; Sun, L.; Liu, X.; Zhao, Y.; Zhang, J. Rational design of Au/Co_3O_4-functionalized $W_{18}O_{49}$ hollow heterostructures with high sensitivity and ultralow limit for triethylamine detection. *Sens. Actuators B Chem.* **2019**, *284*, 202–212. [CrossRef]
11. Xu, Y.; Zheng, W.; Liu, X.; Zhang, L.; Zheng, L.; Yang, C.; Pinna, N.; Zhang, J. Platinum single atoms on tin oxide ultrathin films for extremely sensitive gas detection. *Mater. Horiz.* **2020**, *7*, 1519–1527. [CrossRef]
12. Ayad, M.M.; Torad, N.L. Quartz crystal microbalance sensor for detection of aliphatic amines vapours. *Sens. Actuators B Chem.* **2010**, *147*, 481–487. [CrossRef]
13. Filippo, E.; Manno, D.; Buccolieri, A.; Serra, A. Green synthesis of sucralose-capped silver nanoparticles for fast colorimetric triethylamine detection. *Sens. Actuators B Chem.* **2013**, *178*, 1–9. [CrossRef]
14. Moore, W.M.; Edwards, R.J.; Bavda, L.T. An Improved Capillary Gas Chromatography Method for Triethylamine. Application to Sarafloxacin Hydrochloride and GnRH Residual Solvents Testing. *Anal. Lett.* **1999**, *32*, 2603–2612. [CrossRef]
15. Majder-Łopatka, M.; Węsierski, T.; Dmochowska, A.; Salamonowicz, Z.; Polańczyk, A. The Influence of Hydrogen on the Indications of the Electrochemical Carbon Monoxide Sensors. *Sustainability* **2019**, *12*, 14. [CrossRef]
16. Sun, Y.; Ding, W.; Li, J.; Jia, Y.; Guo, G.; Deng, Z. A novel and simple fluorescent chemical sensor SX based on AIE for relay recognition of Zn^{2+} and Cu^{2+} in aqueous system and analysis in logic gates. *J. Mol. Struct.* **2022**, *1252*, 132219. [CrossRef]

17. Promphet, N.; Ummartyotin, S.; Ngeontae, W.; Puthongkham, P.; Rodthongkum, N. Non-invasive wearable chemical sensors in real-life applications. *Anal. Chim. Acta* **2021**, *1179*, 338643. [CrossRef]
18. Gui, Y.; Zhao, J.; Wang, W.; Tian, J.; Zhao, M. Synthesis of hemispherical WO_3/graphene nanocomposite by a microwave-assisted hydrothermal method and the gas-sensing properties to triethylamine. *Mater. Lett.* **2015**, *155*, 4–7. [CrossRef]
19. Li, Q.; Li, Y.; Zeng, W. Preparation and Application of 2D MXene-Based Gas Sensors: A Review. *Chemosensors* **2021**, *9*, 225. [CrossRef]
20. Zhang, J.; Yu, Y.; Fang, P.; Liu, L.; Yue, H.; Ou, J.; Han, A. Anodization of aluminum in a sealed container. *Electrochem. Commun.* **2021**, *129*, 107086. [CrossRef]
21. Chu, X.; Chen, T.; Zhang, W.; Zheng, B.; Shui, H. Investigation on formaldehyde gas sensor with ZnO thick film prepared through microwave heating method. *Sens. Actuators B Chem.* **2009**, *142*, 49–54. [CrossRef]
22. Dey, A. Semiconductor metal oxide gas sensors: A review. *Mater. Sci. Eng. B* **2018**, *229*, 206–217. [CrossRef]
23. Xu, X.; Ma, S.; Xu, X.; Pei, S.; Han, T.; Liu, W. Transformation synthesis of heterostructured SnS_2/ZnS microspheres for ultrafast triethylamine detection. *J. Alloys Compd.* **2021**, *868*, 159286. [CrossRef]
24. Xu, K.; Zhan, C.; Zhao, W.; Yu, X.; Zhu, Q.; Yang, L. Tunable resistance of MOFs films via an anion exchange strategy for advanced gas sensing. *J. Hazard. Mater.* **2021**, *416*, 125906. [CrossRef] [PubMed]
25. Yang, T.; Yang, X.; Zhu, M.; Zhao, H.; Zhang, M. Coral-like $ZnFe_2O_4$–ZnO mesoporous heterojunction architectures: Synthesis and enhanced sensing properties for triethylamine. *Inorg. Chem. Front.* **2020**, *7*, 1918–1926. [CrossRef]
26. Zeb, S.; Cui, Y.; Zhao, H.; Sui, Y.; Yang, Z.; Khan, Z.U.; Ahmad, S.M.; Ikram, M.; Gao, Y.; Jiang, X. Synergistic Effect of Au-PdO Modified Cu-Doped $K_2W_4O_{13}$ Nanowires for Dual Selectivity High Performance Gas Sensing. *ACS Appl. Mater. Interfaces* **2022**, *14*, 13836–13847. [CrossRef] [PubMed]
27. Gong, F.-L.; Peng, M.-X.; Yue, L.-J.; Chen, J.-L.; Xie, K.-F.; Zhang, Y.-H. Design of p-n heterojunction on mesoporous ZnO/Co_3O_4 nanosheets for triethylamine sensor. *Chem. Phys. Lett.* **2021**, *779*, 138891. [CrossRef]
28. Shao, S.; Xie, C.; Xia, Y.; Zhang, L.; Zhang, J.; Wei, S.; Kim, H.W.; Kim, S.S. Highly conjugated three-dimensional van der Waals heterostructure-based nanocomposite films for ultrahigh-responsive TEA gas sensors at room temperature. *J. Mater. Chem. A* **2022**, *10*, 2995–3008. [CrossRef]
29. Zhang, Y.-H.; Wang, C.-N.; Gong, F.-L.; Chen, J.-L.; Xie, K.-F.; Zhang, H.-L.; Fang, S.-M. Ultra-sensitive triethylamine sensors based on oxygen vacancy-enriched ZnO/SnO_2 micro-camellia. *J. Mater. Chem. C* **2021**, *9*, 6078–6086. [CrossRef]
30. Liu, L.; Zhao, Y.; Song, P.; Yang, Z.; Wang, Q. ppb level triethylamine detection of yolk-shell SnO_2/Au/Fe_2O_3 nanoboxes at low-temperature. *Appl. Surf. Sci.* **2019**, *476*, 391–401. [CrossRef]
31. Li, Z.; Xiong, Y.; Bi, D.; Liu, Q.; Yang, C.; Zhang, J. Continuously improved gas-sensing performance of Zn_2SnO_4 porous octahedrons by structure evolution and further $ZnSnO_3$ nanosheets decoration. *J. Alloys Compd.* **2022**, *901*, 163744. [CrossRef]
32. Xiong, Y.; Liu, W.; Qiao, X.; Song, X.; Wang, S.; Zhang, X.; Wang, X.; Tian, J. Confined synthesis of 2D ultrathin ZnO/Co_3O_4 nanomeshes heterostructure for superior triethylamine detection at low temperature. *Sens. Actuators B Chem.* **2021**, *346*, 130486. [CrossRef]
33. Liang, X.; Zhang, J.; Zhang, K.; Yang, X.; Zhang, M. The modification effect of Fe_2O_3 nanoparticles on ZnO nanorods improves the adsorption and detection capabilities of TEA. *Inorg. Chem. Front.* **2022**, *9*, 259–266. [CrossRef]
34. Jin, Q.; Wen, W.; Wang, Z.-X.; Wang, R.-H.; Zheng, S.; Ye, Z.; Wu, J.-M. Nanoarchitectonics of nest-like MnO_2/TiO_2 thin film for triethylamine sensing. *Sens. Actuators B Chem.* **2022**, *353*, 131137. [CrossRef]
35. Meng, L.; Bu, W.; Li, Y.; Qin, Q.; Zhou, Z.; Hu, C.; Chuai, X.; Wang, C.; Sun, P.; Lu, G. Highly selective triethylamine sensing based on SnO/SnO_2 nanocomposite synthesized by one-step solvothermal process and sintering. *Sens. Actuators B Chem.* **2021**, *342*, 130018. [CrossRef]
36. Meng, F.; Liao, Z.; Xing, C.; Yuan, Z.; Zhang, R.; Zhu, H.; Li, J. Preparation of SnO_2/SiO_2 nanocomposites by sol-gel method for enhancing the gas sensing performance to triethylamine. *J. Alloys Compd.* **2022**, *893*, 162189. [CrossRef]
37. Liu, J.; Zhang, L.; Fan, J.; Yu, J. Semiconductor Gas Sensor for Triethylamine Detection. *Small* **2022**, *18*, e2104984. [CrossRef]
38. Yang, J.; Han, W.; Ma, J.; Wang, C.; Shimanoe, K.; Zhang, S.; Sun, Y.; Cheng, P.; Wang, Y.; Zhang, H.; et al. Sn doping effect on NiO hollow nanofibers based gas sensors about the humidity dependence for triethylamine detection. *Sens. Actuators B Chem.* **2021**, *340*, 129971. [CrossRef]
39. Fan, Y.; Xu, Y.; Wang, Y.; Sun, Y. Fabrication and characterization of Co-doped ZnO nanodiscs for selective TEA sensor applications with high response, high selectivity and ppb-level detection limit. *J. Alloys Compd.* **2021**, *876*, 160170. [CrossRef]
40. Yang, J.; Li, X.; Wu, J.; Han, Y.; Wang, Z.; Zhang, Y.; Xu, Y. Yolk-shell (Cu,Zn)Fe_2O_4 ferrite nano-microspheres with highly selective triethylamine gas-sensing properties. *Dalton Trans.* **2020**, *49*, 14475–14482. [CrossRef]
41. Ma, N.N.; Ma, S.Y.; Wang, L.; Wang, X.P.; Liu, M.M.; Cai, Y.H.; Yang, T.T. Design a $SnWO_4$ coral-like nanostructure for triethylamine (TEA) sensing. *Vacuum* **2022**, *198*, 110913. [CrossRef]
42. Ji, H.; Zeng, W.; Li, Y. Gas sensing mechanisms of metal oxide semiconductors: A focus review. *Nanoscale* **2019**, *11*, 22664–22684. [CrossRef] [PubMed]
43. Kong, D.; Han, J.; Gao, Y.; Gao, Y.; Zhou, W.; Liu, G.; Lu, G. Lower coordination Co_3O_4 mesoporous hierarchical microspheres for comprehensive sensitization of triethylamine vapor sensor. *J. Hazard. Mater.* **2022**, *430*, 128469. [CrossRef]

44. Song, X.; Xu, Q.; Zhang, T.; Song, B.; Li, C.; Cao, B. Room-temperature, high selectivity and low-ppm-level triethylamine sensor assembled with Au decahedrons-decorated porous α-Fe_2O_3 nanorods directly grown on flat substrate. *Sens. Actuators B Chem.* **2018**, *268*, 170–181. [CrossRef]
45. Liu, H.; Cao, X.; Wu, H.; Li, B.; Li, Y.; Zhu, W.; Yang, Z.; Huang, Y. Innovative development on a p-type delafossite $CuCrO_2$ nanoparticles based triethylamine sensor. *Sens. Actuators B Chem.* **2020**, *324*, 128743. [CrossRef]
46. Wang, T.; Wang, X.; Wang, Y.; Yi, G.; Shi, C.; Yang, Y.; Sun, G.; Zhang, Z. Construction of Zn_2SnO_4 decorated ZnO nanoparticles for sensing triethylamine with dramatically enhanced performance. *Mater. Sci. Semicond. Process.* **2022**, *140*, 106403. [CrossRef]
47. Hu, Q.; He, J.; Chang, J.; Gao, J.; Huang, J.; Feng, L. Needle-Shaped WO_3 Nanorods for Triethylamine Gas Sensing. *ACS Appl. Nano Mater.* **2020**, 3, 9046–9054. [CrossRef]
48. Chu, X.; Liang, S.; Sun, W.; Zhang, W.; Chen, T.; Zhang, Q. Trimethylamine sensing properties of sensors based on MoO_3 microrods. *Sens. Actuators B Chem.* **2010**, *148*, 399–403. [CrossRef]
49. Liu, F.; Chen, X.; Wang, X.; Han, Y.; Song, X.; Tian, J.; He, X.; Cui, H. Fabrication of 1D Zn_2SnO_4 nanowire and 2D ZnO nanosheet hybrid hierarchical structures for use in triethylamine gas sensors. *Sens. Actuators B Chem.* **2019**, *291*, 155–163. [CrossRef]
50. Zhai, T.; Xu, H.; Li, W.; Yu, H.; Chen, Z.; Wang, J.; Cao, B. Low-temperature in-situ growth of SnO_2 nanosheets and its high triethylamine sensing response by constructing Au-loaded ZnO/SnO_2 heterostructure. *J. Alloys Compd.* **2018**, *737*, 603–612. [CrossRef]
51. Jin, X.; Li, Y.; Zhang, B.; Xu, X.; Sun, G.; Wang, Y. Temperature-dependent dual selectivity of hierarchical porous In_2O_3 nanospheres for sensing ethanol and TEA. *Sens. Actuators B Chem.* **2021**, *330*, 129271. [CrossRef]
52. Ma, Q.; Chu, S.; Li, H.; Guo, J.; Zhang, Q.; Lin, Z. Flower-like In_2O_3/ZnO heterostructure with accelerated multi-orientation electron transport mechanism for superior triethylamine detection. *Appl. Surf. Sci.* **2021**, *569*, 151074. [CrossRef]
53. Cai, Y.H.; Ma, S.Y.; Yang, T.T.; Cao, P.F.; Wang, L.; Sheng, H.; Liu, M.M. Preparation of YVO_4 octahedral nanomaterials and gas-sensing characteristics to triethylamine. *J. Alloys Compd.* **2022**, *897*, 163167. [CrossRef]
54. Sá, B.S.; Zito, C.A.; Perfecto, T.M.; Volanti, D.P. Porous $ZnSnO_3$ nanocubes as a triethylamine sensor. *Sens. Actuators B Chem.* **2021**, *338*, 129869. [CrossRef]
55. Ji, H.; Qin, W.; Yuan, Z.; Meng, F. Qualitative and quantitative recognition method of drug-producing chemicals based on SnO_2 gas sensor with dynamic measurement and PCA weak separation. *Sens. Actuators B Chem.* **2021**, *348*, 130698. [CrossRef]
56. Ma, Q.; Chu, S.; Li, H.; Guo, J.; Zhang, Q. Cubic-like In_2O_3/α-Fe_2O_3 heterostructures assembled with 2D porous nanoplates for superior triethylamine gas-sensing behavior. *Mater. Lett.* **2021**, *302*, 130452. [CrossRef]
57. Yu, Q.; Kong, W.; Zhang, Y.; Li, X.; Xu, Y. Morphology-controllable synthesis of hierarchical hollow $GaFeO_3$ microcubes with selective triethylamine gas-sensing properties. *Ceram. Int.* **2022**, *48*, 4554–4562. [CrossRef]
58. Meng, X.; Yao, M.; Mu, S.; Wang, Y. Oxygen Vacancies Enhance Triethylamine Sensing Properties of SnO_2 Nanoparticles. *ChemistrySelect* **2019**, *4*, 11268–11274. [CrossRef]
59. Du, J.; Zhao, R.; Xie, Y.; Li, J. Size-controlled synthesis of SnO_2 quantum dots and their gas-sensing performance. *Appl. Surf. Sci.* **2015**, *346*, 256–262. [CrossRef]
60. Li, T.; Zeng, W.; Wang, Z. Quasi-one-dimensional metal-oxide-based heterostructural gas-sensing materials: A review. *Sens. Actuators B Chem.* **2015**, *221*, 1570–1585. [CrossRef]
61. Zhao, S.; Shen, Y.; Yan, X.; Zhou, P.; Yin, Y.; Lu, R.; Han, C.; Cui, B.; Wei, D. Complex-surfactant-assisted hydrothermal synthesis of one-dimensional ZnO nanorods for high-performance ethanol gas sensor. *Sens. Actuators B Chem.* **2019**, *286*, 501–511. [CrossRef]
62. Meng, F.; Shi, X.; Yuan, Z.; Ji, H.; Qin, W.; Shen, Y.; Xing, C. Detection of four alcohol homologue gases by ZnO gas sensor in dynamic interval temperature modulation mode. *Sens. Actuators B Chem.* **2022**, *350*, 130867. [CrossRef]
63. Yang, X.; Zhang, S.; Yu, Q.; Zhao, L.; Sun, P.; Wang, T.; Liu, F.; Yan, X.; Gao, Y.; Liang, X.; et al. One step synthesis of branched SnO_2/ZnO heterostructures and their enhanced gas-sensing properties. *Sens. Actuators B Chem.* **2019**, *281*, 415–423. [CrossRef]
64. Lv, Y.-z.; Li, C.-r.; Guo, L.; Wang, F.-c.; Xu, Y.; Chu, X.-f. Triethylamine gas sensor based on ZnO nanorods prepared by a simple solution route. *Sens. Actuators B Chem.* **2009**, *141*, 85–88. [CrossRef]
65. Liu, B.; Yang, H.; Zhao, H.; An, L.; Zhang, L.; Shi, R.; Wang, L.; Bao, L.; Chen, Y. Synthesis and enhanced gas-sensing properties of ultralong NiO nanowires assembled with NiO nanocrystals. *Sens. Actuators B Chem.* **2011**, *156*, 251–262. [CrossRef]
66. Zou, Y.; Chen, S.; Sun, J.; Liu, J.; Che, Y.; Liu, X.; Zhang, J.; Yang, D. Highly Efficient Gas Sensor Using a Hollow SnO_2 Microfiber for Triethylamine Detection. *ACS Sens.* **2017**, *2*, 897–902. [CrossRef]
67. Lei, G.; Li, Z.; Lu, G.; Hu, J.; Shang, H.; Zhang, X.; Liu, X.; Zhang, J.; Guo, X. Tungsten oxide thin films for highly sensitive triethylamine detection. *J. Mater.* **2022**, *8*, 408–416. [CrossRef]
68. Zhang, L.; Yin, M.; Qiu, J.; Qiu, T.; Chen, Y.; Qi, S.; Wei, X.; Tian, X.; Xu, D. Mesoporous ZnO nanosheet as gas sensor for sensitive triethylamine detection. *Anal Bioanal. Chem.* **2022**, *414*, 2181–2188. [CrossRef]
69. Liu, Y.; Liu, H.; Hu, A.; Wei, Y.; Ou, W.; Deng, X.; Wang, S.; Yu, K. An efficient low-temperature triethylamine gas sensor based on 2D ultrathin SnO_2 nanofilms. *Semicond. Sci. Technol.* **2021**, *36*, 125022. [CrossRef]
70. Zhang, J.; Zhang, B.; Yao, S.; Li, H.; Chen, C.; Bala, H.; Zhang, Z. Improved triethylamine sensing properties of fish-scale-like porous SnO_2 nanosheets by decorating with Ag nanoparticles. *J. Mater.* **2022**, *8*, 518–525. [CrossRef]
71. Zhao, S.; Shen, Y.; Zhou, P.; Zhong, X.; Han, C.; Zhao, Q.; Wei, D. Design of Au@WO_3 core−shell structured nanospheres for ppb-level NO_2 sensing. *Sens. Actuators B Chem.* **2019**, *282*, 917–926. [CrossRef]

72. Meng, F.; Qi, T.; Zhang, J.; Zhu, H.; Yuan, Z.; Liu, C.; Qin, W.; Ding, M. MoS_2-Templated Porous Hollow MoO_3 Microspheres for Highly Selective Ammonia Sensing via a Lewis Acid-Base Interaction. *IEEE Trans. Ind. Electron.* **2022**, *69*, 960–970. [CrossRef]
73. Wang, J.; Pei, C.; Cheng, L.; Wan, W.; Zhao, Q.; Yang, H.; Liu, S. Responses of three-dimensional porous ZnO foam structures to the trace level of triethylamine and ethanol. *Sens. Actuators B Chem.* **2016**, *223*, 650–657. [CrossRef]
74. Zhai, C.; Zhu, M.; Jiang, L.; Yang, T.; Zhao, Q.; Luo, Y.; Zhang, M. Fast triethylamine gas sensing response properties of nanosheets assembled WO_3 hollow microspheres. *Appl. Surf. Sci.* **2019**, *463*, 1078–1084. [CrossRef]
75. Sui, L.-l.; Xu, Y.-M.; Zhang, X.-F.; Cheng, X.-L.; Gao, S.; Zhao, H.; Cai, Z.; Huo, L.-H. Construction of three-dimensional flower-like α-MoO_3 with hierarchical structure for highly selective triethylamine sensor. *Sens. Actuators B Chem.* **2015**, *208*, 406–414. [CrossRef]
76. Yuan, Y.; Wang, Y.; He, X.; Chen, M.; Liu, J.; Liu, B.; Zhao, H.; Liu, S.; Yang, H. Increasing gas sensitivity of Co_3O_4 octahedra by tuning Co-Co_3O_4 surface structure and sensing mechanism of 3-coordinated Co atom as an active center. *J. Mater. Sci. Mater. Electron.* **2020**, *31*, 8852–8864. [CrossRef]
77. Xiang, R.; Zhang, L.; Guo, J.; Wang, Y.; Yuan, Y.; Wang, Y.; Chen, M.; Qi, C.; Liu, B.; Zhao, H.; et al. Enhanced gas sensing performances of hydrogenated MnO octahedrons with {111} facets and the sensing mechanism of unsaturated Mn as a reactive atom. *J. Alloys Compd.* **2021**, *884*, 160872. [CrossRef]
78. Xu, H.-Y.; Chen, Z.-R.; Liu, C.-Y.; Ye, Q.; Yang, X.-P.; Wang, J.-Q.; Cao, B.-Q. Preparation of {200} crystal faced SnO_2 nanorods with extremely high gas sensitivity at lower temperature. *Rare Met.* **2021**, *40*, 2004–2016. [CrossRef]
79. Wang, J.; Zhang, Z.; Rettig, O. Triethylamine sensing characteristics of nonpolar (11–20) and polar (0001) GaN thin films. *Sens. Actuators B Chem.* **2021**, *329*, 129237. [CrossRef]
80. Bai, S.; Zuo, Y.; Zhang, K.; Zhao, Y.; Luo, R.; Li, D.; Chen, A. WO_3-$ZnFe_2O_4$ heterojunction and rGO decoration synergistically improve the sensing performance of triethylamine. *Sens. Actuators B Chem.* **2021**, *347*, 130619. [CrossRef]
81. Guo, L.; Wang, Y.; Shang, Y.; Yang, X.; Zhang, S.; Wang, G.; Wang, Y.; Zhang, B.; Zhang, Z. Preparation of Pd/PdO@ZnO-ZnO nanorods by using metal organic framework templated catalysts for selective detection of triethylamine. *Sens. Actuators B Chem.* **2022**, *350*, 130840. [CrossRef]
82. Li, G.; Ma, Z.; Hu, Q.; Zhang, D.; Fan, Y.; Wang, X.; Chu, X.; Xu, J. PdPt Nanoparticle-Functionalized α-Fe_2O_3 Hollow Nanorods for Triethylamine Sensing. *ACS Appl. Nano Mater.* **2021**, *4*, 10921–10930. [CrossRef]
83. Zeng, J.; Rong, Q.; Xiao, B.; Yu, R.; Zi, B.; Kuang, X.; Deng, X.; Ma, Y.; Zhang, J.; Wu, J.; et al. Single-atom silver loaded on tungsten oxide with oxygen vacancies for high performance triethylamine gas sensors. *J. Mater. Chem. A* **2021**, *9*, 8704–8710. [CrossRef]
84. Feng, B.; Wu, Y.; Ren, Y.; Chen, Y.; Yuan, K.; Deng, Y.; Wei, J. Self-template synthesis of mesoporous Au-SnO_2 nanospheres for low-temperature detection of triethylamine vapor. *Sens. Actuators B Chem.* **2022**, *356*, 131358. [CrossRef]
85. Sun, H.; Tang, X.; Zhang, J.; Li, S.; Liu, L. MOF-derived bow-like Ga-doped Co_3O_4 hierarchical architectures for enhanced triethylamine sensing performance. *Sens. Actuators B Chem.* **2021**, *346*, 130546. [CrossRef]
86. Zhao, Q.; Zhuang, G.; Zhao, Y.; Yang, L.; Zhao, J. Y-doped In_2O_3 hollow nanocubes for improved triethylamine-sensing performance. *New J. Chem.* **2021**, *45*, 6773–6779. [CrossRef]
87. Liu, J.; Zhang, L.; Fan, J.; Zhu, B.; Yu, J. Triethylamine gas sensor based on Pt-functionalized hierarchical ZnO microspheres. *Sens. Actuators B Chem.* **2021**, *331*, 129425. [CrossRef]
88. Zhu, M.; Yang, T.; Zhai, C.; Du, L.; Zhang, J.; Zhang, M. Fast triethylamine gas sensing response properties of Ho-doped SnO_2 nanoparticles. *J. Alloys Compd.* **2020**, *817*, 152724. [CrossRef]
89. Zhang, K.; Xie, K.; Maruf Ahmed, M.; Chai, Z.; Zhao, R.; Li, J.; Du, J. Cr-doped SnO_2 microrods adhering nanoparticles for enhanced triethylamine sensing performance. *Mater. Lett.* **2022**, *312*, 131684. [CrossRef]
90. Bi, W.; Wang, W.; Liu, S. Synthesis of Rh–SnO_2 nanosheets and ultra-high triethylamine sensing performance. *J. Alloys Compd.* **2020**, *817*, 152730. [CrossRef]
91. Yu, Y.; Liao, Z.; Meng, F.; Yuan, Z. Theoretical and Experimental Research on Ammonia Sensing Properties of Sulfur-Doped Graphene Oxide. *Chemosensors* **2021**, *9*, 220. [CrossRef]
92. Wang, X.; Han, W.; Yang, J.; Cheng, P.; Wang, Y.; Feng, C.; Wang, C.; Zhang, H.; Sun, Y.; Lu, G. Conductometric ppb-level triethylamine sensor based on macroporous WO_3-$W_{18}O_{49}$ heterostructures functionalized with carbon layers and PdO nanoparticles. *Sens. Actuators B Chem.* **2022**, *361*, 131707. [CrossRef]
93. Peng, R.; Li, Y.; Liu, T.; Si, P.; Feng, J.; Suhr, J.; Ci, L. Boron-doped graphene coated Au@SnO_2 for high-performance triethylamine gas detection. *Mater. Chem. Phys.* **2020**, *239*, 121961. [CrossRef]
94. Zhang, M.; Zhao, Z.; Hui, B.; Sun, J.; Sun, J.; Tian, W.; Zhang, Z.; Zhang, K.; Xia, Y. Carbonized polymer dots activated hierarchical tungsten oxide for efficient and stable triethylamine sensor. *J. Hazard. Mater.* **2021**, *416*, 126161. [CrossRef]
95. Gai, L.-Y.; Lai, R.-P.; Dong, X.-H.; Wu, X.; Luan, Q.-T.; Wang, J.; Lin, H.-F.; Ding, W.-H.; Wu, G.-L.; Xie, W.-F. Recent advances in ethanol gas sensors based on metal oxide semiconductor heterojunctions. *Rare Met.* **2022**, *41*, 1818–1842. [CrossRef]
96. Liu, Y.; Xiao, S.; Du, K. Chemiresistive Gas Sensors Based on Hollow Heterojunction: A Review. *Adv. Mater. Interfaces* **2021**, *8*, 2002122. [CrossRef]
97. Chen, S.; Gao, N.; Bunes, B.R.; Zang, L. Tunable nanofibril heterojunctions for controlling interfacial charge transfer in chemiresistive gas sensors. *J. Mater. Chem. C* **2019**, *7*, 13709–13735. [CrossRef]
98. Lin, Y.; Fan, Z. Compositing strategies to enhance the performance of chemiresistive CO_2 gas sensors. *Mater. Sci. Semicond. Process.* **2020**, *107*, 104820. [CrossRef]

99. Xu, X.; Liu, W.; Chen, Y.; Wang, S.; Wang, X.; Jiang, H.; Ma, S.; Yuan, F.; Zhang, Q. n-n Heterogeneous MoS_2/SnO_2 Nanotubes and The Excellent Triethylamine (TEA) Sensing Performances. *Mater. Lett.* **2022**, *311*, 131522. [CrossRef]
100. Xu, H.; Ju, J.; Li, W.; Zhang, J.; Wang, J.; Cao, B. Superior triethylamine-sensing properties based on TiO_2/SnO_2 n–n heterojunction nanosheets directly grown on ceramic tubes. *Sens. Actuators B Chem.* **2016**, *228*, 634–642. [CrossRef]
101. Xue, D.; Wang, Y.; Cao, J.; Zhang, Z. Hydrothermal Synthesis of CeO_2-SnO_2 Nanoflowers for Improving Triethylamine Gas Sensing Property. *Nanomaterials* **2018**, *8*, 1025. [CrossRef] [PubMed]
102. Yang, B.; Myung, N.V.; Tran, T.T. 1D Metal Oxide Semiconductor Materials for Chemiresistive Gas Sensors: A Review. *Adv. Electron. Mater.* **2021**, *7*, 2100271. [CrossRef]
103. Gao, H.; Ma, Y.; Song, P.; Yang, Z.; Wang, Q. Three-dimensional reduced graphene oxide/cobaltosic oxide as a high-response sensor for triethylamine gas at room temperature. *Mater. Sci. Semicond. Process.* **2021**, *133*, 105904. [CrossRef]
104. Wang, D.; Gu, K.; Zhang, J.; Xing, D.; Zhang, M. MOF-derived $NiWO_4$@NiO p-p heterostructure for distinguish detection of TEA and xylene by temperature regulation. *J. Alloys Compd.* **2021**, *875*, 160015. [CrossRef]
105. Yang, S.; Lei, G.; Xu, H.; Lan, Z.; Wang, Z.; Gu, H. Metal Oxide Based Heterojunctions for Gas Sensors: A Review. *Nanomaterials* **2021**, *11*, 1026. [CrossRef]
106. Shang, Y.; Shi, W.; Zhao, R.; Ahmed, M.M.; Li, J.; Du, J. Simple self-assembly of 3D laminated CuO/SnO_2 hybrid for the detection of triethylamine. *Chin. Chem. Lett.* **2020**, *31*, 2055–2058. [CrossRef]
107. Zeng, J.; Rong, Q.; Xiao, B.; Zi, B.; Kuang, X.; Deng, X.; Ma, Y.; Song, Z.; Zhang, G.; Zhang, J.; et al. Ultrasensitive ppb-level trimethylamine gas sensor based on p-n heterojunction of Co_3O_4/WO_3. *Nanotechnology* **2021**, *32*, 505511. [CrossRef]
108. Yu, Q.; Jin, R.; Zhao, L.; Wang, T.; Liu, F.; Yan, X.; Wang, C.; Sun, P.; Lu, G. MOF-Derived Mesoporous and Hierarchical Hollow-Structured In_2O_3-NiO Composites for Enhanced Triethylamine Sensing. *ACS Sens.* **2021**, *6*, 3451–3461. [CrossRef]
109. Bai, S.; Tian, Y.; Sun, J.; Tong, Z.; Luo, R.; Li, D.; Chen, A. Heterostructures of polyaniline@SnO_2 loading on flexible PET thin films for triethylamine detection at room temperature. *New J. Chem.* **2016**, *40*, 4595–4600. [CrossRef]
110. Galstyan, V. "Quantum dots: Perspectives in next-generation chemical gas sensors" A review. *Anal. Chim. Acta* **2021**, *1152*, 238192. [CrossRef]
111. Liu, J.; Zhu, B.; Zhang, L.; Fan, J.; Yu, J. 0D/2D CdS/ZnO composite with n-n heterojunction for efficient detection of triethylamine. *J. Colloid Interface Sci.* **2021**, *600*, 898–909. [CrossRef] [PubMed]
112. Liu, J.; Peng, Y.; Zhu, B.; Li, Y.; Zhang, L.; Yu, J. ZIF-8 derived ZnO-$CsPbBr_3$ polyhedrons for efficient triethylamine detection. *Sens. Actuators B Chem.* **2022**, *357*, 131366. [CrossRef]
113. Meng, F.; Zheng, H.; Chang, Y.; Zhao, Y.; Li, M.; Wang, C.; Sun, Y.; Liu, J. One-Step Synthesis of Au/SnO_2/RGO Nanocomposites and Their VOC Sensing Properties. *IEEE Trans. Nanotechnol.* **2018**, *17*, 212–219. [CrossRef]
114. Meng, F.; Li, X.; Yuan, Z.; Lei, Y.; Qi, T.; Li, J. Ppb-Level Xylene Gas Sensors Based on Co_3O_4 Nanoparticle-Coated Reduced Graphene Oxide(rGO) Nanosheets Operating at Low Temperature. *IEEE Trans. Instrum. Meas.* **2021**, *70*, 1–10. [CrossRef]
115. Xiao, Z.; Kong, L.B.; Ruan, S.; Li, X.; Yu, S.; Li, X.; Jiang, Y.; Yao, Z.; Ye, S.; Wang, C.; et al. Recent development in nanocarbon materials for gas sensor applications. *Sens. Actuators B Chem.* **2018**, *274*, 235–267. [CrossRef]
116. Bag, A.; Lee, N.-E. Gas sensing with heterostructures based on two-dimensional nanostructured materials: A review. *J. Mater. Chem. C* **2019**, *7*, 13367–13383. [CrossRef]
117. Liu, B.; Zhu, Q.; Pan, Y.; Huang, F.; Tang, L.; Liu, C.; Cheng, Z.; Wang, P.; Ma, J.; Ding, M. Single-Atom Tailoring of Two-Dimensional Atomic Crystals Enables Highly Efficient Detection and Pattern Recognition of Chemical Vapors. *ACS Sens.* **2022**, *7*, 1533–1543. [CrossRef]
118. Huang, X.; Qi, X.; Boey, F.; Zhang, H. Graphene-based composites. *Chem. Soc. Rev.* **2012**, *41*, 666–686. [CrossRef]
119. Kumar, Y.R.; Deshmukh, K.; Sadasivuni, K.K.; Pasha, S.K.K. Graphene quantum dot based materials for sensing, bio-imaging and energy storage applications: A review. *RSC Adv.* **2020**, *10*, 23861–23898. [CrossRef]
120. Chu, X.; Zhu, X.; Dong, Y.; Zhang, W.; Bai, L. Formaldehyde Sensing Properties of SnO–Graphene Composites Prepared via Hydrothermal Method. *J. Mater. Sci. Technol.* **2015**, *31*, 913–917. [CrossRef]
121. Wang, T.; Huang, D.; Yang, Z.; Xu, S.; He, G.; Li, X.; Hu, N.; Yin, G.; He, D.; Zhang, L. A Review on Graphene-Based Gas/Vapor Sensors with Unique Properties and Potential Applications. *Nanomicro Lett.* **2016**, *8*, 95–119. [CrossRef] [PubMed]
122. Bai, S.; Sun, L.; Sun, J.; Han, J.; Zhang, K.; Li, Q.; Luo, R.; Li, D.; Chen, A. Pine dendritic bismuth vanadate loaded on reduced graphene oxide for detection of low concentration triethylamine. *J. Colloid Interface Sci.* **2021**, *587*, 183–191. [CrossRef] [PubMed]
123. Wei, Q.; Sun, J.; Song, P.; Li, J.; Yang, Z.; Wang, Q. MOF-derived α-Fe_2O_3 porous spindle combined with reduced graphene oxide for improvement of TEA sensing performance. *Sens. Actuators B Chem.* **2020**, *304*, 127306. [CrossRef]
124. Yu, T.; Li, X.; Chen, B.; Yang, Y.; Xu, K.; Liu, Y.; Gong, W.; Yuan, C.; Yang, Y. Enhanced Gas Sensing Performance of rGO Wrapped Crystal Facet-Controlled Co_3O_4 Nanocomposite Heterostructures. *J. Phys. Chem. C* **2022**, *126*, 4879–4888. [CrossRef]
125. Naguib, M.; Kurtoglu, M.; Presser, V.; Lu, J.; Niu, J.; Heon, M.; Hultman, L.; Gogotsi, Y.; Barsoum, M.W. Two-dimensional nanocrystals produced by exfoliation of Ti_3AlC_2. *Adv. Mater.* **2011**, *23*, 4248–4253. [CrossRef]
126. Mehdi Aghaei, S.; Aasi, A.; Panchapakesan, B. Experimental and Theoretical Advances in MXene-Based Gas Sensors. *ACS Omega* **2021**, *6*, 2450–2461. [CrossRef]
127. Bhardwaj, R.; Hazra, A. MXene-based gas sensors. *J. Mater. Chem. C* **2021**, *9*, 15735–15754. [CrossRef]
128. Khazaei, M.; Mishra, A.; Venkataramanan, N.S.; Singh, A.K.; Yunoki, S. Recent advances in MXenes: From fundamentals to applications. *Curr. Opin. Solid State Mater. Sci.* **2019**, *23*, 164–178. [CrossRef]

129. He, T.; Liu, W.; Lv, T.; Ma, M.; Liu, Z.; Vasiliev, A.; Li, X. MXene/SnO_2 heterojunction based chemical gas sensors. *Sens. Actuators B Chem.* **2021**, *329*, 129275. [CrossRef]
130. Tian, X.; Yao, L.; Cui, X.; Zhao, R.; Chen, T.; Xiao, X.; Wang, Y. A two-dimensional $Ti_3C_2T_X$ MXene@TiO_2/MoS_2 heterostructure with excellent selectivity for the room temperature detection of ammonia. *J. Mater. Chem. A* **2022**, *10*, 5505–5519. [CrossRef]
131. Liang, D.; Song, P.; Liu, M.; Wang, Q. 2D/2D SnO_2 nanosheets/$Ti_3C_2T_x$ MXene nanocomposites for detection of triethylamine at low temperature. *Ceram. Int.* **2022**, *48*, 9059–9066. [CrossRef]
132. Anichini, C.; Czepa, W.; Pakulski, D.; Aliprandi, A.; Ciesielski, A.; Samori, P. Chemical sensing with 2D materials. *Chem. Soc. Rev.* **2018**, *47*, 4860–4908. [CrossRef] [PubMed]
133. Ping, J.; Fan, Z.; Sindoro, M.; Ying, Y.; Zhang, H. Recent Advances in Sensing Applications of Two-Dimensional Transition Metal Dichalcogenide Nanosheets and Their Composites. *Adv. Funct. Mater.* **2017**, *27*, 1605817. [CrossRef]
134. Joshi, N.; Hayasaka, T.; Liu, Y.; Liu, H.; Oliveira, O.N., Jr.; Lin, L. A review on chemiresistive room temperature gas sensors based on metal oxide nanostructures, graphene and 2D transition metal dichalcogenides. *Mikrochim. Acta* **2018**, *185*, 213. [CrossRef]
135. Xu, X.L.; Wang, X.P.; Chen, Y.; Liu, W.W.; Wang, S.Y.; Yuan, F.Q.; Ma, S.Y. Design of MoS_2/ZnO bridge-like hetero-nanostructures to boost triethylamine (TEA) sensing. *Vacuum* **2022**, *196*, 110733. [CrossRef]
136. Hou, X.; Wang, Z.; Fan, G.; Ji, H.; Yi, S.; Li, T.; Wang, Y.; Zhang, Z.; Yuan, L.; Zhang, R.; et al. Hierarchical three-dimensional MoS_2/GO hybrid nanostructures for triethylamine-sensing applications with high sensitivity and selectivity. *Sens. Actuators B Chem.* **2020**, *317*, 128236. [CrossRef]
137. Shellaiah, M.; Sun, K.W. Review on Sensing Applications of Perovskite Nanomaterials. *Chemosensors* **2020**, *8*, 55. [CrossRef]
138. Qin, W.; Yuan, Z.; Gao, H.; Zhang, R.; Meng, F. Perovskite-structured $LaCoO_3$ modified ZnO gas sensor and investigation on its gas sensing mechanism by first principle. *Sens. Actuators B Chem.* **2021**, *341*, 130015. [CrossRef]
139. Bulemo, P.M.; Kim, I.-D. Recent advances in ABO_3 perovskites: Their gas-sensing performance as resistive-type gas sensors. *J. Korean Ceram. Soc.* **2019**, *57*, 24–39. [CrossRef]
140. Qin, W.; Yuan, Z.; Shen, Y.; Zhang, R.; Meng, F. Phosphorus-doped porous perovskite $LaFe_{1-x}P_xO_{3-\delta}$ nanosheets with rich surface oxygen vacancies for ppb level acetone sensing at low temperature. *Chem. Eng. J.* **2022**, *431*, 134280. [CrossRef]
141. Yan, Y.; Liu, J.; Zhang, H.; Song, D.; Li, J.; Yang, P.; Zhang, M.; Wang, J. One-pot synthesis of cubic $ZnSnO_3$/ZnO heterostructure composite and enhanced gas-sensing performance. *J. Alloys Compd.* **2019**, *780*, 193–201. [CrossRef]
142. Chu, X.; Zhou, S.; Zhang, W.; Shui, H. Trimethylamine sensing properties of nano-$LaFeO_3$ prepared using solid-state reaction in the presence of PEG400. *Mater. Sci. Eng. B* **2009**, *164*, 65–69. [CrossRef]
143. Zheng, L.; Zhao, Y.; Xu, Y.; Yang, C.; Zhang, J.; Liu, X. Susceptible $CoSnO_3$ nanoboxes with p-type response for triethylamine detection at low temperature. *CrystEngComm* **2020**, *22*, 2795–2805. [CrossRef]
144. Hao, P.; Lin, Z.; Song, P.; Yang, Z.; Wang, Q. rGO-wrapped porous $LaFeO_3$ microspheres for high-performance triethylamine gas sensors. *Ceram. Int.* **2020**, *46*, 9363–9369. [CrossRef]
145. Liu, M.; Wang, C.; Yang, M.; Tang, L.; Wang, Q.; Sun, Y.; Xu, Y. Novel strategy to construct porous Sn-doped ZnO/$ZnFe_2O_4$ heterostructures for superior triethylamine detection. *Mater. Sci. Semicond. Process.* **2021**, *125*, 105643. [CrossRef]
146. Liu, F.; Chu, X.; Dong, Y.; Zhang, W.; Sun, W.; Shen, L. Acetone gas sensors based on graphene-$ZnFe_2O_4$ composite prepared by solvothermal method. *Sens. Actuators B Chem.* **2013**, *188*, 469–474. [CrossRef]
147. Fergus, J.W. Perovskite oxides for semiconductor-based gas sensors. *Sens. Actuators B Chem.* **2007**, *123*, 1169–1179. [CrossRef]
148. Li, Y.; Yuan, Z.; Meng, F. Spinel-Type Materials Used for Gas Sensing: A Review. *Sensors* **2020**, *20*, 5413. [CrossRef] [PubMed]
149. Tian, K.; Zhang, W.; Sun, S.-N.; Xing, L.; Li, Z.-Y.; Zhang, T.-T.; Yang, N.-N.; Kuang, B.-B.; Li, H.-Y. Design and fabrication of spinel nanocomposites derived from perovskite hydroxides as gas sensing layer for volatile organic compounds detection. *Sens. Actuators B Chem.* **2021**, *329*, 129076. [CrossRef]
150. Ma, P.; Li, X.; Zhang, Y.; Han, L.; Xu, Y. Hierarchical spinel-type corn-like MGa_2O_4 (M = Ni, Co) architectures for selective triethylamine gas sensors. *Mater. Sci. Semicond. Process.* **2021**, *133*, 105993. [CrossRef]
151. Yang, C.; Xu, Y.; Zheng, L.; Zhao, Y.; Zheng, W.; Liu, X.; Zhang, J. Hierarchical $NiCo_2O_4$ microspheres assembled by nanorods with p-type response for detection of triethylamine. *Chin. Chem. Lett.* **2020**, *31*, 2077–2082. [CrossRef]
152. Xu, X.; Wang, S.; Liu, W.; Chen, Y.; Ma, S.; Yun, P. An excellent triethylamine (TEA) sensor based on unique hierarchical MoS_2/ZnO composites composed of porous microspheres and nanosheets. *Sens. Actuators B Chem.* **2021**, *333*, 129616. [CrossRef]
153. Zhao, Y.; Yuan, X.; Sun, Y.; Wang, Q.; Xia, X.-Y.; Tang, B. Facile synthesis of tortoise shell-like porous $NiCo_2O_4$ nanoplate with promising triethylamine gas sensing properties. *Sens. Actuators B Chem.* **2020**, *323*, 128663. [CrossRef]
154. Ebrahimifard, R.; Abdizadeh, H.; Golobostanfard, M.R. Controlling the extremely preferred orientation texturing of sol–gel derived ZnO thin films with sol and heat treatment parameters. *J. Sol-Gel Sci. Technol.* **2019**, *93*, 28–35. [CrossRef]
155. Lou, C.; Li, Z.; Yang, C.; Liu, X.; Zheng, W.; Zhang, J. Rational design of ordered porous SnO_2/ZrO_2 thin films for fast and selective triethylamine detection with humidity resistance. *Sens. Actuators B Chem.* **2021**, *333*, 129572. [CrossRef]
156. Yu, H.; Xu, H.; Li, W.; Zhai, T.; Chen, Z.; Wang, J.; Cao, B. Enhanced triethylamine sensing properties by designing Au@SnO_2/ZnO nanosheets directly on alumina tubes. *Surf. Interfaces* **2018**, *10*, 85–92. [CrossRef]
157. Xie, G.; Cheng, G.; Zhu, D.; Yan, J.; Ma, J.; Lv, T.; Zhang, J.; Han, W.; Long, Y.Z. Progress of superconducting nanofibers via electrospinning. *J. Phys. Condens. Matter.* **2021**, *34*, 043002. [CrossRef]
158. Li, Z.; Liu, X.; Zhou, M.; Zhang, S.; Cao, S.; Lei, G.; Lou, C.; Zhang, J. Plasma-induced oxygen vacancies enabled ultrathin ZnO films for highly sensitive detection of triethylamine. *J. Hazard. Mater.* **2021**, *415*, 125757. [CrossRef]

159. Mostafa, A.M. The enhancement of nonlinear absorption of Zn/ZnO thin film by creation oxygen vacancies via infrared laser irradiation and coating with Ag thin film via pulsed laser deposition. *J. Mol. Struct.* **2021**, *1226*, 129407. [CrossRef]
160. Song, X.; Li, L.; Chen, X.; Xu, Q.; Song, B.; Pan, Z.; Liu, Y.; Juan, F.; Xu, F.; Cao, B. Enhanced triethylamine sensing performance of α-Fe_2O_3 nanoparticle/ZnO nanorod heterostructures. *Sens. Actuators B Chem.* **2019**, *298*, 126917. [CrossRef]
161. Yoon, J.; Kim, J.; Park, S.; Jeong, Y.W.; Lee, C.; Oh, S.G. Fabrication of Ag-doped ZnO/PAN composite nanofibers by electrospinning: Photocatalytic and antiviral activities. *Korean J. Chem. Eng.* **2022**, *39*, 1632–1640. [CrossRef]
162. Lee, J.; Lee, K.J. Colloid Syringeless Electrospinning toward Nonwoven Nanofiber Web Containing a Massive Amount of Inorganic Fillers. *Macromol. Mater. Eng.* **2022**, *307*, 2100818. [CrossRef]
163. Ma, Q.; Li, H.; Chu, S.; Liu, Y.; Liu, M.; Fu, X.; Li, H.; Guo, J. In_2O_3 Hierarchical Structures of One-Dimensional Electrospun Fibers with in Situ Growth of Octahedron-like Particles with Superior Sensitivity for Triethylamine at Near Room Temperature. *ACS Sustain. Chem. Eng.* **2020**, *8*, 5240–5250. [CrossRef]
164. Wang, X.-X.; Wang, N.; Qiu, H.; Song, W.; Liu, Q.; Fan, Z.; Yu, M.; Ramakrishna, S.; Long, Y.-Z. Anisotropic nanogenerator for anticounterfeiting and information encrypted transmission. *Nano Energy* **2020**, *71*, 104572. [CrossRef]
165. Xu, D.; Chen, Y.; Qiu, T.; Qi, S.; Zhang, L.; Yin, M.; Ge, K.; Wei, X.; Tian, X.; Wang, P.; et al. Hierarchical mesoporous SnO_2 nanotube templated by staphylococcus aureus through electrospinning for highly sensitive detection of triethylamine. *Mater. Sci. Semicond. Process.* **2021**, *136*, 106129. [CrossRef]
166. Cheng, Y.; Ren, B.; Xu, K.; Jeerapan, I.; Chen, H.; Li, Z.; Ou, J.Z. Recent progress in intrinsic and stimulated room-temperature gas sensors enabled by low-dimensional materials. *J. Mater. Chem. C* **2021**, *9*, 3026–3051. [CrossRef]
167. Sudha, S.; Ramprasath, R.; Cholan, S.; Gokul, B.; Sridhar, S.; Elhosiny Ali, H.; Shkir, M. Enhanced triethylamine gas sensing and photocatalytic performance of Sn doped NiO (SNO) nanoparticles. *Inorg. Chem. Commun.* **2022**, *136*, 109104. [CrossRef]
168. Liu, S.-R.; Guan, M.-Y.; Li, X.-Z.; Guo, Y. Light irradiation enhanced triethylamine gas sensing materials based on ZnO/$ZnFe_2O_4$ composites. *Sens. Actuators B Chem.* **2016**, *236*, 350–357. [CrossRef]
169. Shanmugam, P.; Pushparaj, K.; Sundaramurthy, A.; Sivalingam, Y. Investigation of UV light enhanced gas adsorption properties of CeO_2 Nanoparticles by Scanning Kelvin Probe system. *J. Mol. Struct.* **2022**, *1250*, 131831. [CrossRef]
170. Yang, T.; Gu, K.; Zhu, M.; Lu, Q.; Zhai, C.; Zhao, Q.; Yang, X.; Zhang, M. ZnO-SnO_2 heterojunction nanobelts: Synthesis and ultraviolet light irradiation to improve the triethylamine sensing properties. *Sens. Actuators B Chem.* **2019**, *279*, 410–417. [CrossRef]

chemosensors

Article

One-Step Hydrothermal Synthesis of 3D Interconnected rGO/In_2O_3 Heterojunction Structures for Enhanced Acetone Detection

Xiaoguang San [1], Yue Zhang [1], Lei Zhang [1], Guosheng Wang [1], Dan Meng [1,*], Jia Cui [1] and Quan Jin [2,*]

1 College of Chemical Engineering, Shenyang University of Chemical Technology, Shenyang 110142, China; sanxiaoguang@syuct.edu.cn (X.S.); meat_d@163.com (Y.Z.); 15164038413@163.com (L.Z.); wgsh-lyc@163.com (G.W.); cuijiakj@163.com (J.C.)
2 School of Materials Science and Engineering, Jilin University, Changchun 130012, China
* Correspondence: mengdan0610@hotmail.com (D.M.); qjin@jlu.edu.cn (Q.J.)

Abstract: Acetone detection is of great significance for environmental monitoring or diagnosis of diabetes. Nevertheless, fast and sensitive detection of acetone at low temperatures remains challenging. Herein, a series of rGO-functionalized three-dimensional (3D) In_2O_3 flower-like structures were designed and synthesized via a facile hydrothermal method, and their acetone-sensing properties were systematically investigated. Compared to the pure 3D In_2O_3 flower-like structures, the rGO-functionalized 3D In_2O_3 flower-like structures demonstrated greatly improved acetone-sensing performance at relatively low temperatures. In particular, the 5-rGO/In_2O_3 sensor with an optimized decoration exhibited the highest response value (5.6) to 10 ppm acetone at 150 °C, which was about 2.3 times higher than that of the In_2O_3 sensor (2.4 at 200 °C). Furthermore, the 5-rGO/In_2O_3 sensor also showed good reproducibility, a sub-ppm-level detection limit (1.3 to 0.5 ppm), fast response and recovery rates (3 s and 18 s, respectively), and good long-term stability. The extraordinary acetone-sensing performance of rGO/In_2O_3 composites can be attributed to the synergistic effect of the formation of p-n heterojunctions between rGO and In_2O_3, the large specific surface area, the unique flower-like structures, and the high conductivity of rGO. This work provides a novel sensing material design strategy for effective detection of acetone.

Keywords: In_2O_3 flower-like structures; rGO; composites; gas sensor; acetone

Citation: San, X.; Zhang, Y.; Zhang, L.; Wang, G.; Meng, D.; Cui, J.; Jin, Q. One-Step Hydrothermal Synthesis of 3D Interconnected rGO/In_2O_3 Heterojunction Structures for Enhanced Acetone Detection. *Chemosensors* **2022**, *10*, 270. https://doi.org/10.3390/chemosensors10070270

Academic Editor: Andrea Ponzoni

Received: 6 June 2022
Accepted: 10 July 2022
Published: 11 July 2022

1. Introduction

Acetone, as a volatile organic compound (VOC), is widely used in industries and laboratories [1,2]. However, due to its volatility and toxicity, it is becoming a typical pollution gas and threatening our health. Long-term inhalation of acetone gas at concentrations over 173 ppm may cause serious health problems, such as headache, nausea, vomiting, and central nervous system anesthesia [3]. Additionally, acetone can be used as a biomarker for the diagnosis of type 1 diabetes. It has been found that the exhaled acetone concentration in diabetes patients (1.8 ppm) is significantly higher than that of healthy controls (0.9 ppm) [4]. Therefore, developing an effective sensor operating in a wide acetone concentration range is crucial for environmental monitoring or diagnosis of diabetes.

Recently, many kinds of sensors based on optical, electric, gravimetric transduction techniques, etc., have been developed for detecting VOCs [5,6]. Avramov et al. presented an optimization-mass-sensitive sensor using a surface acoustic wave (SAW)-based two-port resonator as a highly sensitive quartz crystal microbalance (QCM), which exhibited excellent VOC-sensing properties [7]. Kanawade et al. reported a new negative-axicon fiber-optic sensor platform, which exhibits many desirable characteristics for sensing various VOCs at room temperature [8]. However, how to reduce their production costs and limit of detection, and to utilize real-time detection, should be deeply studied. Resistive gas sensors based on metal oxide semiconductors (MOX) are an important type of gas sensor due to their low production cost, ease of fabrication, and good long-term stability [9–11]. Until now, SnO_2 [12], In_2O_3 [13], Fe_2O_3 [14], ZnO [15], WO_3 [16], Co_3O_4 [17], CuO [18], V_2O_5 [19], MoO_3 [20], and other metal oxide semiconductors with micro/nanostructures have been extensively investigated for the detection of various gases. Among them, In_2O_3, owing to its wide band gap, high catalytic activity, and low resistivity, has proven to be an excellent n-type sensing material for detecting VOCs such as methanol, ethanol, formaldehyde, and acetone [21–26]. For example, Yoon et al. [22] synthesized In_2O_3 nanoparticles via a microwave-assisted hydrothermal technology, which showed high responsivity and selectivity to ethanol at 300 °C. Liu et al. [24] reported that Yb-doped porous In_2O_3 nanosheets exhibited excellent sensing performance for acetone. Zhao et al. [25] synthesized hierarchical In_2O_3 nanostructures via a simple hydrothermal method, which showed ultrafast response speed, excellent selectivity, and stability for formaldehyde. Nevertheless, the rapid and accurate detection of trace acetone using In_2O_3 at a low temperatures remains a great challenge and, thus, has been the focus of ongoing research.

In recent years, two-dimensional (2D) graphene-based nanomaterials, including graphene oxide (GO) and reduced graphene oxide (rGO), have been extensively studied as potential gas-sensing materials due to their high specific surface area, outstanding electrical properties, abundant functional groups, and excellent charge-carrier mobility. [27,28]. Unfortunately, the low sensing response, slow response and recovery rates, and poor selectivity of graphene-based sensing materials limit their practical application. Taking into account the advantages of GO/rGO and MOX, the construction of GO/MOX or rGO/MOX hybrid micro–nanostructures is considered an effective way to enhance gas-sensing performances [29–32]. To date, a number of GO/MOX or rGO/MOX hybrids—such as SnO_2/rGO, rGO/Co_3O_4, rGO/WO_3, rGO/In_2O_3, ZnO/rGO, etc.—have been prepared, and demonstrate good sensing performance in detecting both oxidizing and reducing gases [33–38]. For instance, Meng et al. [33] prepared rGO/Co_3O_4 nanocomposites via a one-step hydrothermal method, which could detect ppb level xylene. Cao et al. [37] constructed ZnO/rGO heterostructures through a chemiresistive approach, exhibiting good linearity in the ppb range, high selectivity, and good long-term stability for NO_2. Inspired by the above results, it is hoped that researchers will be able to uniformly distribute 3D In_2O_3 flower-like structures on 2D rGO sheets via a feasible and economic route for developing an acetone gas sensor with high performance.

In the present study, well-designed rGO-functionalized 3D In_2O_3 flower-like structures were constructed via a one-step hydrothermal route, and their acetone-sensing properties were systemically investigated. It was clear that the optimized rGO/In_2O_3 composites demonstrated significantly enhanced acetone-sensing performance in comparison to pure In_2O_3 flower-like structures. The remarkably improved sensing performance of rGO/In_2O_3 composites can be attributed to the synergistic effect of the formation of p-n heterojunctions between rGO and In_2O_3, the large specific surface area, the unique flower-like structures, and the high conductivity of rGO. This work provides a novel sensing material for effective detection of acetone.

2. Experimental

2.1. Materials and Synthesis

Graphene oxide (GO) was synthesized via Hummers' method using graphite flakes as raw materials [39]. Then, the obtained GO was dispersed into ethanol and ultrasonically treated for 1.5 h to obtain a uniform 0.5 mg/mL GO dispersion. The rGO-functionalized 3D In_2O_3 flower-like structures were synthesized by a one-step hydrothermal route, which is schematically illustrated in Figure 1. Typically, 0.13 g of indium nitrate ($In(NO_3)_3{\cdot}4.5H_2O$), 1 g of urea (H_2NCONH_2), and 0.26 g of sodium dodecyl sulfate ($CH_3(CH_2)_{11}OSO_3Na$) were dissolved in 20 mL of distilled water, and then stirred for 0.5 h to obtain a transparent solution. After that, a certain volume of GO solution was added to the above solution and stirred at room temperature for 0.5 h to form a well-dispersed suspension with different mass ratio of rGO/In_2O_3. Diverse volumes of ethanol were added to the above mixtures until the total volume was 40 mL. Subsequently, the mixed solution was transferred into a 50 mL Teflon-lined stainless steel autoclave and then heated in an oven at 160 °C for 3 h, in which structural modification of the In_2O_3 flower-like structures and the reduction of GO occurred simultaneously. After the reaction, the precipitate was collected by centrifugation and washed several times with distilled water and ethanol to remove the impurities, and then dried at 60 °C for 8 h. Finally, the obtained precursor was annealed in a tube furnace at 400 °C for 2 h under a nitrogen atmosphere. In our experiment, the mass ratio of GO in the composites was 1 wt%, 3 wt%, 5 wt%, 7 wt%, and 10 wt%, respectively; consequently, the rGO/In_2O_3 composites were labeled as 1-rGO/In_2O_3, 3-rGO/In_2O_3, 5-rGO/In_2O_3, 7-rGO/In_2O_3, and 10-rGO/In_2O_3, respectively. For comparison, the pure In_2O_3 flower-like structures were synthesized individually by the same procedure without the addition of GO.

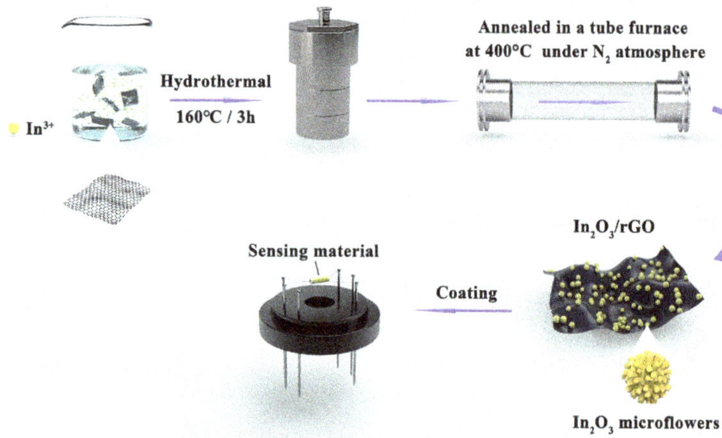

Figure 1. The schematic of the synthesis process of the rGO/In_2O_3 composites.

2.2. *Materials Characterization*

The crystal structure of the rGO/In_2O_3 composites was characterized by X-ray diffraction (XRD, Shimadzu XRD-6100). The surface morphologies of the rGO/In_2O_3 composites were analyzed using field-emission scanning electron microscopy (FE-SEM, ZEISS Ultra Plus). The internal structure of the rGO/In_2O_3 composites was further observed by high-resolution transmission electron microscopy (HRTEM) on a TEM system (JEOL, JEM-2100F). The distribution of elements was obtained by energy-dispersive spectrometry (EDS) mapping on a JEOL-JEM-2100F TEM. Then, the specific surface area and pore size distribution were analyzed by the Brunauer–Emmet–Teller (BET, Quantachrome Autosorb1-C) method and the Barrett–Joyner–Halenda (BJH) model. In addition, the surface compositions and states of the rGO/In_2O_3 composites were investigated by X-ray photoelectron spectroscopy (XPS, JEOL JPS9010MC).

2.3. *Sensor Fabrication and Measurement*

The gas sensor was fabricated as follows: First, an appropriate amount of rGO/In_2O_3 or pure In_2O_3 powder was ground and mixed with ethanol to form a homogeneous slurry, which was then coated on the outer surface of the ceramic tube electrode. The ceramic tube electrode was composed of four Pt wires and two Au electrodes spaced 6 mm apart. After forming a sensing film by evaporating ethanol, a Ni-Cr heated wire was passed through the hollow ceramic tube, and the Ni–Cr wire and Pt wires were subsequently welded on a hexagonal base to fabricate the gas sensor, as shown in Figure 1. Finally, the gas sensor was placed on a TS60 desktop (Winsen Electronics Co., Ltd., Zhengzhou, China) and aged at 400 °C for 48 h to ensure the stability of the sensing device.

Gas-sensitive measurements of the fabricated sensors were performed on a commercial WS-30A system (Winsen Electronics Co., Ltd., Zhengzhou, China) under static laboratory conditions (temperature: 20–25 °C; relative humidity (RH): 20–35%). To obtain the target gas, corresponding volumes of testing liquid were injected into the testing chamber (18 L) using a microsyringe and then evaporated to record and analyze the change in current. The following equation can be used to calculate the target gas concentration:

$$C = \frac{22.4 \times \varphi \times \rho \times V}{V_{chamber} \times M} \tag{1}$$

where V is the volume (μL) of the injected testing liquid, $V_{chamber}$ is the volume (L) of the testing chamber, C is the vapor concentration (ppm) of the target gas, M is the molar mass (g/mol) of the target gas, ρ is the density (g/mL) of the testing liquid, and φ is the volume fraction of the target gas. After the gas-sensing response reached a constant value, the testing gas was released by opening the chamber. Herein, the sensing response to different target gases was defined as Ra/Rg, where Ra and Rg are the resistance of the sensors in air and target gases, respectively. Moreover, the response and recovery time were calculated as the time required to achieve 90% change in resistance after the sensor was exposed to the target gas and fresh air, respectively.

3. Results and Discussion

3.1. Structural and Morphological Characteristics

Figure 2 illustrates the XRD patterns of the In_2O_3 and rGO/In_2O_3 composites. In_2O_3 displayed perfect crystallinity, and the diffraction peaks that appeared at 2θ values of 21.5, 30.6, 35.5, 37.7, 39.8, 45.7, 51.0, and 60.7° were ascribed to the (211), (222), (400), (411), (420), (431), (440), and (622) planes of the cubic phase In_2O_3 (JCPDS No. 06-0416), respectively. As for the rGO/In_2O_3 composites, all of the diffraction peaks were consistent with In_2O_3, indicating that introducing rGO to In_2O_3 has no obvious influence on the phase. Here, there were no obvious diffraction peaks of rGO in the patterns, which might have been due to the low content of rGO in the composites and the low diffraction intensity of rGO. In addition, no other impurity peaks were observed, confirming their high purity.

The morphology of the In_2O_3 and rGO/In_2O_3 composites was investigated by FE-SEM, as shown in Figure 3. The low-magnification FE-SEM image (Figure 3a) clearly demonstrates numerous 3D In_2O_3 flower-like structures with diameters ranging from 1 to 2 μm. These flower-like structures were composed of nanorods with diameters of 100 to 300 nm (Figure 3b), and a specific spacing between each nanorod was clearly observed, illustrating their porous structures. From the FE-SEM images of the 1-rGO/In_2O_3 composites (Figure 3c,d), it is clear that the 3D In_2O_3 flower-like structures remained, and some of them uniformly attached to the surface of the rGO sheets, forming a 3D interconnected structure, and thereby contributing to enhancement of the sensing properties. Moreover, the crumpled rGO sheets with many folds and wrinkles exhibited a clean and transparent film. The morphologies of the 3-rGO/In_2O_3, 5-rGO/In_2O_3, 7-rGO/In_2O_3, and 10-rGO/In_2O_3 composites were similar to that of the 1-rGO/In_2O_3 composites, as shown in Figure 3e–l. It is worth noting that there were no significant changes in the size and the uniformity of the In_2O_3 flower-like structures. Additionally, with increasing the mass ratio of rGO, many ultrathin layered structures were observed, and some of them tend to aggregate with one another, which is unfavorable for the gas-sensing property.

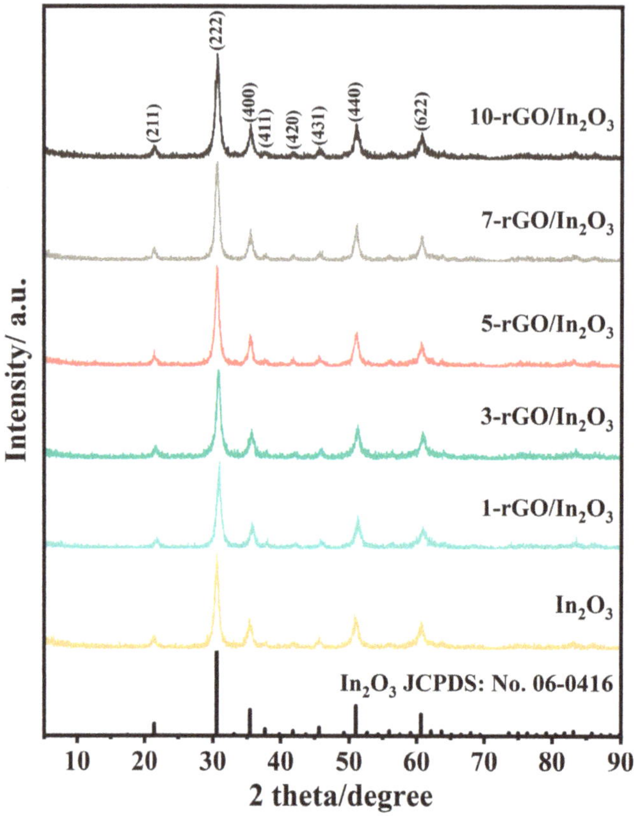

Figure 2. XRD patterns of the pure In_2O_3, 1-rGO/In_2O_3, 3-rGO/In_2O_3, 5-rGO/In_2O_3, 7-rGO/In_2O_3, and 10-rGO/In_2O_3.

Figure 3. FE-SEM images of (**a**,**b**) In_2O_3, (**c**,**d**) 1-rGO/In_2O_3, (**e**,**f**) 3-rGO/In_2O_3, (**g**,**h**) 5-rGO/In_2O_3, (**i**,**j**) 7-rGO/In_2O_3, and (**k**,**l**) 10-rGO/In_2O_3.

The In_2O_3 flower-like structures attached on the rGO sheets were also clearly observed in the TEM image (Figure 4a), consistent with the FE-SEM image. The HRTEM image (Figure 4b) shows that 0.292 and 0.179 nm lattice spacings fit well with the (222) and (440) crystal planes of In_2O_3 [40]. Meanwhile, the EDS elemental mappings (Figure 4c–e) clearly indicate that the In, O, and C are uniformly distributed throughout the whole region, further implying the formation of the rGO/In_2O_3 composites. Taking the results of the FE-SEM and TEM observations into account, it is possible to conclude that the rGO/In_2O_3 composites were successfully prepared.

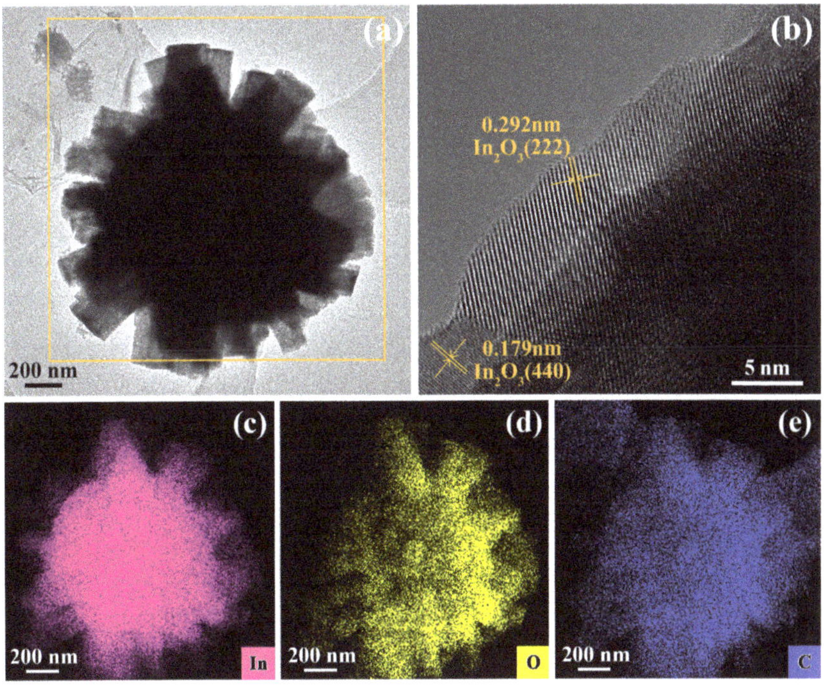

Figure 4. (**a**) TEM image, (**b**) high-resolution TEM image, and (**c**–**e**) its corresponding elemental mapping of 5-rGO/In_2O_3 composites.

XPS analysis was performed to investigate the surface composition and chemical state of the 5-rGO/In_2O_3 composites. The survey spectra shown in Figure 5a show that the rGO/In_2O_3 composites were mainly composed of In, O, and C, confirming the successful preparation of the rGO/In_2O_3 composites. The In 3D symmetric peaks with binding energy of 444.83 eV and 452.37 eV (Figure 5b) were attributed to $In3d_{3/2}$ and $In3d_{5/2}$, respectively, indicating the existence of In^{3+} in the composites [41,42]. Figure 5c depicts three different chemical states of O 1s peaks at 530.30 eV, 531.84 eV, and 533.36 eV, corresponding to lattice oxygen (O_L), oxygen vacancies (O_V), and the surface hydroxyl groups (O_{OH}), respectively [43]. As for the C1s XPS spectrum of the composites (Figure 5d), it can be decomposed into three peaks at 284.60 eV, 285.55 eV, and 289.64 eV, which correspond to C=C–C, C–O/C–In, and C=O bonds in the rGO/In_2O_3 composites, respectively [44]. The above results further demonstrate the successful preparation of rGO/In_2O_3 composites.

The nitrogen adsorption–desorption measurements were performed to further confirm the specific surface area and porous structures of the samples. From Figure 6a,b, it can be seen that both of the curves are recognized as type IV isotherms with obvious hysteresis loops, indicating the formation of mesopores in the samples [45]. The specific surface area of the In_2O_3 and 5-rGO/In_2O_3 composites was calculated to be 35.4 and 64.6 m^2/g, respectively. Obviously, the surface area of the 5-rGO/In_2O_3 composites was significantly higher than that of pure In_2O_3, due to the introduction of rGO. The pore size distribution diagram (inset) shows that the pore size of the In_2O_3 and rGO/In_2O_3 composites was predominantly concentrated at 8.129 and 6.124 nm, respectively, indicating the existence of many mesopores in the products. As a result, the rGO/In_2O_3 composites possess a higher specific surface area and a porous structure, which is advantageous for effective gas absorption and diffusion, contributing to the improved sensing performance.

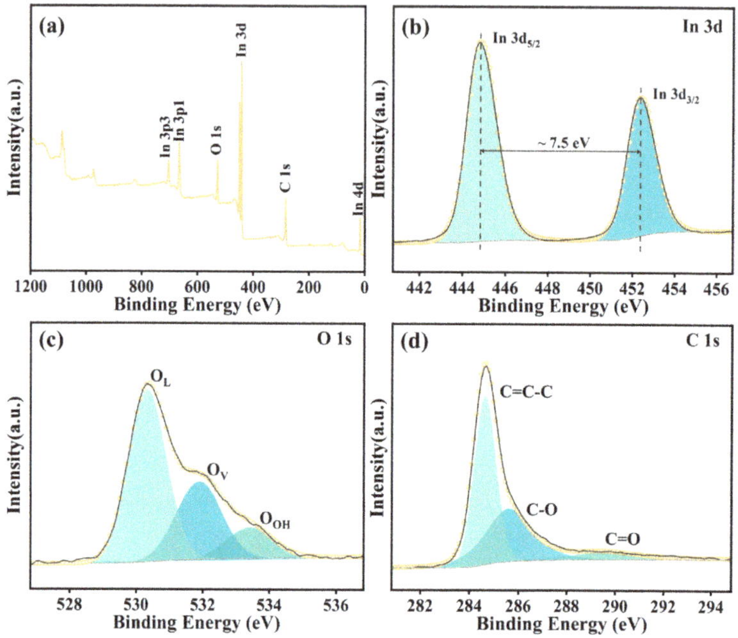

Figure 5. (**a**) Survey XPS spectra of 5-rGO/In_2O_3, (**b**) XPS spectrum of In, (**c**) XPS spectrum of O, and (**d**) XPS spectrum of C.

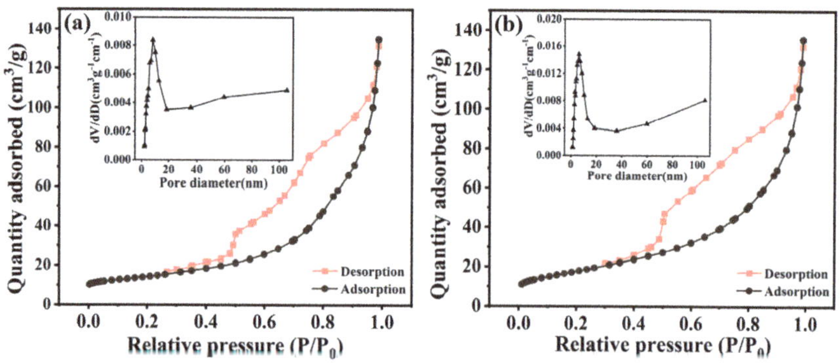

Figure 6. Typical N_2 adsorption–desorption isotherm and pore size distribution plots (inset) of (**a**) In_2O_3 and (**b**) 5-rGO/In_2O_3.

3.2. Gas Sensing Performance

In order to investigate the effects of rGO mass ratios on the sensing performance of In_2O_3 sensors, we fabricated gas sensors made of In_2O_3 flower-like structures and rGO-functionalized In_2O_3 flower-like structures, and tested their sensing performance. It is well known that the operating temperature has a significant impact on the sensing properties of gas sensors. In order to evaluate the effect of operating temperature on the sensing ability and obtain the optimal operating parameters, the sensing response to 10 ppm acetone was first examined in the temperature range of 50 to 250 °C, as shown in Figure 7a. It was clearly found that all response curves showed an "increase-maximum-decrease" trend. This response pattern can be mostly attributed to the kinetic and thermodynamic factors of gas adsorption and desorption on the surface of the sensing materials [46,47]. The optimal operating temperature of the In_2O_3 sensor was 200 °C, while it was 150 °C for all rGO/In_2O_3 sensors, indicating that introducing rGO to In_2O_3 flower-like structures can effectively reduce the optimal operating temperature. The corresponding response values of sensors based on composites with different rGO mass ratios at their optimal operating temperatures are summarized in Figure 7b. Obviously, all rGO/In_2O_3 sensors except the 10-rGO/In_2O_3 sensor exhibited higher responses than that of the In_2O_3 sensor at their optimal operating temperature. In particular, the 5-rGO/In_2O_3 sensor exhibited the highest response value of 5.6, which was about 2.3 times higher than that of the In_2O_3 sensor (2.4 at 200 °C). This suggests that introducing an appropriate amount of rGO not only effectively reduces the optimal operating temperature, but also significantly improves the acetone-sensing response. Herein, considering the high acetone-sensing performance of the 5-rGO/In_2O_3 sensor, this sensor was chosen for further study of other characteristics at its optimal operating temperature of 150 °C.

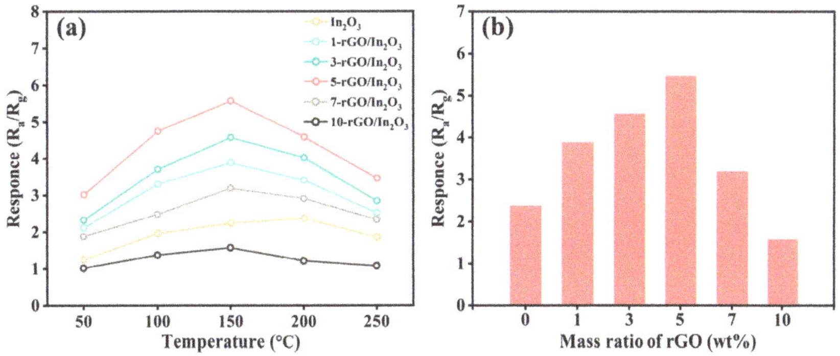

Figure 7. (**a**) The sensing response to 10 ppm acetone at various operating temperatures. (**b**) Correlations between the sensing response to 10 ppm acetone at the optimal operating temperature and rGO mass ratios.

The transient response and recovery curves of the 5-rGO/In_2O_3 sensor to different concentrations of acetone (0.5 to 100 ppm) at 150 °C are displayed in Figure 8a. As can be seen, the resistance dropped rapidly upon exposure to acetone, and quickly returned to its initial resistance value after removing the acetone, demonstrating typical n-type semiconductor sensing features. Moreover, the response and recovery behavior repeated very well, revealing good reversibility and stability. It should be noted that the resistance change is still considerable when exposed to 0.5 ppm acetone, indicating that the detection limit could extend down to sub-ppm levels. To exhibit its response performance, the corresponding response values of the 5-rGO/In_2O_3 sensor as a function of acetone concentration are illustrated in Figure 8b. Apparently, the responses value rapidly increases with increasing acetone concentration in the range of 0.5 ppm to 10 ppm. However, when the concentration is above 10 ppm, the rate of increase in the response value tends to slow down due to the high coverage of acetone molecules on the surface of the sensing material. Moreover, the sensor has an obvious response to sub-ppm levels of acetone (0.5 ppm, 1.3). Figure 8c presents the transient response and recovery curve of the 5-rGO/In_2O_3 sensor when exposed to 10 ppm acetone at 150 °C. The response and recovery times were calculated to be 3 s and 18 s, respectively, revealing quick response and recovery characteristics as the acetone gas is injected and exhausted. As for practical gas-sensing applications, selectivity and stability are of great importance. Figure 8d displays the sensing response of the 5-rGO/In_2O_3 sensor to 10 ppm potential interfering gases (i.e., ethanol, methanol, formaldehyde, trimethylamine, and ammonia) at 150 °C. The sensor exhibited a higher response to acetone compared to other interfering gases, indicating its good selectivity. Additionally, the response values of the 5-rGO/In_2O_3 sensor to different concentrations (0.1 ppm, 0.5 ppm, and 10 ppm) did not decrease significantly after the 30-day testing period (Figure 8e), implying its good long-term stability. As shown in (Figure 8f), the responses of the 5-rGO/In_2O_3 sensor to 10 ppm acetone at 150 °C showed slight decreases under relative humidity (RH) ranging from 25% to 45%, and an obvious decrease at RH levels above 75%. This decrease in sensing responses can be ascribed to the adsorption competition between oxygen species and water molecules on the sensor surface [48]. Furthermore, the acetone-sensing performance in this work was compared with the previous literature, as summarized in in Table 1. Compared with the previous reports, our 5-rGO/In_2O_3 sensor had the noticeable advantages of high response values at a relatively low operating temperature, a sub-ppm-level detection limit, and short response and recovery times, suggesting its potential application as an acetone sensor. However, there are also some drawbacks in this work. For example, how to overcome the influence of water vapor on the gas-sensing characteristics at high RH levels (>75%) should be further investigated for the diagnosis of diabetes in real applications. Moreover, the unsatisfactory selectivity should be also improved by coating functional materials as a "selective gas filter" on the sensing layer surface in future works.

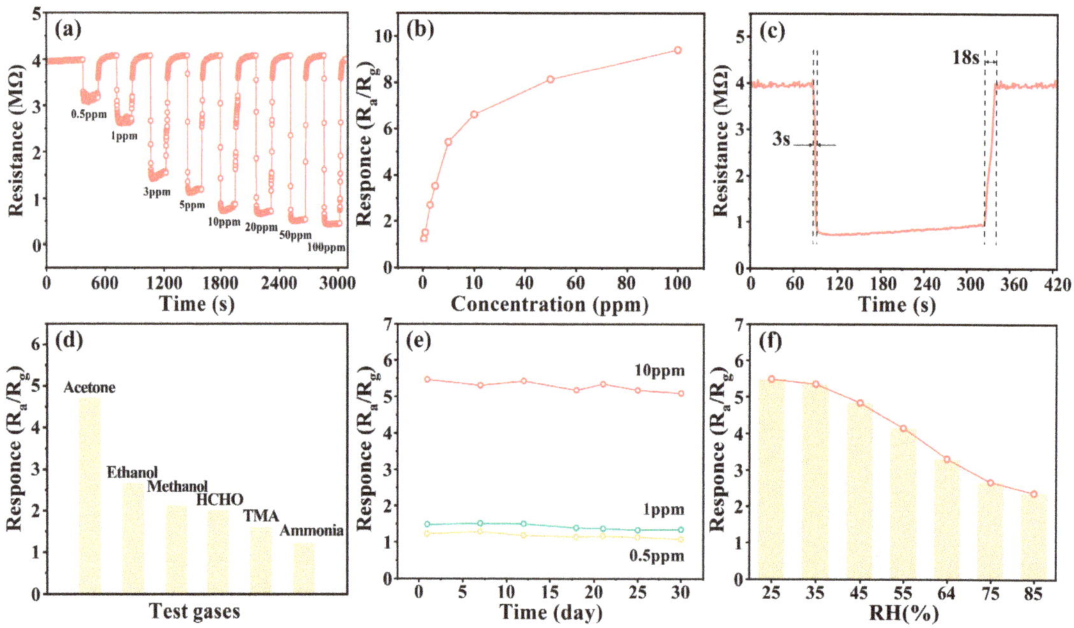

Figure 8. (**a**) The transient response and recovery curves for acetone in the range of 0.5–100 ppm. (**b**) The corresponding sensing response as a function of acetone concentration. (**c**) The dynamic sensing transients to 10 ppm acetone. (**d**) The selectivity and (**e**) long-term stability for 0.5/1/10 ppm acetone. Here, the sensor is made of 5-rGO/In_2O_3 composites, and the operating temperature is 150 °C. (**f**) The sensing responses to 10 ppm acetone under different relative humidity levels at 150 °C.

Table 1. Comparison of sensing properties for acetone detection.

Sensing Materials	Temp. (°C)	Con. (ppm)	Response (Ra/Rg)	Res./Rec. Time (s)	Detection Limit	Ref.
ZnO/NPC	350	100	25.47	3/150	1 ppm	[49]
α-Fe_2O_3-24	220	100	46.6	1/15	1 ppm	[50]
Ag-TiO_2 nanobelts	260	500	28.25	6/8	10 ppm	[51]
Ag@CuO-TiO_2	200	100	6.2	9/56	1 ppm	[52]
WO_3	200	100	28.7	3/113	2 ppm	[53]
SnO_2 nanotubes	325	100	20.3	66/15	5 ppm	[54]
In_2O_3	200	10	2.37	8/44	1 ppm	This work
5-rGO/In_2O_3	150	10	5.57	3/18	0.5 ppm	This work

3.3. Acetone-Sensing Mechanism

Generally, for chemiresistive gas sensors, the sensing characteristics can be attributed to the resistance variation induced by the adsorption and desorption of gas molecules on the surface of the sensing materials, along with the surface redox reaction [55]. For the In_2O_3 sensor, numerous oxygen molecules in the air atmosphere are adsorbed on the surface of the sensing film, and capture free electrons from the In_2O_3 conduction band, forming chemisorbed oxygen species (O_2^- O^- and O^{2-}) via the following reactions [56]:

$$O_{2(ads)} + e^- \rightarrow O_2{}^-{}_{(ads)} \quad (2)$$

$$O_2{}^-{}_{(ads)} + e^- \rightarrow 2O^-{}_{(ads)} \quad (3)$$

$$O^-{}_{(ads)} + e^- \rightarrow O^{2-}{}_{(ads)} \quad (4)$$

Consequently, an electron deletion layer is created at the surface of In_2O_3, leading to a relatively high resistance state. As the reducing gas (acetone) is introduced, redox reactions between acetone molecules and the chemisorbed oxygen species take place, and then the electrons captured by oxygen molecules are released back into the In_2O_3 conduction band. In this process, the electron depletion layer becomes narrow and, thus, the resistance of the sensing materials decreases, ultimately generating a sensing signal. The related reactions are listed as follows [57]:

$$CH_3COCH_3\ (gas) \rightarrow CH_3COCH_3\ (ads) \quad (5)$$

$$CH_3COCH_3\ (ads) + 4O_2{}^-\ (ads) \rightarrow 3CO_2 + 3H_2O + 4e^- \quad (6)$$

$$CH_3COCH_3\ (ads) + 8O^-\ (ads) \rightarrow 3CO_2 + 3H_2O + 8e^- \quad (7)$$

Compared to the In_2O_3 sensor, the rGO/In_2O_3 sensor shows a significant improvement in sensing performance, which can be attributed to the following factors (Figure 9): First of all, the formation of a p-n heterojunction at the interface between n-type In_2O_3 and p-type rGO is one of the main factors improving the sensors' performances [58]. When In_2O_3 and rGO contact one another, the electrons flow from the n-type In_2O_3 with low work function to the p-type rGO with high work function until the Fermi energy level reaches equilibrium [59]; thus, a potential barrier is created at the heterojunctions. Consequently, two different depletion layers coexist in the rGO/In_2O_3 composites: one is the depletion layer formed by the adsorption of oxygen molecules at the surface of In_2O_3, and the other is created at the interface between rGO and In_2O_3 (p-n heterojunctions). As a consequence, the expanded depletion layer and potential barrier simultaneously lead to a high resistance state. The increase in the rGO/In_2O_3 composite's resistance is also confirmed by our experimental results. As shown in Figure 9d, the resistance value of the 5-rGO/In_2O_3 sensor is much higher than that of the In_2O_3 sensor throughout the whole temperature range, indicating that the formation of p-n heterojunctions contributes to raising the resistance value. Upon exposure to acetone, the local heterojunctions between rGO sheets and In_2O_3 flower-like structures offer additional resistance modulation by altering potential barriers and two electron depletion layers. According to the definition of the sensor response (S = Ra/Rg), the total resistance of the rGO/In_2O_3 composites could be distinctly changed, inducing significant enhancement of the sensing response. Moreover, the formation of rGO/In_2O_3 heterojunctions can also generate more active sites, such as including point defects and vacancies, which further adds to the increase in the sensing response [60]. The second aspect originates from the coupling interaction between rGO sheets and In_2O_3 flower-like structures, which creates a 3D interconnected structure and plays an important role in gas-sensing applications. The intercalation of rGO sheets can prevent the agglomeration of In_2O_3 flower-like structures, which not only significantly increase the specific surface area for more efficient gas adsorption reactions, but also offer more pathways for facilitating gas diffusion and transport, thus improving the sensing performance [61]. It has already been confirmed from BET analysis that the specific surface area of the rGO/In_2O_3 composites is higher than that of the In_2O_3 flower-like structures. In addition, rGO, owing to its quantity of dangling bonds and surface defects, is favorable for the adsorption of target gases [62]. Meanwhile, the unique In_2O_3 flower-like structures also facilitate the operative access of gases to the surface. These characteristics greatly contribute to enhancing the sensing response and recovery properties of rGO/In_2O_3 composites. Thirdly, the excellent electron conductivity of rGO is a positive factor for the enhanced sensing performances. From the FE-SEM images, it can be seen that the rGO sheets disperse adequately on the composites, and could act as an electron transfer layer accelerating the electronic transmission rate at the interfaces between In_2O_3 and rGO. This is helpful for reducing the optimal temperature and improving repeatability and stability. Based on the above discussion, the combination of In_2O_3 and rGO contributes to the superior sensing properties, including high response at relatively low temperatures, good stability, and fast response and recovery rates, demonstrating potential application for detecting acetone.

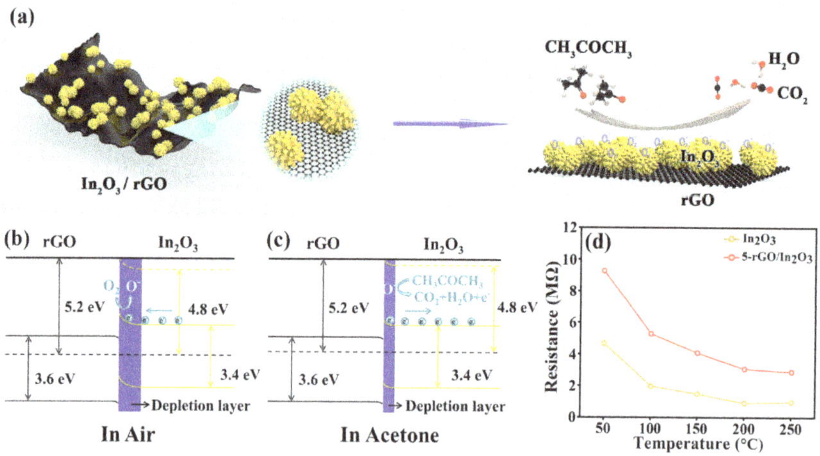

Figure 9. (**a**) Schematic diagram of the mechanism of the acetone gas-sensing process of the rGO/In_2O_3 composites. (**b**,**c**) Schematics of energy band diagrams of rGO/In_2O_3 heterojunctions. (**d**) The resistance of the sensors in air as a function of operating temperature.

4. Conclusions

In summary, a series of well-designed 3D interconnected rGO/In_2O_3 heterojunction structures were successfully constructed using a simple hydrothermal method. The FE-SEM and TEM analyses showed that In_2O_3 flower-like structures with sizes of 1–2 μm uniformly attached on the surface of the rGO sheets. Both the TEM-EDS and XPS results demonstrated the existence of In, O, and C, revealing the successful preparation of the rGO/In_2O_3 composites. In addition, the BET surface area of the 5-rGO/In_2O_3 composites (64.6 m^2/g) was significantly higher than that of pure In_2O_3 (35.4 m^2/g), implying their better sensing properties. As a result, the rGO/In_2O_3 composites achieved much better acetone-sensing properties compared to pure In_2O_3 flower-like structures at the relatively low temperature of 150 °C. Specifically, the 5-rGO/In_2O_3 composites exhibited the highest response value of 5.6 to 10 ppm acetone, which was around 2.3 times higher than that of the In_2O_3 flower-like structures (2.4 at 200 °C). Furthermore, they also displayed good reproducibility, a sub-ppm-level detection limit (1.3 to 0.5 ppm), a fast response/recovery rate (3/18 s), and good long-term stability. The large advancement for rGO/In_2O_3 composites in acetone-sensing performance can be attributed to the synergistic effect of the formation of p-n heterojunctions between rGO and In_2O_3, the large specific surface area and unique flower-like structure, and the high conductivity of rGO. This work not only developed a high-performance gas sensor to detect acetone, but also provides a novel strategy to enhance the sensing performance for MOX gas-sensing materials by reasonably utilizing the synergistic effect of structure and components.

Author Contributions: Conceptualization, Y.Z.; data curation, Y.Z. and J.C.; formal analysis, Y.Z., L.Z., G.W., D.M. and J.C.; funding acquisition, X.S.; investigation, Y.Z.; methodology, X.S., Y.Z. and D.M.; project administration, X.S., D.M. and Q.J.; resources, L.Z. and G.W.; supervision, D.M. and Q.J.; validation, X.S.; visualization, X.S. and Y.Z.; writing—original draft, Y.Z.; writing—review & editing, X.S., D.M. and Q.J. All authors have read and agreed to the published version of the manuscript.

Funding: This research was funded by the National Natural Science Foundation of China (no. 61973223), Support Plan for Innovative Talents in Colleges and Universities in Liaoning Province (no. 2020389), Liao Ning Revitalization Talents Program (no. XLYC2007051), Liaoning Educational Department Foundation (no. LJ2020001), Natural Science Foundation of Liaoning Province (nos. 2019-ZD-0072, 2021-MS-257), and Young and Middle-Aged Scientific and Technological Innovation Talents of Shenyang Science and Technology Bureau (no. RC200352).

Institutional Review Board Statement: Not applicable.

Informed Consent Statement: Not applicable.

Data Availability Statement: Not applicable.

Conflicts of Interest: The authors declare no conflict of interest.

References

1. Zhu, L.; Yuan, K.; Li, Z.; Miao, X.; Wang, J.; Sun, S.; Devi, A.; Lu, H. Highly sensitive and stable MEMS acetone sensors based on well-designed α-Fe_2O_3/C mesoporous nanorods. *J. Colloid Interface Sci.* **2022**, *622*, 156–168. [CrossRef] [PubMed]
2. Deng, L.; Bao, L.; Xu, J.; Wang, D.; Wang, X. Highly sensitive acetone gas sensor based on ultra-low content bimetallic PtCu modified $WO_3·H_2O$ hollow sphere. *Chin. Chem. Lett.* **2020**, *31*, 2041–2044. [CrossRef]
3. Lee, J.; Choi, Y.; Park, B.; Han, J.; Lee, H.; Park, J.; Lee, W. Precise control of surface oxygen vacancies in ZnO nanoparticles for extremely high acetone sensing response. *J. Adv. Ceram.* **2022**, *11*, 769–783. [CrossRef]
4. Ama, O.; Sadiq, M.; Johnson, M.; Zhang, Q.; Wang, D. Novel 1D/2D KWO/$Ti_3C_2T_x$ Nanocomposite-Based Acetone Sensor for Diabetes Prevention and Monitoring. *Chemosensors* **2020**, *8*, 102. [CrossRef]
5. Fanget, S.; Hentz, S.; Puget, P.; Arcamone, J.; Matheron, M.; Colinet, E.; Andreucci, P.; Duraffourg, L.; Myers, E.; Roukes, M.L. Gas sensors based on gravimetric detection—A review. *Sens. Actuators B* **2011**, *160*, 804–821. [CrossRef]
6. Liu, K.; Zhang, C. Volatile organic compounds gas sensor based on quartz crystal microbalance for fruit freshness detection: A review. *Food Chem.* **2021**, *334*, 127615. [CrossRef] [PubMed]
7. Avramov, I.D.; Ivanov, G.R. Layer by Layer Optimization of Langmuir–Blodgett Films for Surface Acoustic Wave (SAW) Based Sensors for Volatile Organic Compounds (VOC) Detection. *Coatings* **2022**, *12*, 669. [CrossRef]
8. Kanawade, R.; Kumar, A.; Pawar, D.; Vairagi, K.; Late, D.; Sarkar, S.; Sinha, R.K.; Mondal, S. Negative axicon tip-based fiber optic interferometer cavity sensor for volatile gas sensing. *Opt. Express* **2019**, *27*, 7277–7290. [CrossRef]
9. Dadkhah, M.; Tulliani, J. Nanostructured Metal Oxide Semiconductors towards Greenhouse Gas Detection. *Chemosensors* **2022**, *10*, 57. [CrossRef]
10. Deng, C.; Zhang, J.; Yu, X.; Zhang, W.; Zhang, X. Determination of acetone in human breath by gas chromatography–mass spectrometry and solid-phase microextraction with on-fiber derivatization. *J. Chromatogr. B* **2004**, *810*, 269–275. [CrossRef]
11. Galstyan, V.; Bhandari, M.P.; Sberveglieri, V.; Sberveglieri, G.; Comini, E. Metal Oxide Nanostructures in Food Applications: Quality Control and Packaging. *Chemosensors* **2018**, *6*, 16. [CrossRef]
12. Li, G.; Shen, Y.; Zhao, S.; Bai, J.; Gao, S.; Liu, W.; Wei, D.; Meng, D.; San, X. Construction of rGO-SnO_2 heterojunction for enhanced hydrogen detection. *Appl. Surf. Sci.* **2022**, *585*, 152623. [CrossRef]

13. Tao, K.; Han, X.; Yin, Q.; Wang, D.; Han, L.; Chen, L. Metal-Organic Frameworks-Derived Porous In_2O_3 Hollow Nanorod for High-Performance Ethanol Gas Sensor. *ChemistrySelect* **2017**, *2*, 10918–10925. [CrossRef]
14. Liu, M.; Song, P.; Zhong, X.; Yang, Z.; Wang, Q. Facile synthesis of Au-decorated α-Fe_2O_3/rGO ternary hybrid structure nanocomposites for enhanced triethylamine gas-sensing properties. *J. Mater. Sci. Mater. Electron.* **2020**, *31*, 22713–22726. [CrossRef]
15. Zhao, S.; Shen, Y.; Yan, X.; Zhou, P.; Yin, Y.; Lu, R.; Han, C.; Cui, B.; Wei, D. Complex-surfactant-assisted hydrothermal synthesis of one-dimensional ZnO nanorods for high-performance ethanol gas sensor. *Sens. Actuators B* **2019**, *286*, 501–511. [CrossRef]
16. Wang, X.; Han, W.; Yang, J.; Jiang, B.; Cheng, P.; Wang, Y.; Sun, P.; Zhang, H.; Sun, Y.; Lu, G. Facet effect on diverse WO_3 morphologies and ideal work function for ppb-level triethylamine detection. *Sens. Actuators B* **2022**, *363*, 131849. [CrossRef]
17. Gao, Y.; Chen, D.; Hou, X.; Zhang, Y.; Yi, S.; Ji, H.; Wang, Y.; Yin, L.; Sun, J. Microwave-assisted synthesis of hierarchically porous Co_3O_4/rGO nanocomposite for low-temperature acetone detection. *J. Colloid Interface Sci.* **2021**, *594*, 690–701. [CrossRef]
18. Bai, H.; Guo, H.; Wang, J.; Dong, Y.; Liu, B.; Xie, Z.; Guo, F.; Chen, D.; Zhang, R.; Zheng, Y. A room-temperature NO_2 gas sensor based on CuO nanoflakes modified with rGO nanosheets. *Sens. Actuators B* **2021**, *337*, 129783. [CrossRef]
19. Meng, D.; Qiao, T.; Wang, G.; San, X.; Meng, F. One-step synthesis of rGO/V_2O_5 flower-like microsphere composites with enhanced trimethylamine sensing properties. *Mater. Lett.* **2021**, *299*, 130023. [CrossRef]
20. Meng, F.; Qi, T.; Zhang, J.; Zhu, H.; Yuan, Z.; Liu, C.; Qin, W.; Ding, M. MoS_2-Templated Porous Hollow MoO_3 Microspheres for Highly Selective Ammonia Sensing via a Lewis Acid-Base Interaction. *IEEE Trans. Ind. Electron.* **2022**, *69*, 960–970. [CrossRef]
21. Choi, J.; Seo, J.; Jeong, H.; Song, K.; Baeck, S.; Shim, S.; Qian, Y. Effects of Field-Effect and Schottky Heterostructure on p-Type Graphene-Based Gas Sensor Modified by n-Type In_2O_3 and Phenylenediamine. *Appl. Surf. Sci.* **2022**, *578*, 152025. [CrossRef]
22. Liang, T.; Kim, D.; Yoon, J.; Yu, Y. Rapid synthesis of rhombohedral In_2O_3 nanoparticles via a microwave-assisted hydrothermal pathway and their application for conductometric ethanol sensing. *Sens. Actuators B* **2021**, *346*, 130578. [CrossRef]
23. Zhang, K.; Qin, S.; Tang, P.; Feng, Y.; Li, D. Ultra-sensitive ethanol gas sensors based on nanosheet-assembled hierarchical ZnO-In_2O_3 heterostructures. *J. Hazard. Mater.* **2020**, *391*, 122191. [CrossRef] [PubMed]
24. Zhang, W.; Ding, S.; Zhang, Q.; Yi, H.; Liu, Z.; Shi, M.; Guan, R.; Yue, L. Rare earth element-doped porous In_2O_3 nanosheets for enhanced gas-sensing performance. *Rare Met.* **2021**, *40*, 1662–1668. [CrossRef]
25. Huang, X.; Tang, Z.; Tan, Z.; Sheng, S.; Zhao, Q. Hierarchical In_2O_3 nanostructures for improved formaldehyde: Sensing performance. *J. Mater. Sci. Mater. Electron.* **2021**, *32*, 11857–11864. [CrossRef]
26. Kohli, N.; Hastir, A.; Kumari, M.; Singh, R.C. Hydrothermally synthesized heterostructures of In_2O_3/MWCNT as acetone gas sensor. *Sens. Actuators A* **2020**, *314*, 112240. [CrossRef]
27. Gupta Chatterjee, S.; Chatterjee, S.; Ray, A.; Chakraborty, A. Graphene–metal oxide nanohybrids for toxic gas sensor: A review. *Sens. Actuators B* **2015**, *221*, 1170–1181. [CrossRef]
28. Liu, X.; Liu, J.; Liu, Q.; Chen, R.; Zhang, H.; Yu, J.; Song, D.; Li, J.; Zhang, M.; Wang, J. Template-free synthesis of rGO decorated hollow Co_3O_4 nano/microspheres for ethanol gas sensor. *Ceram. Int.* **2018**, *44*, 21091–21098. [CrossRef]
29. Ren, H.; Gu, C.; Joo, S.W.; Zhao, J.; Sun, Y.; Huang, J. Effective hydrogen gas sensor based on NiO@rGO nanocomposite. *Sens. Actuators B* **2018**, *266*, 506–513. [CrossRef]
30. de Lima, B.; Komorizono, A.; Ndiaye, A.; Bernardi, M.B.; Brunet, J.; Mastelaro, V. Tunning the Gas Sensing Properties of rGO with In_2O_3 Nanoparticles. *Surfaces* **2022**, *5*, 127–142. [CrossRef]
31. Dong, X.; Han, Q.; Kang, Y.; Li, H.; Huang, X.; Fang, Z.; Yuan, H.; Elzatahry, A.A.; Chi, Z.; Wu, G.; et al. Rational construction and triethylamine sensing performance of foam shaped α-MoO_3@SnS_2 nanosheets. *Chin. Chem. Lett.* **2022**, *33*, 567–572. [CrossRef]
32. Liu, J.-B.; Hu, J.-Y.; Liu, C.; Tan, Y.-M.; Peng, X.; Zhang, Y. Mechanically exfoliated MoS_2 nanosheets decorated with SnS_2 nanoparticles for high-stability gas sensors at room temperature. *Rare Met.* **2021**, *40*, 1536–1544. [CrossRef]
33. Meng, F.; Li, X.; Yuan, Z.; Lei, Y.; Qi, T.; Li, J. Ppb-Level Xylene Gas Sensors Based on Co_3O_4 Nanoparticle-Coated Reduced Graphene Oxide(rGO) Nanosheets Operating at Low Temperature. *IEEE Trans. Instrum. Meas.* **2021**, *70*, 1–10. [CrossRef]
34. Hung, C.; Dat, D.; Van Duy, N.; Van Quang, V.; Van Toan, N.; Van Hieu, N.; Hoa, N. Facile synthesis of ultrafine rGO/WO_3 nanowire nanocomposites for highly sensitive toxic NH_3 gas sensors. *Mater. Res. Bull.* **2020**, *125*, 110810. [CrossRef]
35. Lin, G.; Wang, H.; Lai, X.; Yang, R.; Zou, Y.; Wan, J.; Liu, D.; Jiang, H.; Hu, Y. Co_3O_4/N-doped RGO nanocomposites derived from MOFs and their highly enhanced gas sensing performance. *Sens. Actuators B* **2020**, *303*, 127219. [CrossRef]
36. Zuo, J.; Tavakoli, S.; Mathavakrishnan, D.; Ma, T.; Lim, M.; Rotondo, B.; Pauzauskie, P.; Pavinatto, F.; MacKenzie, D. Additive Manufacturing of a Flexible Carbon Monoxide Sensor Based on a SnO_2-Graphene Nanoink. *Chemosensors* **2020**, *8*, 36. [CrossRef]
37. Cao, P.; Cai, Y.; Pawar, D.; Navale, S.; Rao, C.; Han, S.; Xu, W.; Fang, M.; Liu, X.; Zeng, Y.; et al. Down to ppb level NO_2 detection by ZnO/rGO heterojunction based chemiresistive sensors. *Chem. Eng. J.* **2020**, *401*, 125491. [CrossRef]
38. Mishra, R.K.; Murali, G.; Kim, T.-H.; Kim, J.H.; Lim, Y.J.; Kim, B.-S.; Sahay, P.P.; Lee, S.H. Nanocube In_2O_3@RGO heterostructure based gas sensor for acetone and formaldehyde detection. *RSC Adv.* **2017**, *7*, 38714–38724. [CrossRef]
39. Hummers, W.; Offeman, R. Preparation of Graphitic Oxide. *J. Am. Chem. Soc.* **1958**, *80*, 1339. [CrossRef]
40. Yang, W.; Wan, P.; Zhou, X.; Hu, J.; Guan, Y.; Feng, L. Additive-Free Synthesis of In_2O_3 Cubes Embedded into Graphene Sheets and Their Enhanced NO_2 Sensing Performance at Room Temperature. *ACS Appl. Mater. Interfaces* **2014**, *6*, 21093–21100. [CrossRef]
41. White, J.; Bocarsly, A. Enhanced Carbon Dioxide Reduction Activity on Indium-Based Nanoparticles. *J. Electrochem. Soc.* **2016**, *163*, H410–H416. [CrossRef]

42. Zhang, Q.; Wang, S.; Fu, H.; Wang, Y.; Yu, K.; Wang, L. Facile Design and Hydrothermal Synthesis of In_2O_3 Nanocube Polycrystals with Superior Triethylamine Sensing Properties. *ACS Omega* **2020**, *5*, 11466–11472. [CrossRef] [PubMed]
43. Yang, W.; Feng, L.; He, S.; Liu, L.; Liu, S. Density Gradient Strategy for Preparation of Broken In_2O_3 Microtubes with Remarkably Selective Detection of Triethylamine Vapor. *ACS Appl. Mater. Interfaces* **2018**, *10*, 27131–27140. [CrossRef] [PubMed]
44. Rodwihok, C.; Wongratanaphisan, D.; Thi Ngo, Y.; Khandelwal, M.; Hur, S.; Chung, J. Effect of GO Additive in ZnO/rGO Nanocomposites with Enhanced Photosensitivity and Photocatalytic Activity. *Nanomaterials* **2019**, *9*, 1441. [CrossRef]
45. Yu, L.; Wan, G.; Peng, X.; Dou, Z.; Li, X.; Wang, K.; Lin, S.; Wang, G. Fabrication of carbon-coated NiO supported on graphene for high performance supercapacitors. *RSC Adv.* **2016**, *6*, 14199–14204. [CrossRef]
46. Yan, H.; Song, P.; Zhang, S.; Zhang, J.; Yang, Z.; Wang, Q. Au nanoparticles modified MoO_3 nanosheets with their enhanced properties for gas sensing. *Sens. Actuators B* **2016**, *236*, 201–207. [CrossRef]
47. Liu, B.; Zhu, Q.; Pan, Y.; Huang, F.; Tang, L.; Liu, C.; Cheng, Z.; Wang, P.; Ma, J.; Ding, M. Single-Atom Tailoring of Two-Dimensional Atomic Crystals Enables Highly Efficient Detection and Pattern Recognition of Chemical Vapors. *ACS Sens.* **2022**, *7*, 1533–1543. [CrossRef]
48. Chaloeipote, G.; Prathumwan, R.; Subannajui, K.; Wisitsoraat, A.; Wongchoosuk, C. 3D printed CuO semiconducting gas sensor for ammonia detection at room temperature. *Mater. Sci. Semicond. Process.* **2021**, *123*, 105546. [CrossRef]
49. Chang, X.; Li, K.; Qiao, X.; Xiong, Y.; Xia, F.; Xue, Q. ZIF-8 derived ZnO polyhedrons decorated with biomass derived nitrogen-doped porous carbon for enhanced acetone sensing. *Sens. Actuators B* **2021**, *330*, 129366. [CrossRef]
50. Qin, Q.; Zhu, X.; Zhang, X. Synthesis of α-Fe_2O_3 hollow spheres with rapid response and excellent selectivity towards acetone. *Mater. Sci. Eng. B* **2022**, *275*, 115482. [CrossRef]
51. Zhu, H.; Haidry, A.A.; Wang, Z.; Ji, Y. Improved acetone sensing characteristics of TiO_2 nanobelts with Ag modification. *J. Alloys Comp.* **2021**, *887*, 161312. [CrossRef]
52. Wang, G.; Fu, Z.; Wang, T.; Lei, W.; Sun, P.; Sui, Y.; Zou, B. A rational design of hollow nanocages Ag@CuO-TiO_2 for enhanced acetone sensing performance. *Sens. Actuators B* **2019**, *295*, 70–78. [CrossRef]
53. Wang, Q.; Cheng, X.; Wang, Y.; Yang, Y.; Su, Q.; Li, J.; An, B.; Luo, Y.; Wu, Z.; Xie, E. Sea urchins-like WO_3 as a material for resistive acetone gas sensors. *Sens. Actuators B* **2022**, *355*, 131262. [CrossRef]
54. Cheng, P.; Wang, C.; Wang, Y.; Xu, L.; Dang, F.; Lv, L.; Li, X. Enhanced acetone sensing properties based on in situ growth SnO_2 nanotube arrays. *Nanotechnology* **2021**, *32*, 245503. [CrossRef]
55. Ueda, T.; Boehme, I.; Hyodo, T.; Shimizu, Y.; Weimar, U.; Barsan, N. Enhanced NO_2-Sensing Properties of Au-Loaded Porous In_2O_3 Gas Sensors at Low Operating Temperatures. *Chemosensors* **2020**, *8*, 72. [CrossRef]
56. Gu, D.; Liu, W.; Wang, J.; Yu, J.; Zhang, J.; Huang, B.; Rumyantseva, M.N.; Li, X. Au Functionalized SnS_2 Nanosheets Based Chemiresistive NO_2 Sensors. *Chemosensors* **2022**, *10*, 165. [CrossRef]
57. Lv, L.; Cheng, P.; Zhang, Y.; Zhang, Y.; Lei, Z.; Wang, Y.; Xu, L.; Weng, Z.; Li, C. Ultra-high response acetone gas sensor based on $ZnFe_2O_4$ pleated hollow microspheres prepared by green NaCl template. *Sens. Actuators B* **2022**, *358*, 131490. [CrossRef]
58. Nakate, U.T.; Yu, Y.T.; Park, S. High performance acetaldehyde gas sensor based on p-n heterojunction interface of NiO nanosheets and WO_3 nanorods. *Sens. Actuators B* **2021**, *344*, 130264. [CrossRef]
59. Wan, K.; Wang, D.; Wang, F.; Li, H.; Xu, J.; Wang, X.; Yang, J. Hierarchical In_2O_3@SnO_2 Core-Shell Nanofiber for High Efficiency Formaldehyde Detection. *ACS Appl. Mater. Interfaces* **2019**, *11*, 45214–45225. [CrossRef]
60. Qiu, Y.; Wang, Y.; Song, C. Facile synthesis of $W_{18}O_{49}$/Graphene nanocomposites for highly sensitive ethanol gas sensors. *Colloids Surf. A* **2021**, *616*, 126300. [CrossRef]
61. Cheng, M.; Wu, Z.; Liu, G.; Zhao, L.; Gao, Y.; Zhang, B.; Liu, F.; Yan, X.; Liang, X.; Sun, P.; et al. Highly sensitive sensors based on quasi-2D rGO/SnS_2 hybrid for rapid detection of NO_2 gas. *Sens. Actuators B* **2019**, *291*, 216–225. [CrossRef]
62. Bai, S.; Tian, K.; Zhao, Y.; Feng, Y.; Luo, R.; Li, D.; Chen, A. ZnO/$BiFeO_3$ heterojunction interface modulation and rGO modification for detection of triethylamine. *J. Mater. Chem. C* **2022**, *10*, 8015–8023. [CrossRef]

chemosensors

Article

Influence of Nickel Doping on Ultrahigh Toluene Sensing Performance of Core-Shell ZnO Microsphere Gas Sensor

Zhenhua Li [1], Sijia Li [1], Zijian Song [1], Xueli Yang [1,*], Ziyan Wang [1], Hao Zhang [2], Lanlan Guo [3], Caixuan Sun [1], Hongyan Liu [1], Junkai Shao [1], Yehong Cheng [4,*] and Guofeng Pan [1]

[1] Tianjin Key Laboratory of Electronic Materials and Devices, School of Electronics and Information Engineering, Hebei University of Technology, 5340 Xiping Road, Beichen District, Tianjin 300401, China
[2] Tianjin Key Laboratory of Materials Laminating Fabrication and Interface Control Technology, School of Materials Science and Engineering, Hebei University of Technology, Tianjin 300130, China
[3] School of Physics and Electronic Information Engineering, Henan Polytechnic University, Jiaozuo 454003, China
[4] School of Mechanical Engineering, Hebei University of Technology, Tianjin 300401, China
* Correspondence: xlyang@hebut.edu.cn (X.Y.); chengyehong@hebut.edu.cn (Y.C.)

Abstract: As a volatile organic compound, toluene is extremely harmful to the environment and human health. In this work, through a simple one-step solvothermal method, Ni-doped ZnO sensitive materials (0.5, 1, and 2 at% Ni-doped ZnO) with a core-shell morphology were synthesized for the first time for toluene gas detection. The sensing test results showed that the sensor based on 1 at% Ni-doped ZnO exhibited the best toluene sensing performance. The response was up to 210 to 100 ppm toluene at 325 °C. The sensor exhibited high selectivity, fast response/recovery characteristics (2/77 s), and low detection limit (500 ppb, 3.5). Furthermore, we carried out molecular-level research on the sensitive material prepared in this experiment by various characterization methods. The SEM characterization results showed that ZnO and Ni-doped ZnO possessed the core-shell morphology, and the average grain size decreased with the increase in the Ni doping content. The UV–Vis test showed that the band gap of ZnO became smaller with the increase in the Ni doping amount. The enhanced toluene sensing performance of 1 at% Ni-doped ZnO could be ascribed to the structural sensitization and Ni doping sensitization, which are discussed in detail in the sensing mechanism section.

Keywords: Ni doping; ZnO; gas sensor; toluene; core-shell structure

Citation: Li, Z.; Li, S.; Song, Z.; Yang, X.; Wang, Z.; Zhang, H.; Guo, L.; Sun, C.; Liu, H.; Shao, J.; et al. Influence of Nickel Doping on Ultrahigh Toluene Sensing Performance of Core-Shell ZnO Microsphere Gas Sensor. *Chemosensors* **2022**, *10*, 327. https://doi.org/10.3390/chemosensors10080327

Academic Editors: Fanli Meng, Zhenyu Yuan and Dan Meng

Received: 14 July 2022
Accepted: 10 August 2022
Published: 12 August 2022

1. Introduction

With the development of modern industry, various kinds of poisonous and harmful gases have been produced, which will cause harm to human health and the environment. In particular, volatile organic compounds (VOCs) such as ethanol, acetone, formaldehyde, and toluene, etc., which widely exist in our surroundings. Among them, toluene as a typical aromatic compound is a kind of colorless gas with a special fragrance, and is widely used in many aspects of production and life such as raw materials in the chemical industry, the production of dyes, pharmaceuticals, pesticides, explosives, and other fine chemicals. In addition, it can be used as a food preservative in major food production factories [1]. Meanwhile, toluene is a flammable and harmful gas and can produce an anesthetic effect on the central nervous system, causing great and irreparable harm to the human body. Therefore, toluene has been identified as a highly toxic carcinogen by the International Center for Research on Cancer [2]. Due to the harmfulness of toluene, it is very necessary and urgent to develop a high performance toluene gas sensor to monitor toluene in real-time to ensure human health and environmental safety.

Gas sensors based on semiconductor metal oxides have attracted extensive attention in recent years due to their advantages of high sensitivity, easy fabrication, low cost, and small size. As the most promising sensing material, gas sensors based on semiconductor metal oxides have become a research hotspot. Up to now, many metal oxide semiconductors have been developed and used as gas sensing materials, for instance, ZnO [3,4], In_2O_3 [5–7], SnO_2 [8–10], WO_3 [11–13], NiO [14,15], CuO [16,17], etc. Among the n-type and p-type single semiconductor oxides, ZnO is the most widely studied gas sensing material as ZnO is an n-type semiconductor oxide with a wide band gap energy (3.37 eV) and a large excite binding energy (60 MeV) [18,19]. The physical and chemical properties of ZnO are stable, and the electrical property of ZnO is tunable with transition metal doping [20]. Currently, different morphologies of ZnO have been reported to detect different gases [21–27] including nanoparticles [28], one-dimensional nanowires [29], nanorods and nanofibers, two-dimensional nanosheets [30], nanoribbons and nanorings as well as three-dimensional flower-like structures [31], core-shell structures, and hollow spheres. Studies have shown that the hierarchical structure of ZnO enables better gas sensing properties due to its larger specific surface area, porosity, and permeability.

In addition to constructing the hierarchical structure of ZnO, it is also an effective way to improve the gas sensitive performance by doping. There are many types of doping substances including metal oxides, precious metals, transition metals, carbon-based materials, and polymers [32–36]. Since transition metal ions can be doped into the semiconductor lattice, the lattice distortion of sensitive materials generates a large number of defects and vacancies. The increase in the defect concentration on the surface can improve the adsorption of oxygen anions on the surface of the material, thereby improving the gas-sensing response value of the sensing material. Thus far, many studies have reported that through transition metal doping, improved gas sensing performance can be achieved. Transition metal doping elements including Co, Sn, Fe, and Ni [37–40] etc. have been reported to tune the sensing performance of ZnO. The catalytic properties, surface properties, and charge carriers of pristine semiconductor materials can be improved by doping transition metal elements. Guo [41] et al. successfully synthesized Fe-doped ZnO/rGO nanocomposites via a simple hydrothermal process. The sensor exhibited a response of 12.7 to 5 ppm formaldehyde at 120 °C. Compared with pure ZnO, the response of the doped sensing material was about 3 times higher. The reason for the enhanced sensing performance is that the doping of Fe changes the band gap, making the band gap of ZnO smaller and generating a large number of oxygen vacancies.

Based on that demonstrated above, the gas sensing performance of semiconductor metal oxides can be improved by constructing a hierarchical structure to adjust the surface area and doping transition metal elements to tune the electrical and catalytic properties [42–44]. To our best knowledge, fabricating a core-shell Ni doped ZnO (Ni–ZnO) to achieve an enhanced sensing performance of toluene has rarely been reported.

Here, materials with different ratios of 0, 0.5, 1, and 2 at% Ni–ZnO core-shell morphologies were successfully prepared by a simple solvothermal method and used for toluene gas detection. Gas sensing measurements showed that out of all the four gas sensors, the sensor based on 1 at% Ni–ZnO had the highest toluene response of 210 (100 ppm, 325 °C), which was almost seven times higher than pure ZnO (33.6, 375 °C). Moreover, the sensor also had high selectivity, low detection limit, and fast response/recovery properties. The sensing mechanism of the enhanced sensing properties was analyzed in detail.

2. Materials and Methods

2.1. Synthesis of the Pure ZnO and Ni–ZnO Core-Shell Spheres

First, 20 mL of deionized water and 1 mmol Ni $(CH_3COO)_2{\cdot}4H_2O$ were added into a 50 mL beaker and stirred for 20 min. Then, 7 mL of glycerol and 30 mL of isopropanol were added into a 50 mL beaker with vigorous stirring, and 0.45 mmol Zn $(CH_3COO)_2{\cdot}2H_2O$ was dissolved into the above mixed solution with stirring for 15 min, named as solution A. Then 0, 45, 90, and 180 µL of the above prepared Ni $(CH_3COO)_2{\cdot}4H_2O$ solution were added into solution A, respectively, and stirring was continued for another 20 min. Then, the four solutions were placed in an oven and kept at 180 °C for 24 h. After the reaction was completed, the temperature was naturally cooled to room temperature, and the product was separated by centrifugation, alternately washed with anhydrous ethanol and deionized water, and then dried at 80 °C for 10 h. The dried product was calcined at 500 °C for 2 h (heating rate, 2 °C·min^{-1}). Finally, pure ZnO and 0.5, 1, and 2 at% Ni–ZnO core-shell products were obtained.

2.2. Characterization

In this paper, an X-ray powder diffractometer (XRD; Bruker D8 Discover, Billerica, MA, USA) was used to investigate the composition and crystallinity of pure ZnO and different proportions of the Ni–ZnO samples. Diffraction analysis of sensitive materials was carried out using Cu Kα1 rays (λ = 1.5406 Å) in the range of 2θ from 20 to 80°, and the scanning step was 10°·min^{-1}. The microscopic topography of sensitive materials can be analyzed and studied using scanning electron microscopy (SEM; ZEISS Sigma 500, Jena, Germany). More information about the microscopic morphology and the lattice size of sensitive materials can be obtained by transmission electron microscopy and high-resolution transmission electron microscopy. In addition, energy dispersive X-ray spectrometry (EDS) enables the detection of element mapping distributions. The chemical element analysis was obtained from the X-ray photoelectron spectroscopy (XPS; Escalab 250 XI, Waltham, MA, USA) measurements, and the surface area and pore size distribution were calculated by the Brunauer–Emmett–Teller (BET) and Barrett–Joyner–Halenda (BJH) methods using the nitrogen adsorption–desorption isotherm test (Micromeritics ASAP2000 system, USA). The absorbance and band gap of the sample were measured by using a UV–Visible spectrophotometer (UV–Vis; Lambda 1050+, Waltham, MA, USA) between 250 and 800 nm.

2.3. Fabrication and Measurement Process of the Gas Sensors

As schematically shown in Figure S1a,b, first, the as-synthesized sample (3–5 mg) and an appropriate amount (0.3–0.5 mL) of deionized water were mixed to form a slurry. Then, the slurry was evenly coated on the surface of the ceramic tube (outer diameter 1.2 mm, inner diameter 0.8 mm, length 4 mm) using a small brush. After drying in air for 15 min, the ceramic tube was placed in a Muffle furnace at 400 °C for 120 min with a heating rate of 2 °C min^{-1} to enhance the stability during the test process. After that, a Ni Cr alloy wire, which provides the operating temperature by tuning the constant current, is inserted in the sintered ceramic tube. Finally, the sensor was welded on a six-legged sensor base, and the gas sensing performance was tested after aging for 24 h with 100 mA. Figure S1c is schematically shown as the static gas sensing test system. A constant-current power (Gwinstek GPD-3303S, New Taipei City, Taiwan) was used to supply a constant current for the Ni–Cr alloy wire to control the operating temperature. Two glass cavities with a volume of 1 L were used to hold air and test gas, respectively. A digital multimeter (Fluke 8846a, Everett, WA, USA) and a computer was used to record and monitor the resistance of the gas sensor. During the sensing test process, the sensor was alternately placed in 1 L of fresh air in the glass cavity for a few minutes to obtain a stabilized resistance value recorded as Ra. Then, the gas sensor device was transferred to another glass cavity with 1 L, which was filled with a mixture of fresh air and target gas, the stabilized resistance value in the target gas was recorded as Rg. The sensing test was carried out in a laboratory environment (25 °C, 30 RH%). The volatile organic gas used was an analytical grade liquid, and injected to the glass cavity by a microinjector. The gas preparation process is detailed in the Supplementary Materials. The gas sensing response was defined as S = Ra/Rg, and response/recovery time was the time taken by the sensor to reach 90% of the overall resistance change.

3. Results

The XRD pattern of the pure ZnO, 0.5, 1, and 2 at% Ni–ZnO core-shell spheres are shown in Figure 1a. The diffraction peaks appeared at 34.42°, 36.25°, 47.53°, 56.60°, 62.85°, 68.17°, 69.29°, which corresponded to the (100), (002), (101), (102), (110), (103), and (112) crystal planes of the hexagonal wurtzite phase ZnO (JCPDS No. 36-1451) [45]. However, no peak matched with Ni or NiO in the Ni–ZnO samples. No other diffraction peaks corresponding to the impurities were detected. Figure 1b is the high-resolution peak of the (101) plane of ZnO compared to pure ZnO, and the diffraction peak of Ni–ZnO shifted to a higher angle direction. The shift of 2 at% Ni–ZnO sample was the most obvious, which indicated that Ni^{2+} was incorporated in the ZnO crystal lattice due to the radius of Ni^{2+} being smaller than Zn^{2+}.

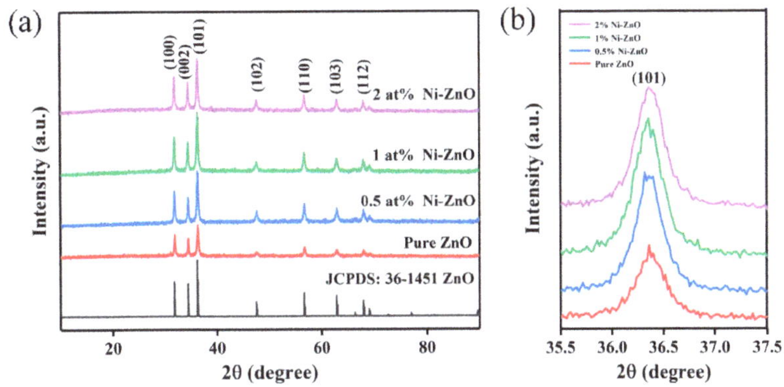

Figure 1. (**a**) The XRD patterns of the pure and Ni–ZnO, (**b**) enlarged spectra of the XRD patterns of the Ni–ZnO.

Figure 2a–d shows the SEM images of the 0, 0.5, 1, and 2 at% Ni–ZnO sample, respectively. The pure ZnO, 0.5, and 1 at% Ni–ZnO samples showed a core-shell spherical hierarchical structure, and there was no obvious core-shell structure observed for 2 at% Ni–ZnO; the morphology was a simple spherical structure. It was found from the SEM images that with the increase in the Ni doping content, the surface of the ZnO microspheres became more and more coarse and the average particle size of the spheres decreased. The average particle size was 15.10, 9.42, 5.62, and 4.66 μm for pure ZnO, 0.5, 1 and 2 at% Ni–ZnO spheres, which suggests that the addition of Ni ions can inhibit the growth of ZnO particles in the hydrothermal reaction process. In addition, the inserted magnified SEM images showed that the core of the core-shell spheres was very large and next to the shell. Proper Ni doping, a rough surface, and loose shell are favorable for gas adsorption and diffusion, with a high gas sensing performance, which will be verified in later sections.

Figure 3a,b are TEM images of different magnifications, whose shape and size are consistent with the SEM characterization results. The inset figure is the selected area electron diffraction (SAED) image and shows that the 1 at% Ni–ZnO sample had a polycrystalline structure. Figure 3c is the HRTEM images of the 1 at% Ni–ZnO sample. The lattice spacings were 0.286 and 0.249 nm, corresponding to the (100) and (101) crystal planes of ZnO [46,47]. Interplanar spacing corresponding to NiO was not found, which is consistent with the results of the XRD. From the element mapping test results in Figure 3d–g, the three elements (Ni, O, and Zn) were evenly distributed in the 1 at% Ni–ZnO sample.

The specific surface area of the pure ZnO and Ni–ZnO samples was determined by the nitrogen adsorption/desorption isotherm. As illustrated in Figure 4a–e, the adsorption–desorption curve is an iv-type isotherm belonging to the H3 hysteresis loop, indicating that Ni–ZnO has a microporous structure [48].The specific surface area increased with the increase in the Ni doping amount, especially for 1 and 2 at% Ni–ZnO, the specific surface areas were 20.43 to 26.41, 37.54 and 40.14 $m^2 \cdot g^{-1}$ for the pure ZnO, 0.5, 1, and 2 at% Ni–ZnO. The results showed that with the proper amount of Ni doping, the specific area of ZnO could increase a lot; when the doping amount reached a certain value, the specific area did not increase too much. A large specific area could supply more surface-active sites for surface adsorbed oxygen species, thus improving the recognition function of the sensing material and contributing to the high gas sensing performance.

Moreover, the inset of Figure 4 shows the nonlocal density functional theory (NLDFT) and grand canonical Monte Carlo (GCMC) pore size distribution of the Ni–ZnO, and through analysis, it was found that the pore sizes of the pure ZnO and 0.5 at%, 1 at%, and 2 at% Ni–ZnO were 46.8, 58.08, 32.72, and 32.71 nm, respectively. Appropriate Ni ion doping can increase the size of the pore size, and combined with the gas sensing data, it can be inferred that the pore size of the sensitive material in this experiment had little effect on the gas sensing performance. The increase in the specific surface area is one of the main factors affecting the gas sensing performance.

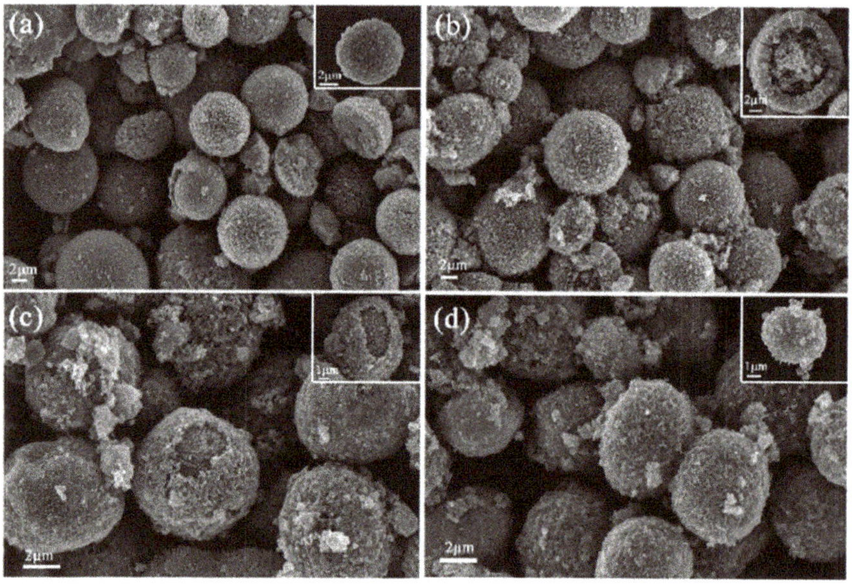

Figure 2. The SEM images of the (**a**) pure (**b**) 0.5 at% (**c**) 1 at%, and (**d**) 2 at% Ni–ZnO.

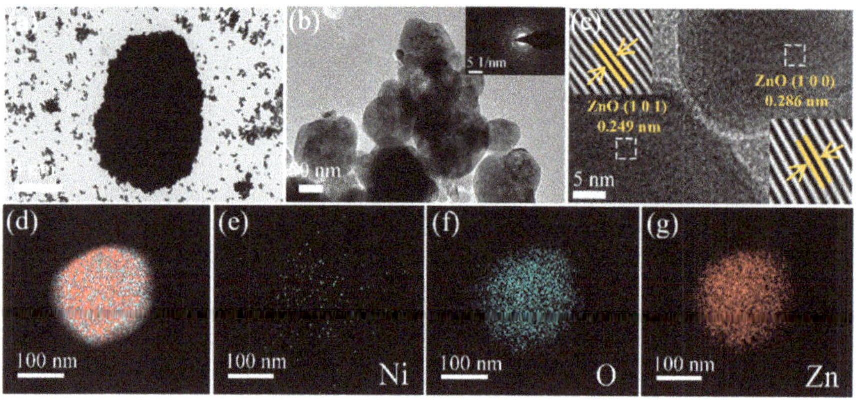

Figure 3. The TEM image of 1 at% Ni–ZnO (**a**,**b**). (**c**) HRTEM images of 1 at% Ni–ZnO. (**d**) The overlapped elemental mapping of Ni, O, Zn in the 1 at% Ni–ZnO sample, and (**e**–**g**) elemental mapping of Ni, O, and Zn.

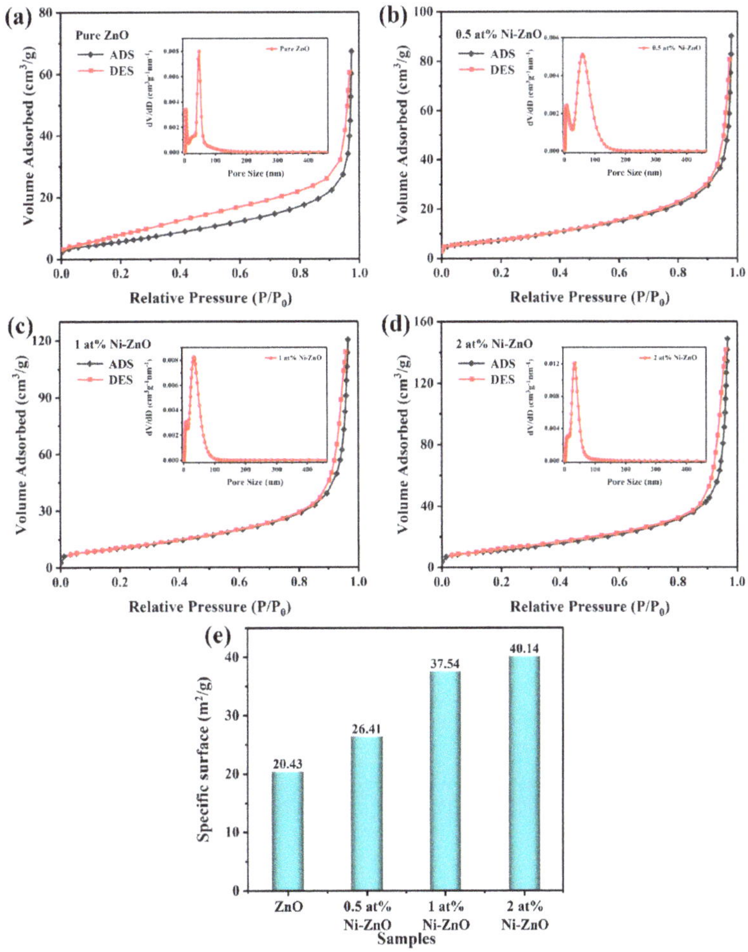

Figure 4. The nitrogen absorption–desorption isotherms and pore size distribution curves (inset) of the (**a**) pure ZnO and (**b**) 0.5 at% (**c**) 1 at%, and (**d**) 2 at% Ni–ZnO. (**e**) Specific surface area of the samples.

The XPS measurement was used to analyze the chemical composition and states of the elements in the pure ZnO and Ni–ZnO samples. Figure 5a exhibits the full spectrum of the ZnO and Ni–ZnO samples, where the full spectrum contained the peaks of Zn, O and Ni, proving the presence of Zn, O, and Ni in the sample. In Figure 5b, the binding energy around 1021.20 and 1044.28 eV corresponded to Zn $2p_{3/2}$ and Zn $2p_{1/2}$, respectively, proving that Zn was in the form of Zn^{2+} [49]. Figure 5c–e shows that the Ni $2p_{3/2}$ orbital can be separated into three peaks of about 854.1, 856.3, and 861.2 eV, corresponding to the Ni^{2+}, Ni^{3+} and satellite peaks, respectively [50], indicating the existence of Ni ions in the ZnO crystals.

Figure 6a–d is the XPS spectra of the O 1s of all samples, the peak shape is asymmetric, and can be allocated into three different oxygen components: chemisorbed gen (O_C), oxygen vacancy (O_V), and lattice oxygen (O_L), which correspond to the peaks at around 532.2 ± 0.5 eV, 531.0 ± 0.5 eV, and 529.5 ± 0.5 eV [51,52], respectively. Table 1 shows the relative proportion of the three oxygen components of O 1s in all of the samples. The relative ratio of O_C did not change a lot, but with Ni doping, the ratio of O_L decreased, while relatively, the proportion of O_V increased, and 1 at% Ni–ZnO showed the highest proportion of O_V with 22.8%. Ni doping led to the formation of an oxygen vacancy in ZnO, and more O_V concentration distributed to more surface-active sites, thus benefiting more surface negatively charged oxygen species.

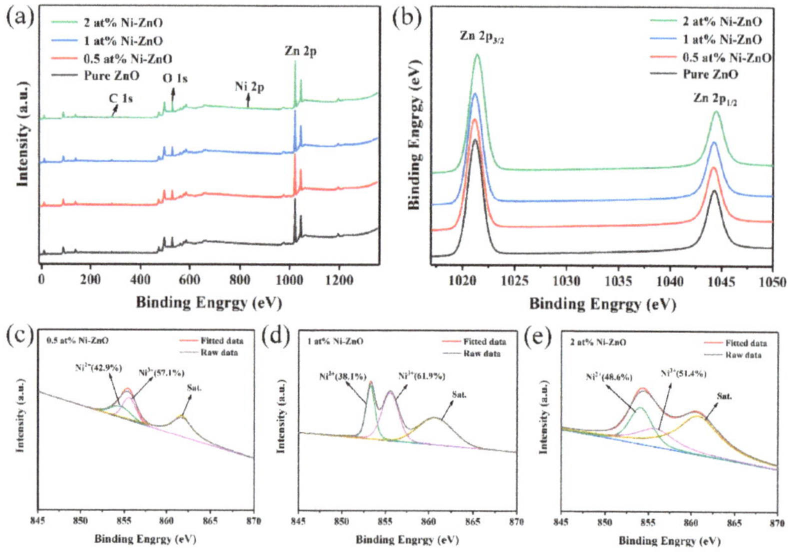

Figure 5. The XPS spectra of the pure and Ni–ZnO for (**a**) the full scan, (**b**) Zn 2p, (**c–e**) Ni $2p_{3/2}$ of 0.5, 1, 2 at% Ni–ZnO.

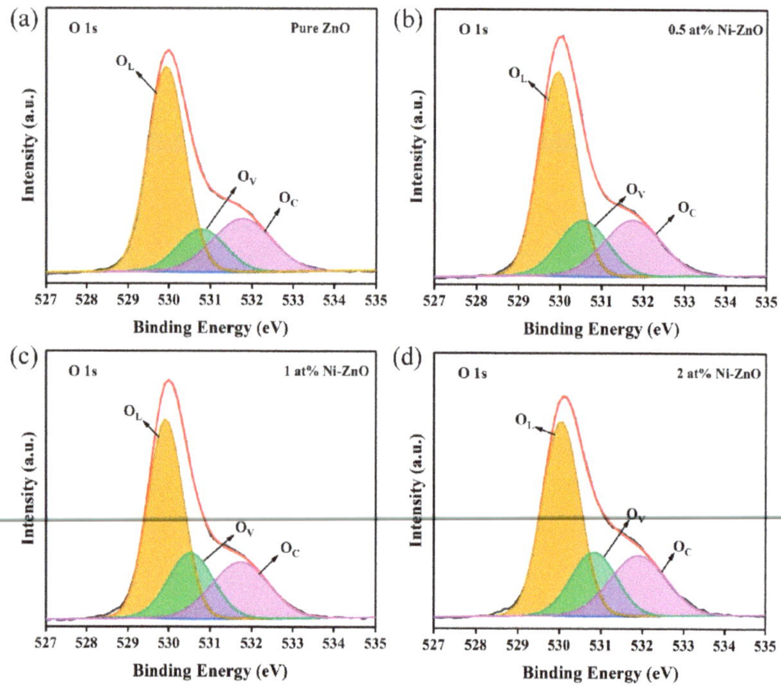

Figure 6. The XPS spectra of O 1s for (**a–d**) pure, 0.5, 1, and 2 at% Ni–ZnO.

Table 1. The relative percentages of O 1s in the XPS spectra ZnO, 0.5 at%, 1 at%, and 2 at% Ni–ZnO.

Species	Peak (eV)	ZnO	0.5 at% Ni–ZnO	1 at% Ni–ZnO	2 at% Ni–ZnO
O_L	529.9	60.0%	55.9%	53.0%	53.7%
O_V	530.8	15.6%	19.9%	22.8%	21.0%
O_C	531.7	24.4%	24.2%	24.2%	25.3%

Gas Sensing Characteristics

The operating temperature can affect the carrier concentration and gas adsorption and desorption of the sensor resistance and gas response [53,54]. Response of the four sensors to 100 ppm toluene at different temperatures (Figure 7a) showed that the response values first increased and then decreased. The maximum response values of the four sensors reached at 375 °C (pure ZnO), 350 °C (0.5 at% Ni–ZnO), and 325 °C (1 and 2 at% Ni–ZnO), and the response values were 33.6, 111.4, 210.0 and 23.1, respectively. The 1 at% Ni–ZnO sample had the highest response of 210.0, which was seven times higher than the pure ZnO. With Ni doping, the operating temperature of 1 at% Ni–ZnO was significantly decreased, which could be explained by the change in the band gap, as illustrated in Figure S2. After Ni doping, the band gap reduced from 3.05 to 2.89 eV, and free electrons were easier to release from the conduction band. Figure 7b displays the responses of the four sensors to the six tested target gases (toluene, ethanol, acetone, methanol, formaldehyde, and xylene) with 100 ppm at their optimal operating temperatures (ZnO, 375 °C; 0.5 at% Ni–ZnO, 350 °C; 1 and 2 at% Ni–ZnO, 325 °C). Obviously, the four sensors had the highest response to toluene. Compared with the other tested gases, the 1 at% Ni–ZnO sensor showed the best toluene sensing performance, and the response to toluene and other gases was 210.0, 52.0, 3.4, 9.4, 3.1, and 1.5. The toluene response was 4–140 times higher than other measured gases, indicating that the sensor had excellent selectivity for toluene.

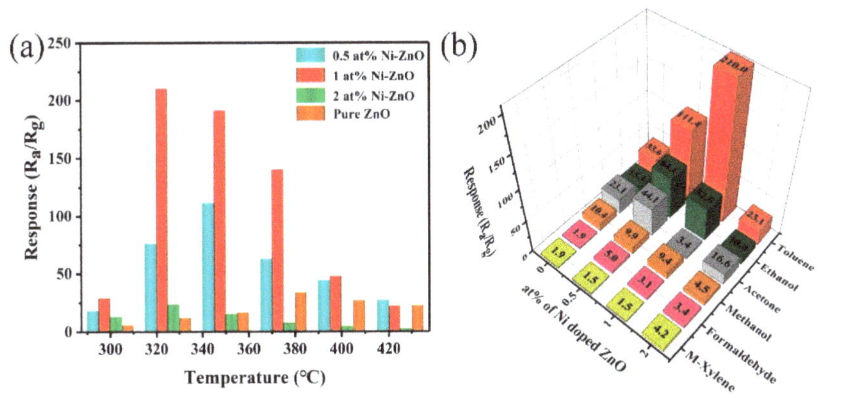

Figure 7. (**a**) The response of the four gas sensors to 100 ppm toluene at different temperatures. (**b**) The selectivity of the four gas sensors to 100 ppm different VOC gases at their optimum operating temperature.

Figure 8a,b shows the real-time response curves of the four sensors at 375 °C (pure ZnO) and 325 °C (Ni–ZnO samples) with different concentrations of toluene. The response value increased with the increase in toluene concentration. The sensor response based on 1 at% Ni–ZnO increased significantly with increasing toluene concentration. The corresponding linear relationship between the toluene response and concentration is listed in Figure 8c,d. It can be observed that the linearity of the sensor was very good in both the low-concentration range and the high-concentration range. The response increased almost linearly with the toluene concentration, and 1 at% Ni–ZnO had the highest sensitivity.

Furthermore, the sensor based on 1 at% Ni–ZnO had a low detection limit of 0.5 ppm with a response value of 3.5, which makes it promising in practical application.

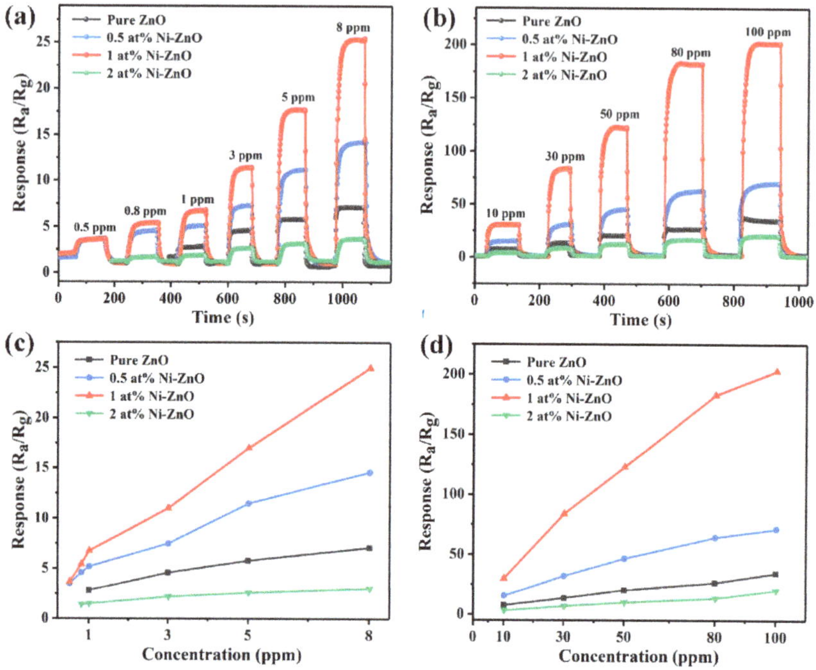

Figure 8. (**a**,**b**) The dynamic response curves of the four sensors to toluene with different concentrations at 375 °C (pure ZnO) and 325 °C (Ni–ZnO samples). (**c**,**d**) The response of the pure, 0.5, 1, and 2 at% Ni–ZnO sensors to (0.5–100 ppm) toluene.

The response and recovery characteristics are also important parameters of gas sensors. Thus, the response and recovery transients of the sensor based on 1 at% Ni–ZnO to 100 ppm toluene at 325 °C are shown in Figure 9a. The response time and recovery time of the sensor based on 1 at% Ni–ZnO are 2 and 77 s, respectively, illustrating that the sensor had a relatively fast response and recovery property. Table 2 is a comparison of the toluene sensing performance of this work and the sensors reported in the literature [55–59]. The operating temperature was relatively higher than that in the literature, but the response and recovery time were fast and the toluene response was much higher than the reported literature.

Figure 9b is the resistance in air as a function of operating temperature. With the increase in Ni doping, the resistance of the sensor increased significantly. The resistance versus temperature plot of the sensors showed an abnormal PTCR (positive temperature coefficient of resistance) behavior during 300–350 °C. Based on the reported literature [60], this phenomenon was caused by Ni doping, which introduced defects in ZnO, thus forming oxygen vacancy-like defects in this temperature region.

Figure 9c is a seven-cycle test curve of the 1 at% Ni–ZnO sample to 100 ppm toluene at 325 °C. The resistance in air and toluene was within the allowable fluctuation range, proving that the sensor had good repeatability. The long-term stability was also studied in this work; the sensing response to 100 ppm toluene in the air for one month is shown in Figure 9d. During the test days, the response values varied and slightly decreased but did not show an obvious fluctuation. The results indicate that the 1 at% Ni–ZnO gas sensor has good long-term stability.

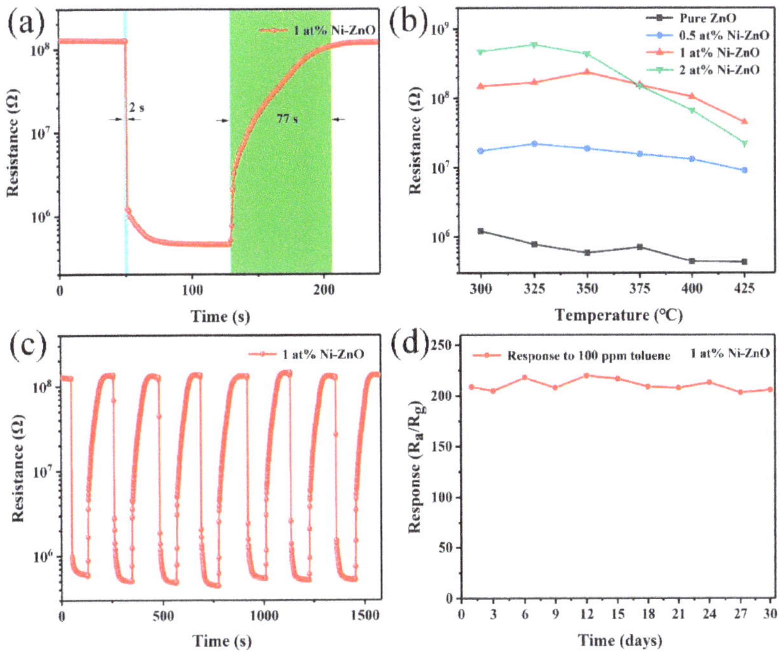

Figure 9. (**a**) The response/recovery time of 1 at% Ni–ZnO gas sensor to 100 ppm toluene at 325 °C. (**b**) Resistance in the air of the four gas sensors at different operating temperatures. (**c**) Seven reversible cycles of 1 at% Ni–ZnO gas sensors. (**d**) The stability testing curves of 1 at% Ni–ZnO gas sensor to 100 ppm toluene in 30 days.

Table 2. A comparison of the toluene gas sensing performance of this work and other sensing material in previously reported works.

Sensing Material	Conc. in ppm	Operating Temperature (°C)	Response	Res/Rec (s)	LOD (ppm)	Ref.
WO_3–SnO_2	10	340	5.6	14.5/406	-	[55]
SnO_2–ZnO	50	200	7.5	90/150	0.1	[56]
$NiGa_2O_4$–NiO	100	230	12.7	60/70	0.5	[57]
NiO/$NiGa_2O_4$	5	200	10.54	600/-	-	[58]
CuO–SnO_2	75	400	540	100/36	10	[59]
Ni–ZnO	100	320	210	2/77	0.5	Present work

4. Discussion

The gas sensing mechanism can be explained by the interaction of multiple factors. In this work, toluene sensing mechanisms with exceptional response are discussed in detail. The gas sensing mechanism has been widely explored in many studies [61,62], and the extensively accepted theory is the chemical reaction between the target gas and chemically adsorbed oxygen species (O_2^- (T < 100 °C), O^- (100 °C < T < 300 °C), and O^{2-} (T > 300 °C)) [63,64]. When redox reactions take place on the surface of the semiconductors, charge transfer will occur during this process, which changes the resistance of the gas sensor.

As illustrated in Figure 10, when the Ni–ZnO gas sensor is exposed to air, O_2 molecules will capture electrons in the conduction band of the material to form negatively charged oxygen species on the surface of the Ni–ZnO material, and a depletion layer will form on the surface at the same time, leading to a high resistance of the sensor in air. Then, when the sensor is exposed to a reducing gas (toluene) environment, the chemically adsorbed oxygen can react with the target gas molecules on the surface of the sensing material, the electron depletion layer becomes smaller. As a result, electrons that are released back into the conduction band induce lower resistance values in reducing gases. The relevant reactions are seen in Equations (1)–(3) (T > 300 °C) [65]:

$$O_2 \rightarrow O_{2(ads)} \tag{1}$$

$$O_{2(ads)} + 4e \rightarrow 2O^{2-} \tag{2}$$

$$C_7H_8 + 18O^{2-} \rightarrow 4H_2O + 7CO_2 + 36e^- \tag{3}$$

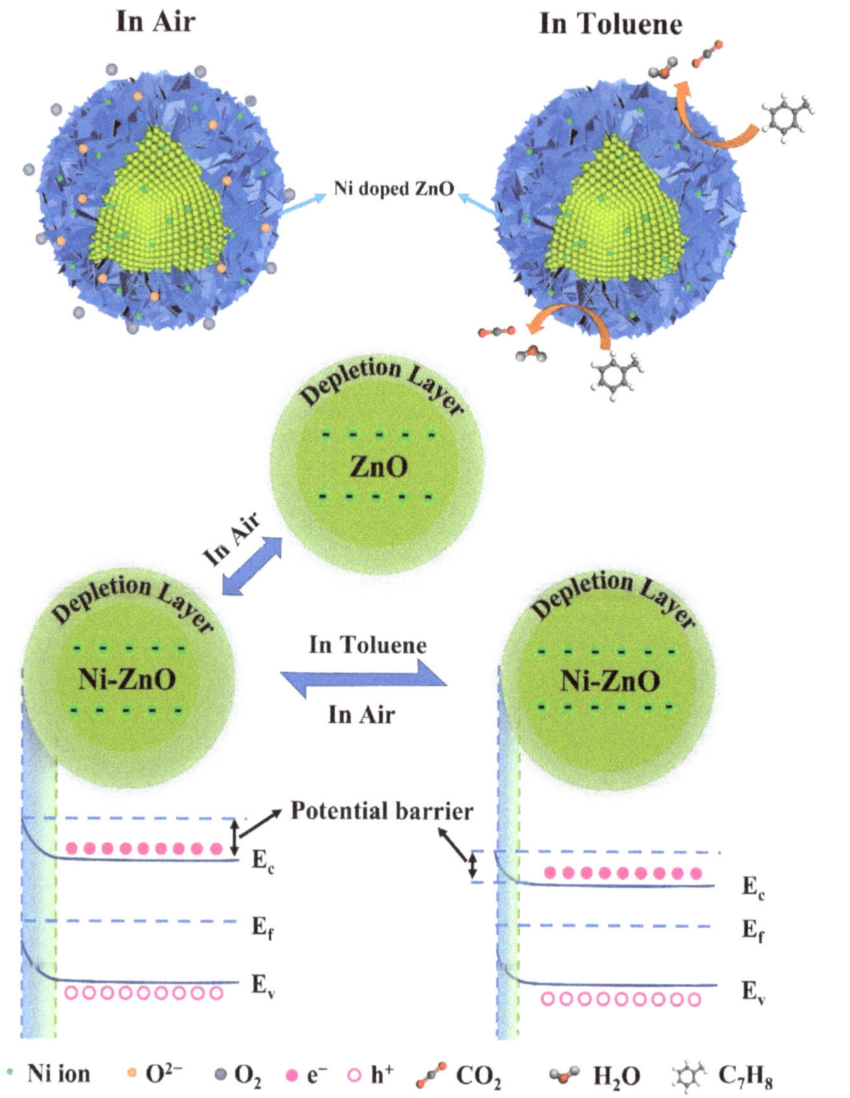

Figure 10. A schematic diagram of the sensing mechanism.

The enhanced sensing performance of the 1 at% Ni–ZnO-based gas sensor can be attributed to the following two aspects including structural sensitization and modified sensitization. The first is structural sensitization. After Ni doping, the hierarchical microstructure of ZnO changed, and the surface of the Ni–ZnO core-shell spheres became rougher. This loose structure is very favorable for the diffusion of gas molecules, which can not only penetrate into the shell at the outer surface of the material, but may also further diffuse to the surface of the core. Therefore, the sensing material has a higher sensing performance [66]. Furthermore, the thickness of the shell of the sensing material becomes thinner with increasing Ni content. According to the literature [67,68], when the thickness of the shell layer is close to the Debye length, the gas sensing material has the best gas sensing performance. Finally, the size of the topography of the gas sensing material is also affected by Ni, resulting in a smaller morphology size. An appropriate reduction in the morphology size can lead to an increase in the specific surface area, which can adsorb more oxygen and toluene molecules and provide more active sites [69].

Modified sensitization includes the following points. First, the specific surface area increased with Ni doping, so the recognition function of the sensing material will be improved. Second, it can be seen from Table 2 that Ni doping increased the relative proportion of oxygen vacancy in ZnO. The oxygen vacancy improved from 15.6% to 22.8 for 1 at% Ni–ZnO, which means that more oxygen species will adsorb on the surface of 1 at% Ni–ZnO and the sensing performance will be enhanced. The third and most important point is that Ni^{2+} ions can be easily oxidized into Ni^{3+} with a higher oxidation state, which will facilitate the redox reaction, due to which Ni^{3+} usually plays the role of the catalyst during the redox process, as many studies have reported [70–72]. Therefore, a greater percentage of Ni^{3+} will obtain a good gas sensing property to a large extent. Figure 5c–e displays the fitted XPS results of Ni^{2+} and Ni^{3+} of the 0.5, 1, and 2 at% Ni–ZnO. The relative percentage of Ni^{2+} and Ni^{3+} varied in the three samples, and the ratio of Ni^{3+}/Ni^{2+} was calculated to be 1.33, 1.62, and 1.05 for the 0.5, 1, and 2 at% Ni–ZnO samples. The 1 at% Ni–ZnO sample had the highest ratio of Ni^{3+}/Ni^{2+} of 1.62, followed by the 0.5 at% Ni–ZnO and 2 at% Ni–ZnO. From the gas sensing tests, the highest sensing performance was obtained by the 1 at% Ni–ZnO, and then the 0.5 at% Ni–ZnO and 2 at% Ni–ZnO samples. The gas sensing results coincided with the inference of the Ni^{3+}/Ni^{2+} ratio.

5. Conclusions

In summary, different Ni–ZnO core-shell spheres were successfully prepared by a one-step solvothermal method and gas sensors based on the prepared materials were prepared. The role of Ni doping in ZnO on the microstructure and gas sensing performance was studied in detail. The results showed that Ni doping changed the hierarchical microstructure of ZnO, and the BET specific area increased after Ni doping. Gas sensing measurements revealed that the operating temperature decreased due to the lower band gap after Ni doping. All of the sensing materials had the highest response to toluene, the best sensing performance was acquired by the 1 at% Ni–ZnO based gas sensor, with high response, excellent selectivity, fast response and recovery time, and relatively long-term stability. The enhanced sensing property can be mainly due to the increase in the specific surface area and relative percentage of oxygen vacancy after Ni doping, and more importantly, is the catalyst effect of Ni ions. This work provides a reasonable way to fabricate a high performance toluene sensing material.

Supplementary Materials: The following supporting information can be downloaded at: https://www.mdpi.com/article/10.3390/chemosensors10080327/s1, Figure S1: The schematic figure of the (a) ceramic tube, (b) sensor device, and (c) the sensing test system. Figure S2: (a) The UV–Vis absorption spectrum and (b) energy band gap of the ZnO samples with different Ni doping amounts. Table S1: The parameter information for all of the gas samples.

Author Contributions: Conceptualization, X.Y. and Z.L.; Methodology, X.Y.; Formal analysis, Z.L.; Investigation, S.L., Z.S., Z.W., H.Z., C.S., H.L. and J.S.; Data curation, Z.L.; Writing—original draft preparation, Z.L.; Writing—review and editing, X.Y., G.P., Y.C. and L.G.; Supervision, X.Y., G.P., Y.C. and L.G.; Funding acquisition, X.Y. and G.P. All authors have read and agreed to the published version of the manuscript.

Funding: This work was supported by the Major National Science and Technology Special Projects (2016ZX02301003-004-007); the National Natural Science Foundation of China (62003123); the National Natural Science Foundation of Hebei Province (F2020202067); the Key Laboratory of Electronic Materials and Devices of Tianjin, China; and the National Demonstration Center for Experimental (Electronic and Communication Engineering) Education (Hebei University of Technology).

Institutional Review Board Statement: Not applicable.

Informed Consent Statement: Not applicable.

Data Availability Statement: Not applicable.

Conflicts of Interest: The authors declare no conflict of interest.

References

1. Mirzaei, A.; Leonardi, S.G.; Neri, G. Detection of hazardous volatile organic compounds (VOCs) by metal oxide nanostructures-based gas sensors: A review. *Ceram. Int.* **2016**, *42*, 15119–15141. [CrossRef]
2. Xu, J.M.; Cheng, J.P. The advances of Co_3O_4 as gas sensing materials: A review. *J. Alloys Compd.* **2016**, *686*, 753–768. [CrossRef]
3. Sharma, B.; Sharma, A.; Joshi, M.; Myung, J.H. Sputtered SnO_2/ZnO heterostructures for improved NO_2 gas sensing properties. *Chemosensors* **2020**, *8*, 67. [CrossRef]
4. Hui, G.Z.; Zhu, M.Y.; Yang, X.L.; Liu, J.J.; Pan, G.F.; Wang, Z.Y. Highly sensitive ethanol gas sensor based on CeO_2/ZnO binary heterojunction composite. *Mater. Lett.* **2020**, *278*, 128453. [CrossRef]
5. Guo, L.L.; Zhang, B.; Yang, X.L.; Zhang, S.S.; Wang, Y.; Wang, G.D. Sensing platform of PdO-ZnO-In_2O_3 nanofibers using MOF templated catalysts for triethylamine detection. *Sens. Actuators B* **2021**, *343*, 130126. [CrossRef]
6. Li, S.H.; Xie, L.L.; He, M.; Hu, X.B.; Luo, G.F.; Chen, C.; Zhu, Z. Metal-Organic frameworks-derived bamboo-like CuO/In_2O_3 Heterostructure for high-performance H_2S gas sensor with Low operating temperature. *Sens. Actuators B* **2020**, *310*, 127828. [CrossRef]
7. Ueda, T.; Boehme, I.; Hyodo, T.; Shimizu, Y.; Weimar, U.; Barsan, N. Enhanced NO_2 sensing properties of Au-loaded porous In_2O_3 gas sensors at low operating temperatures. *Chemosensors* **2020**, *8*, 72. [CrossRef]
8. Yang, X.L.; Zhang, S.F.; Yu, Q.; Zhao, L.P.; Sun, P.; Wang, T.S.; Liu, F.; Yan, X.; Gao, Y.; Liang, X.; et al. One step synthesis of branched SnO_2/ZnO heterostructures and their enhanced gas-sensing properties. *Sens. Actuators B* **2019**, *281*, 415–423. [CrossRef]
9. Hiyoto, K.A.M.; Fisher, E.R. Utilizing plasma modified SnO_2 paper gas sensors to better understand gas-surface interactions at low temperatures. *J. Vac. Sci. Technol. A* **2020**, *38*, 043202. [CrossRef]
10. Zheng, L.; Bi, W.J.; Jin, Z.; Liu, S.T. Synthesis of hierarchical shell-core SnO_2 microspheres and their gas sensing properties. *Chin. Chem. Lett.* **2020**, *31*, 2083–2086. [CrossRef]
11. Ueda, T.; Maeda, T.; Huang, Z.D.; Higuchi, K.; Izawa, K.; Kamada, K.; Hyodo, T.; Shimizu, Y. Enhancement of methylmercaptan sensing response of WO_3 semiconductor gas sensors by gas reactivity and gas diffusivity. *Sens. Actuators B* **2018**, *273*, 826–833. [CrossRef]
12. Buyukkose, S. Highly selective and sensitive WO_3 nanoflakes based ammonia sensor. *Mater. Sci. Semicond. Process.* **2020**, *110*, 104969. [CrossRef]
13. Wang, C.Y.; Li, Y.H.; Qiu, P.P.; Duan, L.L.; Bi, W.; Chen, Y.; Guo, D.; Liu, Y.; Luo, W.; Deng, Y. Controllable synthesis of highly crystallized mesoporous TiO_2/WO_3 heterojunctions for acetone gas sensing. *Chin. Chem. Lett.* **2020**, *31*, 1119–1123. [CrossRef]
14. Mokoena, T.P.; Tshabalala, Z.P.; Hillie, K.T.; Swart, H.C.; Motaung, D.E. The blue luminescence of p-type NiO nanostructured material induced by defects: H_2S gas sensing characteristics at a relatively low operating temperature. *Appl. Surf. Sci.* **2020**, *525*, 146002. [CrossRef]
15. Simion, C.E.; Ghica, C.; Mihalcea, C.G.; Ghica, D.; Mercioniu, I.; Somacescu, S.; Florea, Q.G.; Stanoiu, A. Insights about CO gas-sensing mechanism with NiO-based gas sensors the influence of humidity. *Chemosensors* **2021**, *9*, 244. [CrossRef]
16. Alev, O.; Sarica, N.; Ozdemir, O.; Arslan, L.C.; Buyukkose, S.; Oztur, Z.Z. Cu-doped ZnO nanorods based QCM sensor for hazardous gases. *J. Alloys Compd.* **2020**, *826*, 154177. [CrossRef]
17. Umar, A.; Algadi, H.; Kumar, R.; Akhtar, M.S.; Ibrahim, A.A.; Albargi, H.; Alhamami, M.A.M.; Alsuwian, T.; Zeng, W. Ultrathin leaf-shaped CuO nanosheets based sensor device for enhanced hydrogen sulfide gas sensing application. *Chemosensors* **2021**, *9*, 221. [CrossRef]
18. Reynolds, D.C.; Look, D.C.; Jogai, B.; Hoelscher, J.E.; Sherriff, R.E.; Harris, M.T.; Callahan, M.J. Time-resolved photoluminescence lifetime measurements of the Γ_5 and Γ_6 free excitons in ZnO. *J. Appl. Phys.* **2000**, *88*, 2152–2153. [CrossRef]

19. Norton, D.P.; Heo, Y.W.; Ivill, M.P.; Ip, K.; Pearton, S.J.; Chisholm, M.F. ZnO: Growth, doping & processing. *Mater. Today* **2004**, *7*, 34–40.
20. Modaberi, M.R.; Rooydell, R.; Brahma, S.; Akande, A.A.; Mwakikunga, B.W.; Liu, C.P. Enhanced response and selectivity of H_2S sensing through controlled Ni doping into ZnO nanorods by using single metal organic precursors. *Sens. Acuators B* **2018**, *273*, 1278–1290. [CrossRef]
21. Liang, Y.C.; Liao, W.K.; Deng, X.S. Synthesis and substantially enhanced gas sensing sensitivity of homogeneously nanoscale Pd- and Au-particle decorated ZnO nanostructures. *J. Alloys Compd.* **2014**, *599*, 87–92. [CrossRef]
22. Gardon, M.; Guilemany, J.M. A review on fabrication, sensing mechanisms and performance of metal oxide gas sensors. *J. Mater. Sci. Mater. Electron.* **2013**, *24*, 1410–1421. [CrossRef]
23. Hu, J.; Gao, F.Q.; Sang, S.B.; Li, P.W.; Deng, X.; Zhang, W.D.; Chen, Y.; Lian, K. Optimization of Pd content in ZnO microstructures for high-performance gas detection. *J. Mater. Sci.* **2015**, *50*, 1935–1942. [CrossRef]
24. Wang, W.C.; Tian, Y.T.; Wang, X.C.; He, H.; Xu, Y.R.; He, C.; Li, H. Ethanol sensing properties of porous ZnO spheres via hydrothermal route. *J. Mater. Sci.* **2013**, *48*, 3232–3238. [CrossRef]
25. Khoang, N.D.; Hong, H.S.; Trung, D.D.; van Duy, N.; Hoa, N.D.; Thinh, D.D.; Van Hieu, N. On-chip growth of wafer-scale planar-type ZnO nanorod sensors for effective detection of CO gas. *Sens. Actuators B* **2013**, *181*, 529–536. [CrossRef]
26. Zhang, W.H.; Zhang, W.D.; Zhou, J.F. Solvent thermal synthesis and gas-sensing properties of Fe-doped ZnO. *J. Mater. Sci.* **2010**, *45*, 209–215. [CrossRef]
27. Luo, J.; Ma, S.Y.; Li, F.M.; Li, X.B.; Li, W.Q.; Cheng, L.; Mao, Y.Z.; Gz, D.J. The mesoscopic structure of flower-like ZnO nanorods for acetone detection. *Mater. Lett.* **2014**, *121*, 137–140. [CrossRef]
28. Xiong, H.M. ZnO nanoparticles applied to bioimaging and drug delivery. *Adv. Mater.* **2013**, *25*, 5329–5335. [CrossRef]
29. Huang, J.R.; Xu, X.J.; Gu, C.P.; Yang, M.; Yang, M.; Liu, J.H. Large-scale synthesis of hydrated tungsten oxide 3D architectures by a simple chemical solution route and their gas-sensing properties. *J. Mater. Chem.* **2011**, *21*, 13283–13289. [CrossRef]
30. Li, H.Y.; Wang, X. Three-dimensional architectures constructed using two-dimensional nanosheets. *Sci. China Chem.* **2015**, *58*, 1792–1799. [CrossRef]
31. Sohila, S.; Rajendran, R.; Yaakob, Z.; Teridi, M.A.M.; Sopian, K. Photoelectrochemical water splitting performance of flower like ZnO nanostructures synthesized by a novel chemical method. *J. Mater. Sci. Mater. Electron.* **2016**, *27*, 2846–2851. [CrossRef]
32. Akamatsu, T.; Itoh, T.; Tsuruta, A.; Masuda, Y. CH_3SH and H_2S Sensing Properties of $V_2O_5/WO_3/TiO_2$ Gas Sensor. *Chemosensors* **2021**, *9*, 113. [CrossRef]
33. Li, Y.M.; Hu, H.J.; Zhang, W.F.; Tian, Z.Q.; Jiang, X.Q.; Wang, Y.H.; Zhang, S.; Zhang, Q.; Jian, J.; Zou, J. Theoretical Study on the Electrochemical Catalytic Activity of Au-Doped Pt Electrode for Nitrogen Monoxide. *Chemosensors* **2022**, *10*, 178. [CrossRef]
34. Guo, W.W.; Jian, L.J.; Wang, X.M.; Zeng, W. Hydrothermal synthesis of Ni-doped hydrangea-like Bi_2WO_6 and the enhanced gas sensing property to n-butanol. *Sens. Actuators B* **2022**, *357*, 131396. [CrossRef]
35. Sun, Q.H.; Wu, Z.F.; Cao, B.B.; Chen, X.; Zhang, C.C.; Shaymurat, T.; Duan, H.; Zhang, J.; Zhang, M. Gas sensing performance of biomass carbon materials promoted by nitrogen doping and p-n junction. *Appl. Surf. Sci.* **2022**, *592*, 153254. [CrossRef]
36. Park, H.; Kim, D.H.; Ma, B.S.; Shin, E.; Kim, Y.; Kim, T.S.; Kim, F.S.; Kim, I.D.; Kim, B.J. High-Performance, Flexible NO_2 Chemiresistors Achieved by Design of Imine-Incorporated n-Type Conjugated Polymers. *Adv. Sci.* **2022**, *9*, 2200270. [CrossRef]
37. Fan, Y.R.; Xu, Y.Y.; Wang, Y.X.; Sun, Y.Q. Fabrication and characterization of Co-doped ZnO nanodiscs for selective TEA sensor applications with high response, high selectivity and ppb-level detection limit. *J. Alloys Compd.* **2021**, *876*, 160170. [CrossRef]
38. Lu, S.H.; Hu, X.F.; Zheng, H.; Qiu, J.W.; Tian, R.B.; Quan, W.J.; Min, X.; Ji, P.; Hu, Y.; Cheng, S.; et al. Highly selective, ppb-level xylene gas detection by Sn^{2+}-doped NiO flower-like microspheres prepared by a one-step hydrothermal. *Method Sens.* **2019**, *19*, 2958. [CrossRef]
39. Lee, C.S.; Li, H.Y.; Kim, B.Y.; Jo, Y.M.; Byun, H.G.; Hwang, I.S.; Abdel-Hady, F.; Wazzan, A.A.; Lee, J.-H. Discriminative detection of indoor volatile organic compounds using a sensor array based on pure and Fe-doped In_2O_3 nanofibers. *Sens. Actuators B* **2019**, *285*, 193–200. [CrossRef]
40. Chacko, L.; Massera, E.; Aneesh, P.M. Enhancement in the selectivity and sensitivity of Ni and Pd functionalized MoS_2 toxic gas sensors. *J. Electrochem. Soc.* **2020**, *167*, 106506. [CrossRef]
41. Guo, W.W.; Zhao, B.Y.; Zhou, Q.L.; He, Y.Z.; Wang, Z.C.; Radacsi, N. Fe-doped ZnO/reduced graphene oxide nanocomposite with synergic enhanced gas sensing performance for the effective detection of formaldehyde. *ACS Omega* **2019**, *4*, 10252–10262. [CrossRef]
42. Lu, Y.Y.; Zhan, W.W.; He, Y.; Wang, Y.T.; Kong, X.J.; Kuang, Q.; Xie, Z.; Zheng, X. MOF-templated synthesis of porous Co_3O_4 concave nanocubes with high specific surface area and their gas sensing properties. *ACS Appl. Mater. Interfaces* **2014**, *6*, 4186–4195. [CrossRef]
43. Li, D.P.; Zhang, Y.; Xu, J.C.; Jin, H.X.; Jin, D.F.; Hong, B.; Peng, X.; Wang, P.; Ge, H.; Wang, X. Nanocasting synthesis and gas-sensing behavior of hematite nanowires. *Phys. E Low-Dimens. Syst. Nanostruct. (Amst. Neth.)* **2016**, *84*, 395–400. [CrossRef]
44. Chen, H.D.; Jin, K.L.; Wang, P.F.; Xu, J.C.; Han, Y.B.; Jin, H.X.; Jin, D.F.; Peng, X.L.; Hong, B.; Li, J.; et al. Highly enhanced gas-sensing properties of indium-doped mesoporous hematite nanowires. *J. Phys. Chem. Solids* **2018**, *120*, 271–278. [CrossRef]
45. Su, C.; Zhang, L.; Han, Y.T.; Ren, C.; Li, B.L.; Wang, T.; Zeng, M.; Su, Y.; Hu, N.; Zhou, Z.; et al. Glucose-assisted synthesis of hierarchical NiO-ZnO heterostructure with enhanced glycol gas sensing performance. *Sens. Actuators B* **2021**, *329*, 129167. [CrossRef]

46. Sun, K.; Zhan, G.H.; Chen, H.D.; Lin, S.W. Low-operating-temperature NO_2 sensor based on a CeO_2/ZnO heterojunction. *Sensors* **2021**, *21*, 8269. [CrossRef] [PubMed]
47. Zhao, Y.M.; Wang, S.; Zhai, X.; Shao, L.; Bai, X.J.; Liu, Y.L.; Wang, T.; Li, Y.; Zhang, L.; Fan, F.; et al. Construction of Zn/Ni bimetallic organic framework derived ZnO/NiO heterostructure with superior N-propanol sensing performance. *ACS Appl. Mater. Interfaces* **2021**, *13*, 9206–9215. [CrossRef] [PubMed]
48. Duan, Z.H.; Zhao, Q.N.; Wang, S.; Huang, Q.; Yuan, Z.; Zhang, Y.J.; Jiang, Y.; Tai, H. Halloysite nanotubes: Natural, environmental-friendly and low-cost nanomaterials for high-performance humidity sensor. *Sens. Actuators B* **2020**, *317*, 128204. [CrossRef]
49. Chen, X.X.; Shen, Y.B.; Zhou, P.F.; Zhong, X.X.; Li, G.D.; Han, C.; Wei, D.; Li, S. Bimetallic Au/Pd nanoparticles decorated ZnO nanowires for NO_2 detection. *Sens. Actuators B* **2019**, *289*, 160–168. [CrossRef]
50. Zhou, C.G.; Meng, F.Q.; Chen, K.; Yang, X.L.; Wang, T.S.; Sun, P.; Liu, F.; Yan, X.; Shimanoe, K.; Lu, G. High sensitivity and low detection limit of acetone sensor based on NiO/Zn_2SnO_4 p-n heterojunction octahedrons. *Sens. Actuators B* **2021**, *339*, 129912. [CrossRef]
51. Wang, C.; Cui, X.B.; Liu, J.Y.; Zhou, X.; Cheng, X.Y.; Sun, P.; Hu, X.; Li, X.; Zheng, J.; Lu, G. Design of superior ethanol gas sensor based on Al-doped NiO nanorod-flowers. *ACS Sens.* **2016**, *1*, 131–136. [CrossRef]
52. Wang, S.; Huang, D.; Xu, S.S.; Jiang, W.K.; Wang, T.; Hu, J.; Hu, N.; Su, Y.; Zhang, Y.; Yang, Z. Two-dimensional NiO nanosheets with enhanced room temperature NO_2 sensing performance via Al doping. *Phys. Chem. Chem. Phys.* **2017**, *19*, 19043–19049. [CrossRef]
53. Wang, L.L.; Lou, Z.; Zhang, R.; Zhou, T.T.; Deng, J.N.; Zhang, T. Hybrid Co3O_4/SnO_2 core-shell nanospheres as real-time rapid-response sensors for ammonia gas. *ACS Appl. Mater. Interfaces* **2016**, *8*, 6539–6545. [CrossRef]
54. Korotcenkov, G.; Cho, B.K. Engineering approaches for the improvement of conductometric gas sensor parameters Part 1. Improvement of sensor sensitivity and selectivity (short survey). *Sens. Actuators B* **2013**, *188*, 709–728. [CrossRef]
55. Li, F.; Gao, X.; Wang, R.; Zhang, T. Design of WO_3-SnO_2 core-shell nanofibers and their enhanced gas sensing performance based on different work function. *Appl. Surf. Sci.* **2018**, *442*, 30–37. [CrossRef]
56. Kim, J.H.; Kim, H.W.; Kim, S.S. Self-heating effects on the toluene sensing of Pt-functionalized SnO_2-ZnO core-shell nanowires. *Sens. Actuators B* **2017**, *251*, 781–794. [CrossRef]
57. Chen, H.; Ao, S.R.; Li, G.D.; Gao, Q.; Zou, X.X.; Wei, C.D. Enhanced sensing performance to toluene and xylene by constructing $NiGa_2O_4$-NiO heterostructures. *Sens. Actuators B* **2019**, *282*, 331–338. [CrossRef]
58. Nie, L.F.; Fan, G.J.; Wang, A.Q.; Zhang, L.; Guan, J.; Han, N.; Chen, Y. Finely dispersed and highly toluene sensitive NiO/$NiGa_2O_4$ heterostructures prepared from layered double hydroxides precursors. *Sens. Actuators B* **2021**, *345*, 130412. [CrossRef]
59. Hermawan, A.; Asakura, Y.; Inada, M.; Yin, S. A facile method for preparation of uniformly decorated-spherical SnO_2 by CuO nanoparticles for highly responsive toluene detection at high temperature. *J. Mater. Sci. Technol.* **2020**, *51*, 119–129. [CrossRef]
60. Jamnik, D.J.S.S.I. Investigation of the PTCR effect in ZnO–NiO two-phase ceramics. *Solid State Ion.* **1997**, *99*, 125–135.
61. Kim, H.J.; Lee, J.H. Highly sensitive and selective gas sensors using p-type oxide semiconductors: Overview. *Sens. Actuators B* **2014**, *192*, 607–627. [CrossRef]
62. Jun, J.H.; Yun, J.; Cho, K.; Hwang, I.S.; Lee, J.H.; Kim, S. Necked ZnO nanoparticle-based NO_2 sensors with high and fast response. *Sens. Actuators B* **2009**, *140*, 412–417. [CrossRef]
63. Zhou, J.Y.; Bai, J.L.; Zhao, H.; Yang, Z.Y.; Gu, X.Y.; Huang, B.Y.; Zhao, C.H.; Cairang, L.; Sun, G.Z.; Zhang, Z.X.; et al. Gas sensing enhancing mechanism via doping-induced oxygen vacancies for gas sensors based on indium tin oxide nanotubes. *Sens. Actuators B* **2018**, *265*, 273–284. [CrossRef]
64. Patil, V.L.; Vanalakar, S.A.; Patil, P.S.; Kim, J.H. Fabrication of nanostructured ZnO thin films based NO_2 gas sensor via SILAR technique. *Sens. Actuators B* **2017**, *239*, 1185–1193. [CrossRef]
65. Zhang, R.; Gao, S.; Zhou, T.T.; Tu, J.C.; Zhang, T. Facile preparation of hierarchical structure based on p-type Co_3O_4 as toluene detecting sensor. *Appl. Surf. Sci.* **2020**, *503*, 144167. [CrossRef]
66. Zhang, R.; Xu, Z.W.; Zhou, T.T.; Fei, T.; Wang, R.; Zhang, T. Improvement of gas sensing performance for tin dioxide sensor through construction of nanostructures. *JCIS* **2019**, *557*, 673–682. [CrossRef]
67. Kim, J.H.; Mirzaei, A.; Kim, H.W.; Kim, S.S. Variation of shell thickness in ZnO-SnO_2 core-shell nanowires for optimizing sensing behaviors to CO, C6H6, and C7H8 gases. *Sens. Actuators B* **2020**, *302*, 127150. [CrossRef]
68. Bai, J.L.; Zhao, C.H.; Gong, H.M.; Wang, Q.; Huang, B.Y.; Sun, G.Z.; Wang, Y.; Zhou, J.; Xie, E.; Wang, F. Debye-length controlled gas sensing performances in NiO@ZnO p-n junctional core-shell nanotubes. *J. Phys. D Appl. Phys.* **2019**, *52*, 285103. [CrossRef]
69. Zhou, T.T.; Zhang, T.; Zeng, Y.; Zhang, R.; Lou, Z.; Deng, J.N.; Wang, L. Structure-driven efficient $NiFe_2O_4$ materials for ultra-fast response electronic sensing platform. *Sens. Actuators B* **2018**, *255*, 1436–1444. [CrossRef]
70. Kim, H.R.; Choi, K.I.; Kim, K.M.; Kim, I.D.; Cao, G.Z.; Lee, J.H. Ultra-fast responding and recovering C_2H_5OH sensors using SnO_2 hollow spheres prepared and activated by Ni templates. *Chem. Commun.* **2010**, *46*, 5061–5063. [CrossRef]
71. Yang, M.; Lu, J.Y.; Wang, X.; Zhang, H.; Chen, F.; Sun, J.B.; Yang, J.; Sun, Y.; Lu, G. Acetone sensors with high stability to humidity changes based on Ru-doped NiO flower-like microspheres. *Sens. Actuators B* **2020**, *313*, 127965. [CrossRef]
72. Luo, Y.B.; An, B.X.; Bai, J.L.; Wang, Y.R.; Cheng, X.; Wang, Q.; Li, J.; Yang, Y.; Wu, Z.; Xie, E. Ultrahigh-response hydrogen sensor based on PdO/NiO co-doped In_2O_3 nanotubes. *J. Colloid Interface Sci.* **2021**, *599*, 533–542. [CrossRef]

chemosensors

Article

Non-Invasive Rapid Detection of Lung Cancer Biomarker Toluene with a Cataluminescence Sensor Based on the Two-Dimensional Nanocomposite Pt/$Ti_3C_2T_x$-CNT

Hongyan Wang [1,*], Xiaoqi Shi [1], Fei Liu [2], Tingmei Duan [1] and Bai Sun [2,3,*]

1 Department of Radiation Oncology, The First Affiliated Hospital of Anhui Medical University, Hefei 230022, China
2 Engineering Research Center of Building Energy Efficiency Control and Evaluation of the Ministry of Education, College of Environment and Energy Engineering, Anhui Jianzhu University, Hefei 230601, China
3 Nano-Materials and Environmental Detection Laboratory, Hefei Institute of Physical Science, Chinese Academy of Sciences, Hefei 230031, China
* Correspondence: whhy5000@163.com (H.W.); bsun@iim.ac.cn (B.S.); Tel.: +86-551-63828252 (B.S.)

Abstract: A novel two-dimensional nanocomposite Pt/$Ti_3C_2T_x$-CNT was synthesized for the non-invasive rapid detection of toluene, a lung cancer biomarker, via cataluminescence (CTL). Pt/$Ti_3C_2T_x$-CNT exhibited a good catalytic performance toward toluene. The CTL sensor based on Pt/$Ti_3C_2T_x$-CNT has the advantage of rapid response: The average response time was about 1 s, and the average recovery time was about 30 s. Moreover, the material has a wide scope of detection for toluene, and the limit of detection defined as 3 S/N was about 2 ppm. The optimal working temperature (150 °C) is lower than common sensors, so it has a broad prospect in the actual detection process. Aside from its weak response to formaldehyde, the sensor only exerted a strong response signal to toluene, and no response was observed to other VOCs, indicating that this CTL sensor has good selectivity for toluene. The possible sensing mechanism of CTL showed that toluene was oxidized to generate excited-state CO_2^*, which emitted a luminescent signal when it returned to the ground state.

Keywords: MXene; nanocomposites; cataluminescence; detection of toluene

Citation: Wang, H.; Shi, X.; Liu, F.; Duan, T.; Sun, B. Non-Invasive Rapid Detection of Lung Cancer Biomarker Toluene with a Cataluminescence Sensor Based on the Two-Dimensional Nanocomposite Pt/$Ti_3C_2T_x$-CNT. *Chemosensors* **2022**, *10*, 333. https://doi.org/10.3390/chemosensors10080333

Academic Editor: Tae Geun Kim

Received: 11 July 2022
Accepted: 8 August 2022
Published: 14 August 2022

1. Introduction

Human respiration contains a large number of gases, including carbon dioxide (CO_2), ammonia (NH_3), ethanol (C_2H_5OH), acetone (CH_3COCH_3), and toluene (C_7H_8) [1]. The human-exhaled breath serves to exhaust the body of its metabolic byproducts. These byproducts may be unique in lung cancer, based on metabolic changes within or around cancer cells or due to metabolic changes in the immunological system of the body [2]. Careful analysis of such breathing gases by using reliable detection equipment (such as gas sensors) provides a non-invasive way to check people's health: a significantly high level of gas can be used as a biomarker for pathogenic development (e.g., toluene: lung cancer; acetone: diabetes; ammonia: kidney problems) [3–5]. The development of toluene gas sensors for biomarker applications has attracted particular attention because this medical device can provide a non-invasive tool for the daily monitoring of lung cancer patients [6,7].

At present, there are many methods and technologies that can be used to detect toluene gas. However, when other respiratory gases coexist, the sensitivity and selectivity for the detection of low concentration toluene is still a challenging topic. The existing methods of toluene detection have some drawbacks and cannot achieve rapid detection [8–12]. For example, the chemical detection method takes a long time, and the interference of homologues cannot be ruled out. The infrared spectroscopy method has low sensitivity and large relative error and needs a large number of representative samples to model before detection [9]. Gas chromatography is expensive and requires professional operation [10].

Therefore, it is necessary to establish a rapid, portable, and sensitive detection method for toluene [8]. CTL refers to the phenomenon according to which when a catalytic oxidation reaction occurs at the gas–solid interface, the substance absorbs energy from the ground state and converts it into an excited state, and the excited state releases energy into the ground state, accompanied by chemiluminescence [13]. The CTL method based on nanomaterials is a new sensor detection method, which is widely studied and applied to the detection of volatile organic compounds (VOCs) in the environment [14–19]. Shi et al. [20] prepared Pt/NU-901 nanocomposites for CTL sensors to detect acetone, a marker of diabetes. The detection method is stable, fast, and accurate. Yu et al. [21] reported a camellia-like NiO and used it in a CTL sensor for the rapid detection of H_2S. The large specific surface area of the material, its developed porous structure, and the effect of surface-adsorbed oxygen make the sensor have high sensitivity and selectivity for H_2S detection. This method has high sensitivity, selectivity, and fast detection speed. There is no need to add other reagents when using a CTL sensor to detect VOCs. In addition, sensing elements are not consumed in the CTL reaction process, so the nanomaterial can be reused as a sensing element, which enables the CTL sensor to monitor VOCs in real time [13,22–25]. Therefore, CTL-based sensors can be applied to the detection of toluene.

MXene is a novel two-dimensional material for transition metal carbides or nitrides obtained by using fluoride etching MAX (where M is an early transition metal carbide, A represents the main group elements, and X represents C and/or N elements [26]) ternary layered materials, which has attracted increasing attention because of its unique structure and excellent electrical and optical properties [27]. A single, large layer and less defective Ti_3C_2Tx sheet can be separated via simple ultrasonication or manual shaking by etching with hydrochloric acid and lithium fluoride [28–30]. Additionally, carbon nanotubes (CNTs) have excellent sensing properties [31,32], and metal decoration can improve sensing performance. In recent years, the preparation and exploration of MXene-based composites have started to become research hotspots [33]. In addition, Pt and other precious metals have been widely studied because of their unique physical and chemical properties. Research shows that Pt has good oxidation resistance and high conductivity, and platinum can provide excitation electrons in CTL sensors. Therefore, adding it to a CTL sensor will further enhance the detection signal of the sensor and speed up its response and recovery time [19,34–36]. Therefore, Pt was used as decoration to prepare Pt/$Ti_3C_2T_x$-CNT catalyst in this paper. Then, the CTL characteristics of toluene on the surface of Pt /Ti_3C_2Tx-CNT were determined, and the optimal detection conditions of toluene were investigated using an ultraweak chemiluminescence analyzer.

2. Experiment

2.1. Reagents and Instruments

The reagents used in this experiment were lithium fluoride (LiF), cetyltrimethylammonium bromide (CTAB), ethanol, chloroplatinic acid ($H_2PtCl_6{\cdot}6H_2O$), sodium borohydride ($NaBH_4$), toluene, formaldehyde, chloroform, cyclohexane, tetrachloroethane, diethyl ether, acetone, and ammonia, which were purchased from Sinopharm Chemical Reagent Co., Ltd., (Shanghai, China) and used as received, without further purification. Ti_3AlC_2 was from Kayene Ceramics Co., Ltd., (Yantai, China). CNT was purchased from Beijing Boyu Co., Ltd., (Beijing, China). Deionized water was prepared using FST-TOP-A24 super-pure water equipment made by Shanghai Fushite Instrument Equipment Co., Ltd., (Shanghai, China).

The instruments used in the experiment included an RCT basic magnetic stirrer (EKA Instrument Equipment Co., Ltd., Guangzhou, China), a TG16K-II table high-speed centrifuge (Shanghai Zhaodi Biotechnology Co., Ltd., Shanghai, China), a DHG-9023A air-drying box (Shanghai YiHeng Scientific Instrument Co., Ltd., Shanghai, China), and a PHB-3 digital pH meter (Shanghai SanXin Instrument Factory, Shanghai, China).

2.2. Preparation of Pt/$Ti_3C_2T_x$-CNT Catalysts

The etching methods of Pt/$Ti_3C_2T_x$-CNT have been reported in previous studies [37–39].

2.2.1. Preparation of $Ti_3C_2T_x$

First, 0.8 g of LiF was added to a 10 mL HCl (9 mol/L) solution; then, 0.5 g of Ti_3AlC_2 was added and mixed into the solution with continuous stirring at 35 °C for 24 h. The reaction precipitate was repeatedly washed with deionized (DI) water until the pH value of the supernatant was greater than 6. Finally, the precipitate was dissolved in the DI water, and the resulting solution was sonicated in an Ar atmosphere for 1 h.

2.2.2. Preparation of CTAB-CNT

Briefly, 50 mg CNT was added to 100 mg CTAB, and then the mixture was ultrasonically dispersed in 50 mL DI water. After the reaction was complete, the precipitate was washed with ethanol several times to remove the residual CTAB. Finally, centrifugation (9000 rpm,15 min) was used to collect the black precipitate; thus, the CTAB-CNT was obtained.

2.2.3. Preparation of Pt/$Ti_3C_2T_x$-CNT

First, 40 mg $Ti_3C_2T_x$ and 40 mg CTAB-CNT were dispersed in 40 mL DI water to obtain the $Ti_3C_2T_x$-CNT. Then, $H_2PtCl_6 \cdot 6H_2O$ was added dropwise to the $Ti_3C_2T_x$-CNT suspension and stirred continuously for 2 h at room temperature. Finally, an excess of $NaBH_4$ solution was added to the above solution to react until no bubbles were produced. After centrifugation, washing, and drying, the Pt/$Ti_3C_2T_x$-CNT catalyst was obtained.

2.3. Characterization

The morphology and structure of Pt/$Ti_3C_2T_x$-CNT were investigated using a scanning electron microscope (SEM, Zeiss Auriga FIB-SEM, 10 kV) and a transmission electron microscope (TEM, JEOL-2010, 200 kV). The chemical composition was determined with an energy-dispersive spectrometer (EDS). X-ray diffraction (XRD) patterns were recorded using a Philips X'Pert PRO MPD diffractometer with a Cu Kα X-ray source (λ = 1.5405 Å).

2.4. Cataluminescence Sensing Measurement

Pt/$Ti_3C_2T_x$-CNT was coated on ceramic rods and then placed in a reaction chamber. An adjustable transformer was used to heat the reaction chamber, and a flow pump was used to adjust the airflow rate. In the same reaction condition, using a syringe, 10 kinds of VOCs (toluene, trichloromethane, tetrachloroethane, cyclohexane, formaldehyde, diethyl ether, acetone, carbon dioxide, ammonia, and ethanol) were injected into the reaction chamber to react with Pt/$Ti_3C_2T_x$ -CNT, and the CTL value in Pt/$Ti_3C_2T_x$-CNT surface was determined using an ultraweak chemiluminescence analyzer (BPCL-1-TIC, Guangzhou Microphotonics Technologies Co., Ltd., Guangzhou, China). The sensing device is shown in Figure 1.

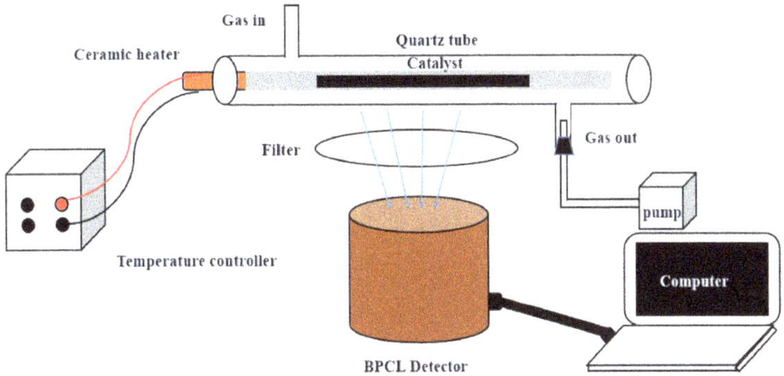

Figure 1. Schematic diagram of CTL sensor experimental device.

The dry air required for the experiment was provided by a compressed gas cylinder and carried out under the condition of constant humidity (40 ± 5%). The temperature was adjusted and detected using the adjustable transformer. The ceramic heating rods coated with nanomaterials were put in a closed quartz tube (500 mL), and the airflow rate was adjusted by adjusting the cylinder valve. The gas flow rate was detected with a rotor gas flowmeter (anti-corrosion LZB-3WBF). Finally, the gas solution to be tested was injected into the cylinder through a micro syringe and then delivered to the reaction chamber. The generated catalytic luminescence signal was detected and processed via a photoelectric detection and data processing system. The measurement wavelength of this experiment was the effective response wavelength (300–650 nm) that can be detected by the photomultiplier tube in the BPCL ultraweak microluminescence instrument, so no specific wavelength was selected. The concentration of the detected gas in the experiment is calculated by the following formula [40]:

$$C = \frac{V_i \times P_0}{V_c \times P_a} \quad (1)$$

where C is the concentration of the gas to be measured, V_i is the volume of the gas to be measured sucked into the syringe, V_c is the volume of the reaction chamber, P_0 is the vapor pressure of the gas to be measured at room temperature, and P_a is the standard atmospheric pressure.

3. Results and Discussions

3.1. Characterization of Pt/$Ti_3C_2T_x$-CNT

Figure 2a–e show the SEM images of Pt/$Ti_3C_2T_x$-CNT and its corresponding EDS mapping. As can be seen from Figure 2a–e, the material consists of Ti, Pt, O, and C. In this composition, Ti and O are $Ti_3C_2T_x$ feature elements, C is a $Ti_3C_2T_x$ and CNT feature element, and Pt is the material surface decoration element. The content of each element is shown in Table 1. Figure 2f,g correspond to the SEM and TEM images of the material. As can be seen from Figure 2f,g, the material has a tubular structure connected by nodes, which shows that the CNT and $Ti_3C_2T_x$ are closely combined to form a stable structure. There is Ti with partial solubility on the surface of $Ti_3C_2T_x$, which makes the Ti cation separate from the $Ti_3C_2T_x$ substrate and freely dissolve into the solution, so $Ti_3C_2T_x$ has a negative charge [26], while the CNT modified with cationic surfactant CTAB has a positive charge [41]. The positively charged CNT and negatively charged $Ti_3C_2T_x$ are combined via electrostatic interaction. Research has shown that the composite of $Ti_3C_2T_x$ and CNT can increase the active sites of the material and improve the cycle life of the material [26,41]. In addition, Pt was attached to the surface of the nanotubes, which indicates that the Pt/$Ti_3C_2T_x$-CNT catalyst was successfully prepared in this experiment.

Table 1. Content of elements in Pt/$Ti_3C_2T_x$-CNT.

Elements	Weight%	Atomic%
C	9.76	93.83
O	0.75	5.39
Ti	0.18	0.44
Pt	0.59	0.34
Totals	11.28	1

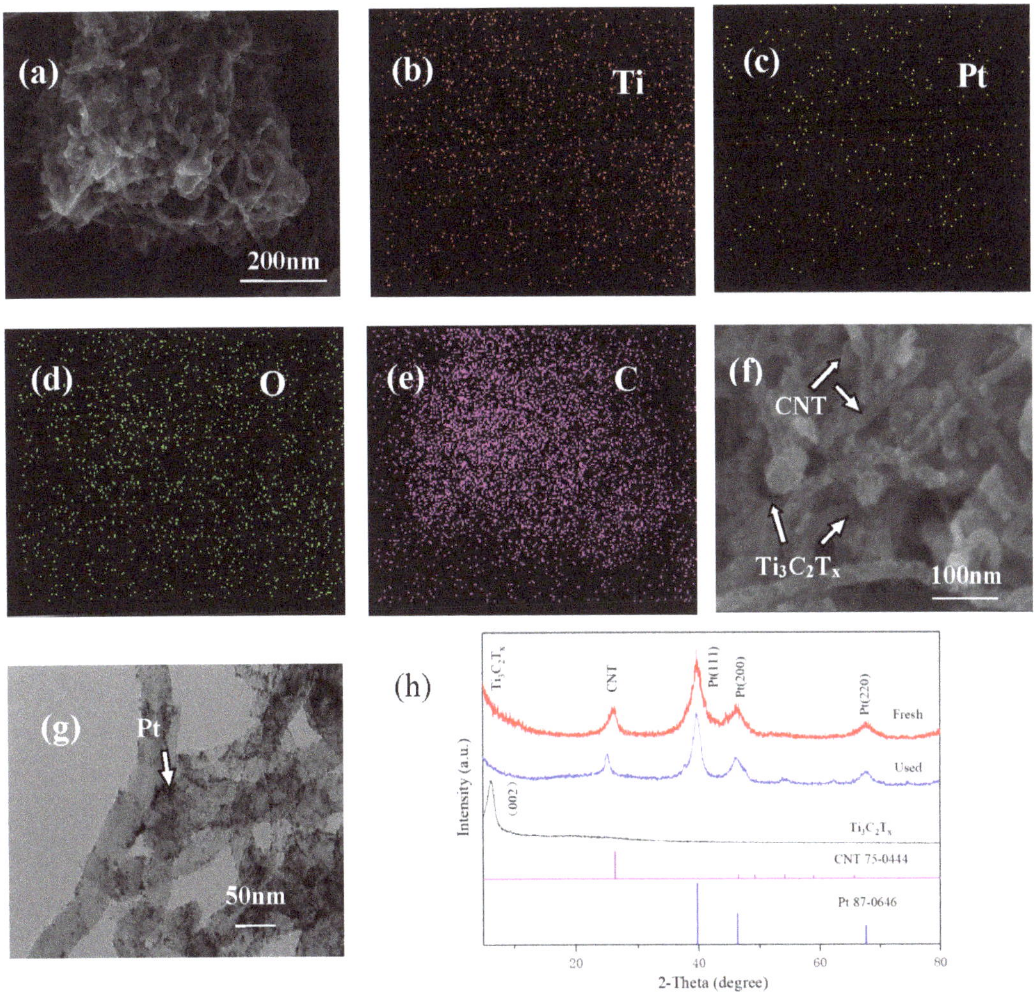

Figure 2. (**a–e**) SEM images and EDS mapping of Pt/$Ti_3C_2T_x$-CNT; (**f**,**g**) SEM and TEM images of Pt/$Ti_3C_2T_x$-CNT; (**h**) XRD pattern of Pt/$Ti_3C_2T_x$-CNT.

The XRD pattern of Pt/$Ti_3C_2T_x$-CNT is shown in Figure 2h. As can be seen from Figure 2h, the diffraction peaks of the materials appeared at 6.3°, 26.3°, 39.8°, 46.3°, and 67.5°, respectively. Diffraction peaks matched well the corresponding standard cards. Among them, the diffraction peak of 6.2° corresponded to $Ti_3C_2T_x$, and 26.3° corresponded to the characteristic peak of CNT (JCPDS No.75-04444). In addition, the diffraction peaks of 39.8°, 46.3°, and 67.8° corresponded to the basal planes of Pt (111), Pt (200), and Pt (220) (JCPDS No.87-0646). No other impurities were observed in the XRD pattern.

3.2. CTL Response of Toluene on Pt/$Ti_3C_2T_x$-CNT

Figure 3a shows the luminescence intensity upon exposure to toluene vapors of the Pt/$Ti_3C_2T_x$-CNT material. After the sensor was debugged and stabilized, the CTL strength was measured. The CTL strength produced by toluene before adding toluene was 0, and after adding toluene, the CTL strength produced by toluene on the material surface could reach about 175,000, with an average response time of 1 s and average recovery time of 30 s. When the CTL strength was restored to 0, toluene was added again, and the CTL strength immediately produced was basically unchanged, indicating that Pt/$Ti_3C_2T_x$-CNT has a good CTL response to toluene. In addition, under the same experimental conditions, after four times of repeated tests, no significant change was observed in signal strength, which shows that the CTL sensor has good reproducibility and high stability. Figure 3b shows the luminous intensity of $Ti_3C_2T_x$-CNT when exposed to toluene vapor. From the comparison of Figure 3a,b, it can be seen that the luminous intensity of $Ti_3C_2T_x$-CNT exposed to toluene vapor was significantly lower than that of Pt/$Ti_3C_2T_x$-CNT exposed to toluene vapor, indicating that the addition of Pt was conducive to the CTL signal detection of toluene and accelerated the corresponding recovery time. The CTL-based method has obvious advantages for the detection of VOCs, such as sensitive and rapid detection. However, research on the detection of toluene via CTL is relatively scarce [19,42,43], so the material has great potential application value.

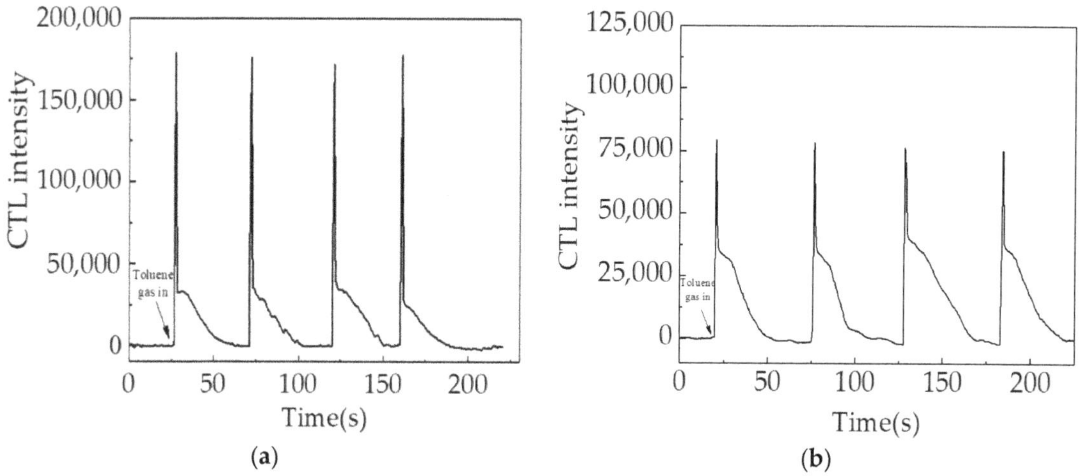

Figure 3. (**a**) The luminescence intensity upon exposure to toluene vapors of Pt/$Ti_3C_2T_x$-CNT material; (**b**) the luminescence intensity upon exposure to toluene vapors of $Ti_3C_2T_x$-CNT material. (detecting concentration: 87 ppm. Working temperature: 150 °C. Airflow rate: 200 mL/min).

3.3. Optimization of CTL Sensor

Temperature is an important factor for CTL reactions, and the CTL value generally increases with the temperature. However, with the increase in temperature, the noise of the detection device also increases. In addition, the high temperature may affect the structure of the material [44]. Therefore, it is necessary to study the optimal working temperature.

The effects of operating temperature on CTL intensity and the S/N [45] value are shown in Figure 4a. The CTL intensity of toluene on the Pt/$Ti_3C_2T_x$-CNT surface increased with the increase in temperature under certain concentrations and airflow rates. At the same time, the S/N value first increased with the increase in temperature and then decreased after 150 °C, which implies that 150 °C is the optimal temperature for the detection of toluene. Therefore, the temperature of 150 °C, which occurs at a large luminescent signal area for toluene, was chosen in subsequent experiments.

The airflow rate has also an obvious influence on CTL strength. A slow air flow rate leads to a low effective contact concentration between the detector and the material, and a high airflow rate causes the detector to be taken out of the reaction chamber before it can react with the material.

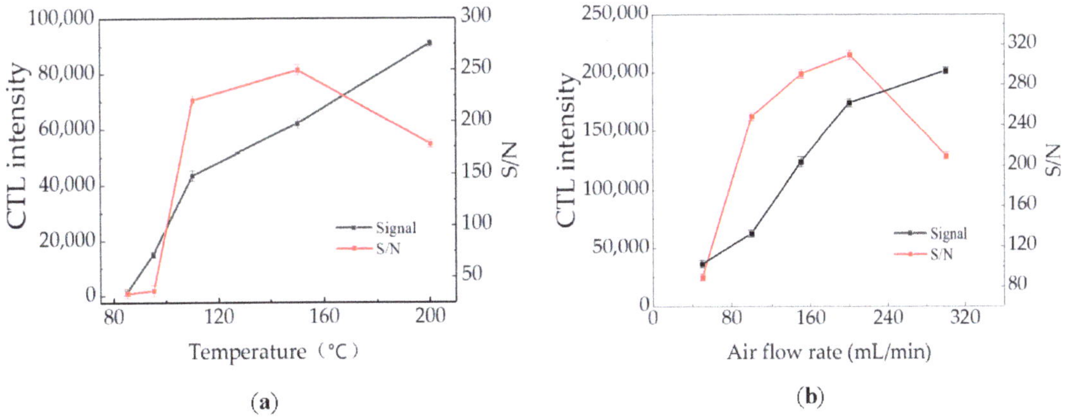

Figure 4. Effects of (**a**) working temperature and (**b**) airflow rate on CTL intensity and the S/N value: (**a**) detecting concentration: 87 ppm; airflow rate: 100 mL/min; (**b**) detecting concentration: 87 ppm; detecting temperature: 150 °C.

At different gas flow rates, the CTL strength between toluene and the material was detected via BPCL-1-TIC, and the noise under the corresponding CTL value was recorded at the same time. The results are shown in Figure 4b. From Figure 4b, the CTL intensity of toluene increased with an increase in the gas flow rate below 300 mL/min, and the S/N first increased with the increase in gas flow rate and then decreased after 200 mL/min. Therefore, the gas flow rate of 200 mL/min was chosen for detection because of the resulting strong and steady CTL signals.

3.4. Analytical Characteristics

Under the optimal experimental conditions, the analytical characteristics of the CTL sensor based on Pt/$Ti_3C_2T_x$-CNT toward toluene were studied. A response curve obtained from the different concentrations of toluene is shown in Figure 5, indicating a good linear relationship within the concentration range of 2.5~87 ppm. The limit of detection defined as 3 S/N was about 2 ppm. The linear regression equation was y = 2034x–4858.3 ($2.5 \leq x \leq 87$ ppm, $R^2 = 0.9228$), where y represents the average relative intensity of CTL after three parallel experiments, and x represents the concentration of toluene.

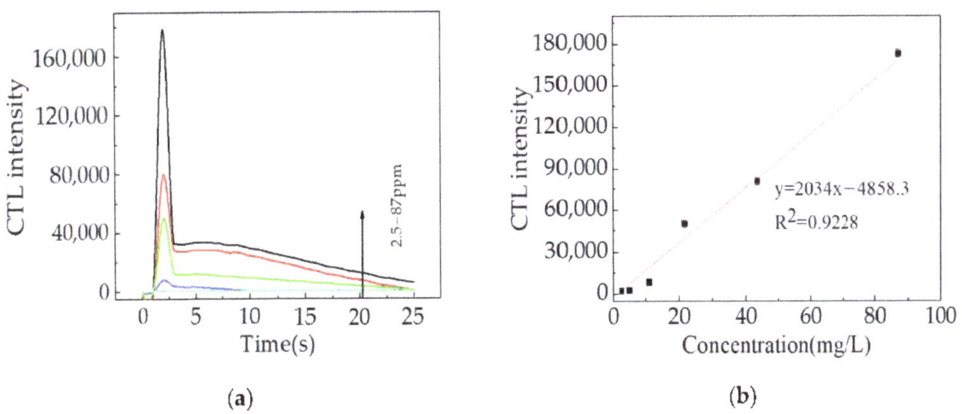

Figure 5. (**a**) Effects of analyte concentration on CTL intensity; (**b**) correlation curve of response signal intensity of different concentrations of toluene on the surface of Pt/$Ti_3C_2T_x$-CNT materials. (working temperature: 150 °C; airflow rate: 200 mL/min).

3.5. Selectivity and Stability of the Pt/$Ti_3C_2T_x$-CNT for Toluene

In addition to good reproducibility, excellent selectivity is also critical for the sensor, especially for practical detection. Under suitable conditions, 10 VOCs (toluene, trichloromethane, tetrachloroethane, cyclohexane, formaldehyde, diethyl ether, acetone, carbon dioxide, ammonia, and ethanol) were introduced into the reaction chamber to prove the selectivity of Pt/$Ti_3C_2T_x$-CNT. The CTL intensity of seven VOCs on the surface of the material is shown in Figure 6a. From Figure 6a, except for toluene and formaldehyde, the other eight VOCs had no obvious response on the Pt/$Ti_3C_2T_x$-CNT surface. There was no evident interference in the detection of toluene. Therefore, Pt/$Ti_3C_2T_x$-CNT has good selectivity in the detection of toluene.

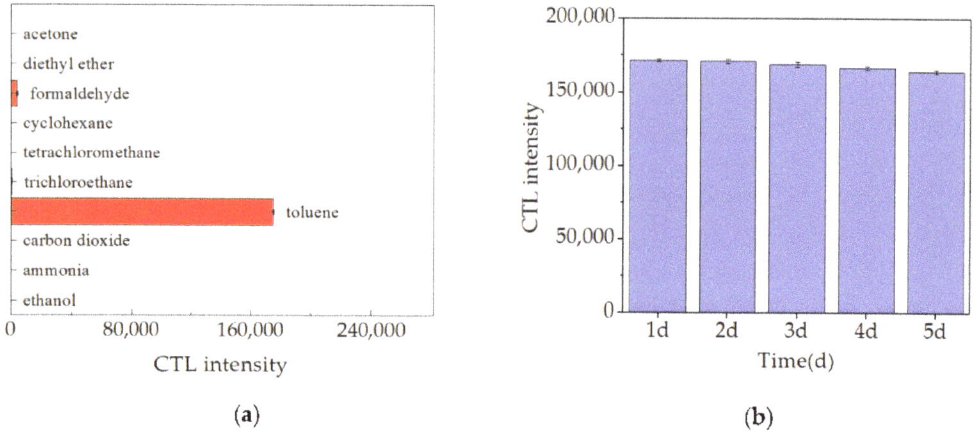

Figure 6. (**a**) Selectivity of the Pt/$Ti_3C_2T_x$-CNT for toluene; (**b**) stability of the Pt/$Ti_3C_2T_x$-CNT for toluene (detecting concentration: 87 ppm; working temperature: 150 °C; airflow rate: 200 mL/min).

In order to verify the stability of the continuous operation of Pt/$Ti_3C_2T_x$-CNT at the optimal temperature (150 °C), the sensing performance of Pt/$Ti_3C_2T_x$-CNT to toluene was measured working for 8 h every day. The average value of the measurement was taken as the measurement result of the day, working continuously for 5 days, and the error is shown by the error bar. As shown in Figure 6b, the results revealed that the CTL intensity produced by toluene on the surface of the material was 171,500 on the first day, and it slightly decreased over the next few days. After 5 days, the CTL intensity produced by toluene on the surface of the material was 164,000, and the signal intensity of toluene decreased by 4.37% after 5 consecutive days. In addition, the XRD pattern of Pt/$Ti_3C_2T_x$-CNT after its continuous operation is shown in Figure 2b. There was no obvious difference between the fresh Pt/$Ti_3C_2T_x$-CNT and the one after detection. This indicates that the CTL sensor has high stability.

3.6. The Advantages of Pt/$Ti_3C_2T_x$-CNT Sensor

Table 2 shows the analytical characteristics of the sensor presented here and some previously reported sensors. As shown in Table 2, the Pt/$Ti_3C_2T_x$-CNT sensor has a faster response, shorter recovery time, and lower working temperature compared with previous reports. The fast response and recovery time of this sensor greatly improve its detection efficiency, and its lower working temperature can reduce the energy consumption of the detection process. Therefore, Pt/$Ti_3C_2T_x$-CNT has great advantages in detecting toluene. In addition, Pt/$Ti_3C_2T_x$-CNT contains the advantages from each component, enabling a possible coordination effect to achieve high catalytical performance. CNT provides a large sensing area for the CTL reaction. $Ti_3C_2T_x$ makes the material junction more stable.

Moreover, the Pt attached to the surface can provide excited electrons according to the principle of catalytic luminescence detection [19].

Table 2. Reports on the detection of toluene based on CTL in recent years.

Principle	Sensing Material	Response Time (s)	Recovery Time (s)	Temperature (°C)	Reference
CTL	TiO_2/SnO_2	ND	ND	270	[42]
CTL	γ-Al_2O_3/PtO_2	1.5	60	236	[43]
CTL	Al_2O_3/Pt	2	38	395	[19]
CTL	γ-Al_2O_3/Eu_2O_3	3	30	432	[16]
CTL	$Pt/Ti_3C_2T_x$-CNT	1	30	150	This work

3.7. Possible Mechanism

Different metal-based catalysts have been found for the detection of VOCs, and their presence in nanomaterials can enhance the sensing characteristics of sensors. Firstly, due to the adsorption of methyl, the adsorption of toluene on the sensor surface is enhanced. Secondly, in our study, considering the interaction between -CH_3 and Pt, electrons more easily transition from the π_{CH_3} level to the Fermi level and from the Fermi level to the π_{CH_3*} level. Therefore, the adsorption barrier of toluene becomes lower, and the electronic effect between methyl and Pt is enhanced [46,47]. From the perspective of the through-space effect, there is through-space repulsion between -CH_3 and Pt, which is not conducive to the adsorption of $-CH_3$ and Pt [48]. Therefore, this indicates that $Ti_3C_2T_x$ and CNT accelerate the conduction of electrons, change the working function and adsorption barriers, and make the electronic effect more dominant than the spatial effect, thus improving the sensor response. Moreover, the CNT-binding $Ti_3C_2T_x$ provides greater contact space for gas molecules [26,30].

It is worth noting that only the materials located at the surface of the sensing layer can make contact with the gas analytes effectively. Hence, it is assumed that the produced CTL signal is the result of the reaction of toluene and oxygen in the surface layer of the material. The whole process of CTL response is described as follows:

$$O_2 + 2e^- \rightarrow 2O^- \text{ or } O_2 + 2e^- \rightarrow O_2^- \quad (2)$$

$$C_7H_8 + O^-(O_2^-) \rightarrow C_{7-x}H_{8-2x} + xCO_2^* + xH_2O + e^- (\text{x represents the number of } H_2O \text{ and } CO_2^*) \quad (3)$$

$$CO_2^* \rightarrow CO_2 + hv \quad (4)$$

Firstly, toluene and oxygen in the air are adsorbed on the surface of $Pt/Ti_3C_2T_x$-CNT. Oxygen adsorbed on the surface of catalyst at appropriate temperature receives free electrons to form oxygen ions (O^- and O_2^-) [39,49,50]. Secondly, toluene adsorbed on the surface of $Pt/Ti_3C_2T_x$-CNT is catalytically oxidized with O^- and O_2^-, and the reaction products then diffuse into the gas phase. In the process of catalytic oxidation reaction, excited CO_2 (CO_2^*) is generated, and the returning process of the released CO_2^* photons to the ground state (CO_2) is accompanied by the generation of optical signals. The main mechanism of CTL signal generation is that the catalyst catalyzes the generation of excited states and converts energy into optical signals [51–54].

4. Conclusions

In this study, a sensor based on $Pt/Ti_3C_2T_x$-CNT was developed for the detection of toluene without the aid of biological reagent identification or instrumental separation. $Pt/Ti_3C_2T_x$-CNT exhibited the best catalytic performance toward toluene due to its large specific surface area and stable structure. The gas sensor based on $Pt/Ti_3C_2T_x$-CNT has the advantage of rapid response. Additionally, the material has a wide scope of detection for toluene, and the optimal working temperature (150 °C) is lower compared with common sensors, so it has a broad prospect in the actual detection process. The limit of detection defined as 3 S/N was about 2 ppm. The possible sensing mechanism of CTL showed that toluene was oxidized to generate excited-state CO_2*, which emitted a luminescent signal when it returned to the ground state.

Author Contributions: Conceptualization, H.W. and B.S.; validation, B.S.; formal analysis, X.S. and F.L.; writing—original draft preparation, H.W. and B.S.; writing—review and editing, B.S.; supervision, B.S.; project administration, T.D.; funding acquisition, B.S. All authors have read and agreed to the published version of the manuscript.

Funding: This work was supported by the Project of the National Key Research and Development Program (2019YFC0408505), the Natural Science Research Project of the Higher Education Institutions of Anhui Province (KJ2021A0303, KJ2021A0616), the Science and Technology Major projects of Anhui Province (18030801106), the teaching research project of Anhui Province (2019jyxm0994), the first batch of natural science projects supported by surplus funds in 2021 of Anhui Jianzhu University (JZ202129, JZ202134), and the Scientific Research Start-up Foundation for Introduction of Talent, Anhui Jianzhu University (2016QD113).

Institutional Review Board Statement: Not applicable.

Informed Consent Statement: Not applicable.

Data Availability Statement: Not applicable.

Conflicts of Interest: The authors declare no conflict of interest.

References

1. Kim, K.H.; Jahan, S.A.; Kabir, E. A review of breath analysis for diagnosis of human health. *TrAC Trends Anal. Chem.* **2012**, *33*, 1–8. [CrossRef]
2. Amann, A.; Corradi, M.; Mazzone, P.; Mutti, A. Lung cancer biomarkers in exhaled breath. *Expert Rev. Mol. Diagn.* **2011**, *11*, 207–217. [CrossRef] [PubMed]
3. Konvalina, G.; Haick, H. Sensors for breath testing: From nanomaterials to comprehensive disease detection. *Acc. Chem. Res.* **2014**, *47*, 66–76. [CrossRef] [PubMed]
4. Zhang, Z.; Zhu, L.; Wen, Z.; Ye, Z. Controllable synthesis of Co_3O_4 crossed nanosheet arrays toward an acetone gas sensor. *Sens. Actuators B Chem.* **2017**, *238*, 1052–1059. [CrossRef]
5. Qin, W.; Yuan, Z.; Shen, Y.; Zhang, R.; Meng, F. Phosphorus-doped porous perovskite $LaFe_{1-x}P_xO_3$-δ nanosheets with rich surface oxygen vacancies for ppb level acetone sensing at low temperature. *Chem. Eng. J.* **2022**, *431*, 134280. [CrossRef]
6. Xiong, Y.; Zhu, Z.; Ding, D.; Lu, W.; Xue, Q. Multi-shelled $ZnCo_2O_4$ yolk-shell spheres for high-performance acetone gas sensor. *Appl. Surf. Sci.* **2018**, *443*, 114–121. [CrossRef]
7. David, S.S.; Veeralakshmi, S.; Sandhya, J.; Nehru, S.; Kalaiselvam, S. Room temperature operatable high sensitive toluene gas sensor using chemiresistive Ag/Bi_2O_3 nanocomposite. *Sens. Actuators B Chem.* **2020**, *320*, 128410. [CrossRef]
8. Healy, R.M.; Wang, J.M.; Karellas, N.S.; Todd, A.; Sofowote, U.; Su, Y.; Munoz, A. Assessment of a passive sampling method and two on-line gas chromatographs for the measurement of benzene, toluene, ethylbenzene and xylenes in ambient air at a highway site. *Atmos. Pollut. Res.* **2019**, *10*, 1123–1127. [CrossRef]
9. Douberly, G.E.; Ricks, A.M.; Schleyer PV, R.; Duncan, M.A. Infrared spectroscopy of gas phase benzenium ions: Protonated benzene and protonated toluene, from 750 to 3400 cm^{-1}. *J. Phys. Chem. A* **2008**, *112*, 4869–4874. [CrossRef]
10. Ji, J.; Deng, C.; Shen, W.; Zhang, X. Field analysis of benzene, toluene, ethylbenzene and xylene in water by portable gas chromatography-microflame ionization detector combined with headspace solid-phase microextraction. *Talanta* **2006**, *69*, 894–899. [CrossRef]
11. Lin, H.; Jang, M.; Suslick, K.S. Preoxidation for colorimetric sensor array detection of VOCs. *J. Am. Chem. Soc.* **2011**, *133*, 16786. [CrossRef] [PubMed]
12. Iwaki, T.; Covington, J.A.; Udrea, F.; Gardner, J.W. Identification and quantification of different vapours using a single polymer chemoresistor and the novel dual transient temperature modulation technique. *Sens. Actuators B Chem.* **2009**, *141*, 370–380. [CrossRef]
13. Breysse, M.; Claudel, B.; Faure, L.; Guenin, M.; Williams, R.J.; Wolkenstein, T. Chemiluminescense during the catalysis of carbon monoxide oxidation on a thoria surface. *J. Catal.* **1976**, *45*, 137–144. [CrossRef]
14. Zhou, K.; Xu, J.; Gu, C.; Liu, B.; Peng, Z. Identification and determination of formaldehyde, benzene and ammonia in air based on cross sensitivity of cataluminescence on single catalyst. *Sens. Actuators B Chem.* **2017**, *246*, 703–709. [CrossRef]
15. Almasian, M.R.; Na, N.; Wen, F.; Zhang, S.; Zhang, X. Development of a plasma-assisted cataluminescence system for benzene, toluene, ethylbenzene, and xylenes analysis. *Anal. Chem.* **2010**, *82*, 3457. [CrossRef] [PubMed]
16. Lu, J.; Cao, X.; Pan, C.; Yang, L.; Lai, G.; Chen, J.; Wu, C. Studies of the Cataluminescence of Benzene Homologues on Nanosized γ-Al_2O_3/Eu_2O_3 and the Development of a Gas Sensorfor Benzene Homologue Vapors. *Sensors* **2006**, *6*, 1827–1836. [CrossRef]

17. Zeng, B.; Zhang, L.; Wan, X.; Song, H.; Lv, Y. Fabrication of α-Fe_2O_3/g-C_3N_4 composites for cataluminescence sensing of H_2S. *Sens. Actuators B Chem.* **2015**, *211*, 370–376. [CrossRef]
18. Leghrib, R.; Felten, A.; Demoisson, F.; Reniers, F.; Pireaux, J.J.; Llobet, E. Room-temperature, selective detection of benzene at trace levels using plasma-treated metal-decorated multiwalled carbon nanotubes. *Carbon* **2010**, *48*, 3477–3484. [CrossRef]
19. Wang, N.J.; Cao, X.A.; He, R.W.; Liu, Y.H.; Huang, Y.J. A cataluminescence-based sensor for detecting benzene, toluene and xylene vapors utilizing the catalytic reduction on the surface of nanosized Al_2O_3/Pt. *Adv. Mat. Res.* **2013**, *663*, 335–342.
20. Shi, Z.; Li, G.; Hu, Y. Cataluminescence sensor based on Pt/NU-901 nanocomposite for rapid capture, catalysis and detection of acetone in exhaled breath. *Anal. Chim. Acta* **2022**, *1206*, 339787. [CrossRef]
21. Zhang, R.K.; Wang, D.; Wu, Y.J.; Hu, Y.H.; Chen, J.Y.; He, J.C.; Wang, J.X. A Cataluminescence Sensor Based on NiO Nanoparticles for Sensitive Detection of Acetaldehyde. *Molecules* **2020**, *25*, 1097. [CrossRef] [PubMed]
22. Shi, G.; Sun, B.; Jin, Z.; Liu, J.; Li, M. Synthesis of SiO_2/Fe_3O_4 nanomaterial and its application as cataluminescence gas sensor material for ether. *Sens. Actuators B Chem.* **2012**, *171–172*, 699–704. [CrossRef]
23. Rushi, A.D.; Datta, K.P.; Ghosh, P.; Mulchandani, A.; Shirsat, M.D. Exercising substituents in porphyrins for real time selective sensing of volatile organic com pounds. *Sens. Actuators B Chem.* **2018**, *257*, 389–397. [CrossRef]
24. Bao, Y.; Xu, P.; Cai, S.; Yu, H.; Li, X. Detection of volatile-organic-compounds (VOCs) in solution using cantilever-based gas sensors. *Talanta* **2018**, *182*, 148–155. [CrossRef] [PubMed]
25. Zhang, R.; Li, G.; Hu, Y. Simple and excellent selective chemiluminescence based CS_2 on line detection system for rapid analysis of sulfur-containing com pounds in complex samples. *Anal. Chem.* **2015**, *87*, 5649–5655. [CrossRef]
26. Xu, M.; Liang, L.; Qi, J.; Wu, T.; Zhou, D.; Xiao, Z. Intralayered Ostwald Ripening-Induced Self-Catalyzed Growth of CNTs on MXene for Robust Lithium–Sulfur Batteries. *Small* **2021**, *17*, 2007446. [CrossRef]
27. Lin, H.; Wang, X.; Yu, L.; Chen, Y.; Shi, J. Two-Dimensional Ultrathin MXene Ceramic Nanosheets for Photothermal Conversion. *Nano Lett.* **2017**, *17*, 384–391. [CrossRef]
28. Anasori, B.; Lukatskaya, M.R.; Gogotsi, Y. 2D metal carbides and nitrides (MXenes) for energy storage. *Nat. Rev. Mater.* **2017**, *2*, 16098. [CrossRef]
29. Tao, Q.; Dahlqvist, M.; Lu, J.; Kota, S.; Meshkian, R.; Halim, J.; Palisaitis, J.; Hultman, L.; Barsoum, M.W.; Persson, P.O.; et al. Two-dimensional Mo1.33C MXene with divacancy ordering prepared from parent 3D laminate with in-plane chemical ordering. *Nat. Commun.* **2017**, *8*, 14949. [CrossRef]
30. Zhou, J.; Zha, X.; Zhou, X.; Chen, F.; Gao, G.; Wang, S.; Shen, C.; Chen, T.; Zhi, C.; Eklund, P.; et al. Synthesis and Electrochemical Properties of Two-Dimensional Hafnium Carbide. *ACS Nano* **2017**, *11*, 3841–3850. [CrossRef]
31. Zhou, Y.; Jiang, Y.; Xie, G.; Du, X.; Tai, H. Gas sensors based on multiple-walled carbon nanotubes-polyethylene oxide films for toluene vapor detection. *Sens. Actuators B Chem.* **2014**, *191*, 24–30. [CrossRef]
32. Crescitelli, A.; Consales, M.; Penza, M.; Aversa, P.; Giordano, M.; Cusano, A. Toluene detection in aqueous phase by optical fiber sensors integrated with single-walled carbon nanotubes. *Open Environ. Biol. J.* **2008**, *1*, 26–32.
33. Firouzjaei, M.D.; Karimiziarani, M.; Moradkhani, H.; Elliott, M.; Anasori, B. MXenes: The two-dimensional influencers. *Mater. Today Adv.* **2022**, *13*, 100202. [CrossRef]
34. Yu, L.; Li, N. Noble Metal Nanoparticles-Based Colorimetric Biosensor for Visual Quantification: A Mini Review. *Chemosensors* **2019**, *7*, 53. [CrossRef]
35. Kim, H.W.; Lee, J.W. GeO2 nanostructures fabricated by heating of Ge powders: Pt-catalyzed growth, structure, and photoluminescence. *Phys. E Low-Dimens. Syst. Nanostruct.* **2008**, *40*, 2499–2503. [CrossRef]
36. Shi, Z.; Xia, L.; Li, G.; Hu, Y. Pt/Au Nanoparticles@Co_3O_4 Cataluminescence Sensor for Rapid Analysis of Methyl Sec-Butyl Ether Impurity in Methyl Tert-Butyl Ether Gasoline Additive. *Chemosensors* **2022**, *10*, 260. [CrossRef]
37. Alhabeb, M.; Maleski, K.; Mathis, T.S.; Sarycheva, A.; Hatter, C.B.; Uzun, S.; Levitt, A.; Gogotsi, Y. Selective etching of silicon from Ti_3SiC_2 (MAX) to obtain 2D titanium carbide (MXene). *Angew. Chem. Int. Edit.* **2018**, *57*, 5444–5448. [CrossRef]
38. Xie, X.; Zhao, M.Q.; Anasori, B.; Maleski, K.; Ren, C.E.; Li, J.; Byles, B.W.; Pomerantseva, E.; Wang, G.; Gogotsi, Y. Porous heterostructured MXene/carbon nanotube composite paper with high volumetric capacity for sodium-based energy storage devices. *Nano Energy* **2016**, *26*, 513–523. [CrossRef]
39. Xu, C.; Fan, C.; Zhang, X.; Chen, H.; Liu, X.; Fu, Z.; Wang, R.; Hong, T.; Cheng, J. MXene ($Ti_3C_2T_x$) and carbon nanotube hybrid-supported platinum catalysts for the high-performance oxygen reduction reaction in PEMFC. *ACS Appl. Mater. Interfaces* **2020**, *12*, 19539–19546. [CrossRef]
40. Meng, F.; Qi, T.; Zhang, J.; Zhu, H.; Yuan, Z.; Liu, C.; Qin, W.; Ding, M. MoS_2-Templated Porous Hollow MoO_3 Microspheres for Highly Selective Ammonia Sensing via a Lewis Acid-Base Interaction. *IEEE Trans. Ind. Electron.* **2022**, *69*, 960–970. [CrossRef]
41. Wu, M.; Park, H.; Cho, K.M.; Kim, J.Y.; Kim, S.J.; Choi, S.; Kang, Y.; Kim, J.; Jung, H.T. Formation of toroidal Li_2O_2 in non-aqueous Li–O_2 batteries with Mo_2CTx MXene/CNT composite. *RSC Adv.* **2019**, *9*, 41120–41125. [CrossRef]
42. Sha, W.; Ni, S.; Zheng, C. Development of cataluminescence sensor system for benzene and toluene determination. *Sens. Actuators B Chem.* **2015**, *209*, 297–305. [CrossRef]
43. Sun, Y.; Cao, X.; Liu, Y.; Wang, N.; He, R. Research on benzene, toluene and dimethylbenzene detection based on a cataluminescence sensor. *Luminescence* **2014**, *29*, 122–126. [CrossRef]
44. Sharma, C.S.; Katepalli, H.; Sharma, A.; Madou, M. Fabrication and electrical conductivity of suspended carbon nanofiber arrays. *Carbon* **2011**, *49*, 1727–1732. [CrossRef]

45. Huang, X.; Song, H.; Zhang, L.; Deng, D.; Lv, Y. ZnO Nanoparticle-Decorated CeO_2 Nanospheres for Cataluminescence Sensing of H_2S. *ACS Appl. Nano Mater.* **2021**, *4*, 9557–9565. [CrossRef]
46. Minot, C.; Gallezot, P. Competitive hydrogenation of benzene and toluene: The oretical study of their adsorption on ruthenium, rhodium, and palladium. *J. Catal.* **1990**, *123*, 341–348. [CrossRef]
47. Kwon, Y.J.; Na, H.G.; Kang, S.Y.; Choi, S.W.; Kim, S.S.; Kim, H.W. Selective detection of low concentration toluene gas using Pt-decorated carbon nanotubes sensors. *Sens. Actuators B Chem.* **2016**, *227*, 157–168. [CrossRef]
48. Orozco, J.M.; Webb, G. The adsorption and hydrogenation of benzene and toluene on alumina-and silica- supported palladium and platinum catalysts. *Appl. Catal.* **1983**, *6*, 67–84. [CrossRef]
49. Qin, W.; Yuan, Z.; Gao, H.; Zhang, R.; Meng, F. Perovskite-structured $LaCoO_3$ modified ZnO gas sensor and investigation on its gas sensing mechanism by first principle. *Sens. Actuators B Chem.* **2021**, *341*, 130015. [CrossRef]
50. Meng, F.; Li, X.; Yuan, Z.; Lei, Y.; Qi, T.; Li, J. Ppb-Level Xylene Gas Sensors based on Co_3O_4 Nanoparticles coated Reduced Graphene Oxide(rGO) Nanosheets Operating at Low Temperature. *IEEE Trans. Instrum. Meas.* **2021**, *70*, 9511510. [CrossRef]
51. Yu, K.; Hu, J.; Li, X.; Zhang, L.; Lv, Y. Camellia-like NiO: A novel cataluminescence sensing material for H_2S. *Sens. Actuators B Chem.* **2019**, *288*, 243–250. [CrossRef]
52. Cao, X.; Zhang, Z.; Zhang, X. A novel gaseous acetaldehyde sensor utilizing cataluminescence on nanosized $BaCO_3$. *Sens. Actuators B Chem.* **2004**, *99*, 30–35. [CrossRef]
53. Ye, Q.; Gao, Q.; Zhang, X.R.; Xu, B.Q. Cataluminescence and catalytic reactions of ethanol oxidation over nanosized $Ce_{1-x}Zr_xO_2$ ($0 \leq x \leq 1$) catalysts. *Catal. Commun.* **2006**, *7*, 589–592. [CrossRef]
54. Wang, S.; Yuan, Z.; Zhang, L.; Lin, Y.; Lu, C. Recent advances in cataluminescence-based optical sensing systems. *Analyst* **2017**, *142*, 1415–1428. [CrossRef]

chemosensors

Article

High-Performance Sulfur Dioxide Gas Sensor Based on Graphite-Phase Carbon-Nitride-Functionalized Tin Diselenide Nanorods Composite

Hao Zhang [1,2], Qiannan Pan [3], Yating Zhang [3], Yanting Zhang [1] and Dongzhi Zhang [3,*]

[1] College of Mechanical and Electrical Engineering, China University of Petroleum (East China), Qingdao 266580, China
[2] Technology Inspection Center, Shengli Oilfield, Dongying 257000, China
[3] College of Control Science and Engineering, China University of Petroleum (East China), Qingdao 266580, China
* Correspondence: dzzhang@upc.edu.cn

Abstract: In this paper, a composite of tin diselenide ($SnSe_2$) functionalized by graphite-phase carbon nitride (g-C_3N_4) was successfully prepared by a hydrothermal method, and was characterized by X-ray diffraction (XRD), scanning electron microscope (SEM) and X-ray photoelectron spectroscopy (XPS). These microstructure characterization results verified the successful synthesis of a multilayer g-C_3N_4/rod-shaped $SnSe_2$ composite. The gas sensitivity results showed that when the g-C_3N_4 ratio was 30%, the g-C_3N_4/$SnSe_2$ composite sensor had the highest response (28.9%) at 200 °C to 20 ppm sulfur dioxide (SO_2) gas, which was much higher than those of pristine g-C_3N_4 and $SnSe_2$ sensors at the optimum temperature. A series of comparative experiments proved that the g-C_3N_4/$SnSe_2$ composite sensor demonstrated an excellent response, strong reversibility and good selectivity for ppm-level SO_2 gas detection. The possible SO_2 sensing mechanism was ascribed to the heterostructure between the n-type $SnSe_2$ and n-type g-C_3N_4 nanomaterials. Furthermore, we also proposed the influence of the special structure of the g-C_3N_4 functionalized $SnSe_2$ composite on the gas-sensing characteristics.

Keywords: tin diselenide; carbon nitride; hydrothermal; heterostructure; SO_2 sensors

Citation: Zhang, H.; Pan, Q.; Zhang, Y.; Zhang, Y.; Zhang, D. High-Performance Sulfur Dioxide Gas Sensor Based on Graphite-Phase Carbon-Nitride-Functionalized Tin Diselenide Nanorods Composite. *Chemosensors* **2022**, *10*, 401. https://doi.org/10.3390/chemosensors10100401

Academic Editor: Andrea Ponzoni

Received: 3 September 2022
Accepted: 2 October 2022
Published: 8 October 2022

1. Introduction

As an important indicator of air pollution, sulfur dioxide (SO_2) is an irritating, highly toxic and colorless gas, which mainly comes from factory exhaust and automobile exhaust emissions [1,2]. When the emitted concentration of sulfur dioxide in air is too high, it is oxidized to sulfur trioxide and combined with water to form acid rain, which can destroy buildings, pollute the environment and reduce soil fertility [3,4]. In addition, human health is seriously threatened by some harmful gas species including sulfur dioxide gas. Respiratory diseases such as bronchitis and asthma can result from the excessive inhalation of sulfur dioxide gas [5]. There are reports that the human-permissible exposure limit for SO_2 gas is 5 ppm and the long-term exposure limit is 2 ppm [6]. Therefore, the development of portable, efficient and reliable gas sensors can be used to monitor the composition and concentration of environmental gas pollutants in the atmosphere, which can help people to deal with dangerous gases in time, and further protect social safety and human health.

Recently, two-dimensional (2D) layered inorganic materials (such as graphene and transition metal materials) have attracted attention due to their unique crystal structures and characteristics. Among them, two-dimensional transition metal dichalcogenides (TMDs) are ideal materials for preparing field effect transistors [7], photodetectors [8] and electronic devices [9] because of their unique electronic, magnetic, optical and mechanical properties. Due to their planar crystal structure, high specific surface area and physical affinity, they also have unique advantages in sensing applications [10]. According to the reports from a few studies, two-dimensional transition metal compounds such as molybdenum disulfide (MoS_2) [11,12], tungsten disulfide (WS_2) [13,14] and tin diselenide ($SnSe_2$) [15,16] offer good gas sensing characteristics for detecting dangerous gases. $SnSe_2$ is an n-type two-dimensional transition chalcogenide, and its gas-sensitive properties have been widely reported [15–17]. The results of a study reported by Moreira et al. on the ammonia (NH_3) and nitrogen dioxide (NO_2) gas sensitivity of tin-diselenide-based sensors revealed that these sensors had steady repeatability and long-term stability under UV radiation [17]. Zhang et al. fabricated coral-like tin diselenide/metal-organic frameworks (MOFs)-derived nanoflower-like tin dioxide (SnO_2) heteronanostructures via a hydrothermal method. The $SnSe_2/SnO_2$ nanocomposite sensor exhibited excellent NO_2-sensing performance at room temperature, which was significantly improved under UV illumination. The enhanced NO_2 sensing performance was attributed to the formation of an n-n heterostructure and light-excited electrons [18]. By using the template-sacrificial approach, Wang et al. created rod-shaped $SnSe_2$ and polyhedral zinc oxide (ZnO) composite nanostructures, and the $ZnO/SnSe_2$ heterostructures exhibited an enhancement of carbon monoxide (CO)-sensing properties at room temperature [19]. Pan et al. reported a coral-like Au-modified $SnSe_2$ Schottky-junction-based ammonia gas sensor and demonstrated good gas sensitivity to ammonia gas detection. In addition, the effect of Au modification on ammonia gas molecules adsorption was also investigated using a first-principles density functional theory (DFT). Because of these studies, it can be concluded that $SnSe_2$ is a feasible material as building block for constructing high-performance gas sensors [20]. However, the use of a single $SnSe_2$ material for gas sensitivity research has certain limitations, and the test results may show poor selectivity and low sensitivity. According to previous studies, the sensing characteristics of gas sensors can be improved by forming a heterojunction [21,22]. Graphite-phase carbon nitride (g-C_3N_4), as a common two-dimensional material, has a structure like graphene and provides more active sites for gas adsorption. The material has excellent chemical stability, good electron mobility and low cost. In addition, g-C_3N_4 can also promote the uniform dispersion of active ingredients, so it can be used as a stable catalyst carrier [23,24].

In this paper, a composite of $SnSe_2$ functionalized by g-C_3N_4 was prepared via a hydrothermal method and served as the sensitive nanomaterial for SO_2 gas sensing. The nanostructure of the g-$C_3N_4/SnSe_2$ composite was characterized by X-ray diffraction (XRD), scanning electron microscope (SEM) and X-ray photoelectron spectroscopy (XPS). The characterization results verified the successful synthesis of a multilayer g-C_3N_4/rod-shaped $SnSe_2$ composite. The as-prepared g-$C_3N_4/SnSe_2$ composite sensor showed an excellent gas response and rapid adsorption/desorption ability towards SO_2 under the optimal temperature of 200 °C, which was much higher than those of pristine g-C_3N_4 and $SnSe_2$ sensors. A series of comparative experiments proved that the g-$C_3N_4/SnSe_2$ composite sensor demonstrated an excellent response, strong reversibility and good selectivity for ppm-level SO_2 gas detection. In this paper, two innovative two-dimensional materials, g-C_3N_4 and $SnSe_2$, were used to discuss in detail the mechanisms that may improve the SO_2 sensing performance, such as the synergistic interaction between g-C_3N_4 and $SnSe_2$, and the effective structural features.

2. Experimental Section

2.1. Materials

Tin chloride dihydrate ($SnCl_2 \cdot 2H_2O$), hydrazine hydrate ($N_2H_4 \cdot H_2O$), selenium dioxide (SeO_2), ethanol (CH_3CH_2OH) and graphitic carbon nitride (g-C_3N_4) were all supplied by Sinopharm Chemical Reagent Co. Ltd. (Shanghai, China).

2.2. Material Synthesis and Sensor Fabrication

The preparation process of the materials is shown in Figure 1. The $SnSe_2$ was prepared by a hydrothermal reduction method. A total of 0.01 mol $SnCl_2 \cdot 2H_2O$ and 0.02 mol SeO_2

were added into 70 mL deionized water and stirred for 30 min. A total of 10 mL of hydrazine hydrate was added into the above mixture solution, and then transferred to a 100 mL Teflon-lined stainless-steel autoclave, and hydrothermally treated at 180 °C for 24 h. The resulting precipitate was filtered, washed and dried at 60 °C for 8 h to obtain a black tin diselenide product.

Figure 1. Synthesis process of $g-C_3N_4/SnSe_2$ nanomaterials.

A total of 280 mg of $SnSe_2$ powder and 120 mg of $g-C_3N_4$ powder were dissolved in 20 mL of DI water. After vigorously stirring for 1 h, the $g-C_3N_4$ was effectively anchored on the surface of $SnSe_2$. The resulting product was dried at 60 °C for overnight. Finally, the $g-C_3N_4/SnSe_2$ composite was obtained. The composite materials with different ratios (0, 20, 30, and 50%) of $g-C_3N_4$ and $SnSe_2$ were prepared by adjusting the quality ratio of $g-C_3N_4$ and $SnSe_2$. In addition, the two materials were dispersed in deionized water in a certain proportion, and after a period of ultrasonic treatment, they were effectively combined through strong physical effects and interactions between charges.

2.3. Gas Sensor Fabrication

The structure illustration of the ceramic tube based SO_2 gas sensor is shown in Figure 2. It consisted of an Al_2O_3 ceramic tube and a base. The ceramic tube was 4 mm long and 1.2 mm in diameter, and its surface was equipped with gold electrodes and two pairs of platinum wires for electrical signals. The heating resistor of the Ni–Cr alloy coil passed through a hollow ceramic tube for heating. The sensing layer materials were coated on the surface of the ceramic tube, and the electrodes were led out to complete the preparation of the sensor. After preparing the sensing film, the sensor was dried at 60 °C for 6 h and then aged at 200 °C for 24 h before the test to obtain good resistance stability. The SO_2-sensing measurement was performed in a home-made gas sensing detection system [25] as shown in Figure 3. The sensor was placed in a home-made chamber, and the SO_2 gas with different concentrations of 1–200 ppm was obtained by diluting 1000 ppm SO_2 standard gas with high-purity air. The sensor resistance was measured with an Agilent 34970A digital multimeter and connected with a computer through RS-232 for data acquisition. The operation temperature of the sensor was controlled by an applied voltage to the Ni–Cr heating resistor. A steady power supply of GPD-4303S was employed for applying voltage for heating. The response of the sensor is defined as $S = (R_A - R_G)/R_A \times 100\%$ (R_A: resistance in air; R_G: resistance in SO_2 gas).

Figure 2. Structure illustration of the ceramic-tube-based SO_2 gas sensor.

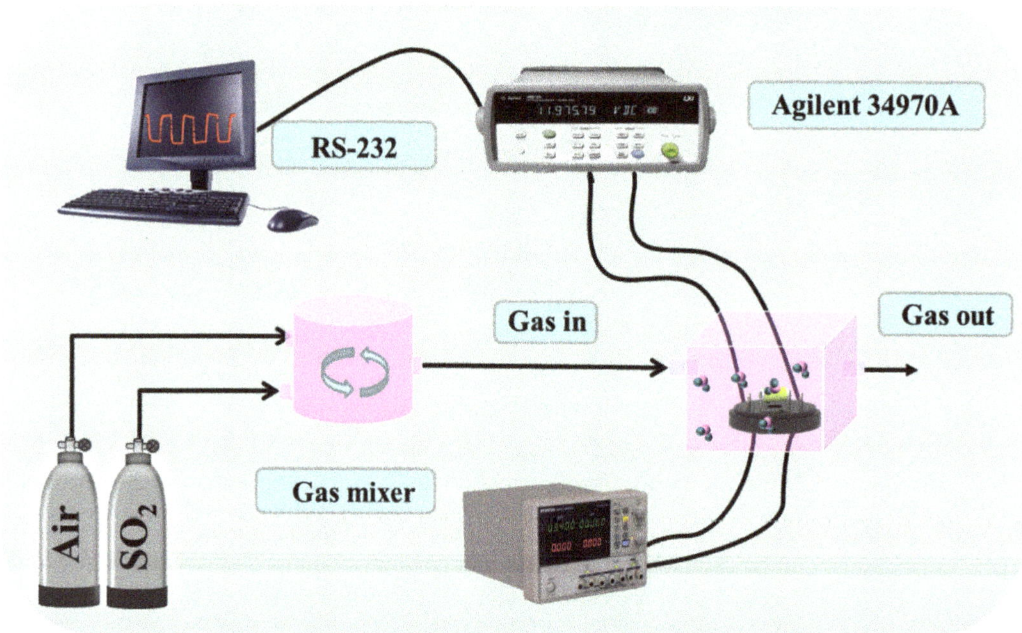

Figure 3. Schematic illustration of sensor performance test platform.

3. Results and Discussion

3.1. Structure Characterization

The crystal structures of the $SnSe_2$, g-C_3N_4 and g-C_3N_4/$SnSe_2$ were characterized by X-ray powder diffraction (XRD, Rigaku D/Max-2550, Rigaku, Japan) with Cu Kα radiation (λ = 0.15418 nm). X-ray photoelectron spectroscopy (XPS, Thermo Scientific K-Alpha XPS spectrometer, Thermo Scientific, Waltham, MA, USA) was used to detect the chemical composition of the samples, and the morphology of pristine $SnSe_2$, g-C_3N_4 and g-C_3N_4/$SnSe_2$ nanocomposites were observed by a scanning electron microscope (SEM, Hitachi S-4800, Hitachi, Japan). The g-C_3N_4/$SnSe_2$ sample with a ratio of 30% g-C_3N_4 was used for characterization.

The crystal phases of the as-prepared materials were identified by XRD analysis. As shown in Figure 4a, the peaks of $SnSe_2$ (JCPDS card number 23-0602) located at 14.38°, 26.98°, 30.88°, 40.17°, 47.77°, 50.19°, 52.58°, 57.92°, 60.34°, 63.96° and 78.22° corresponded to the (001), (100), (011), (012), (110), (111), (103), (201), (004), (202) and (121) diffraction planes of the orthorhombic-phase $SnSe_2$ structures, respectively [26]. The XRD pattern of g-C_3N_4 was well in accordance with the hexagonal crystal g-C_3N_4 (JCPDS Card no. 36-1451) [27]. The formation of the two diffraction peaks of (100) and (002) can be attributed to the in-plane structure stacking of the tri-s-triazine part and the in-plane stacking of the conjugated aromatic hydrocarbon system, respectively. There were corresponding characteristic peaks of g-C_3N_4 and $SnSe_2$ in the XRD pattern of the g-C_3N_4/$SnSe_2$ nanocomposite, and there were no other characteristic peaks, indicating the successful preparation of the $SnSe_2$/g-C_3N_4 composite.

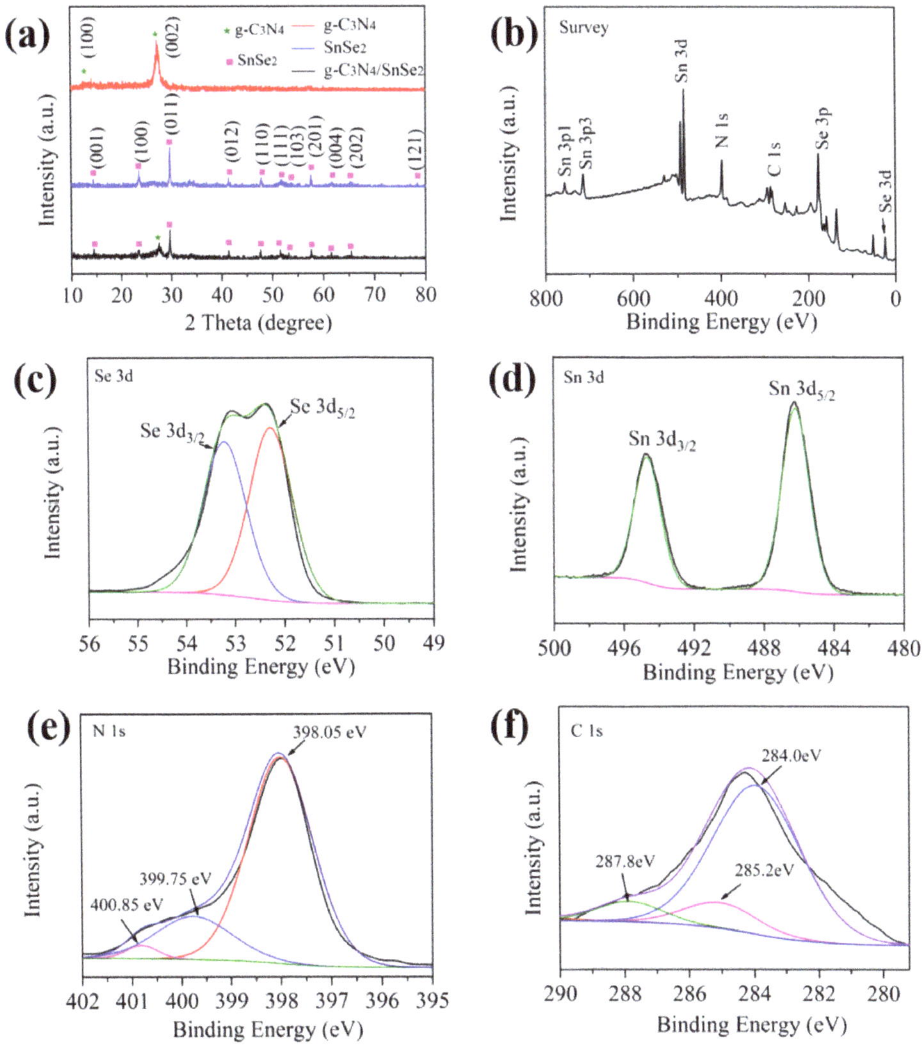

Figure 4. (**a**) XRD patterns of g-C_3N_4, $SnSe_2$ and g-C_3N_4/$SnSe_2$. XPS spectra of g-C_3N_4/$SnSe_2$ product: (**b**) full spectrum, (**c**) Se 3d, (**d**) Sn 3d, (**e**) N 1s and (**f**) C 1s.

XPS is a measurement technique for detecting the elemental composition and chemical valence of materials. Figure 4b is the total spectrum of the g-C_3N_4/$SnSe_2$ composite, and the existence of the four elements Se, Sn, C and N was confirmed from the peak positions in the figure. The Se 3D spectrum shown in Figure 4c is represented by two peaks with binding energies of 53.4 and 52.2 eV, corresponding to Se $3d_{5/2}$ and Se $3d_{3/2}$, respectively [28–30]. The Sn 3d spectrum shown in Figure 4d shows two distinctive peaks at energies of 486.3 and 495.2 eV, which correspond to Sn $3d_{5/2}$ and $3d_{3/2}$, respectively.

Since there was an energy difference of 8.4 eV between the two peaks, the Sn^{4+} state can be confirmed [31]. Figure 4e is the characteristic peak of N 1s. Three asymmetric peaks at 398.3, 400.0 and 401.0 eV were attributed to C=N-C, N-$(C)_3$ and C-N-H, respectively. Figure 4f shows the XPS spectrum of C 1s, where the characteristic peak at 284.6 eV was due to the formation of indefinite carbon absorption on the surface of the material, and the diffraction peak at 287.8 eV may be formed by N=C-N coordination [32].

SEM is a detection technology used to observe the micro-morphology of nanomaterials. Figure 5 shows the SEM characterization of the g-C_3N_4/$SnSe_2$-sensitive material. Figure 5a shows the SEM characterization of the intrinsic $SnSe_2$. The $SnSe_2$ nanorods were composed of numerous nanoparticles, the cross-sectional radius and the length of which were about 300−500 nm and 3−5 μm, respectively. Figure 5b shows the SEM characterization image of the graphite-like-phase carbon nitride. The exfoliated g-C_3N_4 nanosheets had a 2D layered morphology, which was manifested as multiple thin layers stacking together. Figure 5c,d show the surface morphology of the g-C_3N_4/$SnSe_2$ composite. It can be clearly observed that the nanorod-shaped $SnSe_2$ and multilayer g-C_3N_4 grew and aggregated together well, which show s that the g-C_3N_4/$SnSe_2$ composite was successfully synthesized.

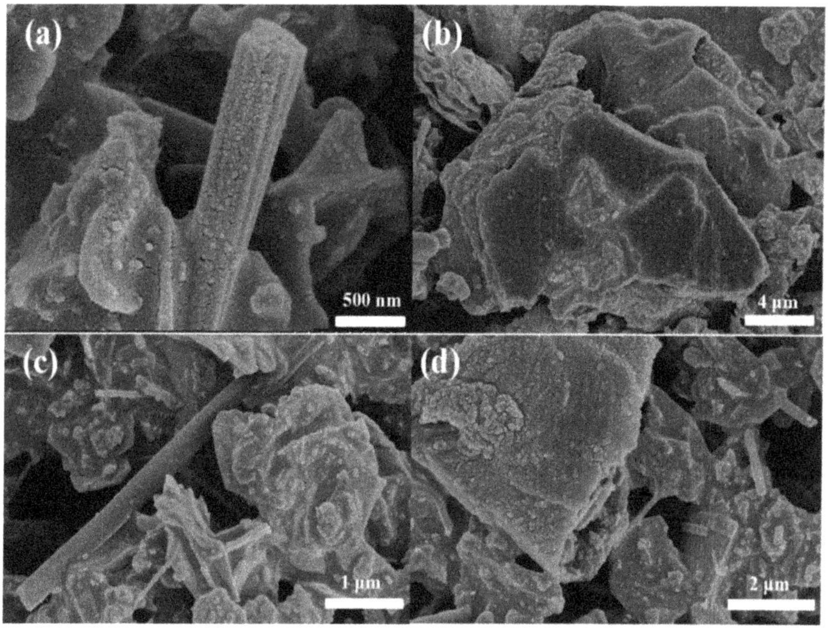

Figure 5. SEM images of (**a**) $SnSe_2$, (**b**) g-C_3N_4, and (**c**,**d**) g-C_3N_4/$SnSe_2$.

3.2. SO_2-Sensing Properties

The working temperature is an important key variable related to the sensing characteristics of the g-C_3N_4/$SnSe_2$ sensor [33–35]. It affects the chemical adsorption and surface reaction of gas molecules. Therefore, to determine the optimal operating temperature as well as the optimum ratio of the two materials, we studied the responses of pristine g-C_3N_4, pristine $SnSe_2$ and the x% g-C_3N_4/$SnSe_2$ (x = 0, 20, 30, 50) film sensors to 20 ppm SO_2 gas at different operating temperatures. As shown in Figure 6a, the operating temperatures and responses of the sensors presented a "triangular" shape. The optimal temperatures of g-C_3N_4, $SnSe_2$ and g-C_3N_4/$SnSe_2$ were 250 °C, 200 °C and 200 °C, respectively. When the g-C_3N_4 ratio was 30%, the response (28.9%) of g-C_3N_4/$SnSe_2$ at 200 °C was the highest, which was much higher than those of the pristine g-C_3N_4 and $SnSe_2$. The gas-sensing properties of the $SnSe_2$, g-C_3N_4 and 30% g-C_3N_4/$SnSe_2$ sensors were investigated by recording the changes in resistance when they were exposed to different concentrations (1−200 ppm) of SO_2 gas at 200 °C, as shown in Figure 6b. When switched to air, the resistances of the three sensors returned to the base value in air. The response values of 30% g-C_3N_4/$SnSe_2$ towards 1, 5, 10, 20, 50, 100, 200 ppm SO_2 gas were, respectively, 8.82%, 15.59%, 20.22%, 28.52%, 35.09%, 41.54% and 44.34%, higher than those of the pristine $SnSe_2$ and g-C_3N_4. Figure 6c shows the fitting curves of the three sensors between sensitivity (Y) and concentration of SO_2 (X). The fitting functions of g-C_3N_4, $SnSe_2$ and g-C_3N_4/$SnSe_2$ sensors were $Y = 14.89 - 11.82 \times 0.98^X$, $Y = 22.76 - 17.85 \times 0.95^X$ and $Y = 42.16 - 33.15 \times 0.96^X$, respectively. The R^2 of the fitting curve of the g-C_3N_4/$SnSe_2$ sensor was 0.975. Figure 6d shows the response/recovery curves of the g-C_3N_4/$SnSe_2$ sensor toward the desired concentrations of SO_2 gas. The sensor exhibited stable response and recovery behaviors.

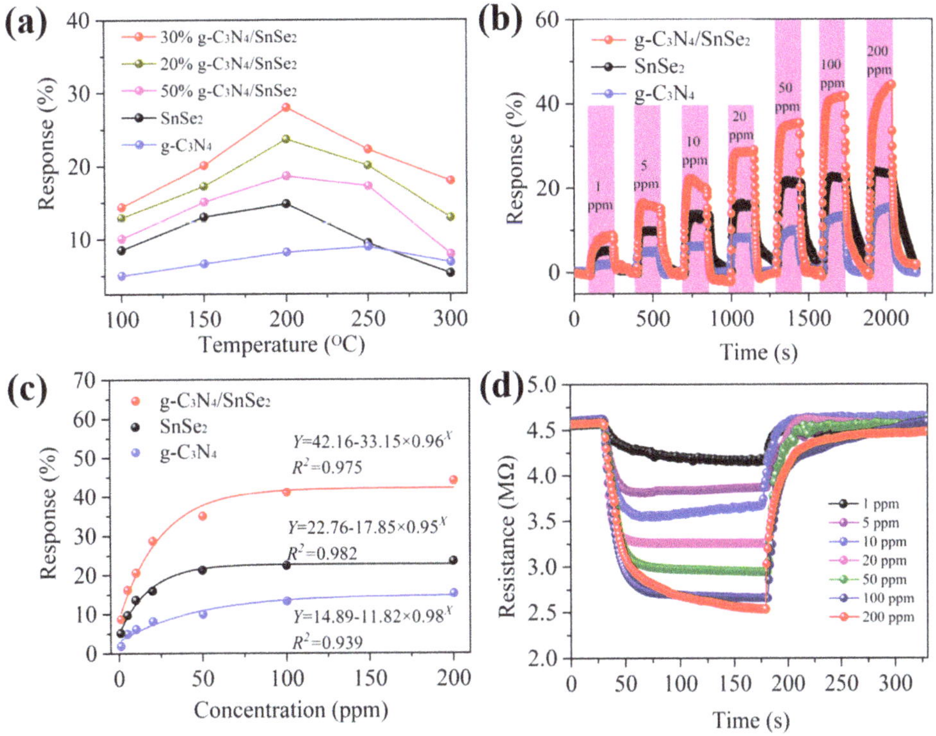

Figure 6. (**a**) Responses of g-C_3N_4/$SnSe_2$ composites with different quality ratios to 20 ppm SO_2 gas at various temperatures. (**b**) Sensing properties of pristine g-C_3N_4, $SnSe_2$ and g-C_3N_4/$SnSe_2$ composite sensors at different concentrations of SO_2 gas at 200 °C. (**c**) Response fitting curves of three sensors versus SO_2 concentration at 200 °C. (**d**) Typical response–recovery curves for various concentrations of SO_2 gas at 200 °C.

The repeatability of the g-C_3N_4/$SnSe_2$ composite sensor against a gas with a concentration of 50 ppm SO_2 at 200 °C is examined in Figure 7a. For each run, the resistance could fully recover its initial state and changes from 4.5 MΩ to 2.9 MΩ, showing a good reproducibility. Figure 7b shows the response/recovery curves of the g-C_3N_4, $SnSe_2$, and g-C_3N_4/$SnSe_2$ composite sensors exposed to 50 ppm SO_2 gas. The response/recovery time of the g-C_3N_4/$SnSe_2$ sensor was 22/24 s, while the values of the pristine g-C_3N_4 and $SnSe_2$ sensors were 46/55 s and 26/82 s, respectively, suggesting that the g-C_3N_4/$SnSe_2$ sensor had a faster detection rate towards SO_2 gas compared with the pristine g-C_3N_4 and $SnSe_2$ sensors at 200 °C. Excellent selectivity is also an important factor for nanomaterial-based gas sensors [36,37]. Therefore, we further studied the SO_2 selectivity of the g-C_3N_4/$SnSe_2$ composite sensor at 200 °C. The sensor was exposed to 50 ppm of various interfering gases, such as LPG, CO, CH_4, H_2S and H_2. As shown in Figure 7c, the g-C_3N_4/$SnSe_2$ composite sensor had the highest response to SO_2 gas, indicating that the selectivity to SO_2 gas was excellent. Additionally, the long-term stability of the g-C_3N_4/$SnSe_2$ composite sensor for various SO_2 gas concentrations (1, 10 and 50 ppm) at a working temperature of 200 °C was examined. The sensor's response had no obvious changes in a period under the same experimental conditions and exhibited a good stability, as shown in Figure 7d.

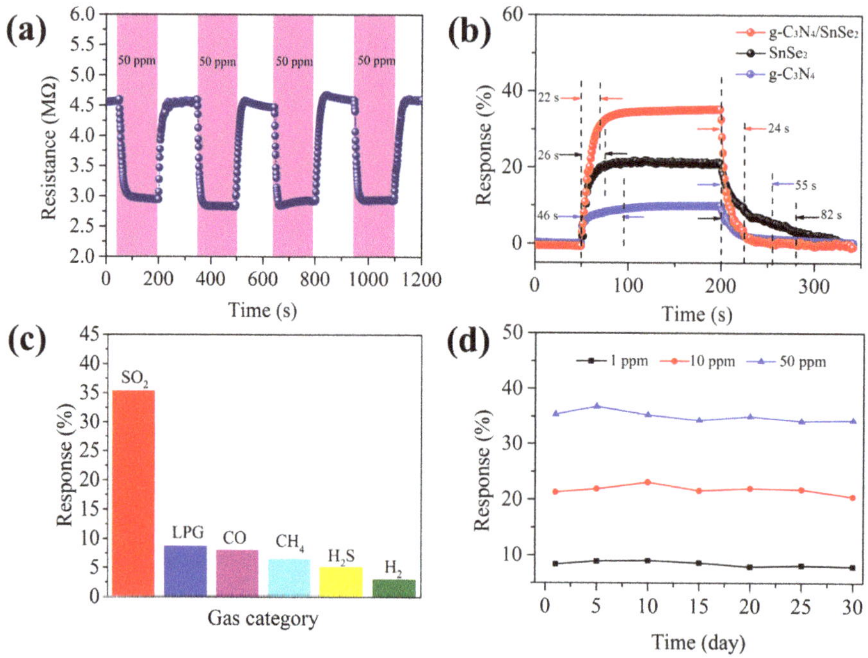

Figure 7. At the optimal temperature of 200 °C (**a**) repeatability of g-C_3N_4/$SnSe_2$ composite sensor. (**b**) Response/recovery curves of g-C_3N_4, $SnSe_2$ and g-C_3N_4/$SnSe_2$ composite sensors exposed to 50 ppm SO_2 gas. (**c**) Selectivity and (**d**) long-term stability of g-C_3N_4/$SnSe_2$ composite sensor.

Numerous studies have described the detection of SO_2 gases utilizing a variety of sensitive materials up to this point. As far as we know, there have been no studies using g-C_3N_4/$SnSe_2$ for SO_2 gas detection. Table 1 lists the comparison of this work with previously reported works. The different SO_2 gas sensors are compared in terms of sensing environment, response value and gas concentration. The comparison results showed that the as-prepared g-C_3N_4/$SnSe_2$ sensor featured a higher response value and a lower operating temperature.

Table 1. Comparison of the SO_2-sensing performance between the present work and previous reported studies.

Sensing Material	Sensing Environment	Response	Concentration	Ref.
AlGaN/ZnO/rGO	RT	2.5%	120 ppb	[38]
SnO_2/rGO	RT/UV	1.7%	5 ppm	[39]
Pt/rGO	120 °C	5%	100 ppm	[40]
g-C_3N_4/rGO	RT	3.2%	100 ppm	[41]
SnO_2/MWCNT	60 °C	6	500 ppm	[42]

Table 1. *Cont.*

Sensing Material	Sensing Environment	Response	Concentration	Ref.
$V_2O_5/WO_3/TiO_2$	400 °C	5%	20 ppm	[43]
WO_3	350 °C	5%	1 ppm	[44]
Cu–SnO_2	400 °C	1.1%	20 ppm	[45]
SnO_2–PANI	RT	3.1%	4 ppm	[46]
TiO_2/rGO	RT	11.14%	5 ppm	[47]
NiO–SnO_2	180 °C	8.3%	50 ppm	[48]
TiO_2	200 °C	11%	10 ppm	[49]
ZnO	RT	0.2%	100 ppm	[50]
WO_3–PANI	RT	4.3%	5 ppm	[51]
Ni–MoS_2	RT	7.4%	5 ppm	[52]
V_2O_5/SnO_2	350 °C	45%	5 ppm	[53]
V-doped TiO_2	400 °C	10%	10 ppm	[54]
GO	RT	6%	5 ppm	[55]
PANI	RT	4.2%	10 ppm	[56]
g-$C_3N_4/SnSe_2$	200 °C	28.9%	20 ppm	This work

3.3. SO_2 Gas-Sensing Mechanism

The sensing mechanism has been widely explained using the surface charge caused by adsorbed oxygen. For the pristine $SnSe_2$ gas sensor, when exposed to air, oxygen molecules in air adsorb on the surface of the sensing material to form adsorbed oxygen molecules. The adsorbed oxygen molecules extract electrons from the conduction band of $SnSe_2$ to form chemically adsorbed oxygen O^- at 200 °C. On the surface of the g-$C_3N_4/SnSe_2$ sensing material, an electron depletion layer consequently forms. The following is a description of the reactions [57,58]:

$$O_2\ (gas) \rightarrow O_2\ (ads) \tag{1}$$

$$O_2\ (ads) + 2e^- \rightarrow 2O^-\ (ads) \tag{2}$$

When the reducing gas SO_2 is introduced, the SO_2 molecules adsorbed on the surface of the sensing material will further react with O^- ions and release electrons to the $SnSe_2$ conduction band to form SO_3, thereby increasing the number of charge carriers and reducing the resistance of the sensor. The specific reaction process is illustrated as follows [59]:

$$SO_2\ (ads) + O^-\ (ads) \rightarrow SO_3 + e^- \tag{3}$$

In our experiment, it is worth noting that the g-$C_3N_4/SnSe_2$ sensor exhibited an improved response to SO_2 gas than the pristine C_3N_4 and $SnSe_2$ sensors. This phenomenon could be explained by two positive factors of g-C_3N_4 decoration. Firstly, the morphology, specific surface area and electrical properties of the material are important factors affecting its gas-sensing performance [60]. The addition of the layered two-dimensional structure of g-C_3N_4 increases the specific surface area of the composite material and provides more active sites, increasing the production of adsorbed oxygen. Another factor influencing the effectiveness of the gas sensing system is the n-n heterojunction created between g-C_3N_4 and $SnSe_2$ [61–63]. Figure 8a shows the energy band structure of the heterojunction formed by the n-type $SnSe_2$ nanorods and n-type g-C_3N_4 layered nanosheets in air, where the band gaps of $SnSe_2$ and g-C_3N_4 were 1.37 and 2.7 eV, respectively. Their Fermi levels were different, and the work function (4.3 eV) of $SnSe_2$ was lower than that (4.67 eV) of g-C_3N_4 [64]. When the two materials contact with different work functions, electrons will transfer from $SnSe_2$ to g-C_3N_4 to reach a Fermi energy balance. Therefore, the n-n heterojunction is formed at the interface between $SnSe_2$ and g-C_3N_4 in air [65]. When exposed to SO_2 gas, SO_2 molecules absorbed on the surface of the composite material react with O^- to generate electrons [66]. Therefore, the carrier concentration in the heterojunction increases by receiving electrons, resulting in a narrowing of the depletion layer at the interface of the two materials [61,67,68], which reduces the resistance of the sensor (Figure 8b). The built-in electric field generated by the nano-scale heterojunction can accelerate the separation process of electrons and facilitate electron transfer.

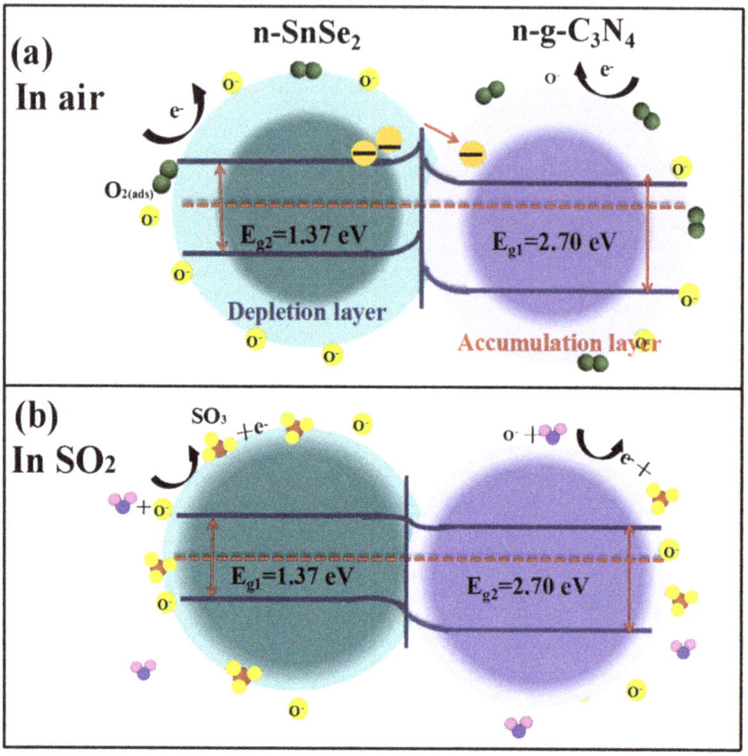

Figure 8. Schematic diagram of microscopic sensing mechanism of g-C_3N_4/$SnSe_2$ composite sensor.

4. Conclusions

In this paper, a g-C_3N_4 functionalized $SnSe_2$ composite thin film sensor was successfully prepared. XRD, XPS and SEM techniques were used to characterize the elements and structure of the g-C_3N_4/$SnSe_2$ composite material. The optimal temperature (200 °C) of the g-C_3N_4/$SnSe_2$ composite sensor was determined through gas-sensing experiments under different working temperatures. The SO_2 gas-sensing performance for the g-C_3N_4/$SnSe_2$ composite sensor was investigated at the optimal temperature. The experimental results showed that under the optimal temperature, the 30% g-C_3N_4/$SnSe_2$ composite sensor gained the highest sensitivity to SO_2 gas. The two-dimensional layered structure of the g-C_3N_4-modified n-type $SnSe_2$ nanorods not only increased the specific surface area and the gas adsorption site of the g-C_3N_4/$SnSe_2$ sensor, but also formed an n-n heterojunction between the g-C_3N_4 nanosheets and $SnSe_2$ nanorods, which improved the sensing performance of the g-C_3N_4/$SnSe_2$ sensor toward SO_2 gas.

Author Contributions: Conceptualization, H.Z. and Q.P.; methodology, D.Z.; validation, H.Z. and Q.P.; formal analysis, H.Z. and Q.P.; investigation, Y.Z. (Yanting Zhang); data curation, H.Z. and Q.P.; writing—original draft preparation, Q.P.; writing—review and editing, Y.Z. (Yating Zhang) and D.Z.; visualization, Q.P. and Y.Z. (Yating Zhang); supervision, D.Z.; funding acquisition, D.Z. All authors have read and agreed to the published version of the manuscript.

Funding: This research was funded by the Original Innovation Special Project of Science and Technology Plan of Qingdao West Coast New Area, China (2020-85).

Institutional Review Board Statement: Not applicable.

Informed Consent Statement: Not applicable.

Data Availability Statement: Not applicable.

Conflicts of Interest: The authors declare no conflict of interest.

References

1. Das, S.; Rana, S.; Mursalin, S.; Rana, P.; Sen, A. Sonochemically prepared nanosized $BiFeO_3$ as novel SO_2 sensor. *Sens. Actuators B* **2015**, *218*, 122–127. [CrossRef]
2. Jung, G.; Jeong, Y.; Hong, Y.; Wu, M.; Hong, S.; Shin, W.; Park, J.; Jang, D.; Lee, J. SO_2 gas sensing characteristics of FET- and resistor-type gas sensors having WO_3 as sensing material. *Solid-State Electron.* **2020**, *165*, 107747. [CrossRef]
3. Ma, C.; Hao, X.; Yang, X.; Liang, X.; Liu, F.; Liu, T.; Yang, C.; Zhu, H.; Lu, G. Sub-ppb SO_2 gas sensor based on NASICON and $La_xSm_{1-x}FeO_3$ sensing electrode. *Sens. Actuators B Chem.* **2018**, *256*, 648–655. [CrossRef]
4. Liu, L.; Liu, S. Oxygen vacancies as an efficient strategy for promotion of low concentration SO_2 gas sensing: The case of Au-modified SnO_2. *ACS Sustain. Chem. Eng.* **2018**, *6*, 13427–13434. [CrossRef]
5. Liu, F.; Wang, J.; Jiang, L.; You, R.; Wang, Q.; Wang, C.; Lin, Z.; Yang, Z.; He, J.; Liu, A.; et al. Compact and planar type rapid response ppb-level SO_2 sensor based on stabilized zirconia and $SrMoO_4$ sensing electrode. *Sens. Actuators B Chem.* **2020**, *307*, 127655. [CrossRef]
6. Tong, P.; Hoa, N.; Nha, H.; Duy, N.; Hung, C.; Hieu, N. SO_2 and H_2S sensing properties of hydrothermally synthesized CuO nanoplates. *J. Electron. Mater.* **2018**, *47*, 7170–7178. [CrossRef]
7. Goel, N.; Kumar, R.; Jain, S.; Rajamani, S.; Roul, B.; Gupta, G.; Kumar, M.; Krupanidhi, S. A high-performance hydrogen sensor based on a reverse-biased MoS_2/GaN heterojunction. *Nanotechnology* **2019**, *30*, 314001. [CrossRef]
8. Xu, T.; Han, Y.; Lin, L.; Xu, J.; Fu, Q.; He, H.; Song, B.; Gai, Q.; Wang, X. Self-power position-sensitive detector with fast optical relaxation time and large position sensitivity basing on the lateral photovoltaic effect in tin diselenide films. *J. Alloys Compd.* **2019**, *790*, 941–946. [CrossRef]
9. Javed, Y.; Mirza, S.; Li, C.; Xu, X.; Rafiq, M. The role of biaxial strain and pressure on the thermoelectric performance of $SnSe_2$: A first principles study. *Semicond. Sci. Technol.* **2019**, *34*, 055009. [CrossRef]
10. Gu, D.; Wang, X.; Liu, W.; Li, X.; Lin, S.; Wang, J.; Rumyantseva, M.; Gaskovc, A.; Akbar, S. Visible-light activated room temperature NO_2 sensing of SnS_2 nanosheets based chemiresistive sensors. *Sens. Actuators B Chem.* **2020**, *305*, 127455. [CrossRef]
11. Jaiswal, J.; Sanger, A.; Tiwari, P.; Chandra, R. MoS_2 hybrid heterostructure thin film decorated with CdTe quantum dots for room temperature NO_2 gas sensor. *Sens. Actuators B Chem.* **2020**, *305*, 127437. [CrossRef]
12. Shen, J.; Yanga, Z.; Wang, Y.; Xu, L.; Liu, R.; Liu, X. The gas sensing performance of borophene/MoS_2 heterostructure. *Appl. Surf. Sci.* **2020**, *504*, 144412. [CrossRef]
13. Liu, D.; Tang, Z.; Zhang, Z. Comparative study on NO_2 and H_2S sensing mechanisms of gas sensors based on WS_2 nanosheets. *Sens. Actuators B Chem.* **2020**, *303*, 127114. [CrossRef]
14. Zhou, Q.; Zhu, L.; Zheng, C.; Wang, J. Nanoporous Functionalized WS2/MWCNTs Nanocomposite for Trimethylamine Detection Based on Quartz Crystal Microbalance Gas Sensor. *ACS Appl. Mater. Interfaces* **2021**, *13*, 41339–41350. [CrossRef] [PubMed]
15. Pawar, M.; Kadam, S.; Late, D. High-performance sensing behavior using electronic ink of 2D $SnSe_2$ nanosheets. *Chem. Sel.* **2017**, *2*, 4068–4075. [CrossRef]
16. Wang, X.; Liu, Y.; Dai, J.; Chen, Q.; Huang, X.; Huang, W. Solution-processed p-SnSe/n-$SnSe_2$ hetero-structure layers for ultrasensitive NO_2 detection. *Chem. Eur. J.* **2020**, *26*, 3870–3876. [CrossRef]
17. Moreira, O.; Cheng, W.; Fuh, H.; Chien, W.; Yan, W.; Fei, H.; Xu, H.; Zhang, D.; Chen, Y.; Zhao, Y.; et al. High Selectivity Gas Sensing and Charge Transfer of $SnSe_2$. *ACS Sens.* **2019**, *4*, 2546–2552. [CrossRef] [PubMed]
18. Li, T.; Zhang, D.; Pan, Q.; Tang, M.; Yu, S. UV enhanced NO_2 gas sensing at room temperature based on coral-like tin diselenide/MOFs-derived nanoflower-like tin dioxide heteronanostructures. *Sens. Actuators B Chem.* **2022**, *355*, 131049. [CrossRef]
19. Wang, D.; Zhang, D.; Pan, Q.; Wang, T.; Chen, F. Gas sensing performance of carbon monoxide sensor based on rod-shaped tin diselenide/MOFs derived zinc oxide polyhedron at room temperature. *Sens. Actuators B Chem.* **2022**, *371*, 132481. [CrossRef]
20. Pan, Q.; Li, T.; Zhang, D. Ammonia gas sensing properties and density functional theory investigation of coral-like Au-$SnSe_2$ Schottky junction. *Sens. Actuators B Chem.* **2021**, *332*, 129440. [CrossRef]
21. Sun, K.; Zhan, G.; Chen, H.; Lin, S. Low-Operating-Temperature NO_2 Sensor Based on a CeO_2/ZnO Heterojunction. *Sensors* **2022**, *21*, 8269. [CrossRef] [PubMed]
22. Liu, K.; Liu, H.; Wang, J.; Feng, L. Synthesis and characterization of $SnSe_2$ hexagonal nanoflakes. *Mater. Lett.* **2009**, *63*, 512–514. [CrossRef]
23. David, S.P.S.; Veeralakshmi, S.; Priya, M.S.; Nehru, S.; Kalaiselvam, S. Room-temperature chemiresistive g-C_3N_4/Ag_2ZrO_3 nanocomposite gas sensor for ethanol detection. *J. Mater. Sci.-Mater. El.* **2022**, *33*, 11498–11510. [CrossRef]
24. Pan, W.; Zhang, Y.; Yu, S.; Liu, X.; Zhang, D. Hydrogen sulfide gas sensing properties of metal organic framework-derived α-Fe_2O_3 hollow nanospheres decorated with $MoSe_2$ nanoflowers. *Sens. Actuators B Chem.* **2021**, *344*, 130221. [CrossRef]
25. Yang, Z.; Zhang, D.; Chen, H. MOF-derived indium oxide hollow microtubes/MoS_2 nanoparticles for NO_2 gas sensing. *Sens. Actuators B Chem.* **2019**, *300*, 127037. [CrossRef]

26. Liu, C.; Huang, Z.; Wang, D.; Wang, X.; Miao, L.; Wang, X.; Wu, S.; Toyama, N.; Asaka, T.; Chen, J.; et al. Dynamic Ag^+-intercalation with $AgSnSe_2$ nanoprecipitates in Cl-doped polycrystalline $SnSe_2$ toward ultra-high thermoelectric performance. *J. Mater. Chem. A* **2019**, *7*, 9761–9772. [CrossRef]
27. Hang, N.; Zhang, S.; Yang, W. Efficient exfoliation of g-C_3N_4 and NO_2 sensing behavior of graphene/g-C_3N_4 nanocomposite. *Sens. Actuators B Chem.* **2017**, *248*, 940–948. [CrossRef]
28. Zhang, D.; Yang, Z.; Li, P.; Pang, M.; Xue, Q. Flexible self-powered high-performance ammonia sensor based on Au-decorated $MoSe_2$ nanoflowers driven by single layer MoS_2-flake piezoelectric nanogenerator. *Nano Energy* **2019**, *65*, 103974. [CrossRef]
29. Yan, J.; Song, Z.; Xu, H.; Lee, L. Carbon-mediated electron transfer channel between SnO_2 QDs and g-C_3N_4 for enhanced photocatalytic H_2 production. *Chem. Eng. J.* **2021**, *425*, 131512. [CrossRef]
30. Yan, J.; Wang, T.; Qiu, S.; Song, Z.; Zhu, W.; Liu, X.; Liang, J.; Sun, C.; Li, H. Insights into the efficient charge separation over Nb_2O_5/2D-C_3N_4 heterostructure for exceptional visible-light driven H_2 evolution. *J. Energy Chem.* **2022**, *65*, 548–555. [CrossRef]
31. Xiong, Y.; Lu, W.; Ding, D.; Zhu, L.; Li, X.; Ling, C.; Xue, Q. Enhanced room temperature oxygen sensing properties of LaOCl-SnO_2 hollow spheres by UV light illumination. *ACS Sens.* **2017**, *2*, 679–686. [CrossRef] [PubMed]
32. Zeng, B.; Zhang, L.; Wan, X. Fabrication of α-Fe_2O_3/g-C_3N_4 composites for cataluminescence sensing of H_2S. *Sens. Actuators B Chem.* **2015**, *211*, 370–376. [CrossRef]
33. Guo, J.; Zhang, J.; Zhu, M.; Ju, D.; Xu, H.; Cao, B. High-performance gas sensor based on ZnO nanowires functionalized by Au nanoparticles. *Sens. Actuator B* **2014**, *199*, 339–345. [CrossRef]
34. Zhou, X.; Feng, W.; Wang, C.; Hu, X.; Li, X.; Sun, P.; Shimanoe, K.; Yamazoe, N.; Lu, G. Porous ZnO/$ZnCo_2O_4$ hollow spheres: Synthesis, characterization, and applications in gas sensing. *J. Mater. Chem. A* **2014**, *2*, 17683–17690. [CrossRef]
35. Bai, S.; Chen, C.; Luo, R.; Chen, A.; Li, D. Synthesis of MoO_3/reduced graphene oxide hybrids and mechanism of enhancing H_2S sensing performances. *Sens. Actuator B* **2015**, *216*, 113–120. [CrossRef]
36. Zhang, D.; Cao, Y.; Wu, J.; Zhang, X. Tungsten trioxide nanoparticles decorated tungsten disulfide nanoheterojunction for highly sensitive ethanol gas sensing application. *Appl. Surf. Sci.* **2020**, *503*, 144063. [CrossRef]
37. Zhang, D.; Yang, Z.; Yu, S.; Mi, Q.; Pan, Q. Diversiform metal oxide-based hybrid nanostructure for gas sensing with versatile prospects. *Coord. Chem. Rev.* **2020**, *413*, 213272. [CrossRef]
38. Triet, N.M.; Duy, L.T.; Hwang, B.U.; Hanif, A.; Siddiqui, S.; Park, K.H.; Cho, C.Y.; Lee, N.E. High-performance Schottky diode gas sensor based on the heterojunction of three-dimensional nanohybrids of reduced graphene oxide-vertical ZnO vanorods on an AlGaN/GaN layer. *ACS Appl. Mater. Interfaces* **2017**, *9*, 30722–30732. [CrossRef]
39. Li, W.; Guo, J.; Cai, L.; Qi, W.; Sun, Y.; Xu, J.; Sun, M.; Zhu, H.; Xiang, L.; Xie, D.; et al. UV light irradiation enhanced gas sensor selectivity of NO_2 and SO_2 using rGO functionalized with hollow SnO_2 nanofibers. *Sens. Actuators B Chem.* **2019**, *290*, 443–452. [CrossRef]
40. Chen, D.; Tang, J.; Zhang, X.; Fang, J.; Li, Y.; Zhuo, R. Detecting decompositions of sulfur hexafluoride using reduced graphene oxide decorated with Pt nanoparticles. *J. Phys. D Appl. Phys.* **2018**, *51*, 185304. [CrossRef]
41. Chen, A.; Liu, R.; Peng, X.; Chen, Q.; Wu, J. 2D hybrid nanomaterials for selective detection of NO_2 and SO_2 using "light on and off" strategy. *ACS Appl. Mater. Interfaces* **2017**, *9*, 37191–37200. [CrossRef] [PubMed]
42. Tyagi, P.; Sharma, A.; Tomar, M.; Gupta, V. A comparative study of RGO-SnO_2 and MWCNT-SnO_2 nanocomposites based SO_2 gas sensors. *Sens. Actuators B Chem.* **2017**, *248*, 980–986. [CrossRef]
43. Chaudhary, V.; Kaur, A. Enhanced room temperature sulfur dioxide sensing behaviour of in situ polymerized polyaniline–tungsten oxide nanocomposite possessing honeycomb morphology. *RSC Adv.* **2015**, *5*, 73535–73544. [CrossRef]
44. Stankova, M.; Vilanova, X.; Calderer, J. Detection of SO_2 and H_2S in CO_2 stream by means of WO_3-based micro-hotplate sensors. *Sens. Actuators B Chem.* **2004**, *2*, 219–225. [CrossRef]
45. Ghimbeu, M.; Lumbreras, M.; Schoonman, J. Electrosprayed metal oxide semiconductor films for sensitive and selective detection of hydrogen sulfide. *Sensors* **2009**, *9*, 9122–9132. [CrossRef] [PubMed]
46. Shen, F.; Wang, D.; Liu, R. Edge-tailored graphene oxide nanosheet-based field effect transistors for fast and reversible electronic detection of sulfur dioxide. *Nanoscale* **2012**, *5*, 537–540. [CrossRef] [PubMed]
47. Zhang, D.; Liu, J.; Jiang, C.; Li, P.; Sun, Y. High-performance sulfur dioxide sensing properties of layer-by-layer self-assembled titania-modified graphene hybrid nanocomposite. *Sens. Actuators B Chem.* **2017**, *245*, 560–567. [CrossRef]
48. Tyagi, P.; Sharma, A.; Tomar, M.; Gupta, V. Metal oxide catalyst assisted SnO_2 thin film based SO_2 gas sensor. *Sens. Actuators B Chem.* **2016**, *224*, 282–289. [CrossRef]
49. Zhang, X.; Tie, J.; Zhang, J. A Pt-doped TiO_2 nanotube arrays sensor for detecting SF_6 decomposition products. *Sensors* **2013**, *13*, 14764–14776. [CrossRef]
50. Chaudhary, V.; Kaur, A. Solitary surfactant assisted morphology dependent chemiresistive polyanilne sensors for room temperature monitoring of low ppm sulphur dioxide. *Polym. Int.* **2015**, *64*, 1475–1481. [CrossRef]
51. Lupan, O.; Chow, L.; Chai, G. A single ZnO tetrapod-based sensor. *Sens. Actuators B Chem.* **2009**, *141*, 511–517. [CrossRef]
52. Zhang, D.; Wu, J.; Peng, L. Room-temperature SO_2 gas-sensing properties based on a metal-doped MoS_2 nanoflower: An experimental and density functional theory investigation. *J. Mater. Chem. A* **2017**, *5*, 20666–20677. [CrossRef]
53. Das, S.; Chakraborty, S.; Parkash, O. Vanadium doped tin dioxide as a novel sulfur dioxide sensor. *Talanta* **2008**, *75*, 385–389. [CrossRef] [PubMed]
54. Morris, D.; Egdell, R. Application of V-doped TiO_2 as a sensor for detection of SO_2. *J. Mater. Chem.* **2001**, *11*, 3207–3210. [CrossRef]

55. Kumar, R.; Avasthi, D.; Kaur, A. Fabrication of chemiresistive gas sensors based on multistep reduced graphene oxide for low parts per million monitoring of sulfur dioxide at room temperature. *Sens. Actuators B Chem.* **2017**, *242*, 461–468. [CrossRef]
56. Nisar, J.; Topalian, Z.; de Sarkar, A.; Osterlund, L.; Ahuja, R. TiO_2 based gas sensor: A possible application to SO_2. *ACS Appl. Mater. Interfaces* **2013**, *5*, 8516–8522. [CrossRef]
57. Zhang, S.; Song, P.; Liu, L.; Yang, Z.; Wang, Q. In_2O_3 nanosheets array directly grown on Al_2O_3 ceramic tube and their gas sensing performance. *Ceram. Int.* **2017**, *43*, 7942–7947. [CrossRef]
58. Si, W.; Du, W.; Wang, F.; Wu, L.; Liu, J.; Liu, W.; Cui, P.; Zhang, X. One-pot hydrothermal synthesis of nano-sheet assembled NiO/ZnO microspheres for efficient sulfur dioxide detection. *Ceram. Int.* **2020**, *46*, 7279–7287. [CrossRef]
59. Zhuo, Q.; Zeng, W.; Chen, W.; Xu, L.; Kumar, R.; Umar, A. High sensitive and low-concentration sulfur dioxide (SO_2) gas sensor application of heterostructure NiO-ZnO nanodisks. *Sens. Actuators B Chem.* **2019**, *298*, 126870. [CrossRef]
60. Ye, H.; Liu, L.; Xu, Y.; Wang, L.; Chen, X.; Zhang, K.; Liu, Y.; Koh, S.; Zhang, G. SnSe monolayer: A promising candidate of SO_2 sensor with high adsorption quantity. *Appl. Surf. Sci.* **2019**, *484*, 33–38. [CrossRef]
61. Seif, A.; nikfarjam, A.; Ghassem, H. UV enhanced ammonia gas sensing properties of PANI/TiO_2 core-shell nanofibers. *Sens. Actuators B Chem.* **2019**, *298*, 126906.
62. Bai, J.; Zhao, C.; Gong, H.; Wang, Q.; Huang, B.; Sun, G.; Wang, Y.; Zhou, J.; Xie, E.; Wang, F. Debye-length controlled gas sensing performances in NiO@ZnO p-n junctional core-shell nanotubes. *J. Phys. D Appl. Phys.* **2019**, *52*, 285103. [CrossRef]
63. Su, P.; Zheng, Y. Room-temperature ppb-level SO_2 gas sensors based on RGO/WO_3 and MWCNTs/WO_3 nanocomposites. *Anal. Methods* **2021**, *13*, 782–788. [CrossRef] [PubMed]
64. Veeralingam, S.; Sahatiya, P.; Badhulika, S. Low cost, flexible and disposable $SnSe_2$ based photoresponsive ammonia sensor for detection of ammonia in urine samples. *Sens. Actuators B Chem.* **2019**, *297*, 126725. [CrossRef]
65. Yang, Z.; Zhang, D.; Wang, D. Carbon monoxide gas sensing properties of metal-organic frameworks-derived tin dioxide nanoparticles/molybdenum diselenide nanoflowers. *Sens. Actuators B Chem.* **2020**, *304*, 127369. [CrossRef]
66. Guo, J.; Li, Y.; Jiang, B.; Gao, H.; Wang, T.; Sun, P.; Liu, F.; Yan, X.; Liang, X.; Gao, Y.; et al. Xylene gas sensing properties of hydrothermal synthesized SnO_2-Co_3O_4 microstructure. *Sens. Actuators B Chem.* **2020**, *310*, 127780. [CrossRef]
67. Beniwai, A.; Sunny. Electrospun SnO_2/PPy nanocomposite for ultra-low ammonia concentration detection at room temperature. *Sens. Actuators B Chem.* **2019**, *296*, 126660. [CrossRef]
68. Xue, Q.; Chen, H.; Li, Q.; Yan, K.; Besenbacher, F.; Dong, M. Room-temperature high-sensitivity detection of ammonia gas using the capacitance of carbon/silicon heterojunctions. *Energy Environ. Sci.* **2010**, *3*, 288–291. [CrossRef]

chemosensors

Article

Nanostructuring of SnO_2 Thin Films by Associating Glancing Angle Deposition and Sputtering Pressure for Gas Sensing Applications

Achraf El Mohajir [1], Mohammad Arab Pour Yazdi [2,3], Anna Krystianiak [4], Olivier Heintz [4], Nicolas Martin [1], Franck Berger [1] and Jean-Baptiste Sanchez [1,*]

1 Institut FEMTO-ST, UMR 6174, CNRS, Université Bourgogne Franche-Comté, 15B, Avenue des Montboucons, 25030 Besançon, France
2 Institut FEMTO-ST, UMR 6174, CNRS, Université Bourgogne Franche-Comté, UTBM, Site de Montbéliard, 90010 Belfort, France
3 Anton Paar TriTec SA, Les Vernets 6, 2035 Corcelles, Switzerland
4 Laboratoire ICB, UMR 6303, CNRS, Université Bourgogne Franche-Comté, 9, Avenue Alain Savary, CEDEX BP 47 870, 21078 Dijon, France
* Correspondence: jbsanche@femto-st.fr

Abstract: SnO_2 thin films were prepared by conventional and Glancing Angle Deposition reactive sputtering, and their gas sensing properties were investigated. The porosity of the as-prepared films was widely assessed using optical methods, and the sensing performances of these active layers were correlated with the evolution of surface and film porosity as a function of deposition conditions and annealing treatment. The sensor made of inclined columns grown at high sputtering pressure (6×10^{-3} mbar) and annealed at 500 °C in air exhibited the best response to benzene, with a limit of detection of 30 ppb. In addition, successful BTEX (i.e., benzene, toluene, ethylbenzene, and xylenes) discrimination was achieved by combining the sensing signals of four nanostructured tin-oxide-based gas sensors.

Keywords: GLAD; SnO_2; nanocolumns; BTEX gas sensor; porous architecture

Citation: Mohajir, A.E.; Yazdi, M.A.P.; Krystianiak, A.; Heintz, O.; Martin, N.; Berger, F.; Sanchez, J.-B. Nanostructuring of SnO_2 Thin Films by Associating Glancing Angle Deposition and Sputtering Pressure for Gas Sensing Applications. *Chemosensors* **2022**, *10*, 426. https://doi.org/10.3390/chemosensors10100426

Academic Editors: Fanli Meng, Zhenyu Yuan and Dan Meng

Received: 21 September 2022
Accepted: 16 October 2022
Published: 17 October 2022

1. Introduction

Metal-oxide-semiconductor (MOS) gas sensors are the dominant gas sensing devices for monitoring the presence of air pollutants [1,2]. Despite poor selectivity, these chemical gas sensors show many advantages such as low cost, flexibility of production, and reduced power consumption, as well as high sensitivity and minimal performance drift over time in a controlled atmosphere (laboratory conditions). The sensing performances of these metal-oxide-based gas sensors are directly controlled by the grain size and the morphology of the sensing materials [3–8]. In order to improve MOS gas sensor performances, most studies have been focused on the structuring of sensitive thin films. In particular, the nanostructuring of the sensitive material led to an improvement in detection performance, especially to reach very low detection limits in the ppb level or lower. Several innovative fabrication methods have been developed to produce high-sensitive chemical gas sensors with metal oxide nanostructures, including bottom-up (chemical vapor deposition (CVD), sol–gel processes, thermal evaporation, hydrothermal route, and reactive sputtering) and also top-down approaches (electron-beam lithography, reactive ion etching, field ion beam technology) [9–14]. Recently, the Glancing Angle Deposition method (GLAD) has been applied in reactive sputtering for the growth of different nanosized columnar films with controlled porosities and shapes [15–17]. Thin films with very original architectures such as normal and inclined columns, zigzag or spiral structures can be achieved with this deposition method [18–20]. Although GLAD technique offers a fully three-dimensional control of the films and almost no limitation on materials that can be fabricated, only a few studies address the use of the GLAD deposition technique for the development of sensitive layers for chemical gas sensor applications. Nevertheless, we can cite works concerning the use of copper oxide films [21], ZnO nanospiral thin films [22], and nanocolumnar SnO_2 and NiO metal oxides [23] or, more recently, a sensor array composed of different nanocolumnar structures made of metal oxides [24]. All these studies have shown that the interaction mechanisms between gas phase and metal oxide are fully dependent on the open porous microstructure associated with GLAD films. Thus, to fabricate efficient gas sensors, it is interesting to produce materials with a large surface in contact with the gas phase, as target analytes can easily interact with the high surface area microstructure of thin films.

In this study, the GLAD method has been used to produce tin oxide films exhibiting normal and inclined columns. The porosity of tin oxide thin films was controlled by changing both the deposition flux of particles onto substrates and the sputtering pressure. To evaluate the influence of the porosity on the detection performances of these sensitive layers, benzene was chosen as pollutant gas. Among the numerous Volatile Organic Compounds (VOCs) that can be found in the air, BTEX compounds (i.e., benzene, toluene, ethylbenzene, and xylenes) have long been known to severely threaten human health. Especially, benzene is considered as one of the most toxic compounds due to its carcinogenic effects even at very low concentrations near the ppb level [25].

In this manuscript, SnO_2 thin films were thoroughly characterized in order to obtain information about texture, morphology, and chemical composition. We used optical approaches to study the impact of deposition conditions on the films' porosity, inside the film and on the surface. Eventually, detection performances of the SnO_2-based gas sensors fabricated were evaluated in the presence of low concentrations of a representative aromatic indoor air pollutant, i.e., benzene. In addition, we used Linear Discriminant Analysis (LDA) to evaluate the sensing performance of a system combining the responses of four nanostructured thin-film-sensitive surfaces for the discrimination of BTEX. Here, we demonstrated the interest in using the GLAD deposition technique and the role of sputtering pressure to obtain very porous sensitive thin films. In particular, we discussed relationships between deposition parameters of tin oxide thin films and gas sensing performances.

2. Materials and Methods

2.1. Film Growth

DC magnetron sputtering was used to deposit SnO_2 thin films in a 40 L homemade vacuum chamber at a base pressure below 10^{-8} mbar. A tin metallic target with a purity of 99.9 % and a diameter of 51 mm was used. The distance between target and substrate was 65 mm, and the current of the tin target (I_{Sn}) was fixed at 30 mA. During the magnetron sputtering deposition, oxygen-to-argon flow rate ratios were fixed at 0.6 and 0.8, leading to argon sputtering pressures of 3×10^{-3} and 6×10^{-3} mbar, respectively. SnO_2 thin films were deposited onto glass and (100) silicon substrates as well as alumina sensor platforms at two different angles ($\alpha = 0°$ and $80°$ for normal and oblique films, respectively) and at two different argon sputtering pressures (3×10^{-3} and 6×10^{-3} mbar). A tilting angle of $\alpha = 80°$ was chosen based on previous studies [20,26] considering that films deposited at a tilting angle of $\alpha > 70°$ offered higher porosities compared with conventional sputtering angle. The deposition time was adjusted to obtain a film thickness close to 250 nm. The experimental setup used to deposit SnO_2 thin films can be found in previous works [26,27].

In this study, we labelled the nanostructured thin films "C" for normal angle ($\alpha = 0°$), while "I" was used for glancing angle films ($\alpha = 80°$). Additionally, the labels "3" and "6" were used for sputtering pressures of 3×10^{-3} and 6×10^{-3} mbar, respectively. For example, "I6" stands for an inclined film ($\alpha = 80°$) deposited at a sputtering pressure of 6×10^{-3} mbar.

2.2. Characterization

The surface and cross-section views of SnO_2 thin films were observed by scanning electron microscopy (SEM) (Field Electron and Ion Company, Hillsboro, OR, USA) in a

Dual-Beam SEM/FIB FEI Helios 600i microscope. Crystal structure of each film was characterized by X-ray diffraction (XRD) (Bruker, Karlsruhe, Germany) using a Bruker D8 focus diffractometer with a Cobalt X-ray tube (Co $_{\lambda K\alpha 1}$ = 0.178897 nm) with a $\theta/2\theta$ configuration. Scans were performed with a step of 0.02° per 0.2s and a 2θ angle ranging from 20 to 80°. X-ray photoelectron spectroscopy (XPS) data were obtained with a PHI VersaProbe I system (PHI, Chanhassen, MN, USA) using a non-monochromatized Al Kα X-ray source (hυ = 1486.7 eV, power of 50 W). Energy calibrations were performed on adventitious carbon (CC/CH bonds) at 248.8 eV. CasaXPS software (Casa Software Ltd.) was used for data treatment [28]. Optical transmission spectra of the thin film deposited on glass substrates were recorded with a Lambda 365 UV/Vis optical spectrometer (PerkinElmer, Wellesley, MA, USA). The reflectance and refractive index were determined using the Filmetrics F50 EXR reflectometer (Jobin Yvon, Longjumeau, France) on the films deposited on (100) silicon substrates.

2.3. Sensing Tests

For sensing tests, SnO_2 thin films were deposited onto 0.5 cm^2 alumina platforms (Heraeus MSP 632) equipped with interdigitated electrodes and allowed both the heating control and conductance measurement of the sensitive material as a function of time. Before exposure to pollutants, a conditioning sequence was applied to each sensor by aging the sensitive surface under a synthetic air flow (relative humidity (RH) 8% at 25 °C, 100 mL min 1) at 500 °C for 48 h to ensure material crystallization. Concentrations of pollutants (BTEX) were obtained by adjusting the ratio between pollutant vapors and air. The normalized response of each sensor S was defined as S = (G − G°)/G°, where G and G° are the conductance under synthetic air mixed with BTEX and under pure synthetic air, respectively.

3. Results and Discussion

3.1. Morphology, Structure, and Composition of the Films

Figure 1 displays the different architectures obtained with the GLAD technique before and after a thermal treatment at 500 °C for 48 h in ambient air. SnO_2 thin film deposited with a conventional incident angle (α = 0°) and under low argon sputtering pressure (3×10^{-3} mbar), i.e., C3 film, exhibits a dense surface and a poorly defined cross-section morphology. On the other hand, for oblique films (α = 80°), the shadowing effect occurring during the early island nucleation on the surface causes voids formation between the small islands, subsequently creating columnar structures [29]. Films deposited at a higher pressure and/or with an angle of α = 80 ° exhibit an apparently more porous morphology. Basically, increasing the pressure in the deposition chamber induces multiple collisions in the gas phase, which makes it possible to cover areas which are not initially in the incident path of the sputtered particle flux. Thus, the surface becomes rougher with higher porosity. Despite a thermal treatment at 500 °C for several hours, the orientation of the nanostructured columns is preserved. In addition, it is worth noting that for I6 film, the columns tend to straighten out because of higher collisions due to higher pressure in the vacuum chamber, and thus a less directional particle flux. The tilting angle of the columns for I3 and I6 films are 34° and 14°, respectively.

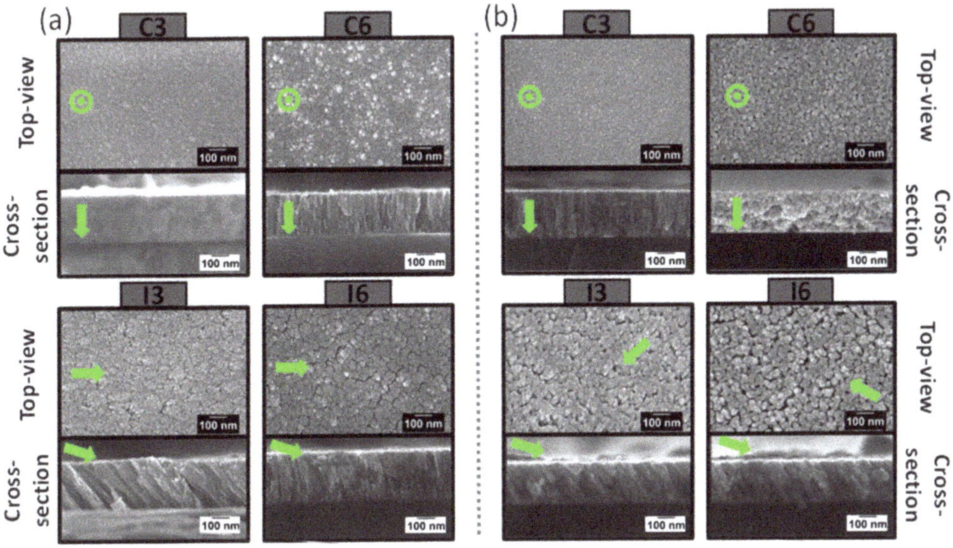

Figure 1. Top and cross-section views of SnO_2 thin films sputter deposited on silicon (**a**) before and (**b**) after an annealing treatment in air for 48 h at 500 °C. Arrows indicate the incoming particle flux during the deposition.

XRD patterns of the nanostructured SnO_2 thin films are shown in Figure 2a. As-deposited SnO_2 thin films with a pressure of 6×10^{-3} mbar remain amorphous, whatever the incidence angle (C6 and I6 thin films). Very weak peaks appear after an annealing treatment of 350°C for 48 h. In contrast, all films deposited with a lower pressure (3×10^{-3} mbar, i.e., C3 and I3 thin films) exhibit diffracted signals prior to a post-thermal treatment. After a thermal treatment of 500 °C for 48 h, XRD peaks of all films become sharper and stronger, indicating a better crystallinity. XRD patterns exhibit peaks between 20 and 70° corresponding to the (110), (101), (200), (211), and (220) reflections, which are assigned to the standard rutile-like crystalline structure of SnO_2 (cassiterite, space group: P42/mnm, JCPDS file no. 41-1445). No peaks of impurities were observed, demonstrating the high purity of the as-prepared SnO_2 thin films. The corresponding tetragonal lattice constants are reported in Table 1.

Despite a slight evolution of the lattice constants with annealing temperature, these data agree with values already reported for SnO_2 (a = b = 4.737 Å and c = 3.186 Å) [30].

In addition, the average crystallite size of SnO_2 films was also calculated using the Scherrer's formula and the Williamson–Hall method (see Table 1) [31]. For both methods, the results show that the crystallite size increases with the annealing temperature, and films deposited with a pressure of 6×10^{-3} mbar exhibit the lower crystallite size after a thermal treatment at 500 °C, i.e., 1.7 and 1.4 $\pm$ 0. 1 nm for C6 and I6 thin films, respectively.

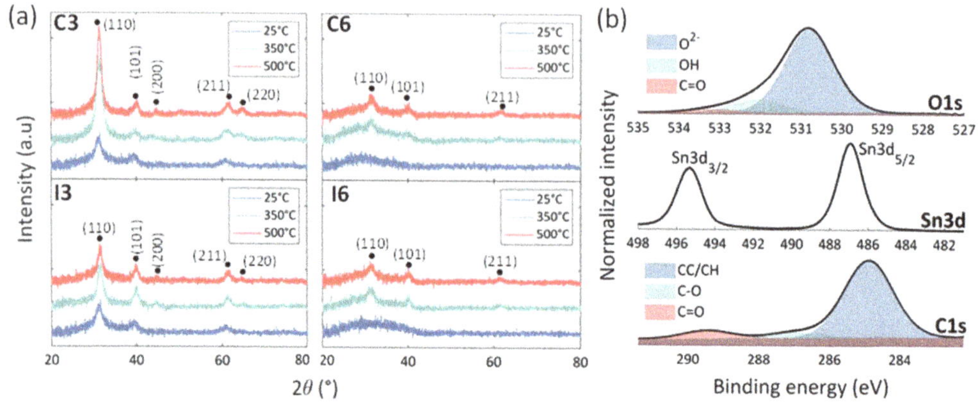

Figure 2. (**a**) XRD patterns of SnO_2 thin films sputter deposited on glass substrates before and after an annealing treatment in ambient air for 48 h at 2 different temperatures (350 and 500 °C). (**b**) XPS O1s, Sn3d, and C1s spectral windows of the I6 SnO_2 thin film sputter deposited on (100) Si substrate after an annealing treatment in ambient air at 500 °C for 48 h.

Table 1. Crystallographic properties, surface chemical composition, and porosity data of SnO_2 thin film.

		Lattice Parameters (±0.003 Å)		Average Crystallite Size (±0.1 nm)		Surface Concentration (±0.5 at. %)		Refractive Index at 470 nm (±0.01)	Packing Density (±2%)
		a = b	c	Scherrer	Williamson-Hall	O	Sn		
C3	25 °C	4.783	3.217	3.0	2.8	-	-	1.97	100
	350 °C	4.780	3.217	4.8	4.6	-	-	1.91	97
	500 °C	4.761	3.186	6.9	6.7	71.1	28.9	1.81	93
C6	25 °C	-	-	-	-	-	-	1.91	-
	350 °C	-	-	-	-	-	-	1.93	-
	500 °C	4.782	3.189	1.7	1.3	67.6	32.4	1.85	92
I3	25 °C	4.774	3.203	4.6	5.1	-	-	1.91	97
	350 °C	4.766	3.196	5.6	5.7	-	-	1.89	95
	500 °C	4.756	3.191	5.4	5.1	67.3	32.7	1.83	91
I6	25 °C	-	-	-	-	-	-	1.80	-
	350 °C	4.766	3.190	-	-	-	-	1.79	89
	500 °C	4.749	3.192	1.4	1.4	67.6	32.4	1.73	84

The surface chemistry of SnO_2 thin films deposited on (100) Si substrates was also investigated by XPS after a thermal treatment at 500 °C for 48 h in ambient air. All XPS spectra showed the presence of oxygen, tin, and carbon. A similar surface chemistry was observed for all films (see Figure S1a in the Supplementary Materials file). Figure 2b presents an example of the Sn3d, O1s, and C1s core line decomposition obtained from the I6 thin-film sample. The evident contribution of the XPS C1s peak is attributed to the carbon contamination adsorbed on the surface of the films from the ambient air and during the annealing treatment. The decomposition of the O1s confirmed that it is built-up essentially as a mixture of three components. The main component at 530.81 eV is related to O-Sn^{2+} and O-Sn^{4+} bindings. Two additional components can be noticed at a binding energy of 533.22 and 532.09 eV, corresponding to O=C and -O-H bindings, respectively. These species at the near surface of SnO_2 materials and after exposure to air come from partially ionized –OH groups originating from dissociated atmospheric water and adsorbed CO/CO_2 from the ambient air [32]. The XPS Sn3d spectral line is narrow and symmetrical. The difference between the binding energy of 3d5/2 level for SnO and SnO_2 is very low (486.6 and 486.9 eV, respectively) but, for all samples, the shape of the MNN Auger transition for Sn is close to that observed for SnO_2 (see Figure S1b) [33]. It is worth noting that there is no contribution of metallic Sn (Sn^0) at the binding energy about 485 eV. This finding indicates that all films were completely oxidized after a thermal treatment at 500 °C in ambient air for 48 h.

The relative surface concentrations of Sn and O, expressed in at. %, agree with the formula of SnO_2 composition (see Table 1). It is worth noting that the oxygen concentration is probably slightly overestimated due to the presence of contaminants at the near surface of the films.

3.2. *Film Growth*

In this section, we present results concerning the influence of experimental deposition conditions on the porosity of SnO_2 thin films, through the films, and on their surface.

3.2.1. Through the Films

Figure 3a shows the transmittance and reflectance spectra of films deposited on glass and (100) Si substrates, respectively, before and after an annealing treatment at 350 and 500 °C for 48 h. All films show a sharp absorption around 300 nm and exhibit a high transmission in the visible and near-infrared region. Films deposited at 3×10^{-3} mbar (C3 and I3) show distinct interference fringe patterns in the optical transmittance spectra, indicating homogeneous growth. On the other hand, films deposited at 6×10^{-3} mbar (C6 and I6) show minor interference fringes, indicating inhomogeneity in the films. As mentioned in the work of Ohlídal et al. [34], this inhomogeneity is due to the surficial and inner roughness of the films. From Figure 3a, it can be underlined that there is no absorption edge shift for the C3 film whatever the annealing temperature. This indicates that there are less defects in C3 thin film in comparison with other films. The absence of interference fringes makes the determination of refractive indices from methods based on the adjustment of envelopes impossible [35].

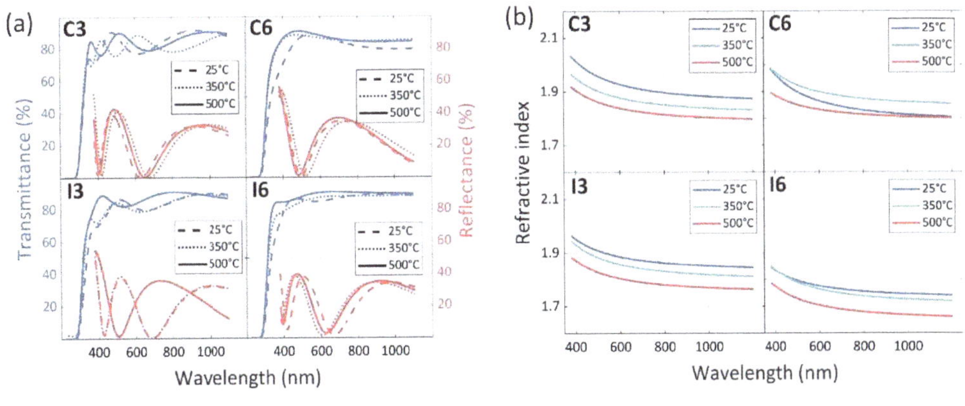

Figure 3. (**a**) Transmittance and reflectance spectra of SnO_2 thin films sputter deposited on glass and (100) Si substrates before and after an annealing treatment in air for 48 h and at 2 different temperatures (350 and 500 °C). (**b**) Evolution of the calculated refractive index vs. wavelength for all films at various temperatures.

Consequently, using the three-phase Cauchy dispersion model (k = 0) showed in Equation (1), refractive indices can be calculated from the measured reflectance spectra (see again Figure 3a) [36].

$$n(\lambda) = A + \frac{B}{\lambda^2} + \frac{C}{\lambda^4} \tag{1}$$

Fitting constants A, B, and C were calculated from the transmittance spectrum of C3 film and are estimated at 1.592, 7.44×10^4 nm^2, and -4.558×10^9 nm^4, respectively. These fitting constants were used for the determination of the refractive indices for all films.

Figure 3b shows the evolution of the refractive indices as a function of the wavelength for each thin film, before and after the annealing treatment. The values of the refractive indices at 470 nm for all films annealed at every temperature are reported in Table 1. It is worth noting that these values were found to be lower than the corresponding value for bulk tin oxide at $\lambda = 470$ nm (n = 2 for bulk SnO_2 [37]). The refractive index was found to be lower for inclined columns ($\alpha = 80°$) and for films deposited at high pressure (6×10^{-3} mbar), which indicates higher porosity present inside these films [38]. In addition, after a thermal treatment, the value of the refractive index decreases for all films, indicating that the porosity of the films evolves with the annealing temperature. From the determined refractive indices at 470 nm and lattice constants previously calculated from XRD results (see Table 1), the packing density p was evaluated for all SnO_2 thin films using Equation (2) [39]. The calculated values are gathered in Table 1.

$$n_f = \frac{(1-p)n_v^2 + (1+p)n_v^4 n_b^2}{(1+p)n_v^2 + (1-p)n_b^2} + \frac{5}{2}\left(\frac{n_b^2 - 1}{n_b C_b}\right)(C_b - C_f) \quad (2)$$

n_f is the film refractive index at 470 nm, n_b the refractive index of bulk SnO_2 ($n_b = 2$), n_v the refractive index of void ($n_v = 1$), C_b the lattice parameter for bulk SnO_2 ($C_b = 3.186$ Å), C_f the lattice parameter for porous SnO_2, and p the packing density of the film.

For all studied films, the packing density p decreases with the annealing temperature, which indicates an increase in the porosity inside the films. The evolution of the packing density is in line with the tendencies observed for the refractive index, i.e., the inner porosity gradually rises between normal and inclined columns. The values of the inner porosities obtained in this work are in agreement with those found in the study of Bagga et al. [40], where the porosity of nanostructured SnO_2 thin films varied from 23% to 9%. Moreover, the porosity is favored with a higher deposition pressure, whatever the films deposited (normal or inclined columns). For example, the packing index is estimated to be around 91% (i.e., 9% inner porosity) for the inclined columns deposited at 3×10^{-3} mbar compared with 84% (i.e., 16% inner porosity) for the same structure at 6×10^{-3} mbar.

3.2.2. Surface Porosity

Gas sensing being surface state dependent, the surface porosity of each thin film was also investigated using SEM images. Pore characterization, including surface porosity and pore-size distribution of each film, was evaluated after a post-treatment of binary SEM images using both ImageJ and MATLAB software. Figure 4 depicts, for each thin film, the surface porosity determined by ImageJ versus the annealing temperature. The inserts show examples of the binary images obtained after the post-image treatment for the I6 film, as-deposited, and after two different annealing temperatures for 48 h (350 and 500 °C).

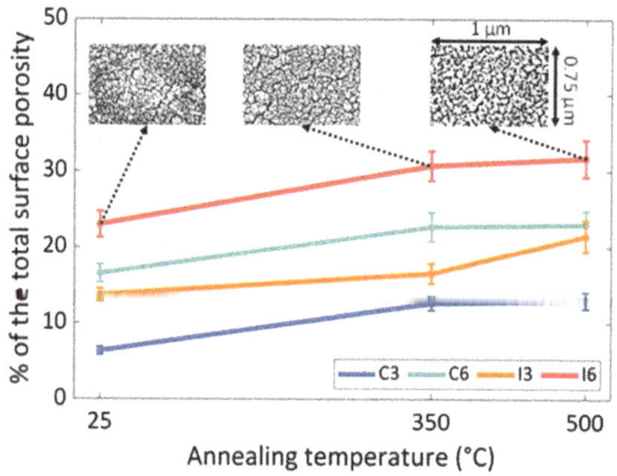

Figure 4. Percent of the total surface porosity vs. annealing temperature. The inserts show binary SEM images for I6 films (analyzed surface around 0.75 µm^2).

For all films, the analyzed surface was around 0.75 μm^2, which constitutes a good representation of the total surface. A comparison between all films as deposited, whatever the annealing temperature, shows that by increasing the deposition pressure and tilting the substrate holder, the surface porosity increases significantly. To summarize, the surface porosity increases according to the series: C3 < I3 < C6 < I6. By annealing the films at 500 °C, with an intermediate temperature at 350 °C, the growth of crystallites causes tensile stress, which creates a more porous surface. Moreover, many studies showed that the annealing procedure increases the length and width of already existing pores. They become interconnected and form bigger pores that affect the surface morphology massively [41,42]. In addition, the cooling phase occurring when the temperature drops from 500 to 25 °C in a short time (a few minutes) induces supplementary strain due to the mismatch of the expansion coefficients between the film and the substrate, leading to the formation of additional cracks on the surface [43].

The pore-size distribution (PSD) was also investigated using Otsu's method. This algorithm allows automatic multilevel thresholding of binarized SEM images. The pore segmentation was then obtained based on the lowest standard deviation of intensity in each segment. In addition, we used the MATLAB code developed by Rabbani et al. to determine the pore sizes [44]. This code was adapted to our analysis by adjusting the different parameters used for the thresholding method. Figure 5a displays an example of the PSD and the corresponding boxplot showing the interquartile range median, minimum, and maximum for the I6 thin film annealed at 500 °C for 48 h in ambient air.

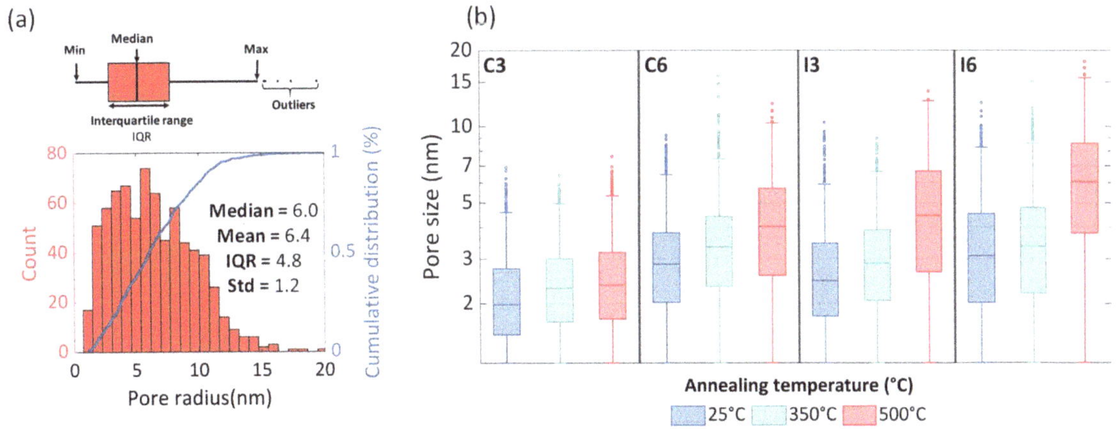

Figure 5. (**a**) Pore-size distribution and boxplot calculated for the I6 film annealed at 500 °C. (**b**) Boxplots showing the evolution of the pore size for all SnO_2 thin films sputter deposited before and after an annealing treatment in air for 48 h and at 2 different temperatures (350 and 500 °C).

The pore-size distribution of I6 exhibits a multimodal distribution extending from ca. 1 to 20 nm, with the majority of the pores being in the mesopore range (i.e., width > 2 nm). Figure S2 demonstrates that the PSDs of all films are right-skewed, regardless of the deposition conditions and the annealing temperature. After a thermal treatment at 500 °C for 48 h, C6, I3, and I6 films exhibit multimodal distributions, indicating wider pore formation in the mesopore range. On the contrary, irrespective of the annealing treatment, C3 film presents a narrow unimodal PSD, with the larger pores being less than 7 nm. Figure 5b summarizes the corresponding boxplots calculated from the PSDs of all films. It was noticed that the pore-size median increases continuously with the annealing temperature, especially for oblique films. This rise is noticeably higher for I6 film, with a pore-size median going from 3.4 nm to 6.0 nm at 350 °C and 500 °C, respectively.

This porosity assessment shows that a simple study of the total surface porosity can be misleading, as in some cases the porosity consists mostly of small and non-accessible pores.

By combining the porosity results obtained from the inner and surface characterization, we obtained a better understanding of the evolution of the morphology of each tin oxide thin film. Comparatively, I6 thin film annealed at 500 °C offers the highest inner porosity (16%) and surface porosity (around 31%), as well as a wide range of pore sizes (1–20 nm). These textural properties prove to be of great interest for gas sensing applications since the diffusion of pollutants inside the material will be favored. It should be noted that the growth of the nanostructured columns, which is correlated to the porous structure of the films, is preserved regardless of the substrate used (Si, glass or alumina platforms).

After having evaluated the porous structure of the films, gas sensing performances were investigated.

3.3. *Sensing Performances*

3.3.1. Benzene Detection

The gas sensing properties of metal-oxide-based gas sensors are highly affected by the operating temperature, which generally ranges from 300 to 500 °C. Figure S3 represents the typical electrical responses of the four sensors at various sensing temperatures and under 900 ppb of benzene for 2 min. Each benzene exposition was replicated three times. Figure 6a displays the evolution of the relative conductance versus the sensitive surface temperature for each gas sensor (under 900 ppb of benzene for 2 min).

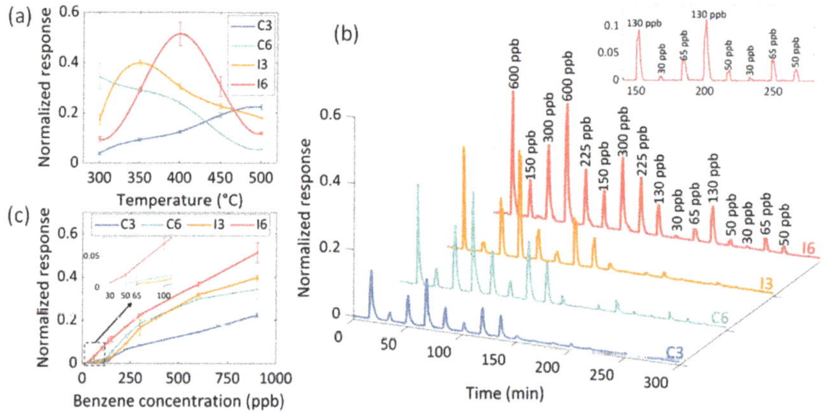

Figure 6. (**a**) Evolution of the normalized response as a function of the sensitive surface temperature for each gas sensor under 900 ppb of benzene. (**b**) Real-time response of each gas sensor for various benzene concentrations (sensing temperatures equal to 500, 300, 350, and 400 °C for C3, C6, I3, and I6 sensors, respectively). (**c**) Normalized response of the SnO_2-based gas sensors as a function of the benzene concentration. For all tests, 2 min. exposition time, 15 min. recovery time, and 100 mL min^{-1} flow rate were applied.

It is evident that the I6-based gas sensor showed a substantial change in the response as a function of metal-oxide temperature. The optimized temperatures, corresponding to the sensor's temperature showing a maximal conductance under pollutant, were estimated at 500, 300, 350, and 400 °C for C3, C6, I3, and I6 sensors, respectively.

The detection mechanism of benzene with SnO_2-based gas sensors can be explained by considering that the chemisorbed oxygen species in the form O^- and O^{2-} are dominating for sensing layer temperatures above 200 °C [45]. Since SnO_2 is an n-type semiconductor, and benzene a reducing compound, an oxidation mechanism of benzene molecules occurs at the sensor surface according to the successive reactions [46]:

$$C_6H_{6\,(gas)} \leftrightarrow C_6H_{6\,(ads)} \quad (3)$$

$$C_6H_{6\,(ads)} + 15\,O^-{}_{(ads)} \rightarrow 6\,CO_{2(gas)} + 3\,H_2O_{(gas)} + 15\,e^-{}_{c.b.} \tag{4}$$

$$C_6H_{6\,(ads)} + 15\,O^{2-}{}_{(ads)} \rightarrow 6\,CO_{2(gas)} + 3\,H_2O_{(gas)} + 30\,e^-{}_{c.b.} \tag{5}$$

c.b. is the conduction band of the material.

These reactions explain the increase in the conductance when SnO_2 is in contact with benzene. Figure 6b plots the impulse response/recovery curves of each sensor at their respective optimized temperature for various concentrations of benzene mixed in air. The insert shows a zoom of the impulse response/recovery curves for I6 sensor at very low benzene concentrations. The sensors were stabilized in nearly dry air (8% RH at 25 °C) for 15 min. and then exposed to different benzene concentrations for 2 min. Each benzene concentration was chosen randomly and replicated twice in order to evaluate both the repeatability and the stability of each sensor. From Figure 6b, it can be easily found that there is a good correspondence between benzene concentrations and the dynamic response of each sensor. All sensors present an excellent repeatability and stability over time since the amplitude of the responses is in line with the benzene concentrations. The corresponding responses calculated from the change in the conductance of each sensor as a function of the benzene concentration are plotted in Figure 6c. It can be noticed that the I6 sensor exhibits the highest response in comparison with the other ones, even though the difference in the deposition conditions (incidence angle and argon sputtering pressure), C6, and I3 sensors show similar responses to benzene.

The limit of detection (LOD) was defined by the concentration at which the amplitude of the signal was three times higher than the signal/noise ratio (S = 0.002/N > 3). For benzene, the LOD was estimated at 100 ppb for C3, 65 ppb for I3, and 50 ppb for C6 sensors, while the LOD achieved for I6 sensor was around 30 ppb. From these results, it is interesting to note that not only the architecture of the film but also the deposition pressure are directly connected to the sensing performances. According to results obtained on the morphology and structure of the synthesized materials, the good sensing performances of the I6 coating is mainly due to the high inner and surface porosity, the wide range of pore sizes, and also to the small crystallite size (1.4 nm). The sensitivity of MOS gas sensors increases when the crystallite size is less than the double of the thickness of the space-charge layer (L) produced around the surface of the crystallites due to the chemisorbed oxygens. For SnO_2-based material, it is admitted that the space-charge layer L is equal to 3 nm [47]. In the present study, the average crystallite size obtained by XRD for all annealed thin films is lower than 2L and, more particularly for the I6-sensitive surface, the crystallite size is far lower.

3.3.2. BTEX Discrimination

Since benzene is not the only air pollutant, we also studied the cross-sensitivity of sensors for various VOCs (i.e., toluene, ethylbenzene, and ortho- and para-xylene). Figure 7a depicts the normalized responses of each sensor to around 100 ppb of each compound diluted in air and analyzed separately (i.e., not in a mixture).

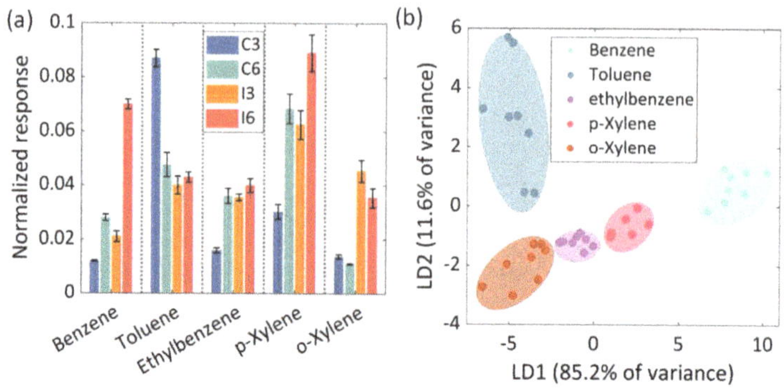

Figure 7. (**a**) Sensors' response to various BTEX at a concentration of 100 ppb. (**b**) Linear Discriminant Analysis (LDA) scores using responses of the 4 SnO_2-based gas sensors. For all tests, 2 min. exposition time, 15 min. recovery time, and 100 mL min^{-1} flow rate were applied.

For these experiments, sensing temperates were maintained to 500, 300, 350, and 400 °C for C3, C6, I3, and I6 sensors, respectively. As expected, manufactured sensors are sensitive to all VOCs as long as each compound can react with the sensitive surface according to the following redox reactions:

$$C_xH_{y\,(gas)} \leftrightarrow C_xH_{y\,(ads)} \tag{6}$$

$$C_xH_{y\,(ads)} + (2x + \frac{y}{2})O^-_{(ads)} \rightarrow x\,CO_{2(gas)} + \frac{y}{2}\,H_2O_{(gas)} + (2x + \frac{y}{2}).\,e^-_{c.b.} \tag{7}$$

$$C_xH_{y\,(ads)} + (2x + \frac{y}{2})\,O^{2-}_{(ads)} \rightarrow x\,CO_{2(gas)} + \frac{y}{2}\,H_2O_{(gas)} + (4x + y)\,e^-_{c.b.} \tag{8}$$

In other words, this confirms that metal-oxide-based gas sensors are not selective, and it is not possible to identify one chemical compound in a mixture using a single gas sensor.

Nevertheless, classifying the species can be achieved by using a Linear Discriminant Analysis (LDA) that can map multidimensional data onto two- or three-dimensional axes [48]. This method uses linear combinations of sensors' response to predict the class of a given observation (chemical nature of compound). In order to obtain greater precision on the discrimination of these VOCs compounds, we combined the normalized response and sensing kinetics (response and recovery times) of each sensor (i.e., C3, C6, I3, and I6) after exposition to VOCs (Figure S4). Figure 7b represents the LDA scores using the experimental data measured using the responses of the four sensors manufactured in this study and under BTEX. Plots confirm the ability to cluster points according to the chemical nature of species. The within-class distance of each VOC compound class in LDA are well-gathered. This indicates that combining sensing responses of the four gas sensors improves the discrimination of BTEX with concentrations ranging from 50 to 900 ppb.

Table 2 reports some of the literature's data on the gas sensing performances of various metal-oxide-based gas sensor arrays towards BTEX detection in synthetic air.

Table 2. A comparison of MOS-based gas sensor arrays for the detection of BTEX in air.

Materials	Number of Sensors Used	BTEX Concentrations Tested	Ref.
SnO_2	4	50–900 ppb (BTEX)	This study
NiO, WO_3, SnO_2	3	30–80 ppm (BTX)	[49]
SnO_2 NPs/cobalt-porphyrin, SnO_2 NPs/zinc-porphyrin, SnO_2 NPs/nickel-porphyrin and ZnO NPs/cobalt-porphyrin	4	1–9 ppm (BTEX)	[50]
SnO_2 with several additives, including Pt, Pd, CuO, LaO, ScO, TiO, WO or ZnO	10	benzene 50 ppm toluene 500 ppm	[51]

It should be also kept in mind that all results were obtained by analyzing compounds separately and in nearly dry air. By combining the responses from the four sensors fabricated in the present work, we obtain better sensing performances compared with those reported in the literature. It is worth noting that all sensor arrays reported in Table 2 are composed of several sensitive layers made with different doped or undoped materials (SnO_2, WO_3, etc.). The use of sensitive layers made of various materials makes it difficult to manufacture sensor arrays. Furthermore, the limit of detection of these arrays is around a few ppm, which is much higher than the concentrations of BTEX found in air (a few ppb). In our study, we showed that by modifying only the deposition conditions of the films (sputtering pressure and deposition angle), it is possible to obtain an efficient sensor array made of the same sensitive material (undoped SnO_2 here) for the discrimination of BTEX even at low concentrations (50 ppb). Consequently, this leads to an easier manufacturing procedure.

4. Conclusions

In this study, we used the conventional and Glancing Angle Deposition (GLAD) reactive sputtering method to prepare nanostructured tin-oxide-based sensitive films for gas sensor applications. The effect of the deposition parameters of SnO_2 films was studied in detail, and it was found that the intergranular porosity can be tuned by adjusting the pressure and deposition angle. All sensors exhibited excellent stability and repeatability to benzene at a concentration of 900 ppb. The excellent sensing performances of the I6 sensor (deposition angle of 80° with a sputtering pressure of 6×10^{-3} mbar) were mainly attributed to high inner and surface porosity (16 and 31 %, respectively), a wide range of pore sizes (1–20 nm), and small crystallite size (1.4 nm). With these deposition conditions, it is possible to reach a limit of detection close to 30 ppb with an operating temperature of 400 °C. Despite the lack of intrinsic selectivity of each gas sensor taken individually, a chemical signature was obtained for the discrimination of a BTEX mixture by collectively considering the gas sensing responses from the four gas sensors developed in this study. We showed that it is possible to obtain a sensor array by using diverse undoped nanostructured SnO_2-sensitive films. In summary, the GLAD method seems to be very promising for the development of gas sensor arrays and has great potential for exploring high-performance gas sensors for VOC detection.

Supplementary Materials: The following supporting information can be downloaded at: https://www.mdpi.com/article/10.3390/chemosensors10100426/s1, Figure S1: (a) XPS O1s, Sn3d, and C1s spectral windows of the C3, C6, and I3 SnO_2 thin films and (b) experimental Sn MNN Auger spectra for C3, C6, I3, and I6 SnO_2 thin films. All films were sputter deposited on (110) Si substrates and annealed in ambient air at 500 °C for 48 h; Figure S2: Pore-size distributions (PSDs) of all films sputter deposited on (100) Si substrates, before and after annealing treatments at 350 and 500 °C for 48 h in ambient air. All PSDs were calculated using MATLAB software; Figure S3: Dynamic normalized response of the SnO_2-based gas sensors (C3, C6, I3, and I6) as a function of the sensing temperature under 900 ppb of benzene. For all tests, 2 min. exposition time, 15 min. recovery time, and 100 mL min^{-1} flow rate were applied. Each benzene exposition was replicated 3 times; Figure S4: Dynamic normalized response of the SnO_2-based gas sensors (C3, C6, I3, and I6) for different concentrations of (a) toluene, (b) ethylbenzene, (c) p-xylene, and (d) o-xylene. For all tests, 2 min. exposition time, 15 min. recovery time, and 100 mL min^{-1} flow rate were applied.

Author Contributions: Investigation, methodology, data curation, and software, A.E.M.; investigation, M.A.P.Y., A.K. and O.H.; formal analysis and validation, N.M.; resources, F.B.; funding acquisition, supervision, conceptualization, and writing—review and editing, J.-B.S. All authors have read and agreed to the published version of the manuscript.

Funding: This research was funded by the Bourgogne Franche-Comté Region, the European fund FEDER, and the EUR EIPHI (contract ANR-17-EURE-0002) through the project DECOLAIR.

Institutional Review Board Statement: Not applicable.

Informed Consent Statement: Not applicable.

Data Availability Statement: Not applicable.

Acknowledgments: The authors would like to express their gratitude to the Bourgogne Franche-Comté Region, the European fund FEDER, and the EUR EIPHI (contract ANR-17-EURE-0002) for financial support through the project DECOLAIR. This work was partially supported by the French RENATECH network and its FEMTO-ST technological facility.

Conflicts of Interest: The authors declare no conflict of interest.

References

1. Spinelle, L.; Gerboles, M.; Kok, G.; Persijn, S.; Sauerwald, T. Review of Portable and Low-Cost Sensors for the Ambient Air Monitoring of Benzene and Other Volatile Organic Compounds. *Sensors* **2017**, *17*, 1520. [CrossRef] [PubMed]
2. Dhall, S.; Mehta, B.R.; Tyagi, A.K.; Sood, K. A Review on Environmental Gas Sensors: Materials and Technologies. *Sens. Int.* **2021**, *2*, 100116. [CrossRef]
3. Shimizu, Y.; Jono, A.; Hyodo, T.; Egashira, M. Preparation of Large Mesoporous SnO_2 Powder for Gas Sensor Application. *Sens. Actuators B Chem.* **2005**, *108*, 56–61. [CrossRef]
4. Korotcenkov, G.; Han, S.-D.; Cho, B.K.; Brinzari, V. Grain Size Effects in Sensor Response of Nanostructured SnO_2- and In_2O_3-Based Conductometric Thin Film Gas Sensor. *Crit. Rev. Solid State Mater. Sci.* **2009**, *34*, 1–17. [CrossRef]
5. Kong, Y.; Li, Y.; Cui, X.; Su, L.; Ma, D.; Lai, T.; Yao, L.; Xiao, X.; Wang, Y. SnO_2 Nanostructured Materials Used as Gas Sensors for the Detection of Hazardous and Flammable Gases: A Review. *Nano Mater. Sci.* **2021**, *in press*. [CrossRef]
6. Sriram, S.R.; Parne, S.; Vaddadi, V.S.C.S.; Edla, D.; Nagaraju, P.; Avala, R.R.; Yelsani, V.; Sontu, U.B. Nanostructured WO3 Based Gas Sensors: A Short Review. *Sens. Rev.* **2021**, *41*, 406–424. [CrossRef]
7. Pant, B.R.; Jayatissa, A.H. Gas Sensor Application of Zinc Oxide. In *Chemical Methods for Processing Nanomaterials*; CRC Press: Boca Raton, FL, USA, 2021; ISBN 978-0-429-02318-7.
8. Tian, X.; Cui, X.; Lai, T.; Ren, J.; Yang, Z.; Xiao, M.; Wang, B.; Xiao, X.; Wang, Y. Gas Sensors Based on TiO_2 Nanostructured Materials for the Detection of Hazardous Gases: A Review. *Nano Mater. Sci.* **2021**, *3*, 390–403. [CrossRef]
9. Seekaew, Y.; Wisitsoraat, A.; Phokharatkul, D.; Wongchoosuk, C. Room Temperature Toluene Gas Sensor Based on TiO_2 Nanoparticles Decorated 3D Graphene-Carbon Nanotube Nanostructures. *Sens. Actuators B Chem.* **2019**, *279*, 69–78. [CrossRef]
10. Subbiah, D.K.; Mani, G.K.; Babu, K.J.; Das, A.; Balaguru Rayappan, J.B. Nanostructured ZnO on Cotton Fabrics—A Novel Flexible Gas Sensor & UV Filter. *J. Clean. Prod.* **2018**, *194*, 372–382. [CrossRef]
11. Nemec, P.; Hotový, I.; Andok, R.; Kostič, I. Increased Sensitivity of a Gas Sensor by Controlled Extension of TiO_2 Active Area. *AIP Conf. Proc.* **2018**, *1996*, 020032. [CrossRef]
12. Atanasova, G.; Dikovska, A.O.; Dilova, T.; Georgieva, B.; Avdeev, G.V.; Stefanov, P.; Nedyalkov, N.N. Metal-Oxide Nanostructures Produced by PLD in Open Air for Gas Sensor Applications. *Appl. Surf. Sci.* **2019**, *470*, 861–869. [CrossRef]
13. Singh, A.; Sharma, A.; Tomar, M.; Gupta, V. Tunable Nanostructured Columnar Growth of SnO_2 for Efficient Detection of CO Gas. *Nanotechnology* **2018**, *29*, 065502. [CrossRef] [PubMed]
14. Nemec, P.; Hotovy, I.; Rehacek, V.; Andok, R. TiO_2 Sensoric Structures with Controlled Extension of Their Active Area by Electron-Beam Lithography and Reactive Ion Etching Techniques. *AIP Conf. Proc.* **2021**, *2411*, 060003. [CrossRef]
15. Bairagi, S.; Järrendahl, K.; Eriksson, F.; Hultman, L.; Birch, J.; Hsiao, C.-L. Glancing Angle Deposition and Growth Mechanism of Inclined AlN Nanostructures Using Reactive Magnetron Sputtering. *Coatings* **2020**, *10*, 768. [CrossRef]
16. El Beainou, R.; Garcia-Valenzuela, A.; Raschetti, M.; Cote, J.-M.; Alvarez, R.; Palmero, A.; Potin, V.; Martin, N. A 4-View Imaging to Reveal Microstructural Differences in Obliquely Sputter-Deposited Tungsten Films. *Mater. Lett.* **2020**, *264*, 127381. [CrossRef]
17. Shin, S.J.; Bayu Aji, L.B.; Bae, J.H.; Engwall, A.M.; Nielsen, M.H.; Hammons, J.A.; Zuo, X.B.; Lee, B.; Lepro, X.; Mirkarimi, P.B.; et al. Oblique Angle Deposition of Boron Carbide Films by Magnetron Sputtering. *J. Appl. Phys.* **2021**, *130*, 125305. [CrossRef]

18. Liedtke, S.; Grüner, C.; Lotnyk, A.; Rauschenbach, B. Glancing Angle Deposition of Sculptured Thin Metal Films at Room Temperature. *Nanotechnology* **2017**, *28*, 385604. [CrossRef]
19. Pedrosa, P.; Ferreira, A.; Martin, N.; Yazdi, M.A.P.; Billard, A.; Lanceros-Méndez, S.; Vaz, F. Nano-Sculptured Janus-like TiAg Thin Films Obliquely Deposited by GLAD Co-Sputtering for Temperature Sensing. *Nanotechnology* **2018**, *29*, 355706. [CrossRef]
20. Sakkas, C.; Rauch, J.-Y.; Cote, J.-M.; Tissot, V.; Gavoille, J.; Martin, N. Tuning the Optical Properties of WO3 Films Exhibiting a Zigzag Columnar Microstructure. *Coatings* **2021**, *11*, 438. [CrossRef]
21. Rydosz, A.; Dyndał, K.; Andrysiewicz, W.; Grochala, D.; Marszałek, K. GLAD Magnetron Sputtered Ultra-Thin Copper Oxide Films for Gas-Sensing Application. *Coatings* **2020**, *10*, 378. [CrossRef]
22. Luo, P.; Xie, M.; Luo, J.; Kan, H.; Wei, Q. Nitric Oxide Sensors Using Nanospiral ZnO Thin Film Deposited by GLAD for Application to Exhaled Human Breath. *RSC Adv.* **2020**, *10*, 14877–14884. [CrossRef] [PubMed]
23. Song, Y.G.; Shim, Y.-S.; Han, S.D.; Lee, H.R.; Ju, B.-K.; Kang, C.Y. Metal Oxide Nanocolumns for Extremely Sensitive Gas Sensors. *J. Sens. Sci. Technol.* **2016**, *25*, 184–188. [CrossRef]
24. Kang, M.; Cho, I.; Park, J.; Jeong, J.; Lee, K.; Lee, B.; Henriquez, D.D.O.; Yoon, K.; Park, I. High Accuracy Real-Time Multi-Gas Identification by a Batch-Uniform Gas Sensor Array and Deep Learning Algorithm. *ACS Sens.* **2022**, *7*, 430–440. [CrossRef]
25. Davidson, C.J.; Hannigan, J.H.; Bowen, S.E. Effects of Inhaled Combined Benzene, Toluene, Ethylbenzene, and Xylenes (BTEX): Toward an Environmental Exposure Model. *Environ. Toxicol. Pharmacol.* **2021**, *81*, 103518. [CrossRef] [PubMed]
26. Xu, X.; Arab Pour Yazdi, M.; Sanchez, J.-B.; Billard, A.; Berger, F.; Martin, N. Exploiting the Dodecane and Ozone Sensing Capabilities of Nanostructured Tungsten Oxide Films. *Sens. Actuators B Chem.* **2018**, *266*, 773–783. [CrossRef]
27. Pedrosa, P.; Ferreira, A.; Cote, J.-M.; Martin, N.; Yazdi, M.A.P.; Billard, A.; Lanceros-Mendez, S.; Vaz, F. Influence of the Sputtering Pressure on the Morphological Features and Electrical Resistivity Anisotropy of Nanostructured Titanium Films. *Appl. Surf. Sci.* **2017**, *420*, 681–690. [CrossRef]
28. Fairley, N.; Fernandez, V.; Richard-Plouet, M.; Guillot-Deudon, C.; Walton, J.; Smith, E.; Flahaut, D.; Greiner, M.; Biesinger, M.; Tougaard, S.; et al. Systematic and Collaborative Approach to Problem Solving Using X-Ray Photoelectron Spectroscopy. *Appl. Surf. Sci. Adv.* **2021**, *5*, 100112. [CrossRef]
29. Chen, L.-C.; Chen, C.-C.; Sung, Y.-T.; Hsu, Y.-Y. Oblique-Angle Sputtering Effects on Characteristics of Nanocolumnar Structure Anisotropic Indium Tin Oxide Films. *J. Electrochem. Soc.* **2009**, *156*, H471. [CrossRef]
30. Bolzan, A.A.; Fong, C.; Kennedy, B.J.; Howard, C.J. Structural Studies of Rutile-Type Metal Dioxides. *Acta Crystallogr. B* **1997**, *53*, 373–380. [CrossRef]
31. Mustapha, S.; Ndamitso, M.M.; Abdulkareem, A.S.; Tijani, J.O.; Shuaib, D.T.; Mohammed, A.K.; Sumaila, A. Comparative Study of Crystallite Size Using Williamson-Hall and Debye-Scherrer Plots for ZnO Nanoparticles. *Adv. Nat. Sci. Nanosci. Nanotechnol.* **2019**, *10*, 045013. [CrossRef]
32. Güzelçimen, F.; Tanören, B.; Çetinkaya, Ç.; Kaya, M.D.; Efkere, H.İ.; Özen, Y.; Bingöl, D.; Sirkeci, M.; Kınacı, B.; Ünlü, M.B.; et al. The Effect of Thickness on Surface Structure of Rf Sputtered TiO_2 Thin Films by XPS, SEM/EDS, AFM and SAM. *Vacuum* **2020**, *182*, 109766. [CrossRef]
33. Kövér, L.; Kovács, Z.; Sanjinés, R.; Moretti, G.; Cserny, I.; Margaritondo, G.; Pálinkás, J.; Adachi, H. Electronic Structure of Tin Oxides: High-Resolution Study of XPS and Auger Spectra. *Surf. Interface Anal.* **1995**, *23*, 461–466. [CrossRef]
34. Ohlídal, I.; Vohánka, J.; Čermák, M. Optics of Inhomogeneous Thin Films with Defects: Application to Optical Characterization. *Coatings* **2020**, *11*, 22. [CrossRef]
35. Sánchez-González, J.; Díaz-Parralejo, A.; Ortiz, A.L.; Guiberteau, F. Determination of Optical Properties in Nanostructured Thin Films Using the Swanepoel Method. *Appl. Surf. Sci.* **2006**, *252*, 6013–6017. [CrossRef]
36. Wang, Z.-Y.; Zhang, R.-J.; Lu, H.-L.; Chen, X.; Sun, Y.; Zhang, Y.; Wei, Y.-F.; Xu, J.-P.; Wang, S.-Y.; Zheng, Y.-X.; et al. The Impact of Thickness and Thermal Annealing on Refractive Index for Aluminum Oxide Thin Films Deposited by Atomic Layer Deposition. *Nanoscale Res. Lett.* **2015**, *10*, 46. [CrossRef] [PubMed]
37. Raoult, E.; Bodeux, R.; Jutteau, S.; Rives, S.; Yaiche, A.; Coutancier, D.; Rousset, J.; Collin, S. Optical characterizations and modelling of semitransparent perovskite solar cells for tandem applications. In Proceedings of the 36th European Photovoltaic Solar Energy Conference and Exhibition, Marseille, France, 9–13 September 2019; pp. 757–763. [CrossRef]
38. Robbie, K.; Brett, M.J. Sculptured Thin Films and Glancing Angle Deposition: Growth Mechanics and Applications. *J. Vac. Sci. Technol. Vac. Surf. Films* **1997**, *15*, 1460–1465. [CrossRef]
39. Sharma, A.; Tomar, M.; Gupta, V. SnO_2 Thin Film Sensor with Enhanced Response for NO_2 Gas at Lower Temperatures. *Sens. Actuators B Chem.* **2011**, *156*, 743–752. [CrossRef]
40. Bagga, S.; Akhtar, J.; Mishra, S. Influence of Porosity on the Properties of Nanostructured Tin Oxide Thin Film. *Mater. Res. Express* **2018**, *5*, 116406. [CrossRef]
41. Tang, Z.; Chan, P.C.H.; Sharma, R.K.; Yan, G.; Hsing, I.-M.; Sin, J.K.O. Investigation and Control of Microcracks in Tin Oxide Gas Sensing Thin-Films. *Sens. Actuators B Chem.* **2001**, *79*, 39–47. [CrossRef]
42. Sonder, E.; Levinson, L.M.; Katz, W. Role of Short-circuiting Pathways in Reduced ZnO Varistors. *J. Appl. Phys.* **1985**, *58*, 4420–4425. [CrossRef]
43. Filipovic, L.; Selberherr, S. Performance and Stress Analysis of Metal Oxide Films for CMOS-Integrated Gas Sensors. *Sensors* **2015**, *15*, 7206–7227. [CrossRef]

44. Rabbani, A.; Salehi, S. Dynamic Modeling of the Formation Damage and Mud Cake Deposition Using Filtration Theories Coupled with SEM Image Processing. *J. Nat. Gas Sci. Eng.* **2017**, *42*, 157–168. [CrossRef]
45. Kohl, D. Surface Processes in the Detection of Reducing Gases with SnO_2-Based Devices. *Sens. Actuators* **1989**, *18*, 71–113. [CrossRef]
46. Mirzaei, A.; Kim, J.-H.; Kim, H.W.; Kim, S.S. Resistive-Based Gas Sensors for Detection of Benzene, Toluene and Xylene (BTX) Gases: A Review. *J. Mater. Chem. C* **2018**, *6*, 4342–4370. [CrossRef]
47. Xu, C.; Tamaki, J.; Miura, N.; Yamazoe, N. Grain Size Effects on Gas Sensitivity of Porous SnO_2-Based Elements. *Sens. Actuators B Chem.* **1991**, *3*, 147–155. [CrossRef]
48. Akbar, M.A.; Ait Si Ali, A.; Amira, A.; Bensaali, F.; Benammar, M.; Hassan, M.; Bermak, A. An Empirical Study for PCA- and LDA-Based Feature Reduction for Gas Identification. *IEEE Sens. J.* **2016**, *16*, 5734–5746. [CrossRef]
49. Liu, H.; Meng, G.; Deng, Z.; Nagashima, K.; Wang, S.; Dai, T.; Li, L.; Yanagida, T.; Fang, X. Discriminating BTX Molecules by the Nonselective Metal Oxide Sensor-Based Smart Sensing System. *ACS Sens.* **2021**, *6*, 4167–4175. [CrossRef]
50. Cho, B.; Lee, K.; Pyo, S.; Kim, J. Fabrication and Characterization of VOC Sensor Array Based on SnO_2 and ZnO Nanoparticles Functionalized by Metalloporphyrins. *Micro Nano Syst. Lett.* **2018**, *6*, 10. [CrossRef]
51. Lee, D.-S.; Kim, Y.T.; Huh, J.-S.; Lee, D.-D. Fabrication and Characteristics of SnO_2 Gas Sensor Array for Volatile Organic Compounds Recognition. *Thin Solid Films* **2002**, *416*, 271–278. [CrossRef]

chemosensors

MDPI

Article

Synthesis of Graphene-Oxide-Decorated Porous ZnO Nanosheet Composites and Their Gas Sensing Properties

Jie Li [1,2], Zhen Jin [1,2,*], Yang Chao [2], Aijing Wang [2], Decai Wang [2], Shaohua Chen [1,2] and Quan Qian [2]

[1] Anhui Advanced Building Materials Engineering Laboratory, Anhui JianZhu University, Hefei 230601, China
[2] School of Materials and Chemical Engineering, Anhui JianZhu University, Hefei 230601, China
* Correspondence: ftbjin@hotmail.com

Abstract: In this work, graphene-oxide-decorated porous ZnO nanosheets were prepared using a hydrothermal method. The graphene oxide/porous ZnO nanosheet (GO/ZnO nanosheet) composites were characterized with SEM, HRTEM, XRD, Raman spectroscopy, XPS and BET. The results indicate that the ZnO nanosheets have a porous, single-crystal structure. Thin GO nanosheets closely cover the surface of porous ZnO nanosheets. The sensing performance of GO/ZnO nanosheet composites is investigated. At the optimized temperature of 300 °C, the GO/ZnO nanosheet composites exhibit a superior sensing performance in n-propanol detection. In a wide range of 5–200 ppm, the composites exhibit a linear response to n-propanol. Moreover, the sensing performance of the GO/ZnO nanosheet composites to n-propanol is largely higher than that to other VOC gases, indicating a high selectivity in n-propanol detection. This can be ascribed to the higher electron separation efficiency and larger depletion layer brought by the modification of the GO on ZnO nanosheets. It is considered that the GO/ZnO nanosheet composites have a great application potential in n-propanol detection.

Keywords: graphene oxide; ZnO nanosheet; n-propanol; sensing property

1. Introduction

The demand for accurate detection for volatile organic compounds (VOCs) is increasing alongside the development of industrialization [1,2]. Therefore, many methods have been developed for the efficient detection of VOCs. Among various techniques, metal oxide semiconductor (MOS) gas sensors with benefits including fast response, low energy consumption and compact size are widely used [3–5]. Among MOS materials, ZnO has attracted great attention in the field of gas detection due to its excellent electronic and photonic properties [6–9]. However, owing to the large resistance value and the aggregation during the process of detection, pure ZnO sensing materials always suffer imperfections such as poor sensitivity and selectivity in VOC detection [10–12]. Numerous efforts have been devoted to enhance the performance of ZnO sensing materials [13–15]. Composite ZnO with a secondary material is a rational method to enhanced its sensing property, since the composite materials always combines the advantages of each component [16,17]. Aubekerov et al. synthesized $ZnFe_2O_4$-modified ZnO nanowires, which show better sensitivity than pure ZnO nanowires in isopropyl alcohol detection [18]. Xu et al. reported Eu(III)-functionalized ZnO@MOF heterostructures with excellent selectivity and sensitivity for aldehyde detection [19]. Nevertheless, it is still a big challenge to obtain ZnO-based sensing materials with superior sensing performance.

Two-dimensional materials, such as graphene and graphene oxide (GO), have attracted much attention due to their unique structures and excellent physical and chemical properties [20,21]. The unique two-dimensional network of graphene facilitates the exposure of all carbon atoms to air, allowing the maximum contact with the target molecules [17,22]. Wu et al. [23] found that a graphene/ZnO composite with low resistance exhibits better gas-sensing performance. GO, with abundant oxygen-containing functional groups, has a

Citation: Li, J.; Jin, Z.; Chao, Y.; Wang, A.; Wang, D.; Chen, S.; Qian, Q. Synthesis of Graphene-Oxide-Decorated Porous ZnO Nanosheet Composites and Their Gas Sensing Properties. *Chemosensors* **2023**, *11*, 65. https://doi.org/10.3390/chemosensors11010065

Academic Editor: Boris Lakard

Received: 3 November 2022
Revised: 4 January 2023
Accepted: 6 January 2023
Published: 13 January 2023

similar structure to graphene, and can also be used as an ideal modifier to ZnO [24,25]. GO can not only effectively prevent the aggregation of ZnO nanomaterials, but also improve the electrical conductivity of ZnO materials, thus enhancing the sensing performance of the composites [26,27]. Singh et al. [28] prepared GO/ZnO composites using a one-step method, and these composites could efficiently detect NH_3 (1 ppm), NO (5 ppm) and CO (22 ppm) gases. Although many reports mention GO-modified sensing materials, it is still a big challenge to promote their sensing performance.

As one of the most widely used chemical raw materials, n-propanol is used in various industries, such as preservatives, detergents, cosmetics and pharmaceuticals [29]. However, n-propanol does harm to human health, and is known as a toxic volatile organic compound (VOC). When the concentration of n-propanol exceeds 400 ppm, it may impair the upper respiratory tract, retina and optic nerve. Therefore, the high-performance detection of n-propanol has attracted great interest in recent years [30]. In this work, graphene-oxide-decorated porous single-crystalline ZnO nanosheet (GO/ZnO nanosheet) composites were synthesized using a hydrothermal method. In the composites, the ZnO nanosheets of hexagonal wurtzite are closely wrapped by GO lamellae, and this unique structure makes the GO/ZnO nanosheet composites exhibit a high response, excellent selectivity and stability in n-propanol detection. At the optimized temperature of 300 °C, the GO/ZnO nanosheet composites exhibit a linear response in n-propanol detection. The sensing performance of the GO/ZnO nanosheet composites to n-propanol is much higher than to other VOC gases. The sensing mechanism is also discussed. It is believed that the GO/ZnO nanosheet composites have great potential for practical usage.

2. Materials and Methods

2.1. Materials

Zinc acetate, urea, sodium nitrate, potassium permanganate, graphite powder and concentrated sulfuric acid of analytical grade were purchased from Sinopharm Chemical Reagent Co., Ltd. (Shanghai, China) and used without further purification. Deionized water was used in all experiments.

2.2. Preparation of GO and Porous Single-Crystal ZnO Nanosheet Composites

GO was prepared using the modified Hummer method using graphite, sodium nitrate, concentrated sulfuric acid and potassium permanganate as raw materials. Graphite powder (0.5 g) and concentrated sulfuric acid (23 mL) were added to a 250 mL three necked round-bottomed flask. The flask was sealed and stirred in an ice bath continuously overnight. $NaNO_3$ (0.5 g) was quickly added into the mixture and $KMnO_4$ (3 g) was slowly added for 30 min. Then the mixture was transferred to a 35 °C water bath under vigorous stirring for 1 h. Deionized water (40 mL) was slowly added and the system was heated up to 90 °C and stirred continuously for 2 h. Then a mixture of deionized water (10 mL) and hydrogen peroxide (30%) (3 mL) was slowly added and stirred for a few minutes. GO was prepared by filtering and washing the solution until it was neutral. GO (20 mg) was dissolved in 1 mL deionized water and ultrasonicated to form a 20 g/L GO dispersion.

The GO dispersion was added to a mixture of zinc acetate (1 g), urea (3 g) and 40 mL deionized water in a conical flask under ultrasound for 30 min. The conical flask was tightly corked and heated in an oven at 100 °C for 6 h, and then cooled to room temperature. Then, the complexes of GO and ZnO precursor were prepared. The product was washed with deionized water and ethanol by centrifugation and dried at 60 °C for 24 h. Finally, the GO-modified porous single crystal ZnO nanosheet composites were prepared after calcinating the obtained samples at 300 °C for 2 h in air.

2.3. Characterization

The morphologies and structure of the samples were examined using field emission scanning electron microscopy (FESEM, FEI Sirion-200, Eindhoven, Netherlands) at 5 kV of accelerating voltage, and high resolution transmission electron microscopy (HRTEM,

JEOL JEM-2010, Tokyo, Japan) at 200 kV of accelerating voltage. The crystal structure of the samples was determined using X-ray diffraction (XRD, Philips X′ pert PRO, Amsterdam, Netherlands) with Cu Kα radiation, λ = 0.15418 nm. The Raman spectroscopy was processed with a Raman spectrometer (invia, Renishaw, 532 nm wavelength, Gloucestershire, UK). X-ray photoelectron spectra (XPS) were recorded on an AXIS ULTRADLD spectrometer(Manchester, UK), with monochromatized Al Kα X-rays as the excitation source and C 1s chosen as the reference line. N_2 adsorption/desorption isotherm was carried out at 77 K using an Autosorb-1-C analyzer (Quantachrome Instruments, Boynton Beach, FL, USA), and the corresponding pore size distribution was derived from the adsorption branch of the isotherm using the BJH method.

2.4. Fabrication of the Gas Sensor and the Gas-Sensing Measurement System

An illustration of the operating procedure of the gas sensor and its measurement system is shown in Figure 1. Two gold electrodes with a distance of 10 mm were added onto a ceramic tube. Next, two pairs of gold wires were closely bonded to those two gold electrodes. Then, a nichrome wire (with a resistance of 30 Ω) was placed in the interior of the ceramic tube as a heating wire. For the fabrication of the gas sensor, the prepared GO/ZnO nanosheet composite and porous ZnO nanosheet powders were dispersed in ethanol solvent and directly coated on the surface of the ceramic tube and then dried naturally at room temperature. The final thickness of the sensing film of the GO/ZnO nanosheet composite and porous ZnO nanosheet were both approximately 200 μm.

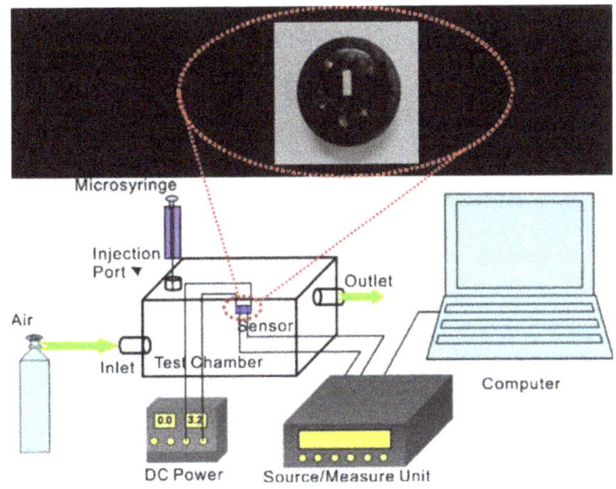

Figure 1. Schematic diagram of the experimental setup. The inset is the structure of the sensor.

The gas-sensing measurements were performed in a closed test chamber (1 L) equipped with an inlet and an outlet for gas. A Keithley 6487 Source/Measure Unit was used to record the change in current. When working, a constant voltage was applied onto the two gold electrodes between the sensing films, and the corresponding current was measured. In the typical gas-sensing test, certain amount of target gas was first collected from the headspace vapor of the organic solvent (e.g., methanol, ethanol, n-propanol, isopropanol, ammonia, methanal, benzene, water, etc., all of analytical grade), and then introduced into the test chamber using a microsyringe. The saturated vapor pressure of the organic vapor was recorded to calculate the concentration of the target gas under a standard atmospheric pressure. The injection rate of the target gas was about 1 mL/s. When the test was over, the target gas in the testing chamber was released by inputting fresh compressed air into the air bottle.

The response of the sensor is defined as:

$$\text{Response} = R_a/R_g = I_g/I_a \tag{1}$$

Here, R_a and R_g are the electric resistance of the sensor in air and the target gas, respectively. I_a and I_g are the electric current of the sensor in air and the target gas, respectively.

3. Results

3.1. Characterization

The porous ZnO nanosheets were first synthesized through a hydrothermal method, then the GO nanosheets were modified onto their surface. The morphologies of the porous ZnO nanosheets and the GO/porous ZnO nanosheet (GO/ZnO nanosheet) composites were investigated with SEM, and the results are shown in Figure 2. Figure 2a,b are the SEM images of the porous ZnO nanosheets at different magnifications. In these images, a great many nanosized pores can be clearly seen on the surface of the ZnO nanosheets. After the modification of GO nanosheets, from Figure 2c, it can be seen that the edges of the modified ZnO nanosheets are blurred. Furthermore, in the magnified image (Figure 2d), a thin and wrinkled structure can be observed on the surface of the nanosheets, indicating the successful modification of the lamellar GO on the porous ZnO nanosheets. The statistical distributions of the length and width of the GO/ZnO nanosheet composites are presented in Figure S1. The average length and width of the GO/ZnO nanosheet composites are 5.0 and 3.5 μm, respectively.

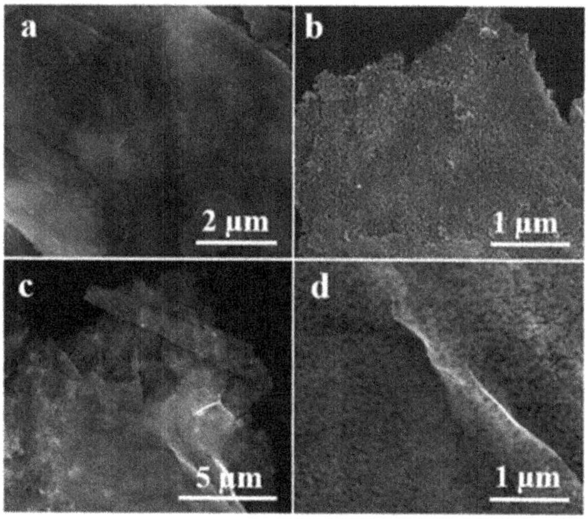

Figure 2. (**a**,**b**) SEM images of porous ZnO nanosheets, (**c**,**d**) GO/ZnO nanosheet composites, (**b**,**d**) are the magnification of (**a**,**c**), respectively.

Figure 3a,b are TEM images of the GO/ZnO nanosheets at different magnifications. It can be seen that the pores on the ZnO nanosheet are very rich, and the average diameter of the pores is about 30 nm. When looking closely at the surface of the ZnO nanosheets, it can be found that the ZnO nanosheet is covered with a thin layer of GO. In addition, at the edge of the ZnO nanosheet, the uncovered GO nanosheets also can be clearly observed. Figure 3c is the corresponding SAED pattern of the GO/ZnO nanosheet, which consists of well-ordered dots, indicating the single crystallinity of the porous ZnO nanosheets. The (112), (110) and (002) planes of wurtzite ZnO can also be indexed from the SAED pattern. Due to its small amount and poor crystallinity, the SAED pattern of GO cannot be observed. From the HRTEM image of GO/ZnO nanosheets (Figure 3d), it can been seen that the clear and coherent lattice fringes go throughout the nanosheet. The lattice spacing is 0.26 nm, which can be ascribed to the (002) plane of the hexagonal phase ZnO. Those results definitively prove the formation of the GO/ZnO nanosheet composites.

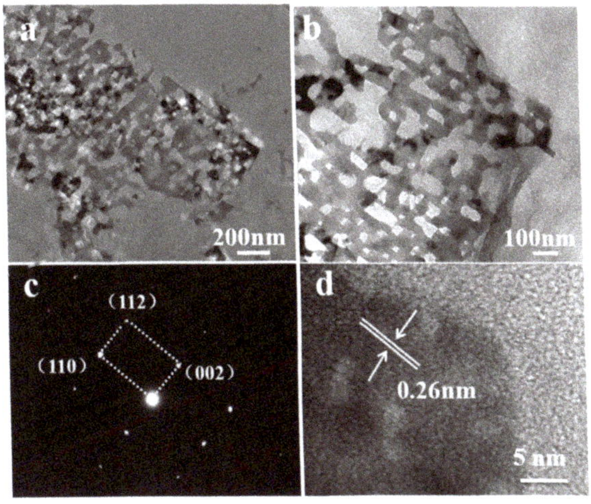

Figure 3. (**a**,**b**) TEM images of GO/ZnO nanosheets, (**c**) SAED pattern of the GO/ZnO nanosheets, (**d**) HRTEM image of GO/ZnO nanosheets.

The XRD patterns of pure porous ZnO nanosheets and GO/ZnO nanosheet composites are shown in Figure 4. All peaks in the porous ZnO nanosheets and GO/ZnO nanosheet composites correspond to the characteristic peaks of hexagonal wurtzite zinc oxide (JCPDS No. 36-1451), and no other peaks can be observed, indicating that there are no other impurities in the GO/ZnO nanosheet composites. The disappearance of the diffraction peak of GO in the composite material is due to its low amount and poor crystallinity.

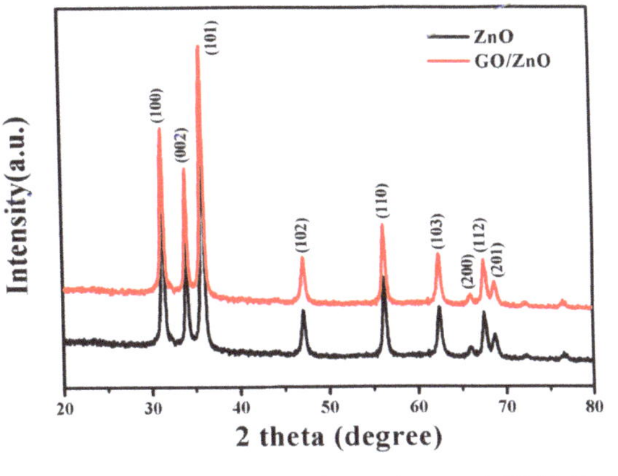

Figure 4. XRD patterns of the ZnO nanosheets and GO/ZnO nanosheet composites.

Raman spectroscopy is an effective method to characterize the structure of GO; thus, the GO nanosheets, porous single-crystal ZnO nanosheets and GO/ZnO nanosheet composites were characterized with Raman spectroscopy, and the results are presented in Figure 5. From the Raman spectra of the GO and GO/ZnO nanosheet composites, two obvious peaks at 1362 cm^{-1} and 1590 cm^{-1} can be observed, which correspond to the D and G bands of graphene, respectively. Furthermore, the Raman spectra of the GO/ZnO nanosheet composites also contain three small peaks at 322.2, 432.3 and 562.2 cm^{-1}, which can also be observed in the Raman spectrum of the porous ZnO nanosheets. This result further confirms the combination of the ZnO nanosheets and GO.

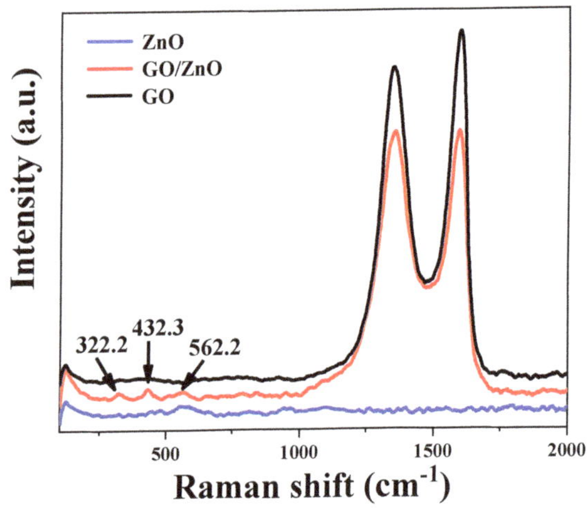

Figure 5. Raman spectra of pure GO, porous single-crystal ZnO nanosheets and GO/ZnO nanosheet composites.

The structure of GO/ZnO nanosheet composites was further investigated with XPS, and the results are shown in Figure 6. Figure 6a is the survey XPS spectrum of the composites, while Figure 6b presents the Zn 2p spectrum of the composites. The two characteristic peaks at 1044.6 and 1021.5 correspond to Zn 2p1/2 and Zn 2p3/2, respectively, which are associated with the Zn-O bonding in ZnO. In Figure 6c, the C 1s peak of the GO/ZnO nanosheet composites can be deconvoluted into three peaks observed at 284.8, 286.2 and 288.7 eV, corresponding to functional groups C-C, C-O and C-COOH, respectively [21,31]. The O 1s spectrum (Figure 6d) can be deconvoluted into three peaks at 529.7, 531.2 and 532.5 eV, which correspond to the O^{2-} valance state (O-Zn bonding) in GO/ZnO nanosheet composites, oxygen vacancies and defects, and chemisorbed oxygen species, respectively [26,31]. The greater amount of oxygen vacancies and defects of the GO/ZnO nanosheet composites can help to trap electrons, improve the electron-hole recombination rate and keep a high carrier mobility, resulting in the enhanced sensing performance of the GO/ZnO nanosheet composites. The results are consistent with the sensing performance of the composites.

The N_2 adsorption/desorption isotherm of GO/ZnO nanosheet composites was studied and the result is presented in Figure 7. The N_2 adsorption/desorption isotherm of GO/ZnO nanosheet composites displays a type IV isotherm with an H3 hysteresis loop, which is typical of mesoporous materials. Figure 7b presents the pore size distribution of the composites. It can be seen that the average pore diameter is around 18.8 nm, which is consistent with the TEM result. The BET surface area of the GO/ZnO nanosheet composites is 25.1 m^2g^{-1}, which is larger than that of the ZnO nanosheet (21.3 m^2g^{-1}) [32]. The larger surface area and abundant pore structure of the GO/ZnO nanosheet composite make it a promising candidate for VOC detection.

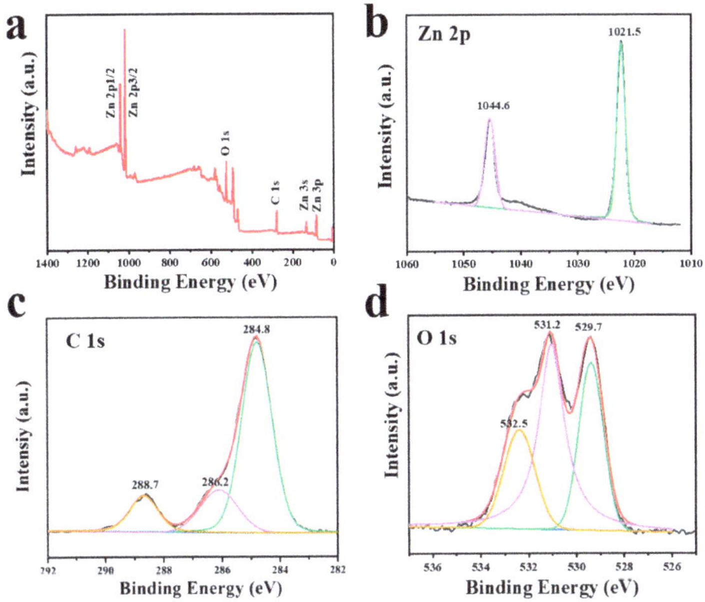

Figure 6. (**a**) Survey XPS spectrum of GO/ZnO nanosheet composites, and the corre-sponding elements' high-resolution XPS spectra: (**b**) Zn 2p, the purple line corresponds to Zn 2p1/2 and the green line corresponds to 2p3/2, respectively, (**c**) C 1s, the green, purple and yellow line correspond to functional groups C-C, C-O and C-COOH, respectively, (**d**) O 1s, the green, purple and yellow line correspond to O-Zn bonding, oxygen vacancies and defects, and chemisorbed oxygen species, respectively.

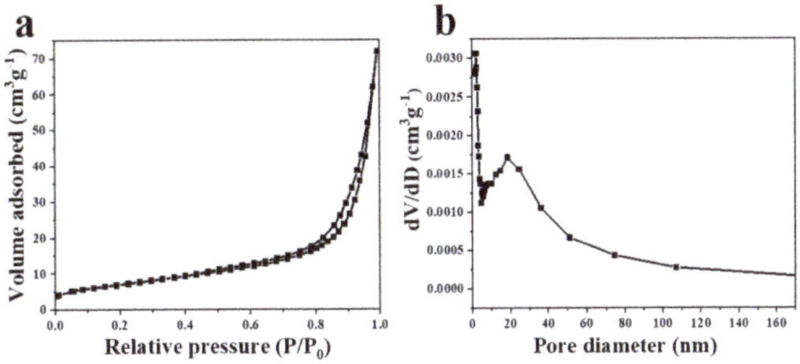

Figure 7. (**a**) N_2 adsorption-desorption isotherm of the GO/ZnO nanosheet composites, (**b**) the corresponding pore size distribution.

3.2. Sensing Performance

The electrical behavior of GO/ZnO nanosheet composites was investigated, and Figure S2 presents the corresponding current-voltage (I–V) curves. It can be seen that the almost linear characteristics of the ZnO/GO samples indicate the p-type ohmic contact of the sensing films.

The sensing performance of MOS sensors usually depends on the operating temperature. The responses of the GO/ZnO nanosheet composite sensor to 50 ppm n-propanol and isopropanol at different operating temperatures were investigated. As illustrated in Figure 8, the responses of GO/ZnO nanosheet composites to both n-propanol and isopropanol first increase with the increasing working temperature. However, when the temperature increases beyond 300 °C, the responses decrease. At low temperatures, there are not enough O^{2-} anions on the surface of the GO/ZnO nanosheet composites to react with n-propanol and isopropanol molecules, and when the working temperature is too high, the desorption rate of molecules is far greater than the adsorption rate, thus resulting in a decrease in the response of the GO/ZnO nanosheet composites. Thus, the optimum operating temperature of the GO/ZnO nanosheet composites is 300 °C, and the subsequent gas-sensing tests were carried out at this temperature.

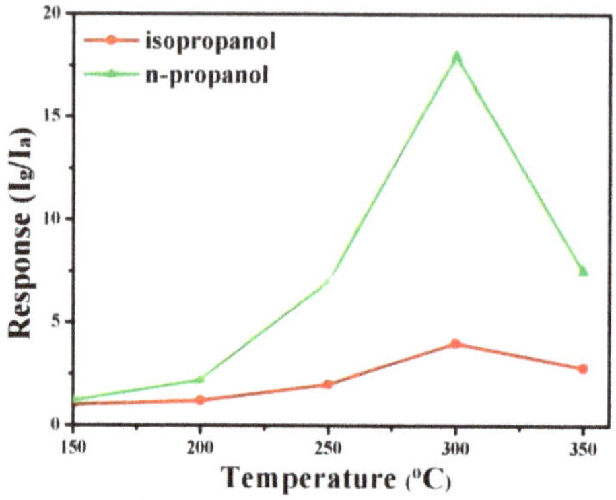

Figure 8. Responses of the GO/ZnO nanosheet composite sensor to 50 ppm isopropanol and n-propanol at different operating temperatures.

The real-time responses of the GO/ZnO nanosheet composites to n-propanol at different concentrations were also investigated, and the results are presented in Figure 9a. With an increase in the n-propanol concentration from 5 to 200 ppm, the response currents increase rapidly, and then immediately drop as the n-propanol gas is removed from the testing chamber. Figure 9b displays the plot of the sensitivity as a function of the concentration of n-propanol, revealing a linear range from 5 to 200 ppm. The least-squares fitting is y = 5.48 + 0.26x, and the regression coefficient (R^2) is 97.6%. The real-time response curves of GO/ZnO nanosheet composites to 50 ppm n-propanol and isopropanol at 300 °C are shown in Figure 10. The sensitivity of the GO/ZnO nanosheet composites to 50 ppm n-propanol and isopropanol is 17.2 and 4.4, respectively. In addition, the responses and recovery of the composites to both n-propanol and isopropanol are very fast, and the corresponding response times are 45 and 36 s, while the recovery times are 15 and 19 s, respectively. The results indicate that the GO/ZnO nanosheet composites have a good response to n-propanol.

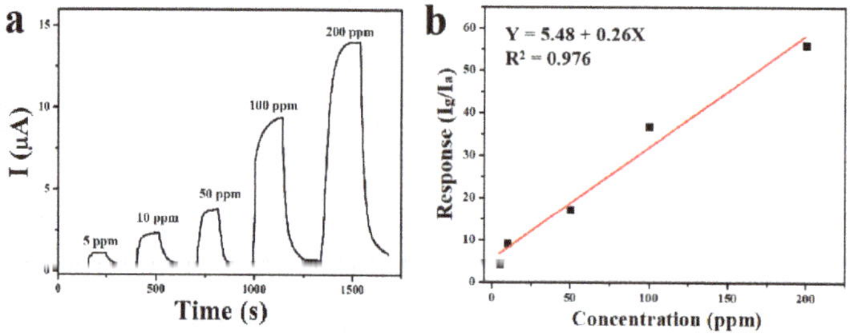

Figure 9. (**a**) Real-time responses of the GO/ZnO nanosheet composites to n-propanol at different concentrations at 300 °C, (**b**) the corresponding plots of the response vs. concentration.

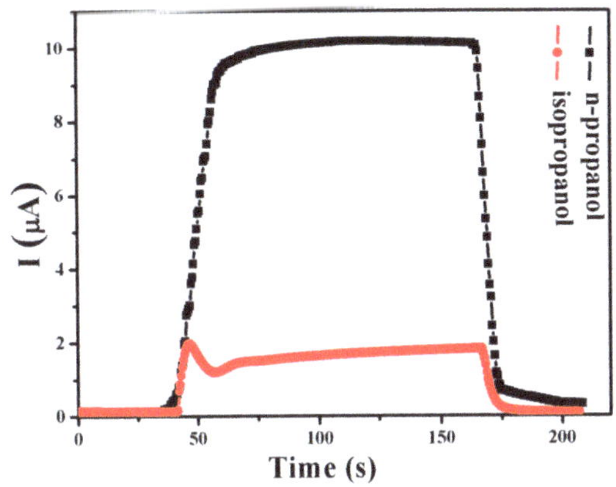

Figure 10. Real-time responses of GO/ZnO nanosheet composites to 50 ppm n-propanol and 50 ppm isopropanol at 300 °C.

The sensing performance of the GO/ZnO nanosheet composites was also compared with that of the porous ZnO nanosheets, and the results are presented in Figure 11. As shown, the responses of the ZnO nanosheets to 50 ppm n-propanol and isopropanol are 2.58 and 2.39, respectively. Nevertheless, when the ZnO nanosheets are coupled with GO nanosheets, the sensing performance of the composites is largely enhanced, especially to n-propanol. Obviously, this huge enhancement in sensing performance is due to the modification with GO.

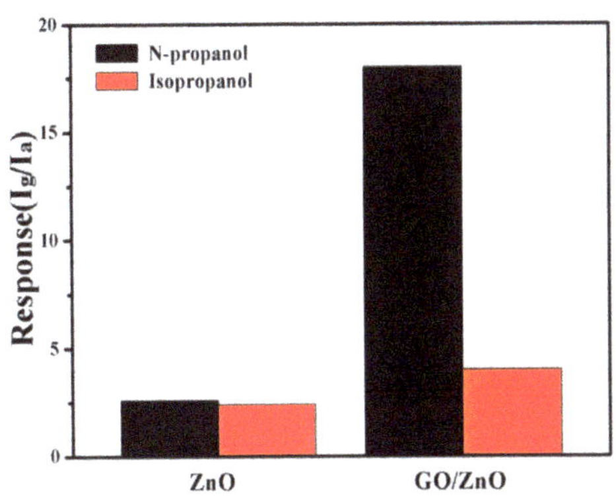

Figure 11. Response sensitivity of the porous ZnO nanosheets and GO/ZnO nanosheet composites to 50 ppm n-propanol and 50 ppm isopropanol at 300 °C.

The selectivity of the gas sensor is an important parameter for its application. Thus, the sensing properties of the GO/ZnO nanosheet composites to 50 ppm of methanol, ethanol, n-propanol, isopropanol, ammonia, methanal, benzene and hydrogen were investigated at 300 °C, and the results are presented in Figure 12. Obviously, the responses of the GO/ZnO nanosheet composites to ammonia, methanal and benzene are not high, namely 2.2, 1.5 and 1.2, respectively. For alcohols, the responses of the composites increase significantly. As well, the sensitivity of the GO/ZnO nanosheet composites largely increases with the growth of the carbon chain. The sensitivities of the GO/ZnO nanosheet composites to methanol, ethanol and n-propanol are 2.7, 7.0 and 17.2, respectively. Furthermore, the sensitivity of the composites to isopropanol, the isomer of n-propanol, largely reduces to 4.0. Thus, it is believed that the GO/ZnO nanosheet composites can identify n-propanol in a complex environment. The sensing performance of the GO/ZnO nanosheet composites was also compared with other values from the literature, and the results are presented in

Table 1. Obviously, the GO/ZnO nanosheet composites exhibit good sensitivity towards n-propanol.

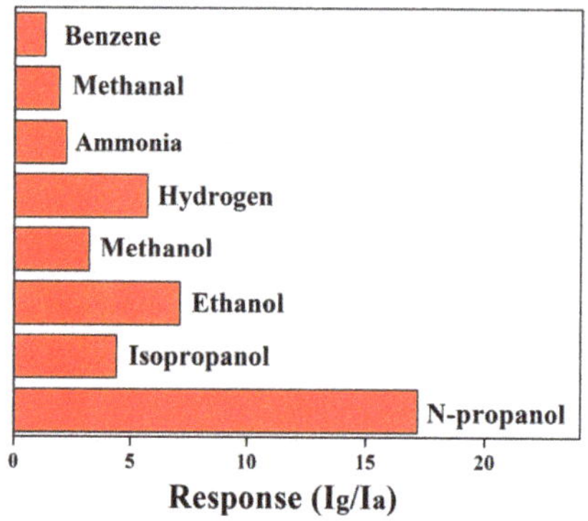

Figure 12. Responses of the GO-decorated porous single-crystalline ZnO nanosheets to 50 ppm of various gases at 300 °C.

Table 1. N-propanol sensors based on metal oxide nanostructures.

Sensing Materials	Temperature (°C)	Response @ Concentration (ppm)	Selectivity	Response Time (s)	References
ZnO nanowires	300	30.1@50 (2-propanol)	General	5	[33]
CuO nanofiber	200	4.66@100	General	4.66	[34]
TeO_2 nanowires	50	3.15@500	Poor	20	[35]
$CuO/CuCo_2O_4$ nanotubes	Room temperature	14@100	General	6.3	[36]
Cu_2O hollow microspheres	187	11@100	General	50	[37]
GO/ZnO nanosheet composites	300	17.2@50	Good	45	This work

The sensing mechanism of the GO/ZnO nanosheet composites was also deduced. As shown in Figure 13, GO nanosheets exhibit a p-type semiconductor behavior, while ZnO is an n-type semiconductor; thus, the p-n heterojunction of GO and ZnO is formed. When working, both of the GO and ZnO nanosheets are hot excited. As GO (4.5 eV) and ZnO (4.22 eV) have different work functions, the electrons in the higher energy level will be transferred to the lower energy level [21,26]. In the GO/ZnO heterojunction, the electrons will transfer from p-GO to n-ZnO, and the holes move in the opposite direction until equilibrium. The abundant oxygen vacancies and efficient electron mobility at the GO/ZnO heterojunction will provide more active sites for gas adsorption by increasing the specific surface area. Furthermore, they supply more channels to facilitate the diffusion of gas molecules, and consequently, favor the enhancement of n-propanol detection. Moreover, the unique structure of the ZnO nanosheet is also beneficial for the sensing properties of the composites. First, the porous structure provides more active sites. Due to the porous structure, there are more surface defects on the surface of the nanosheet, and those defects can effectively adsorb VOC molecules. In addition, these nanoholes could be the ideal transmission channel for the target gases, effectively facilitating the diffusion of gas molecules, which can greatly increase the sensing performance of the composite. Second, the single-crystalline structure of individual nanosheets can also increase the sensitivity. During transportation, electrons need not jump cross-grain boundaries, which is beneficial for the stability of the sensing materials. Third, the ZnO nanosheets are ultra-thin, which makes the space charge layers cover the whole nanosheet. Fourth, in the composite, both ZnO and GO are lamellar structures, which grant the composite a larger boundary area; thus, electrons can be effectively transferred between the GO/ZnO nanosheets heterogeneously. Thus, the composites exhibit superior n-propanol sensing performance.

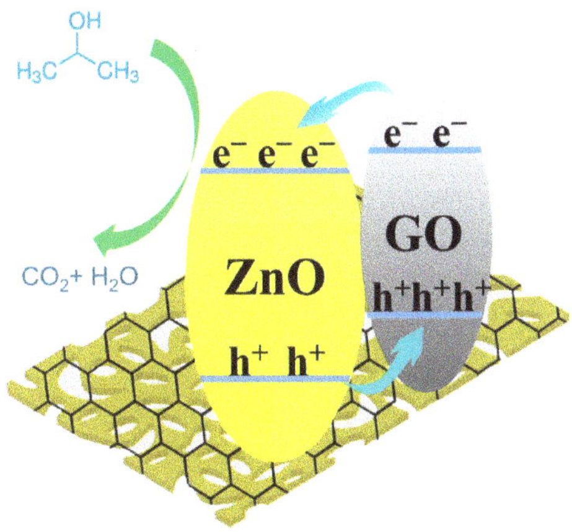

Figure 13. Sensing mechanism of the GO/ZnO nanosheet composites.

The stability of the GO/ZnO nanosheet composite sensor was further studied. As shown in Figure 14, there is no significant reduction in the response of the composite during a 14-day test. It is believed that the good stability of the GO/ZnO nanosheet composite can be attributed to the unique structure of the porous ZnO nanosheet.

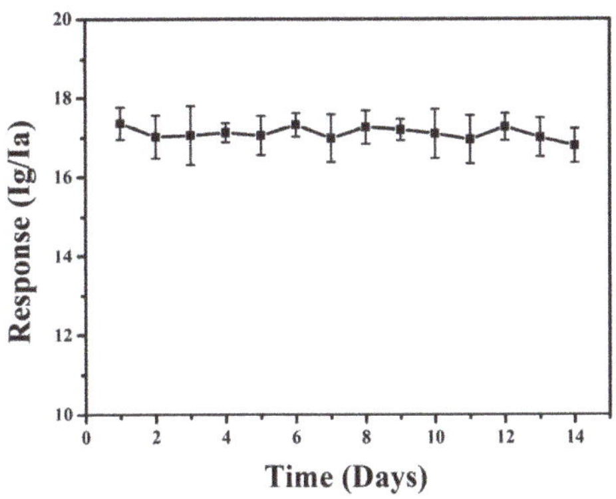

Figure 14. Response curve of the GO/ZnO nanosheet composites to 50 ppm n-propanol after 14 days of testing.

4. Conclusions

In conclusion, the GO/ZnO nanosheet composites were successfully synthesized through a hydrothermal method. In the composites, the ZnO nanosheets of hexagonal wurtzite are closely wrapped by GO lamellae, and the unique structure makes the GO/ZnO nanosheet composites exhibit a high response, excellent selectivity and stability in n-propanol detection. At the optimized temperature of 300 °C, the GO/ZnO nanosheet composites exhibit a linear response in n-propanol detection in a large range. Moreover, the sensing performance of the composites to n-propanol is largely higher than to other VOC gases, indicating a high sensitivity and selectivity in n-propanol detection. This can be ascribed to the higher electron-separation efficiency and larger depletion layer brought by the modification of the GO on ZnO nanosheets. It is considered that the GO/ZnO nanosheet composites have a great application potential in n-propanol detection.

Supplementary Materials: The following supporting information can be downloaded at: https://www.mdpi.com/article/10.3390/chemosensors11010065/s1, Figure S1: The statistical distributions of the length and width of the GO/ZnO nanosheet composites; Figure S2: The corresponding current–voltage (I–V) curves of the GO/ZnO nanosheet composites.

Author Contributions: Conceptualization, J.L.; methodology, Y.C.; software, A.W.; validation, S.C. and D.W.; formal analysis, Z.J.; investigation, J.L.; writing—original draft preparation, J.L.; writing—review and editing, Q.Q.; supervision, Z.J.; funding acquisition, Z.J. All authors have read and agreed to the published version of the manuscript.

Funding: This research was funded by the National Key Research and Development (No. 2019YFC0408504), Natural Science Foundation of Anhui Province (No. 2008085MB44), Natural Science Research Project of Anhui Educational Committee (No. 2022AH040046, KJ2019A0772) and Young Talents and PhD Start-up Fund Project from Anhui Jianzhu University (2018QD52, 2019QDZ30).

Institutional Review Board Statement: Not applicable.

Informed Consent Statement: Not applicable.

Data Availability Statement: The data presented in this study are openly available.

Conflicts of Interest: We declare that we have no financial and personal relationships with other people or organizations that can inappropriately influence our work, and there is no professional or other personal interest of any nature or kind in any product, service and/or company that could be construed as influencing the position presented in, or the review of, the manuscript entitled.

References

1. Han, B.; Liu, X.; Xing, X.; Chen, N.; Xiao, X.; Liu, S.; Wang, Y. A high response butanol gas sensor based on ZnO hollow spheres. *Sensors Actuators Chem.* **2016**, *237*, 423–430. [CrossRef]
2. Meng, F.L.; Zheng, H.X.; Chang, Y.L.; Zhao, Y.; Li, M.Q.; Wang, C.; Sun, Y.F.; Liu. J., H. One-step synthesis of Au/SnO_2/RGO nanocomposites and their VOC sensing properties. *IEEE Trans. Nanotechnol.* **2018**, *172*, 212–219. [CrossRef]
3. Ma, A.; Park, H.J.; Seo, J.H.; Jang, K.Y.; Lee, H.K.; Kim, D.Y.; Lee, J.E.; Nam, K.M.; Lee, D.S. Phase transition of non-equilibrium wurtzite CoO: Spontaneous deposition of sensing material for ultrasensitive detection of acetone. *Sensors Actuators B Chem.* **2020**, *308*, 127698. [CrossRef]
4. Qin, W.B.; Yuan, Z.Y.; Gao, H.L.; Zhang, R.; Meng, F.L. Perovskite-structured $LaCoO_3$ modified ZnO gas sensor and investigation on its gas sensing mechanism by first principle. *Sens. Actuators B Chem.* **2021**, *341*, 130015. [CrossRef]
5. Ji, H.Y.; Qin, W.B.; Yuan, Z.Y.; Meng, F.L. Qualitative and quantitative recognition method of drug-producing chemicals based on SnO2 gas Sensor with dynamic measurement and PCA weak separation. *Sens. Actuators B Chem.* **2021**, *348*, 130698. [CrossRef]
6. Caicedo, N.; Leturcq, R.; Raskin, J.-P.; Flandre, D.; Lenoble, D. Detection mechanism in highly sensitive ZnO nanowires network gas sensors. *Sensors Actuators B Chem.* **2019**, *297*, 126602. [CrossRef]
7. Bhati, V.S.; Hojamberdiev, M.; Kumar, M. Enhanced sensing performance of ZnO nanostructures-based gas sensors: A review. *Energy Rep.* **2020**, *6*, 46–62. [CrossRef]
8. Wang, C.N.; Li, Y.L.; Gong, F.L.; Zhang, Y.H.; Fang, S.M.; Zhang, H.L. Advances in doped ZnO nanostructures for gas sensor. *Chem. Rec.* **2020**, *20*, 1553–1567. [CrossRef] [PubMed]
9. Meng, F.L.; Shi, X.; Yuan, Z.Y.; Ji, H.Y.; Qin, W.B.; Shen, Y.B.; Xing, C.Y. Detection of Four Alcohol Homologue Gases by ZnO Gas Sensor in Dynamic Interval Temperature Modulation Mode. *Sens. Actuators B Chem.* **2022**, *350*, 130867. [CrossRef]
10. Nundy, S.; Eom, T.Y.; Kang, J.G.; Suh, J.; Cho, M.; Park, J.S.; Lee, H.J. Flower-shaped ZnO nanomaterials for low-temperature operations in NO_x gas sensors. *Ceram. Int.* **2020**, *46*, 5706–5714. [CrossRef]
11. Alev, O.; Sarıca, N.; Özdemir, O.; Arslan, L.C.; Büyükköse, S.; Öztürk, Z.Z. Cu-doped ZnO nanorods based QCM sensor for hazardous gases. *J. Alloy. Compd.* **2020**, *826*, 154177. [CrossRef]
12. Zhang, L.; Yin, M.; Qiu, J.; Qiu, T.; Chen, Y.; Qi, S.; Wei, X.; Tian, X.; Xu, D. Mesoporous ZnO nanosheet as gas sensor for sensitive triethylamine detection. *Anal. Bioanal. Chem.* **2022**, *414*, 2181–2188. [CrossRef] [PubMed]
13. Campos, A.C.; Paes, S.C.; Correa, B.S.; Cabrera-Pasca, G.A.; Costa, M.S.; Costa, C.S.; Otubo, L.; Carbonari, A.W. Growth of long ZnO nanowires with high density on the ZnO surface for gas sensors. *ACS Appl. Nano Mater.* **2019**, *3*, 175–185. [CrossRef]
14. Kim, J.W.; Porte, Y.; Ko, K.Y.; Kim, H.; Myoung, J.M. Micropatternable double-faced ZnO nanoflowers for flexible gas sensor. *ACS Appl. Mater. Inter.* **2017**, *9*, 32876–32886. [CrossRef]

15. Choi, S.M.; Kim, Y.M.; Mirzaei, A.; Kim, H.S.; Kim, S.I.; Baek, S.H.; Chun, D.W.; Jin, C.; Lee, H.K. Selective, sensitive, and stable NO_2 gas sensor based on porous ZnO nanosheets. *Appl. Surf. Sci.* **2021**, *568*, 150910. [CrossRef]
16. Liu, J.; Zhang, L.; Fan, J.; Zhu, B.; Yu, J. Triethylamine gas sensor based on Pt-functionalized hierarchical ZnO microspheres. *Sensors Actuators B Chem.* **2021**, *331*, 129425. [CrossRef]
17. Young, S.J.; Chu, Y.L. Platinum nanoparticle-decorated ZnO nanorods improved the performance of methanol gas sensor. *J. Electrochem. Soc.* **2020**, *167*, 147508. [CrossRef]
18. Aubekerov, K.K.; Punegova, N.; Sergeenko, R.; Kuznetsov, A.; Kondratev, V.M.; Kadinskaya, S.A.; Nalimova, S.S.; Moshnikov, V.A. Synthesis and study of gas sensitive $ZnFe_2O_4$-modified ZnO nanowires. *J. Phys.* **2022**, *2227*, 012014. [CrossRef]
19. Xu, X.-Y.; Yan, B. Eu(iii)-functionalized ZnO@MOF heterostructures: Integration of pre-concentration and efficient charge transfer for the fabrication of a ppb-level sensing platform for volatile aldehyde gases in vehicles. *J. Mater. Chem. A* **2017**, *5*, 2215–2223. [CrossRef]
20. Ugale, A.D.; Umarji, G.G.; Jung, S.H.; Deshpande, N.G.; Lee, W.; Cho, H.K.; Yoo, J.B. ZnO decorated flexible and strong graphene fibers for sensing NO_2 and H_2S at room temperature. *Sensors Actuators B Chem.* **2020**, *308*, 127690. [CrossRef]
21. Zhang, J.; Jia, X.; Liu, T.; Yang, J.; Wang, S.; Li, Y.; Shao, D.; Feng, L.; Song, H. Facile strategy to synthesize porous GO/ZnO heterostructure for enhanced acetone gas sensing properties. *Sensors Actuators B Chem.* **2022**, *359*, 131601. [CrossRef]
22. Huang, W.C.; Tsai, H.J.; Lin, T.C.; Weng, W.C.; Chang, Y.C.; Chiu, J.L.; Lin, J.-J.; Lin, C.F.; Lin, Y.-S.; Chen, H. Incorporation of carbon nanotube and graphene in ZnO nanorods-based hydrogen gas sensor. *Ceram. Int.* **2018**, *44*, 12308–12314. [CrossRef]
23. Wu, J.; Shen, X.; Jiang, L.; Wang, K.; Chen, K. Solvothermal synthesis and characterization of sandwich-like graphene/ZnO nanocomposites. *Appl. Surf. Sci.* **2010**, *256*, 2826–2830. [CrossRef]
24. Khosravi, Y.; Sasar, M.; Abdi, Y. Light-induced oxygen sensing using ZnO/GO based gas sensor. *Mat. Sci. Semicon. Proc.* **2018**, *85*, 9–14. [CrossRef]
25. Galstyan, V.; Comini, E.; Kholmanov, I.; Faglia, G.; Sberveglieri, G. Reduced graphene oxide/ZnO nanocomposite for application in chemical gas sensors. *RSC Adv.* **2016**, *6*, 34225–34232. [CrossRef]
26. Kamble, C.; Narwade, S.; Mane, R. Detection of acetylene (C_2H_2) gas using Ag-modified ZnO/GO nanorods prepared by a hydrothermal synthesis. *Mat. Sci. Semicon. Proc.* **2023**, *153*, 107145. [CrossRef]
27. Vessalli, B.A.; Zito, C.A.; Perfecto, T.M.; Volanti, D.P.; Mazon, T. ZnO nanorods/graphene oxide sheets prepared by chemical bath deposition for volatile organic compounds detection. *J. Alloys Compd.* **2017**, *696*, 996–1003. [CrossRef]
28. Singh, G.; Choudhary, A.; Haranath, D.; Joshi, A.G.; Singh, N.; Singh, S.; Pasricha, R. ZnO decorated luminescent graphene as a potential gas sensor at room temperature. *Carbon* **2012**, *50*, 385–394. [CrossRef]
29. Yin, Y.; Shen, Y.; Zhou, P.; Lu, R.; Li, A.; Zhao, S.; Liu, W.; Wei, D.; Wei, K. Fabrication, characterization and n-propanol sensing properties of perovskite-type ZnSnO3 nanospheres based gas sensor. *Appl. Surf. Sci.* **2020**, *509*, 145335. [CrossRef]
30. Mokoena, T.P.; Hillie, K.T.; Swart, H.C.; Leshabane, N.; Tshilongo, J.; Motaung, D.E. Fabrication of a propanol gas sensor using p-type nickel oxide nanostructures: The effect of ramping rate towards luminescence and gas sensing characteristics. *Mater. Chem. Phys.* **2020**, *253*, 123316. [CrossRef]
31. Dai, K.; Lu, L.H.; Liang, C.H.; Dai, J.M.; Zhu, G.P.; Liu, Z.L.; Liu, Q.Z.; Zhang, Y.X. Graphene oxide modified ZnO nanorods hybrid with high reusable photocatalytic activity under UV-LED irradiation. *Mater. Chem. Phys.* **2014**, *143*, 1410–1416. [CrossRef]
32. Jin, Z.; Zhang, Y.X.; Meng, F.L.; Jia, Y.; Luo, T.; Yu, X.Y.; Wang, J.; Liu, J.H.; Huang, X.J. Facile synthesis of porous single crystalline ZnO nanoplates and their application in photocatalytic reduction of Cr(VI) in the presence of phenol. *J. Hazard. Mater.* **2014**, *276*, 400–407. [CrossRef] [PubMed]
33. Huang, J.R.; Ren, H.B.; Sun, P.P.; Gu, C.P.; Sun, Y.F.; Liu, J.H. Facile synthesis of porous ZnO nanowires consisting of ordered nanocrystallites and their enhanced gas-sensing. *Sensors Actuators B Chem.* **2013**, *188*, 249–256. [CrossRef]
34. Dong, C.J.; Xing, X.X.; Chen, N.; Liu, X.; Wang, Y.D. Biomorphic synthesis of hollow CuO fibers for low-ppm-level n-propanol detection via a facile solution combustion method. Sensors Actuators *B Chem.* **2016**, *230*, 1–8.
35. Shen, Y.B.; Fan, A.F.; Wei, D.Z.; Gao, S.L.; Liu, W.G.; Han, C.; Cui, B.Y. A low-temperature n-propanol gas sensor based on TeO_2 nanowires as the sensing layer. *RSC Adv.* **2015**, *5*, 29126. [CrossRef]
36. Alali, K.T.; Lu, Z.T.; Zhang, H.S.; Liu, J.Y.; Liu, Q.; Li, R.M.; Aljebawi, K.; Wang, J. P-p heterojunction $CuO/CuCo_2O_4$ nanotubes synthesized via electrospinning technology for detecting n-propanol gas at room temperature. *Inorg. Chem. Front.* **2017**, *4*, 1219–1230. [CrossRef]
37. Wang, N.; Zhou, Y.; Chen, K.; Wang, T.S.; Sun, P.; Wang, C.G.; Chuai, X.H.; Zhang, S.M.; Liu, X.M.; Lu, G.Y. Double shell Cu_2O hollow microspheres as sensing material for high performance n-propanol sensor. *Sensors Actuators B Chem.* **2021**, *333*, 129540. [CrossRef]

chemosensors

MDPI

Article

A Novel Gas Recognition Algorithm for Gas Sensor Array Combining Savitzky–Golay Smooth and Image Conversion Route

Xi Wang [1,2], Chen Qian [1,2], Zhikai Zhao [1,2,*], Jiaming Li [1,2] and Mingzhi Jiao [1,2,*]

1 National and Local Joint Engineering Laboratory of Internet Application Technology on Mine, China University of Mining and Technology, Xuzhou 221116, China
2 School of Information and Control Engineering, China University of Mining and Technology, Xuzhou 221116, China
* Correspondence: zhikaizhao@163.com (Z.Z.); mingzhijiao@cumt.edu.cn (M.J.); Tel.: +86-13685146374 (Z.Z.); +86-13952204541 (M.J.)

Abstract: In recent years, the application of Deep Neural Networks to gas recognition has been developing. The classification performance of the Deep Neural Network depends on the efficient representation of the input data samples. Therefore, a variety of filtering methods are firstly adopted to smooth filter the gas sensing response data, which can remove redundant information and greatly improve the performance of the classifier. Additionally, the optimization experiment of the Savitzky–Golay filtering algorithm is carried out. After that, we used the Gramian Angular Summation Field (GASF) method to encode the gas sensing response data into two-dimensional sensing images. In addition, data augmentation technology is used to reduce the impact of small sample numbers on the classifier and improve the robustness and generalization ability of the model. Then, combined with fine-tuning of the GoogLeNet neural network, which owns the ability to automatically learn the characteristics of deep samples, the classification of four gases has finally been realized: methane, ethanol, ethylene, and carbon monoxide. Through setting a variety of different comparison experiments, it is known that the Savitzky–Golay smooth filtering pretreatment method effectively improves the recognition accuracy of the classifier, and the gas recognition network adopted is superior to the fine-tuned ResNet50, Alex-Net, and ResNet34 networks in both accuracy and sample processing times. Finally, the highest recognition accuracy of the classification results of our proposed route is 99.9%, which is better than other similar work.

Keywords: gas recognition; gas sensor array; Savitzky–Golay smooth filter; sensor data visualization; Deep Neural Network

Citation: Wang, X.; Qian, C.; Zhao, Z.; Li, J.; Jiao, M. A Novel Gas Recognition Algorithm for Gas Sensor Array Combining Savitzky–Golay Smooth and Image Conversion Route. *Chemosensors* **2023**, *11*, 96. https://doi.org/10.3390/chemosensors11020096

Academic Editor: Corrado Di Natale

Received: 22 December 2022
Revised: 19 January 2023
Accepted: 24 January 2023
Published: 29 January 2023

1. Introduction

Gas recognition technology based on gas sensor array is widely used, playing an important role in many fields such as disease prediction [1–3], food safety [4–6], environmental monitoring [7,8], coal mine risk prediction [9], and so on. For example, Liu et al. [9] adopted the gas sensor array technology to identify and detect the concentration of carbon monoxide and methane released in the process of coal oxidation or spontaneous combustion. By capturing the slight change of gas release in the initial stage of coal oxidation and combining with an artificial neural network, the risk prediction ability in a coal mine has been greatly improved. Methane, ethanol, ethylene, and carbon monoxide are common flammable gases and are often mixed together in practical situations, such as in a chemical industrial park. Therefore, it is of practical significance to study the classification and identification technology of methane, ethanol, ethylene, and carbon monoxide based on gas sensor array.

Since the time series sensing data may contain redundant data or noise, the performance of the classifier largely depends on the input data representation. Efficient input data representation is the key to gas classification, which is helpful for training the classified gas model with efficient input data. For example, Pan et al. [10] proposed a new hybrid convolutional and recurrent neural network method to achieve fast gas recognition, extract valuable transient features contained at the beginning of the response curve, and finally achieve an identification accuracy of 84.06% within a response time as short as 0.5 s, and increase to 98.28% when the response time is 4 s. Pareek et al. [11] proposed a new 3DCN-RDN (3D convolution neural-based regression dual network) to achieve gas quantitation and identification, with a classification accuracy of 94.37%. YongKyung et al. [12] proposed a new classification method for mixed gases, which is based on the representation of simulated images with several sensor specific channels and the Convolutional Neural Network (CNN) classifier. He et al. [13] proposed a new hybrid CNN-Bi-LSTM-AM network model based on Bayesian as an optimization algorithm to realize rapid gas identification. However, in the process of gas classification, they did not consider the possible redundancy or noise information in the time series sensing data and did not conduct smooth filtering and noise reduction pre-processing on the time series sensing data, which would lead to the decline of classification accuracy.

However, time series sensing data are characterized by relationships between different attributes, which makes it difficult to obtain complex information and temporal correlations from sensor responses. Scholars have proposed to encode the time series into different types of images for visual analysis, which enables computer vision technology to be used for time series classification. This idea provides a new perspective for us to solve the classification of time series sensing data. Donner et al. [14] proposed a new method to analyze the structural nature of the time series of complex systems. By constructing an adjacency matrix with predefined recursive functions, the time series can be interpreted as a complex network. Silva et al. [15] use recursive graphs as the representation domain of the time series classification and measure the similarity between the recursive graphs with the Campana-Keogh (CK-1) distance, which is a distance based on Kolmogorov complexity. A video compression algorithm is used to estimate the image similarity, which is a simple and parameterless time series classification method. Javed et al. [16] used different line graph techniques to analyze the performance of multiple time series and introduced a graphic awareness framework for multiple time series. These methods explore how to visualize the topology and intrinsic properties of the time series. However, their visualization process did not consider the integrity of the mapping between time series and visual cues, which would have resulted in a degradation of the classification performance. Wang et al. [17] proposed a new framework based on the Gramian Angular Summation/Difference Field (GASF/GADF) and Markov Transition Fields (MTFs) to encode time series into different types of images, which enables computer vision technology to be used for time series classification and interpolation. Liu et al. [18] used the Markov Transition Fields (MTFs) to visualize sensor responses into images. Combined with the Small-Scale Convolutional Neural Network (SSCNN), gas classification was further realized. Wang et al. [19] took the whole process of gas reaction as the feature map and used the mixed neural network to classify the gas, achieving a 95% classification accuracy. Inspired by this, we consider using the Gramian Angular Summation Field (GASF) method to encode the time series into two-dimensional sensing images. Because it contains time correlation and can maintain time dependence, it can be well applied in our experimental research.

A pattern recognition algorithm [20–22] is another key process of gas recognition. Ha et al. [23] apply the K-Nearest Neighbor (KNN) algorithm in combination with the Multi-Mode Principle Component Analysis (MPCA), which can be used for process monitoring and fault monitoring. Sun et al. [24] proposed a pattern recognition method based on the Local Mean Decomposition (LMD) envelope spectrum entropy and the Support Vector Machine (SVM) for classifying leakage aperture categories. However, all the above methods require feature extraction (typical features include response maximum value, response time, integral area under response curve, etc.) before pattern recognition, which undoubtedly increases the difficulty of achieving the target task. As artificial intelligence technology has made great progress in multiple fields, Deep Neural Networks (DNN) have achieved remarkable achievements in the field of artificial intelligence, such as computer vision [25], natural language processing [26], malware detection [27–29], etc. For example, Peng et al. [30] have designed a Deep Convolutional Neural Network (DCNN) to classify four gases. The recognition accuracy is 95.2%, and the training time is 154 s. It is far better than that of the Support Vector Machine (SVM) and the Multi-Layer Perception (MLP).

The main contributions of this study can be summarized as follows: We illustrate with examples how Deep Neural Networks can be applied to time series gas sensing data. We use a variety of filtering methods to pre-process the gas sensing response data for more efficient input data representations and to improve the performance of the classifier. After that, we adopted the GASF method to encode the gas sensing response data into two-dimensional images, and converted the classification and recognition based on the time series gas sensing data into the classification and recognition of two-dimensional sensing images, and finally realized the classification of methane, ethanol, ethylene, and carbon monoxide. In addition, the data enhancement technique is applied in the experimental study to further improve the performance of the classifier with relatively few data samples. Combined with fine-tuning the GoogLeNet classification network, gas data samples were trained and tested. We carried out a variety of comparative experiments under different experimental settings, including using the 10-fold cross validation method to verify the accuracy of the algorithm, whether the gas sample data are smoothly pre-processed, whether the data are data enhanced, dividing the dataset into different proportions, using different classification network models to identify the gas, comparing the experiments performed by other researchers on this dataset. The performances of the suggested algorithms in other UCI gas datasets are presented as well. The experimental results show that the Savitzky–Golay smooth filtering algorithm can greatly improve the performance of the neural network gas classification algorithm.

2. Materials and Methods

2.1. Materials

The data used in this experiment are an open-source gas dataset downloaded from UCI Machine Learning Repository public database [31]. The twin gas sensor arrays dataset is a time series gas sensor dataset of 8 sensor arrays. The dataset was collected by five identical sensor arrays, each containing four metal oxide gas sensors, namely TGS2611, TGS2612, TGS2610, and TGS26028. Each gas sensor operates at two heating voltage levels of 5.65 V and 5.00 V, resulting in a total of eight sensor combinations. The eight sensors are integrated in a specially designed circuit board, which is equipped with a temperature control and signal acquisition module to monitor the operation of the sensors. The experimental setup is shown in Figure 1 below.

During the signal acquisition process, the same experimental method was used to measure the gas of 5 sensor arrays, and the measurements were made with different sensor arrays every day. There are four types of gas tested, namely methane, ethanol, ethylene, and carbon monoxide. The duration of a single test experiment is 600 s, and the sampling frequency is 100 Hz. The types of sensors on a single sensor array and the working voltages of each sensor are shown in Table 1.

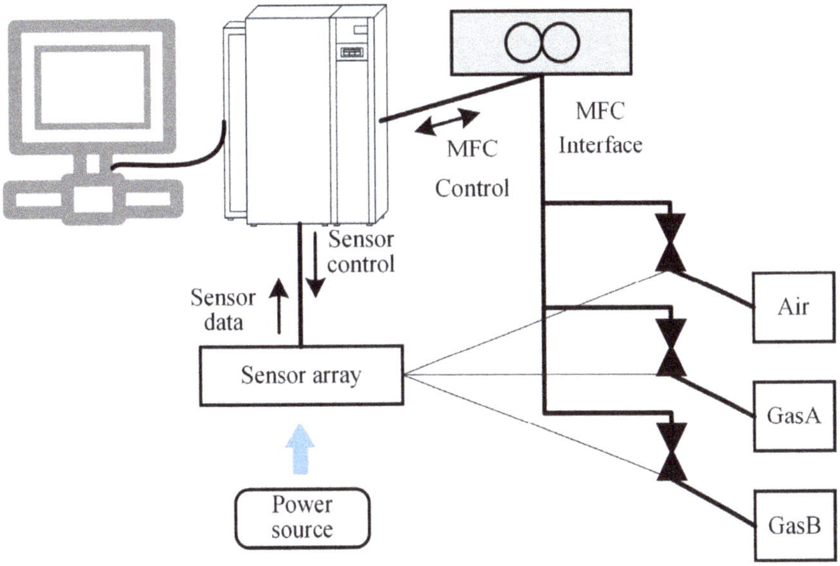

Figure 1. Experimental setup used for data acquisition.

Table 1. Sensor types and operating voltages.

Channel	Category of Sensors	Operating Voltage
0	TGS2611	5.65 V
1	TGS2612	5.65 V
2	TGS2610	5.65 V
3	TGS2602	5.65 V
4	TGS2611	5.00 V
5	TGS2612	5.00 V
6	TGS2610	5.00 V
7	TGS2602	5.00 V

The different concentration levels of the four measured gases are shown in Table 2.

Table 2. Different concentration levels of measured gases (ppm).

Gas Type	Level 1	Level 2	Level 3	Level 4	Level 5	Level 6	Level 7	Level 8	Level 9	Level 10
Methane	25.0	50.0	75.0	100.0	125.0	150.0	175.0	200.0	225.0	250.0
Ethanol	12.5	25.0	37.5	50.0	62.5	75.0	87.5	100.0	112.5	125.0
Ethylene	12.5	25.0	37.5	50.0	62.5	75.0	87.5	100.0	112.5	125.0
Carbon monoxide	25.0	50.0	75.0	100.0	125.0	150.0	175.0	200.0	225.0	250.0

The date of the detection of various gas substances of different concentrations by each sensor array is shown in Table 3.

Table 3. The testing schedule for each batch of sensor arrays.

Sensor Array	The First Test	The Second Test	The Third Test	The Fourth Test
Array 1	4th day	10th day	15th day	21st day
Array 2	1st day	7th day	11th day	16th day
Array 3	2nd day	8th day	14th day	17th day
Array 4	3rd day	9th day	-	-
Array 5	18th day	22nd day	-	-

2.2. The Whole Experimental Process

The whole process of the experiment in this paper is as follows: First, in order to effectively extract the characteristics of the gas to be tested, the response data of the gas sensor were analyzed. The response data of the gas sensor within the range of important response and recovery time were retained, and the redundant and interference information was eliminated. Then, the time series data within the intercepted time range was visualized, and it was found that the response data of the gas sensor had certain data fluctuations and contained noise. Therefore, the smooth filtering method was considered to pre-process the data. In this paper, multiple smooth filtering methods were compared and analyzed, and the experimental results showed that Savitzky–Golay smooth filtering had the best classification effect. Therefore, Savitzky–Golay smooth filtering was selected for the basic experiment. Next, the GASF method was used to convert the pre-processed time series data into two-dimensional sensing images, and the data enhancement technology was used to reduce the impact of small samples on the classifier. In addition, the GoogLeNet classification network was fine-tuned to realize the recognition of four different gases, namely methane, ethanol, ethylene, and carbon monoxide. The flow chart of the whole experimental process is shown in Figure 2 below.

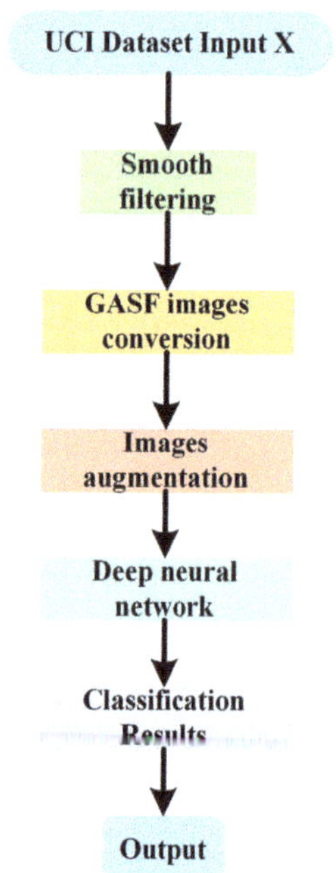

Figure 2. The flow chart of the whole experimental process.

2.3. Data Pre-Processing

According to the literature [31], the specific collection experiment design of this gas sensor dataset is known. The experimental design of the four gases was the same, and the test time of a single experiment was 600 s. The process is as follows: First, constant flow of clean air is circulated through the gas sensor chamber for 50 s to form the initial stabilization phase of the response, which can be used as a baseline for measuring the sensor response. Secondly, the selected gas is passed into the sensor gas chamber and mixed with air according to the required concentration level to produce a gas mixture, which is circulated for 100 s. Finally, clean air is circulated to remove the gas mixture from the sensor chamber for the next 450 s.

The dataset used in this paper is 8 channels time series, $X_{gas} = \{X_{gas1}, X_{gas2}, \ldots, X_{gas8}\}$, where $X_{gas1}, \ldots, X_{gas8}$, respectively, represents gas sensor array (TGS2611,TGS2612,..., andTGS2602) dynamic response data. Gas sensor array data are usually complex, high-dimensional time signals. To process such complex data, efficient input data representation is needed to complete the task of identifying gases. Since gas classification requires a large number of samples for model training, the performance of the classifier largely depends on the input data representation. Efficient input data representation is the key to gas classification and helps train the gas classification network model with input data. Otherwise, without proper input data representation, important information will inevitably be lost, which will degrade the performance of the classification model. In order to effectively extract the features of the gas to be identified, the data of the 8 channel time series are intercepted. In the experiment, the response data within 30–100 s are intercepted as the classification object. On the one hand, the response data within this time range includes both the gas response process and the gas recovery process, without losing the important response data in the gas reaction process. On the other hand, some redundant information is removed from the response data in this time range, which improves the efficiency of network model calculation. The original response data curve of CO gas is shown in Figure 3 below.

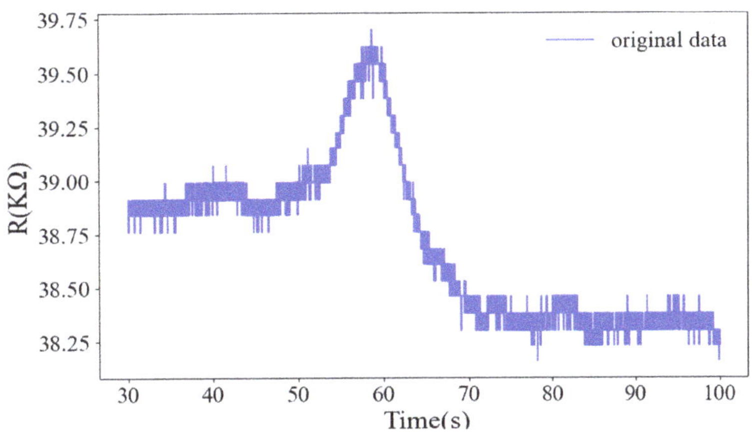

Figure 3. The original response data curve of CO gas.

As can be seen from the above original response data curve of CO gas, the response of CO gas has certain data fluctuations and noise. In order to reduce the fluctuation of response data, a variety of smoothing filtering methods are used to pre-process the original gas response data. Based on the classification experiment results, Savitzky–Golay smoothing filtering algorithm has the best effect. Therefore, Savitzky–Golay smoothing filtering algorithm is mainly used in the following years. It is a function commonly used in curve smoothing processing and is a convolution algorithm based on smoothing time series data and the least squares principle. The core idea is to carry out order polynomial fitting to the data points in a certain length window, so as to get the fitting result. Savitzky–Golay smoothing filtering (SG smoothing filtering) is actually a weighted average algorithm for moving windows, but its weighting coefficient is not simply a constant window, but a least square fitting for a given higher-order polynomial within a sliding window.

The principle of Savitzky–Golay smoothing filtering is as follows: It is assumed that the width of the sliding time window is $n = 2m + 1$, and the data point is $x = (-m, -m + 1, \ldots, m - 1, \text{and } m)$. The sub-polynomial is used to fit the data points in the time window. The $k - 1$ degree polynomial is used to fit the data points in the window, $y = a_0 + a_1x + a_2x^2 + \ldots + a_{k-1}x^{k-1}$. Therefore, n linear equations with k elements are obtained. In order for the system to have a solution, n should be greater than or equal to k. Generally, n greater than k is selected, and the parameters are fitted by the least square method, $A = \{a_0, a_1, \ldots, a_{k-2}, a_{k-1}\}$.

The biggest feature of Savitzky–Golay smooth filtering is that while filtering noise, it can ensure that the shape and width of the signal remain unchanged, which can better retain the information of characteristic peaks and improve the accuracy of input data. Therefore, based on the advantages of the Savitzky–Golay smooth denoising algorithm mentioned above, it can be well applied to the gas sensor dataset. As for the selection of experimental parameters, we compared the influences of different parameter selection on the gas classification results, as shown in Table 4. When window length increases from 29 to 89, the accuracy first increased, and then decreased. The time consumed kept on increasing when the window length increased. Window length of 59 provides better accuracy with little time as well. Based on the experimental results, this paper selects window length as 59 and k value as 3, and performs certain smoothing noise filtering pre-processing on the 8-channel data to improve the quality of input data.

Table 4. The influence of different experimental parameters on experimental results.

Window Length (k = 3)	Accuracy	Time (One Epoch)
29	98.4%	25.6 s
59	99.9%	26 s
89	99.2%	27.8 s

The comparison of gas response data curve with or without Savitzky–Golay smooth filtering is shown in Figure 4 below.

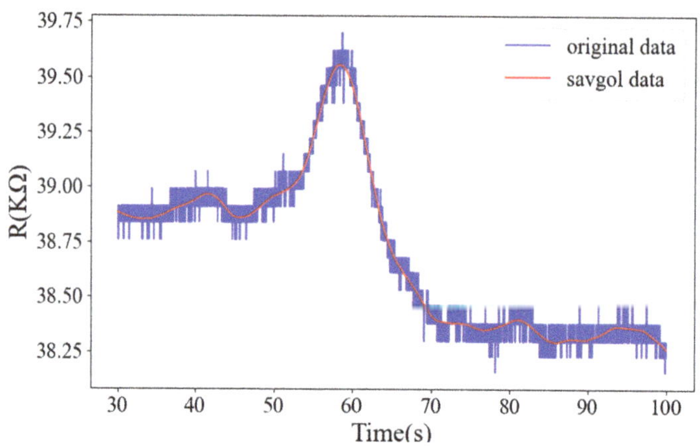

Figure 4. Comparison diagram of gas response data curve with or without filtering.

2.4. Images Conversion

With the development of computer vision, Deep Neural Network has shown outstanding performance in image recognition. According to the time correlation embedded in the gas sensor data, the SG algorithm in the previous section is used to smooth and filter the time series data into a two-dimensional sensor image, so as to be used for the subsequent feature learning. Therefore, gas classification and recognition based on time series data are transformed into gas classification and recognition for two-dimensional sensor images. The Gramian Angular Field (GAF) method was used to convert the response data of the gas sensor into a two-dimensional sensor image. The features in the two-dimensional image were learned by combining with the gas classification network model, and the features were automatically extracted. Finally, the recognition of four gas types was realized.

The specific implementation process of Gramian Angular Field (GAF) is as follows: First, scale the sequence data of each channel in the 8-channel time series data of each test sample, $X = \{X_1, X_2, \ldots X_8\}$ of each test sample, and scale the data range to [0, 1]. The expression is as follows:$\widetilde{X}_i^t = \left[X_i^t - \min(X_i)\right] / \left[\max(X_i) - \min(X_i)\right]$, where $\widetilde{X}_i^t$ is the normalized value at time t, S_i^t is the response value of the ith sensor at time t, and X_i is the response value of the ith sensor within the sampling time range. The comparison of gas response data curves with or without Savitzky–Golay smooth filtering after data normalization is shown in Figure 5a,b below.

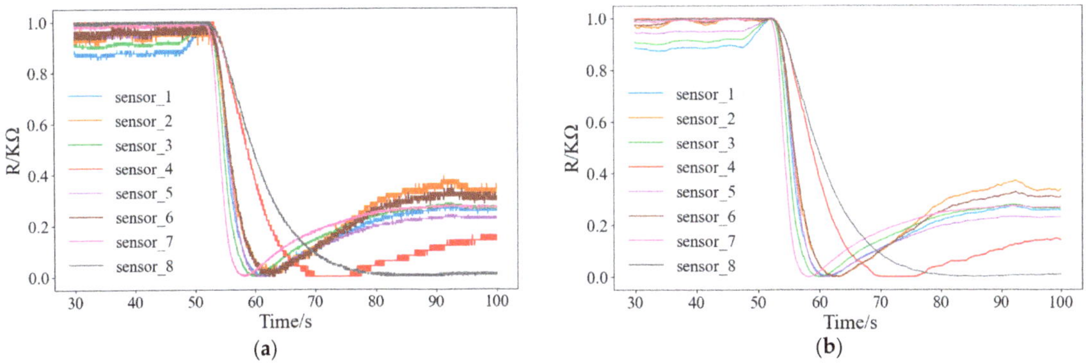

Figure 5. Gas response data curves: (**a**) without filtering after data normalization; and (**b**) with filtering after data normalization.

Next, the normalized time series data X_i^t are converted to the polar coordinate system, that is, the value is regarded as the cosine of the included angle, and the time stamp is regarded as the radius. The formula expression is shown in (1):

$$\begin{cases} \varphi = \arccos(\widetilde{X}_i^t),\ 0 \leq \widetilde{X}_i^t \leq 1,\ \widetilde{X}_i^t \in \widetilde{X} \\ r = \frac{t_i}{N},\ t_i \epsilon\, N \end{cases} \tag{1}$$

Since the response data values are normalized to the interval of [0, 1], the angle range is $[0, \pi/2]$. The polar coordinate transformation through this formula has significant advantages because the encoding is bijective, and for a given time series, the result of a given mapping in the polar coordinate system is unique. Finally, there are two methods to convert GAF image: GASF (Gramian Angular Summation Fields) and GADF (Gramian Angular Difference Fields). The formula expression is shown in (2):

$$\begin{cases} GASF = \left[\cos(\varphi_i + \varphi_j)\right] = \widetilde{X}' \cdot \widetilde{X} - \sqrt{I - \widetilde{X}^2}' \cdot \sqrt{I - \widetilde{X}^2} \\ GADF = \left[\sin(\varphi_i - \varphi_j)\right] = \sqrt{I - \widetilde{X}^2}' \cdot \widetilde{X} - \widetilde{X}' \cdot \sqrt{I - \widetilde{X}^2} \end{cases} \tag{2}$$

For the conversion of single-channel time series gas sensing data into GAF images, GASF method was used to convert the data into two-dimensional sensing images, and the image size of single-channel two-dimensional sensing images was set to 64 × 64. The single-channel time series data are converted into a single-channel two-dimensional sensor image (including GASF and GADF methods), as shown in Figure 6 below.

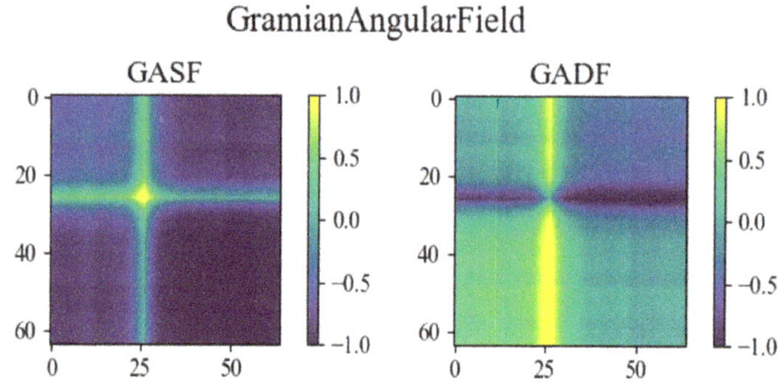

Figure 6. Single-channel 2D sensor image.

Since the gas response data of each test sample is measured by 8 gas sensors, considering that the time series data intervals intercepted by all experimental samples are identical, that is, the time column data of each sample will not affect the performance of the subsequent gas classification network model, a 3 × 3 two-dimensional combined sensor image is constructed. Finally, the gas sensor response data of each test sample are processed by GASF method and converted into a 192 × 192 two-dimensional combination sensor image. The two-dimensional combination sensor images of the four gases are shown in Figure 7 below.

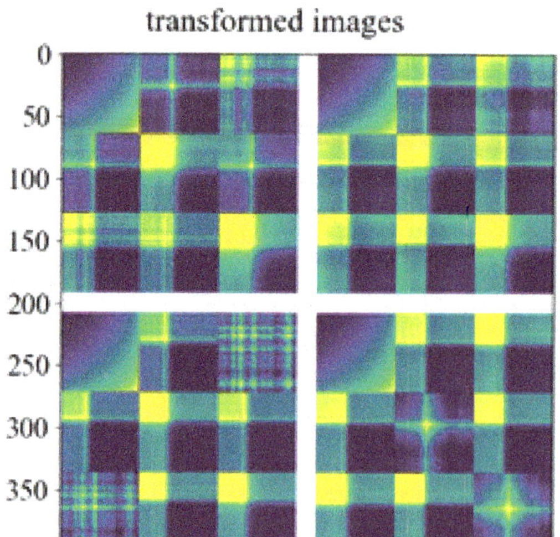

Figure 7. The two-dimensional combination sensor images of the four gases. The x and y axis represent the number of pixels converted from the gas sensing response data to the gas sensing image.

2.5. Data Augmentation Technology

Since the collection and labeling of gas sensor array data are energy-consuming and costly processes, it is a great challenge. The total number of data samples used in this experiment is 640, and the gas recognition network model is easy to over-fit using small samples. Therefore, data augmentation technology [12] is adopted to increase the number of experimental samples, reduce the occurrence of over-fitting phenomenon, expand the decision boundary of the model, and help to improve the generalization ability of the training model, so that the gas classification model has stronger robustness. GASF method is used to convert the time series gas sensing data into two-dimensional combined sensing images. Since the Gram matrix is symmetrical about the diagonal, the sensor response image obtained is also symmetrical about the diagonal. The data augmentation technology of the experimental samples includes image mirroring, transformation brightness (brightening and darkening), and angle rotation (90° and 180°). Finally, the total amount of data samples in this experiment was increased 6 times, from 640 gas data samples to 3840 gas data samples. After that, the two-dimensional combination sensor image is input into the gas recognition network to realize the training and testing process of gas classification. The two-dimensional combined sensor image after the enhancement of single gas sample data are shown in Figure 8 below.

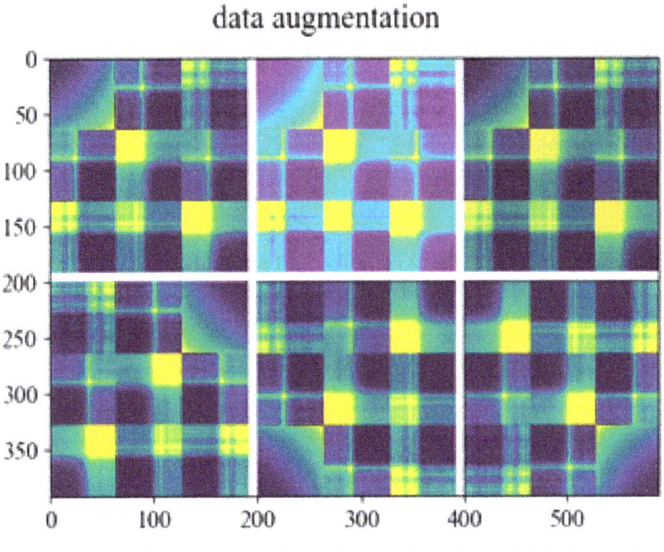

Figure 8. A two-dimensional combined image with enhanced data from a single gas sample. The x and y axis represent the number of pixels converted from the gas sensing response data to the gas sensing image.

2.6. Model Execution

(1) Introduction to the Model

Pytorch framework in deep learning is used and run on GPU to accelerate the calculation and realize the gas recognition model. The fine-tuning of GoogLeNet network structure for gas recognition is shown in Figure 9 below, which is roughly divided into five modules. The specific operation process of each module is as follows: Firstly, in the first module, a convolution layer and a maximum pooling layer are used. In our gas recognition model, the rectified linear units (ReLUs) are required after the convolution operation. In the second module, two convolution layers and one maximum pooling layer are used. Local response normalization is not adopted in the first two modules, because this layer structure does not play a significant role, so it is discarded to simplify the network model structure. In the third module, there are two layers of structure, namely inception (3a) layer and inception (3b) layer, which are divided into four branches. Multi-scale processing is adopted. Finally, the eigenmatrix of four branches of inception layer is connected in parallel to the depth direction. After that, the specific experimental operations of the fourth and fifth modules are similar to those of inception (3a) and inception (3b) in the third module. Finally, the output layer is different from the previous neural network output layer, which adopts three continuous and fully connected layers. This output layer network adopts the adaptive average pooling layer, which plays a role of dimension reduction on the one hand and abstracts the global features of the image by combining low-level features on the other hand. No matter how much the height and width of the input feature matrix are set, both the height and width of the specified feature matrix can be obtained (the convolution layer with both height and width of 1 is finally obtained), and then the dropout with 50% probability of dropping is added. Through the operation of dropout, the number of neurons and connection weights in the network will be randomly reduced, which can improve the numerical performance and prevent over-fitting. Finally, the soft-max layer (activation function) is used as the classifier to identify four different gases: methane, ethylene, ethanol, and carbon monoxide.

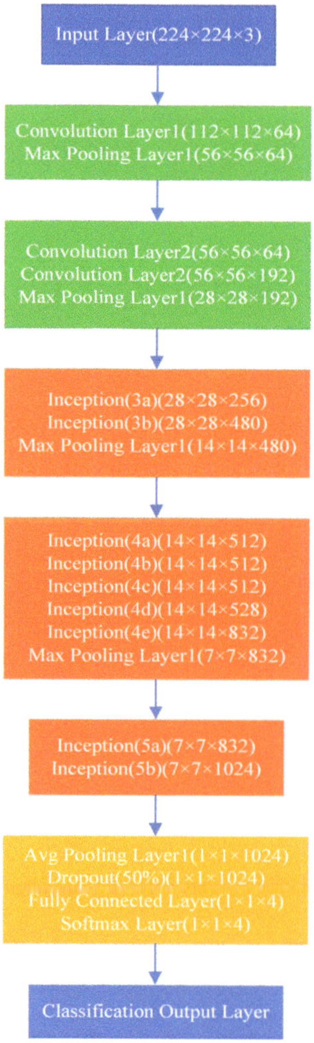

Figure 9. The structure of DNN model.

GoogLeNet network model structure mainly introduces the inception architecture. The improved inception structure diagram is shown in Figure 10 below, which mainly integrates the characteristic information of different scales. On the basis of the original inception structure, a convolution layer with a convolution kernel size of 1 × 1 was added before the 3 × 3 and 5 × 5 convolution layers, and a convolution layer with a convolution kernel size of 1 × 1 was added after the pooling layer, in order to reduce the dimension and model training parameters, so as to reduce the amount of computation.

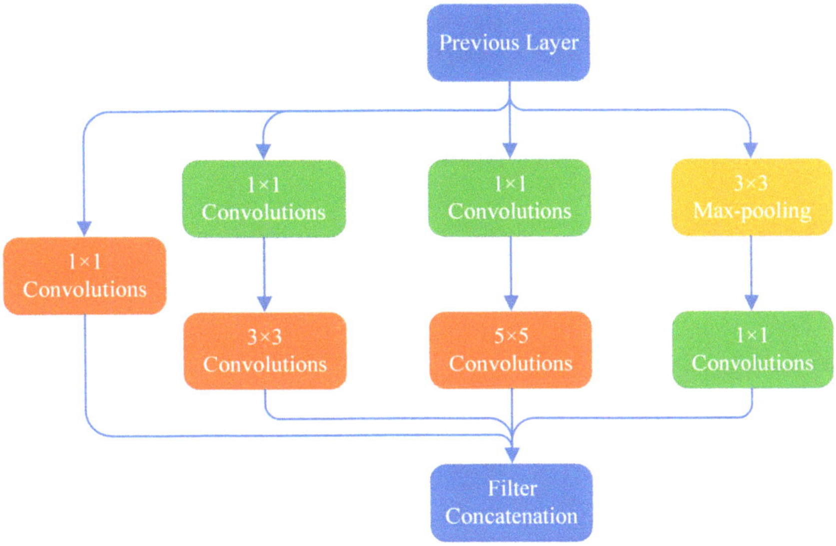

Figure 10. Improved inception structure diagram.

(2) Training and testing procedures

1. In the process of model training, Pytorch framework in deep learning is used and run on GPU to accelerate calculation and realize gas classification and recognition. In the experiment, the final goal is to classify 4 different gases. For the multi-gas classification tasks, cross-entropy loss is used as the loss function. The adaptive moment estimation (Adam) optimization algorithm is used as the optimization algorithm. Batch size is set to 32, the learning rate to 0.0003, and iterations to 100. All 2D sensing images are transformed accordingly, for example, the random aspect ratio is trimmed to 224 × 224 pixels, the horizontal flip is according to probability p = 0.5, transformed into tensors and normalized to 0~1, and the image is normalized. The gas classification is realized in combination with the fine-tuning GoogLeNet network model. It is worth noting that the two auxiliary classifiers of GoogLeNet are applied in the training of the model, the loss of the two auxiliary classifiers is multiplied by the weight and added to the overall loss of the network, and then the back propagation is carried out, which can prevent the over-fitting of the network and improve the discriminant power of the classifier at the lower network layer. In the actual test, the two auxiliary classifiers will not be used.

(3) Process of prediction

In the prediction process, first load the two-dimensional sensor image after gas response data conversion. In order to match the input dimension of the network, add a Batch dimension to the prediction image. By reading the label file generated in the training process, load the model parameters saved in the training process on the basis of the model establishment, and finally realize the recognition and prediction of the image. In other words, it can accurately predict four different gases: methane, ethylene, ethanol, and carbon monoxide. The following Figure 11 shows the prediction results, achieving accurate prediction of carbon monoxide gas.

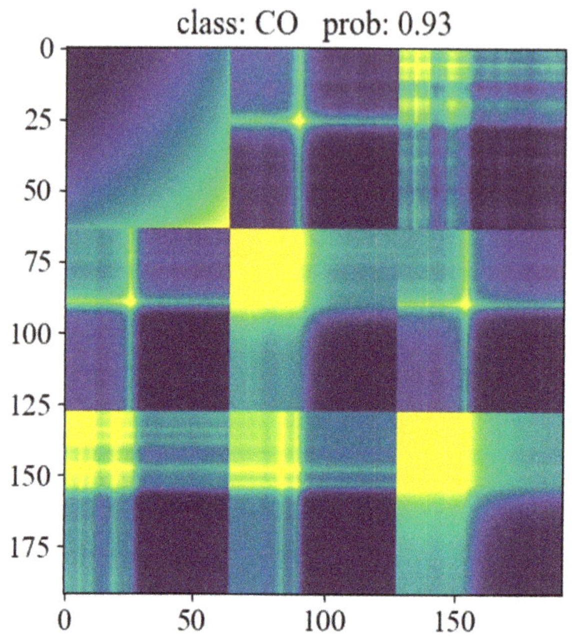

Figure 11. Gas identification and prediction results. The x and y axis represent the number of pixels converted from the gas sensing response data to the gas sensing image.

3. Results

3.1. Basic Experiment

In the basic experiment, the Savitzky–Golay smooth filtering was carried out on the response data of the original gas sample. After that, the GASF method was used to encode the gas sensing response data into two-dimensional sensing images. In addition, data augmentation technology was also adopted in the experiment to reduce the impact of small samples on the classifier. The dataset was randomly divided into the training set and the test set according to the ratio of 8:2. Combined with the fine-tuning GoogLeNet neural network, the classification of four gases, including methane, ethanol, ethylene, and carbon monoxide, was finally realized by taking advantage of its automatic learning of deep-seated sample characteristics. As shown in Figure 12a–c, gas identification achieves a higher precision, and the convergence is completed in fewer iteration cycles and the final loss value converges to around zero. In the process of training and testing, the accuracy has the same upward trend and the two are close. In the process of training and testing, the loss value also has the same downward trend and the two are close. In addition, when the accuracy increases, the loss value shows a downward trend, indicating that there is no over-fitting phenomenon in the experimental method.

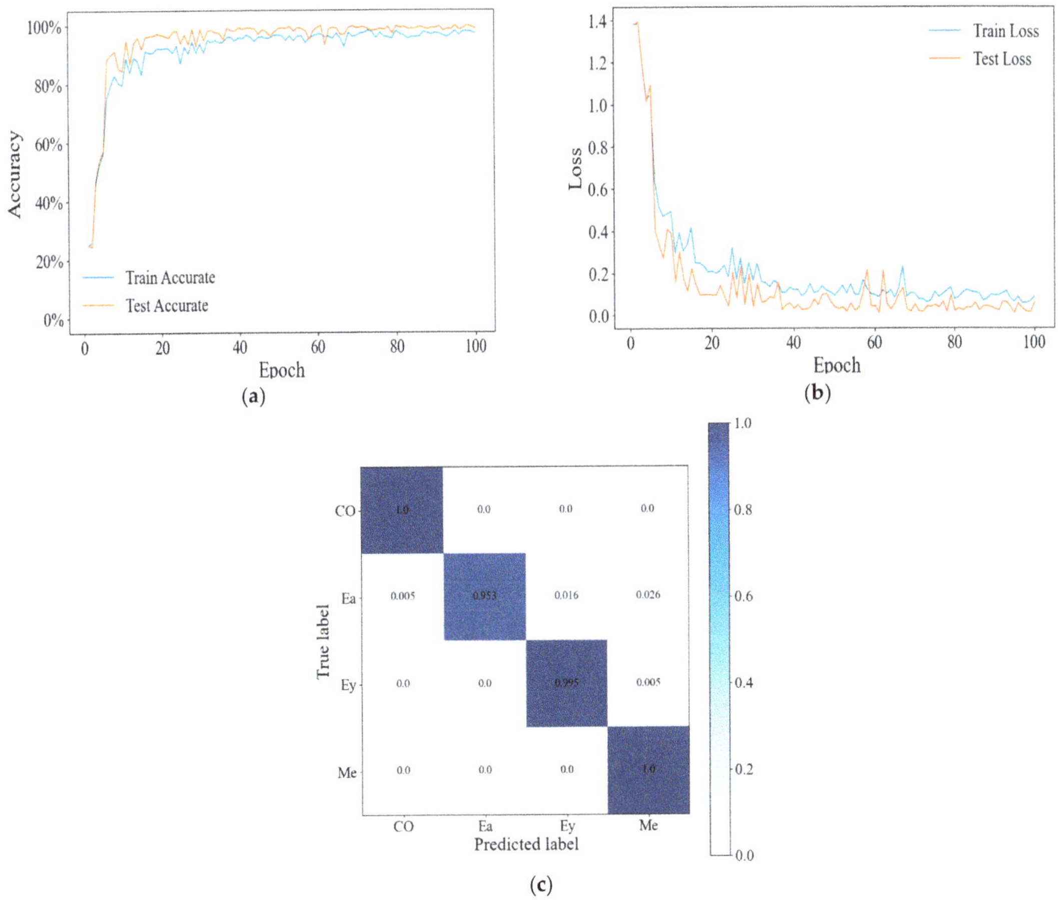

Figure 12. (**a**) The training and test accuracy curves in basic experiment. (**b**) The training and test loss curves in basic experiment. (**c**) Confusion matrix of basic experiment.

3.2. Comparison Experiments

(1) Influence of the 10-fold cross validation method on the experimental classification results

In the experiment, a 10-fold cross validation method was adopted to verify the accuracy of the model recognition. The two-dimensional sensor images were divided into ten parts, nine of which were taken as training data and one as test data for experiment. As shown in Figure 13a,b, the gas recognition accuracy is 99.9%, and the gas recognition accuracy and loss rate curves in the training and testing of the model are basically the same. Despite the small oscillations during the test, the overall accuracy and loss rate curves during the training and testing process tend to smooth out and eventually stabilize. The confusion matrix of the experimental classification results is shown in Figure 13c below. It is clear from the diagram that the model performs well in identifying all four gases.

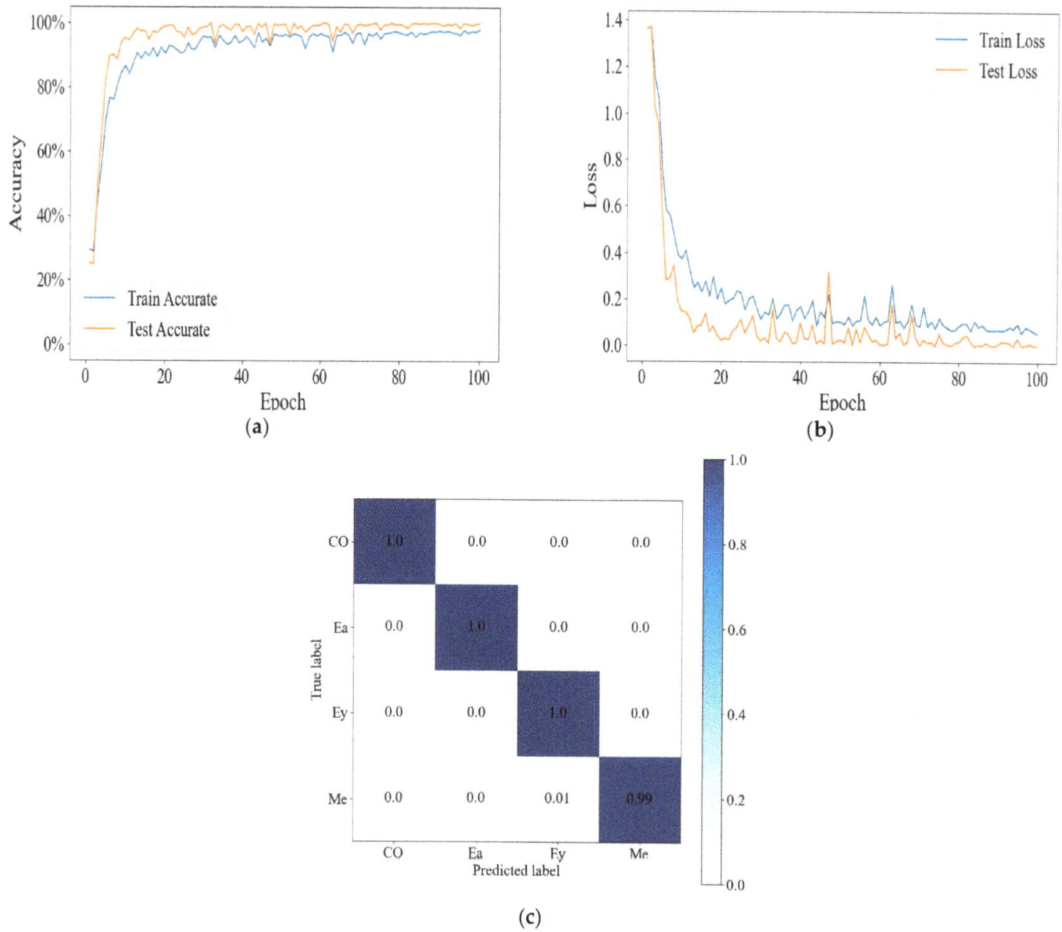

Figure 13. (**a**) The training and test accuracy curves of the comparison experiment. (**b**) The training and test loss curves of the comparison experiment. (**c**) Confusion matrix of the comparison experiment.

(2) The influence of smooth treatment on the experimental classification results.

In order to make the experimental research more persuasive, three smoothing filtering methods, namely Savitzky–Golay smoothing filtering, Kalman filtering, Gaussian filtering and moving average filtering, were adopted in the experiment to filter the original gas response data, respectively. The corresponding accuracy and loss rate results in the test process are shown in Table 5 below. As can be seen from Table 5, smooth filtering has well-improved the classification accuracy of the gas identification network. In particular, the Savitzky–Golay smooth filtering algorithm can better improve the accuracy of gas identification. Therefore, the subsequent comparison experiments all use the Savitzky–Golay algorithm for smooth filtering.

Table 5. The experimental results contrast.

Algorithm	Accuracy	Loss
With SG smooth filtering	99.9%	0.223
With Kalman filtering	98.2%	0.278
With Gaussian filtering	97.8%	0.324
With moving average filtering	97.2%	0.328
Without SG smooth filtering	93.5%	0.652

(3) The influence of data augmentation on the experimental classification results.

In order to reduce the influence of the small sample data size of the gas sensor array on the classification network model, the data augmentation network strategy is adopted in the experiment. The data samples, with or without data augmentation, were trained and tested, respectively, in the experiment, and the number of iterations was randomly set to 200. After several experiments, it was proved that the model based on the fine-tuning GoogLeNet network structure had a fast convergence speed, and when the number of iterations was 60, it was close to convergence. Therefore, it was appropriate to set the number of iterations to 100. The experimental results of accuracy and loss rates in the testing process are shown in Figure 14a,b below. As can be seen from the figure, the gas sample data after data augmentation makes the classification model more robust, the test effect more stable, and the convergence faster, as the highest classification accuracy is 99.9%, and the loss is small. However, for the gas sample data with no data augmentation input into the classification network model, the test effect has certain data fluctuations, without good stability. The highest classification accuracy was 98.4%, and the loss was great. It can be concluded that the data augmentation network strategy can better improve the classification accuracy of the gas identification network model, so that the gas identification network model has better robustness and generalization ability.

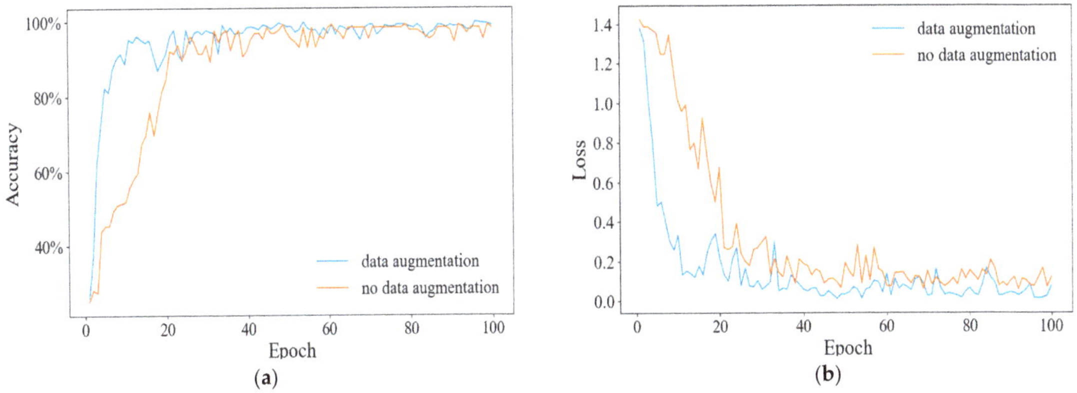

Figure 14. (**a**) Training and accuracy curves with and without data augmentation experiment. (**b**) Training and test loss curves with and without data augmentation experiment.

(4) The influence of different sample data division ratios on the classification results.

In the experiment, the response data of gas samples were divided into different proportions of training sets and test sets. The two-dimensional sensor response image enhanced by the data in the previous section was used as the input data of the gas identification network. During the experiment, the sample data were divided into training sets and test sets according to various proportions, as shown in Table 6 below.

Table 6. Data partition ratio.

Partition Ratio	Divided Sample Number
9:1	3456:384
8:2	3072:768
7:3	2688:1152

The response data of gas samples are divided according to the division proportion of different training sets and test sets. The experimental results of the accuracy and loss rate of the test process are shown in Figure 15a,b below. In general, when dividing the gas sample data with different proportions of training set and test set, the higher the proportion of the training set data, the higher the classification accuracy will be, and the higher the proportion of the test set data, the more stable the classification results of the classifier will be.

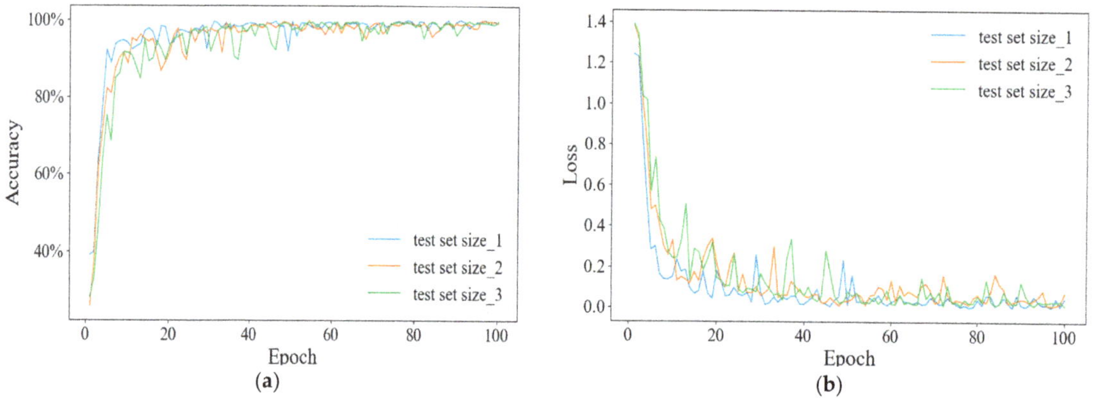

Figure 15. (**a**) Training and test accuracy curves for different proportions of divided datasets. (**b**) Training and test loss curves of different proportions of divided datasets.

(5) The influence of different classification network models on the classification results.

The data enhanced two-dimensional sensor images were input into different classification network models for a comparison experiment. The experimental results are shown in the following Table 7. As can be seen from Table 7, the fine-tuned GoogLeNet network classification model has the highest classification accuracy, followed by the fine-tuned ResNet50 network, Alex-Net network, and finally, the ResNet34 network. In contrast, the fine-tuned GoogLeNet network classification model selected by us is more suitable for the identification of methane, ethanol, ethylene, and carbon monoxide, achieving higher recognition accuracy and more stable experimental results. The variation trends of accuracy and loss rate in the training process and the testing process are similar, which indicates that the method we adopted improves the generalization ability of the classification model.

Table 7. Comparison of experimental results of different classification network models.

Classification Network	Accuracy	Time (One Epoch)
Fine-tune GoogLeNet	99.9%	26 s
Fine-tune Alex-Net	96.8%	30 s
Fine-tune ResNet34	95.4%	33 s
Fine-tune ResNet50	97.4%	32 s

(6) Compare and analyze the experimental results with others.

In order to make the experimental results more comparative and persuasive, this paper will compare them with the classification accuracy obtained by other people's experiments on this public dataset. As can be seen from Table 8, compared with the classification accuracy achieved by other experiments in the table, our gas classification accuracy is higher, and the classification effect is better.

Table 8. Performance of accuracy compared with other people's experiments.

Name	Model	Accuracy
Xiaofang Pan [10]	CRNN	98.28%
Vishakha Pareek [11]	3DCNRDN	94.37%
Juan He [13]	CNN-Bi-LSTM-AM	84.9%
Ruijie Gu [32]	Inception-Time	88.11%
Ruijie Gu [32]	MSRSN	94.84%
In this paper	With Savitzky–Golay smooth-GASF-Fine tune GoogLeNet	99.9%
	Without Savitzky–Golay mooth-GASF-Fine tune GoogLeNet	93.5%

(7) Application to different UCI gas datasets.

In order to apply the gas recognition algorithm proposed in this paper to the recognition of a binary gas mixture, we found the binary gas mixture dataset in the UCI machine learning library. The binary gas mixture dataset is named the gas sensor array under dynamic gas mixtures. The gas identification process uses a mixture of carbon monoxide and ethylene, and finally realizes the identification of air, carbon monoxide, ethylene, and a mixture of carbon monoxide and ethylene. Carbon monoxide ranges from 0 ppm to 533.33 ppm, and ethylene ranges from 0 ppm to 20 ppm.

The data sample processing process is as follows: when the concentration of any gas in the mixture of carbon monoxide and ethylene changes, it is divided into one data sample. Finally, it is divided into 373 binary gas data samples of four types. By using the Gramian Angular Summation Fields (GASF) method, the binary gas mixture data samples are converted into two-dimensional sensing images, and the fine-tune GoogLeNet network model is applied to the binary gas mixture image recognition, which can classify the binary gas mixture. The experimental results of the accuracy and loss rate of the test process are shown in Figure 16a,b below. It can be seen from the curve of experimental results that the gas recognition algorithm can be well applied to a binary gas mixture.

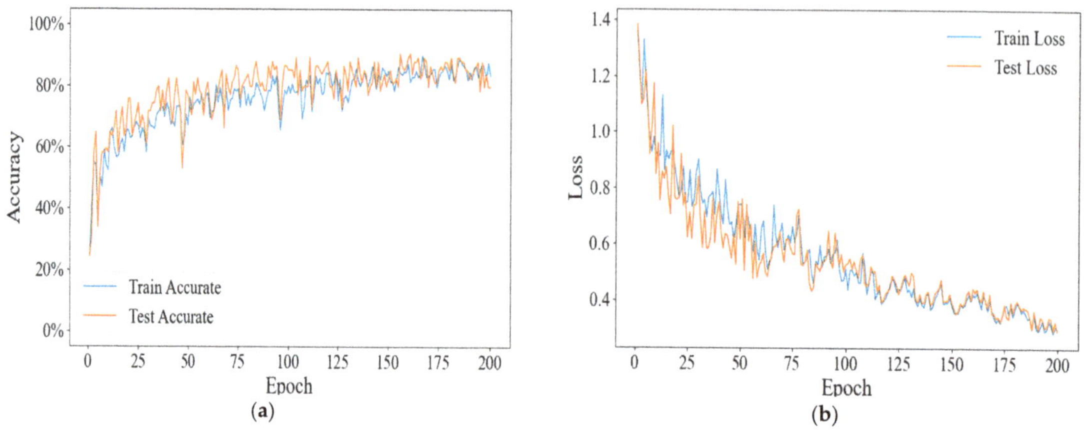

Figure 16. (**a**) Training and test accuracy curves of binary gas mixture datasets. (**b**) Training and test loss curves of binary gas mixture datasets.

4. Discussion

As for the classification experimental results under different comparison experiments obtained in the last section, we will, respectively, discuss as follows:

(1) Firstly, the experiment adopts four different smoothing filtering algorithms to pre-process the original gas sample data. The experimental results show that, firstly, smoothing filtering pre-processing greatly improves the classification and identification accuracy of gas. The highest recognition accuracy is 99.9%. Secondly, among the smoothing filtering algorithms, the Savitzky–Golay smoothing filtering algorithm has the highest accuracy and the best effect. Compared with the other three filtering algorithms, the Savitzky–Golay smoothing filtering algorithm can better retain the information of characteristic peaks and improve the accuracy of input data. In conclusion, the smooth filtering pre-processing can filter the noise existing in the signal sample data, reduce the existence of redundant information without losing important response data information, and obtain efficient input data representation. The classification accuracy of gas verifies the superiority of the smooth filtering pre-processing method.

(2) Secondly, gas sample data collection and labeling are energy-consuming and costly processes. Data augmentation network technology is used to amplify datasets. On the one hand, the number of training samples is increased; on the other hand, the generalization ability and robustness of the model can be improved. We carried out a six-fold data enhancement on the gas sample data used in the experiment, and finally the highest classification accuracy reached 99.9%, while the highest classification accuracy of the gas sample without data enhancement was 98.4%. In addition, it can be seen from the comparison curve of the experimental results of the data classification with or without data enhancement in the previous section that the gas samples after data enhancement perform better than those without data enhancement in terms of accuracy and loss rate. The classification accuracy of gas samples with data enhancement is more stable and converges faster. Therefore, data enhancement network technology can improve the accuracy of gas classification to a certain extent.

(3) Thirdly, the gas sample data are divided into training sets and test sets according to various proportions. The higher the proportion of the training set data, the higher the classification accuracy will be; the higher the proportion of the test set data, the more stable the classification results of the classifier. Because the more training gas samples there are, the more data information they contain, the more features they will learn, and the higher classification accuracy they can achieve in subsequent tests. At the same time, setting a certain number of test gas samples of data can improve the generalization ability

of the model, making the classification results more stable, preventing the occurrence of overfitting phenomenon, and finding the optimal parameters of the classification network.

(4) Fourthly, the comparison experiments of different classification networks in the previous section show that the fine-tuned GoogLeNet gas recognition network model selected is superior to the fine-tuned ResNet50, Alex-Net, and ResNet34 networks, in terms of classification accuracy, sample processing time, and network connection complexity. Since fine-tuning the GoogLeNet gas identification network model simplifies the model structure and has multi-inception structure, which can integrate feature information of different scales, two auxiliary classifiers are added to the network model to help training, and the output layer dismisses the fully connected layer and uses the average pooling layer instead, greatly reducing the model parameters. High gas classification and recognition accuracy is achieved.

(5) Finally, compared with the classification recognition accuracy of other people's experiments on this public dataset, it was found that our experimental research achieved a higher classification accuracy. Different from other scholars' experiments, we adopted different network strategies (data smoothing filtering and data enhancement) to efficiently optimize the input data samples and we fine-tuned the GoogLeNet gas identification network model to automatically learn subsequent features and finally achieve a higher classification accuracy.

(6) In general, we used a variety of smoothing filtering algorithms to pre-process the gas sensor data, and according to the time correlation embedded in the sensor data, we used the GASF method to convert the gas sensor data into a two-dimensional sensor image for the subsequent feature learning. We combined with the fine-tuned GoogLeNet classification network model to automatically learn the features of the sensor image to classify the four gases, and achieve a good classification accuracy.

5. Conclusions and Future Outlook

To identify different kinds of gas, we used a variety of smoothing filtering algorithms to perform data smoothing pretreatments on the multi-channel data representation and obtain efficient input data representation, improving the performance of the classification model. The optimization experiment verifies that the data from the Savitzky–Golay smoothing filtering algorithm is inputted into the gas classification network, and the final gas classification accuracy is higher. The smoothing filter pre-processing method plays a key role in the gas classification experiment, greatly improving the accuracy of the experimental results. Using the time correlation embedded in the sensor data, we used the GASF method to convert the gas sensor data into two-dimensional sensor image representations. Further gas classification is achieved using sensor images rather than time series data. Then, we also use the data enhancement network strategy to reduce the impact of small samples on the classifier, and to improve the robustness and generalization ability of the gas identification network model. The model can automatically and comprehensively extract different features of target gas for subsequent learning to realize the gas classification. Different from traditional methods, this model does not need to carry out tedious steps, such as artificial feature extraction and feature selection for gas sensing data and can directly classify the converted two-dimensional sensor image with a better classification performance. In addition, for the dataset samples in this paper, the fine-tuned GoogLeNet gas identification network model has obvious advantages over the fine-tuned ResNet50, Alex-Net, and ResNet34 networks. Compared with other advanced methods previously reported, this method has more advantages. These features help to identify gases in real time and quickly, with excellent accuracy and robustness, and they are suitable for a wide range of applications.

In our future work, we aim to find the optimal algorithm to directly classify and recognize multivariate time series data through more extensive experiments, especially on multivariate time series datasets. In addition, the time convolution neural network (TCN) and cyclic neural network (RNN) are considered to process multivariate time series data, and a variety of different classification methods are integrated to compare their classification performance. For another important future work, we hope to apply the selected method to different applications and different gas sensor array systems to further evaluate their classification performance and versatility.

Author Contributions: Conceptualization, M.J.; methodology, X.W.; software, Z.Z.; validation, C.Q. and J.L.; formal analysis, X.W.; investigation, X.W. and C.Q.; resources, M.J.; data curation, Z.Z.; writing—original draft preparation, X.W.; writing—review and editing, M.J.; visualization, X.W.; supervision, M.J.; project administration, X.W.; funding acquisition, M.J. All authors have read and agreed to the published version of the manuscript.

Funding: This research was funded by the Ministry of Science and Technology of China (2022CSJGG0703), National Natural Science Foundation of China (62204260).

Institutional Review Board Statement: Not applicable.

Informed Consent Statement: Not applicable.

Data Availability Statement: https://archive.ics.uci.edu/ml/datasets/Twin+gas+sensor+arrays (accessed on19 May 2016).

Conflicts of Interest: The authors declare no conflict of interest.

References

1. Navaneeth, B.; Suchetha, M. PSO optimized 1-D CNN-SVM architecture for real-time detection and classification applications. *Comput. Biol. Med.* **2019**, *108*, 85–92. [CrossRef] [PubMed]
2. Gasparri, R.; Santonico, M.; Valentini, C.; Sedda, G.; Borri, A.; Petrella, F.; Maisonneuve, P.; Pennazza, G.; D'Amico, A.; Natale, C.D. Volatile signature for the early diagnosis of lung cancer. *J. Breath Res.* **2016**, *10*, 016007. [CrossRef] [PubMed]
3. Chapman, E.A.; Thomas, P.S.; Stone, E.; Lewis, C.; Yates, D.S. A breath test for malignant mesothelioma using an electronic nose. *Eur. Respir. J.* **2012**, *40*, 448–454. [CrossRef] [PubMed]
4. Paknahad, M.; Ahmadi, A.; Rousseau, J.; Nejad, H.R.; Hoorfar, M. On-chip electronic nose for wine tasting: A digital microfluidic approach. *IEEE Sens. J.* **2017**, *17*, 4322–4329. [CrossRef]
5. Yin, Y.; Hao, Y.; Bai, Y.; Yu, H.C. A Gaussian-based kernel Fisher discriminant analysis for electronic nose data and applications in spirit and vinegar classification. *J. Food Meas. Charact.* **2017**, *11*, 24–32. [CrossRef]
6. Chen, L.Y.; Wong, D.M.; Fang, C.Y.; Chiu, C.I.; Chou, T.I.; Wu, C.C.; Chiu, S.W.; Tang, K.T. Development of an electronic-nose system for fruit maturity and quality monitoring. In Proceedings of the 2018 IEEE International Conference on Applied System Invention (ICASI), Chiba, Japan, 13–17 April 2018.
7. Men, H.; Yin, C.; Shi, Y.; Liu, X.T.; Fang, H.R.; Han, X.J.; Liu, J.J. Quantification of acrylonitrile butadiene styrene odor intensity based on a novel odor assessment system with a sensor array. *IEEE Access* **2020**, *8*, 33237–33249. [CrossRef]
8. Xu, L.Y.; He, J.; Duan, S.H.; Wu, X.B.; Wang, Q. Comparison of machine learning algorithms for concentration detection and prediction of formaldehyde based on electronic nose. *Sens. Rev.* **2016**, *36*, 207–216. [CrossRef]
9. Liu, M.H.; Li, Y. Application of electronic nose technology in coal mine risk prediction. *Chem. Eng. Trans.* **2018**, *68*, 307–312.
10. Pan, X.F.; Zhang, H.E.; Ye, W.B.; Zhao, X.J. A fast and robust gas recognition algorithm based on hybrid convolutional and recurrent neural network. *IEEE Access* **2019**, *7*, 100954–100963. [CrossRef]
11. Pareek, V.; Chaudhury, S.; Singh, S. Gas Discrimination & Quantification using Sensor Array with 3D Convolution Regression Dual Network. In Proceedings of the 2021 11th IEEE International Conference on Intelligent Data Acquisition and Advanced Computing Systems: Technology and Applications (IDAACS), Cracow, Poland, 22–25 September 2021.
12. Oh, Y.K.; Lim, C.; Lee, J.; Kim, S.; Kim, S. Multichannel convolution neural network for gas mixture classification. *Ann. Oper. Res.* **2022**. [CrossRef]
13. He, J.; Li, M.Y.; Zhou, R.; Li, N.; Liang, Y. Rapid Identification of Multiple Gases. In Proceedings of the 2021 3rd International Conference on Advanced Information Science and System (AISS 2021), Sanya, China, 26–28 November 2021.
14. Donner, R.V.; Zou, Y.; Donges, J.F.; Marwan, N.; Kurths, J. Recurrence networks—A novel paradigm for nonlinear time series analysis. *New J. Phys.* **2010**, *12*, 033025. [CrossRef]
15. Silva, D.F.; De Souza, V.M.A.; Batista, G.E.A.P.A. Time series classification using compression distance of recurrence plots. In Proceedings of the 2013 IEEE 13th International Conference on Data Mining, Dallas, TX, USA, 7–10 December 2013.
16. Javed, W.; McDonnel, B.; Elmqvist, N. Graphical perception of multiple time series. *IEEE Trans. Vis. Comput. Graph.* **2010**, *16*, 927–934. [CrossRef] [PubMed]
17. Wang, Z.; Oates, T. Imaging time-series to improve classification and imputation. In Proceedings of the Twenty-Fourth International Joint Conference on Artificial Intelligence, Buenos Aires, Argentina, 25–31 July 2015.
18. Liu, Y.J.; Meng, Q.H.; Zhang, X.N. Data processing for multiple electronic noses using sensor response visualization. *IEEE Sens. J.* **2018**, *18*, 9360–9369. [CrossRef]

19. Wang, S.H.; Chou, T.I.; Chiu, S.W.; Tang, K.T. Using a hybrid deep neural network for gas classification. *IEEE Sens. J.* **2020**, *21*, 6401–6407. [CrossRef]
20. Tsui, L.; Benavidez, A.; Palanisamy, P.; Evans, L.; Garzon, F. Quantitative decoding of the response a ceramic mixed potential sensor array for engine emissions control and diagnostics. *Sens. Actuators B Chem.* **2017**, *249*, 673–684. [CrossRef]
21. Yi, S.; Tian, S.; Zeng, D.; Xu, K.; Peng, X.L.; Wang, H.; Zhang, S.P.; Xie, C.S. A novel approach to fabricate metal oxide nanowire-like networks based coplanar gas sensors array for enhanced selectivity. *Sens. Actuators B Chem.* **2014**, *204*, 351–359. [CrossRef]
22. Akbar, M.A.; Ali, A.A.S.; Amira, A.; Bensaali, F.; Benammar, M.; Hassan, M.; Bermak, A. An empirical study for PCA-and LDA-based feature reduction for gas identification. *IEEE Sens. J.* **2016**, *16*, 5734–5746. [CrossRef]
23. Ha, D.; Ahmed, U.; Pyun, H.; Lee, C.J.; Baek, K.H.; Han, C. Multi-mode operation of principal component analysis with k-nearest neighbor algorithm to monitor compressors for liquefied natural gas mixed refrigerant processes. *Comput. Chem. Eng.* **2017**, *106*, 96–105. [CrossRef]
24. Sun, J.; Xiao, Q.; Wen, J.; Wang, F. Natural gas pipeline small leakage feature extraction and recognition based on LMD envelope spectrum entropy and SVM. *Measurement* **2014**, *55*, 434–443. [CrossRef]
25. Yang, F.; Zhang, L.; Yu, S.; Prokhorov, D.; Mei, X.; Ling, H. Feature pyramid and hierarchical boosting network for pavement crack detection. *IEEE Trans. Intell. Transp. Syst.* **2019**, *21*, 1525–1535. [CrossRef]
26. Ahmad, F.; Abbasi, A.; Li, J.; Dobolyi, D.G.; Netemeyer, R.G.; Clifford, G.D.; Chen, H. A deep learning architecture for psychometric natural language processing. *ACM Trans. Inf. Syst. (TOIS)* **2020**, *38*, 1–29. [CrossRef]
27. Mahindru, A.; Sangal, A.L. MLDroid—Framework for Android malware detection using machine learning techniques. *Neural Comput. Appl.* **2021**, *33*, 5183–5240. [CrossRef]
28. Mahindru, A.; Sangal, A.L. Deepdroid: Feature selection approach to detect android malware using deep learning. In Proceedings of the 2019 IEEE 10th International Conference on Software Engineering and Service Science (ICSESS), Beijing, China, 18–20 October 2019.
29. Mahindru, A.; Sangal, A.L. Perbdroid: Effective malware detection model developed using machine learning classification techniques. In *A Journey towards Bio-Inspired Techniques in Software Engineering*; Springer: Cham, Switzerland, 2020; pp. 103–139.
30. Peng, P.; Zhao, X.J.; Pan, X.F.; Ye, W.B. Gas classification using deep convolutional neural networks. *Sensors* **2018**, *18*, 157. [CrossRef] [PubMed]
31. Fonollosa, J.; Fernandez, L.; Gutiérrez-Gálvez, A.; Huerta, R.; Marco, S. Calibration transfer and drift counteraction in chemical sensor arrays using Direct Standardization. *Sens. Actuators B Chem.* **2016**, *236*, 1044–1053. [CrossRef]
32. Ruijie, G. Research on Technologies of Gas Recognition by Electronic Nose Based on Deep Learning. Master's Thesis, Computer Technology, Huazhong University of Science and Technology, Wuhan, China, 2021.

chemosensors

Article

Cross-Linked SnO_2 Nanosheets Modified by Ag Nanoparticles for Formaldehyde Vapor Detection

Huaipeng Weng [1], Xumeng Dong [1], Yufeng Sun [1,*], Haibo Ren [1], Jiarui Huang [2] and Sang Woo Joo [3,*]

1 Key Lab for High Performance Nonferrous Metals of Anhui Province, Anhui Polytechnic University, Wuhu 241000, China
2 Key Laboratory of Functional Molecular Solids of the Ministry of Education, Anhui Laboratory of Molecule-Based Materials, College of Chemistry and Materials Science, Anhui Normal University, Wuhu 241002, China
3 School of Mechanical Engineering, Yeungnam University, Gyeongsan 712749, Gyeoungbuk, Republic of Korea
* Correspondence: sunyufeng118@126.com (Y.S.); swjoo@yu.ac.kr (S.W.J.)

Abstract: Ag@SnO_2 nanosheets were prepared through a hydrothermal method followed by heat treatment and a liquid reduction process. Many Ag nanoparticles (Ag NPs) were dispersed uniformly over the surface of the SnO_2 nanosheets. The thickness of the SnO_2 nanosheets was approximately 10 nm. After decoration with Ag NPs, the Ag@SnO_2 nanosheet sensors exhibited improved gas-sensing behaviors compared to the pure SnO_2 nanosheet sensor. The response of cross-linked SnO_2 nanosheets decorated by Ag NP sensors for 100 ppm formaldehyde vapor was up to 101.4, which was double that (45.5) of the pure SnO_2 nanosheet sensor. The response and recovery times of the Ag@SnO_2 sensor were 21 s and 23 s, respectively. The Ag@SnO_2 nanosheet sensors showed reasonable cycling stability, as demonstrated by testing with 100 ppm formaldehyde 10 times. The superior gas-sensing behaviors of the Ag@SnO_2 sensor were due to the large specific surface area, cross-linked nanostructure, and synergistic effect of the Ag NPs with huge sensitizing active sites and numerous SnO_2 nanosheets.

Keywords: SnO_2; Ag nanoparticles; nanosheet; gas sensor; formaldehyde

Citation: Weng, H.; Dong, X.; Sun, Y.; Ren, H.; Huang, J.; Joo, S.W. Cross-Linked SnO_2 Nanosheets Modified by Ag Nanoparticles for Formaldehyde Vapor Detection. *Chemosensors* **2023**, *11*, 116. https://doi.org/10.3390/chemosensors11020116

Academic Editor: Boris Lakard

Received: 11 January 2023
Revised: 29 January 2023
Accepted: 1 February 2023
Published: 4 February 2023

1. Introduction

Formaldehyde is an important chemical raw material often used in the chemical, textile, and wood industries [1–3]. However, formaldehyde is toxic, flammable, and volatile. Repeated exposure to even low formaldehyde concentrations (e.g., 0.08 ppm) can harm the nervous and immune systems of humans [4,5]. Therefore, a gas sensor with high sensitivity and a low detection limit is needed to detect formaldehyde gas.

In recent years, researchers reported that gas sensors based on metal oxide semiconductors (MOS) have higher sensitivity, better selectivity, and lower manufacturing cost. Common gas-sensing materials include ZnO [6], SnO_2 [7], MoO_3 [8], NiO [9], and In_2O_3 [10]. Among these MOSs, SnO_2 is an *n*-type semiconductor with a bandgap of 3.6 eV, a rapid electron transmission rate (160 cm^2/V s), and desirable physicochemical properties [11–13].

Controlling the morphology of the SnO_2 to improve the sensor response is a viable strategy. Based on the sensing mechanism, the sensing property is also related to its morphology and the size of the nanocrystals. SnO_2 materials with uniform morphology and adequate size would improve the performance of gas sensors. Therefore, SnO_2 materials with different morphologies, such as nanosphere [14], nanowire [15], nanorod [16], nanotube [17], nanosheet [12], and micro/nanoflower [18], were constructed to enhance their sensing properties. Li et al. constructed a sensor using SnO_2 nanospheres prepared through a hydrothermal method [19]. The sensor showed a strong response (38.3) for

100 ppm formaldehyde at 200 °C, and the response/recovery times were 17 s/25 s, respectively. The exceptional sensing performance was ascribed to the favorable porous nanosphere structure providing numerous surface-active sites and convenient channels for the easy adsorption and diffusion of gas molecules. Hu et al. utilized flower-like SnO_2 as a sensing material [20]. The sensor response to 100 ppm formaldehyde reached 34.6 with response/recovery times of 64 s/10 s. The flower-like structure can supply many active sites for accelerating the diffusion and transport of gas molecules and electrons. Among these morphologies, 2D nanosheets have a large specific surface area, suitable surface energy, and stable structure, which can significantly improve the response, shorten the response/recovery times, and reduce the detection limit [21]. Xu et al. prepared SnO_2 nanosheets using a hydrothermal method [12]. The nanosheet sensor displayed a strong response to 79.5 to 100 ppm formaldehyde at 200 °C, with response and recovery times of 42 s and 112 s, respectively. Yu et al. synthesized SnO_2 nanosheets using a solvothermal method [22]. In addition, Meng et al. used the stepwise recognition method of qualitative classification and quantitative regression to improve the recognition rate, the resolution, and the generalization performance of the SnO_2 sensor, which is a new way of on-line rapid sensor detection for drug-producing chemicals [23]. The SnO_2 sensor exhibited a response of 7 to 100 ppm formaldehyde at 200 °C, with short response/recovery times (1 s/6 s). Many researchers have made progress in improving the sensor performance by synthesizing various morphologies of tin dioxide. Nevertheless, it is still necessary to synthesize hierarchical SnO_2 with a large specific surface area to further enhance the sensor performance.

The following methods are often used to improve the practicality of SnO_2-based gas sensors by reducing the working temperature and shortening the response/recovery times: element doping [24], construction of heterojunctions [25,26], and noble metal modification [27]. Among them, the decoration of noble metals on sensing material is an effective strategy to promote the sensing properties. The commonly used noble metals include Pd [28], Au [29], Pt [30], and Ag [31]. For example, Li et al. modified SnO_2 nanosheets with Pd/Au nanoparticles [32]. The Pd NPs-modified SnO_2 sensor exhibited a strong response of 125 and a response time of 68 s toward 100 ppm formaldehyde at 110 °C. Tang et al. synthesized Pt-modified SnO_2 nanospheres with a mean diameter of 30–50 nm [14]. As sensing materials, it achieved a response of 70.1 toward 100 ppm formaldehyde with an extremely short response time (5 s) at 200 °C. The enhanced sensing property was mainly ascribed to two reasons. First, the electron transfer speed can be accelerated because of the electron sensitization effect. Second, the synergistic effect of noble metals and SnO_2 promotes the redox reactions on the surface of the materials [33]. Ag is relatively inexpensive with stable chemical properties compared with other noble metals. Therefore, Ag is a promising alternative to modify sensing materials and improve the sensing performance. For example, Liu et al. [34] attached Ag NPs to the surface of SnO_2 using a liquid reduction method. The resulting sensor exhibited a strong response to volatile organic compounds (VOCs). The response of the Ag@SnO_2 sensor was 14.4 for 10 ppm formaldehyde at the optimal operating temperature (150 °C), while the response of pure SnO_2 was only 3.2 at the same working temperature. Many strategies to decorate SnO_2 with noble metals have been developed. Despite many achievements in the preparation of Ag@SnO_2 sensing materials, there is a considerable focus on SnO_2 hierarchical structures modified by Ag NPs with improved sensing performance for VOCs.

In this study, numerous cross-linked SnO_2 nanosheets loaded with uniform Ag NPs were prepared via a hydrothermal method combined with a heat treatment and liquid reduction process. Compared with pure SnO_2 nanosheets, Ag@SnO_2 nanosheets showed enhanced sensing properties towards VOCs. This paper discusses the effect of Ag NPs on the sensing performance, including the sensing mechanism of the Ag@SnO_2 nanosheets.

2. Experimental Details

2.1. Preparation of Cross-Linked SnO_2 Nanosheets

First, 1.0 g of $SnCl_2·2H_2O$ was dissolved in 20 mL deionized water with stirring, followed by adding 10 mL NaOH (1.5 mol l^{-1}) solution. Subsequently, 0.3 g cetyltrimethyl ammonium bromide (CTAB) was placed into a mixed solution. After stirring for 30 min, 0.8 mL hydrazine hydrate was added to the mixture. Subsequently, the mixture was maintained at 150 °C for 24 h. The precursor was obtained after drying at 70 °C for 10 h. Finally, the cross-linked SnO_2 nanosheets were achieved by annealing the precursor at 700 °C for 30 min at a constant heating rate (5 °C min^{-1}).

2.2. Preparation of Cross-Linked Ag@SnO_2 Nanosheets

The prepared SnO_2 powder (0.5 g) was dispersed in 15 mL of deionized water with constant stirring. Subsequently, 1 mL of a 20 mM $AgNO_3$ solution and 2 mL of a 10 mM ascorbic acid solution were added to the mixture solution with stirring for another 60 min. Finally, the product was collected after washing three times and dried at 70 °C for 5 h.

2.3. Characterization

The related information is shown in the Supporting Information (SI).

3. Results and Discussion

3.1. Characterizations

Figure 1a exhibits the X-ray diffraction (XRD) pattern of the precursor and annealed product. Before calcination, all the XRD peaks of the precursor were assigned to SnO_2 (JCPDS No. 06-0395) and Sn_3O_4 (JCPDS No. 16-0737). After calcining the precursor at 700 °C, all the peaks at 26.1°, 33.9°, 37.9°, 51.8°, 54.7°, 61.8°, 64.7°, 65.9°, and 71.2° 2θ were indexed to the (110), (101), (200), (211), (220), (310), (112), (301), and (202) lattice planes of tetragonal rutile SnO_2 (JCPDS No. 41-1445), respectively. No impurity peaks were detected. Moreover, the high crystallinity of the final product was obtained by observing the sharp and strong XRD peaks. The XRD pattern of Ag@SnO_2 exhibited the peaks for SnO_2 and peaks at 38.1°, 44.2°, 64.4°, and 77.5° 2θ corresponding to the (111), (200), (220), and (311) planes of Ag (JCPDS Card No. 04-0783), respectively, indicating that the product contained Ag and SnO_2. Figure 1b exhibits energy-dispersive X-ray spectroscopy (EDS) analysis, confirming the existence of Sn, O, and Ag in the final product.

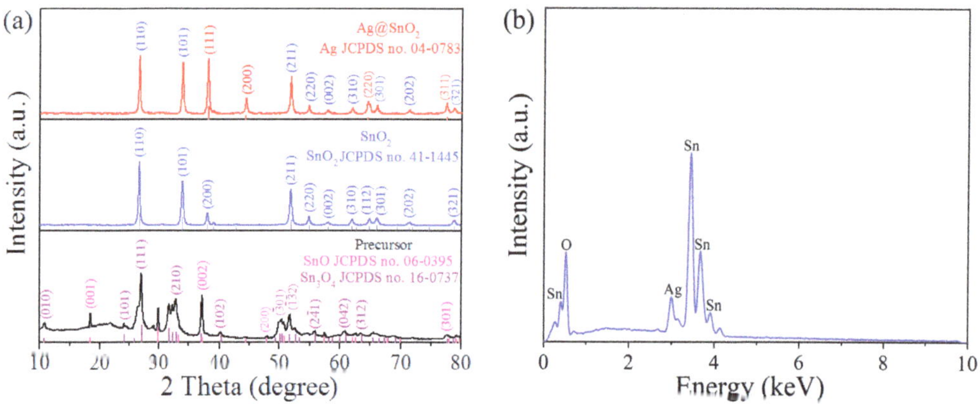

Figure 1. (**a**) XRD pattern; (**b**) EDS analysis of the cross-linked Ag@SnO_2 nanosheets.

The morphologies of the samples were studied by scanning electron microscopy (SEM). Figure 2a,b shows the sheet-linked morphology of the precursor. The nanosheets had a smooth surface. Many cross-linked nanosheets formed a hierarchical structure by self-assembly. Some pores formed among the cross-linked nanosheets. Synthesis of cross-linked SnO_2 nanosheets from an aqueous solution containing Sn^{2+} ions may be explained via the following mechanism:

$$4SnCl_2{\cdot}2H_2O + 6OH^- = Sn_4(OH)_6Cl_2{\downarrow} + 8H_2O + 6Cl^- \quad (1)$$

$$Sn_4(OH)_6Cl_2 + O_2 = 2Sn_2O_3 + 2H_2O + 2HCl \quad (2)$$

$$2Sn_2O_3 + O_2 = 4SnO_2 \quad (3)$$

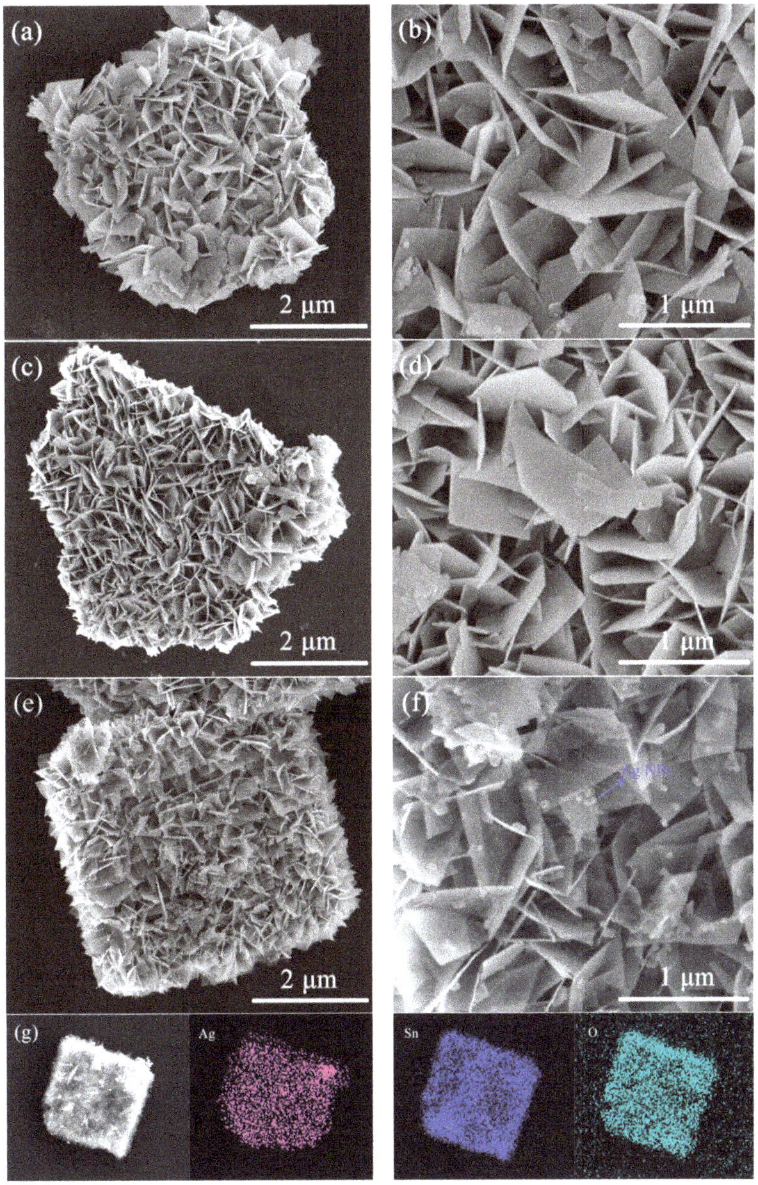

Figure 2. SEM images of (**a**) the precursor; (**b**) the precursor; (**c**) cross-linked SnO_2 nanosheets; (**d**) cross-linked SnO_2 nanosheets; (**e**) cross-linked $Ag@SnO_2$ nanosheets; (**f**) cross-linked $Ag@SnO_2$ nanosheets. (**g**) Elemental mapping images of $Ag@SnO_2$ nanosheets.

The reaction (1) could be easily observed when adding NaOH to the Sn^{2+} solution, resulting in some $Sn_4(OH)_6Cl_2$ microplates. After that, the $Sn_4(OH)_6Cl_2$ microplates were gradually oxidized, forming some Sn_2O_3 nuclei on the surface of the microplates. Upon increasing the time further, Sn_2O_3 nuclei formed from the oxidation of $Sn_4(OH)_6Cl_2$, followed by crystal growth. In liquid medium, the growth habit of Sn_2O_3 crystal is mainly determined by its intrinsic structure. Therefore, the cross-linked nanosheet-like precursor was obtained. Finally, the cross-linked SnO_2 nanosheets were obtained via calcination of the precursor in air. Figure 2c,d presents SEM images of the precursors after high-temperature calcination. The morphology maintains the structure of cross-linked nanosheets. The thickness of the nanosheets was ca. 10 nm, as shown in Figure S3. As displayed in Figure 2e,f, after being decorated by Ag NPs, abundant Ag NPs were attached uniformly to the surface of each nanosheet, and no obvious aggregation of Ag NPs occurred. Figure S4 presents the size of the Ag NPs with 80 nm. The loaded Ag NPs can supply many active sites for gas adsorption and accelerate the transfer of electrons. Figure 2g presents the mapping images of the Ag@SnO_2 nanosheets, confirming the presence of Ag, Sn, and O. The Ag NPs were modified uniformly on the nanosheet surface.

The morphology of the composites was explored further by transmission electron microscopy (TEM). Figure 3a,b shows TEM images (low/high magnification, respectively) of the SnO_2 nanosheets modified by Ag NPs (Ag@SnO_2). The sample maintained a cross-linked nanosheet morphology. Abundant Ag NPs were scattered uniformly over the SnO_2 nanosheet surface, with no serious aggregation of Ag NPs. As shown in Figure 3c, the lattice spacing of 0.176 nm matches the crystal plane (211) of rutile SnO_2 in the TEM image with high-resolution TEM (HRTEM). From Figure 3b,c, the crystallographic orientation of the SnO_2 nanosheet is the (211) plane. The nanosheet grows along the (211) direction, but the basal plane cannot be determined. Moreover, the lattice spacing (0.236 nm) corresponded to the crystal plane (111) of Ag [35]. From the SAED pattern of Ag@SnO_2 (Figure 3d), five distinct diffraction rings corresponded to the (100), (101), (200), (211), and (301) crystal planes of the tetragonal SnO_2, indicating that the nanosheet was polycrystalline [36]. The diffraction ring could be attributed to the crystal plane (311) of Ag [35].

Figure 4a presents the nitrogen adsorption/desorption isotherms of the Ag@SnO_2 nanosheets. The specific surface area of the nanocomposites is 11.9 $m^2\ g^{-1}$. The pore size ranged from 0.86 nm to 24.65 nm, with an average pore size of 12.86 nm. By comparison, the specific surface area of pure SnO_2 nanosheets was only 8.1 $m^2\ g^{-1}$. The pore sizes were between 0.89 and 28.12 nm, averaging 11.78 nm. The cross-linked nanosheets contributed to the formation of numerous pores. The Ag@SnO_2 nanosheets had a larger specific surface area than the pure SnO_2 nanosheets. The higher specific surface area was favorable for providing more available active sites for the gas molecules. The porous structure helped improve the diffusion efficiency of gas molecules by shortening the diffusion distance.

Figure 5a presents the UV-vis spectra of cross-linked Ag@SnO_2 and pure cross-linked SnO_2 nanosheets. The maximum absorption peak of the as-prepared samples occurred around 300 nm. The absorption band (350–280 nm) of Ag-SnO_2 composite is red-shifted compared with pristine SnO_2, which can be attributed to the charge transition between Ag and SnO_2, leading to the enhancement of gas-sensing performance [34]. Figure 5b presents the calculated bandgap of Ag@SnO_2 and the SnO_2 using the Tucker model. The calculated bandgaps for SnO_2 and Ag@SnO_2 were 2.81 eV and 2.66 eV, respectively. The bandgap of Ag@SnO_2 was smaller than that of SnO_2, possibly due to Ag NPs accelerating the electron transfer rate, improving the surface electron concentration of Ag@SnO_2, so that more electrons were injected into the conduction band of Ag@SnO_2, inducing the Burstein–Moss effect [37,38].

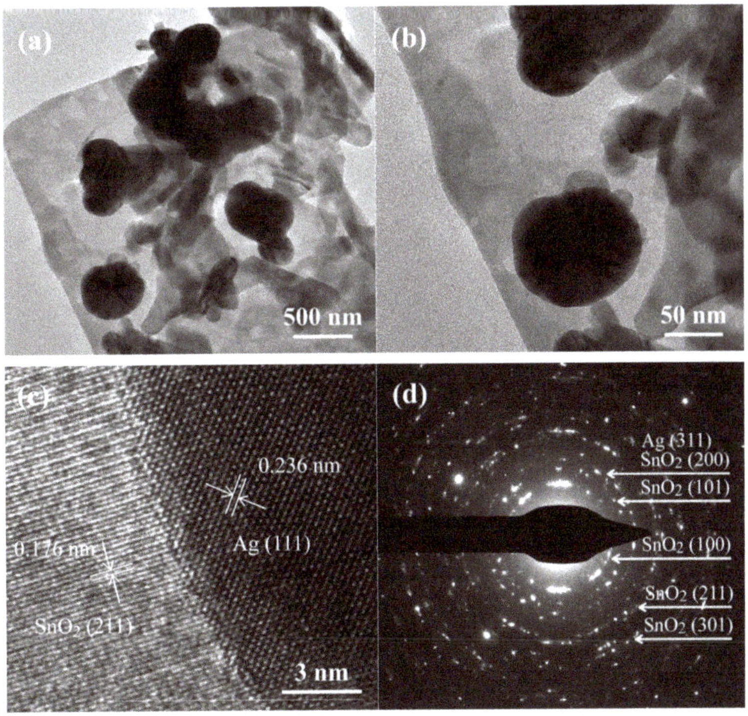

Figure 3. (**a**) TEM image; (**b**) HRTEM image; (**c**) lattice-resolved HRTEM image; (**d**) selected area electron diffraction (SAED) pattern of Ag@SnO_2 nanosheets.

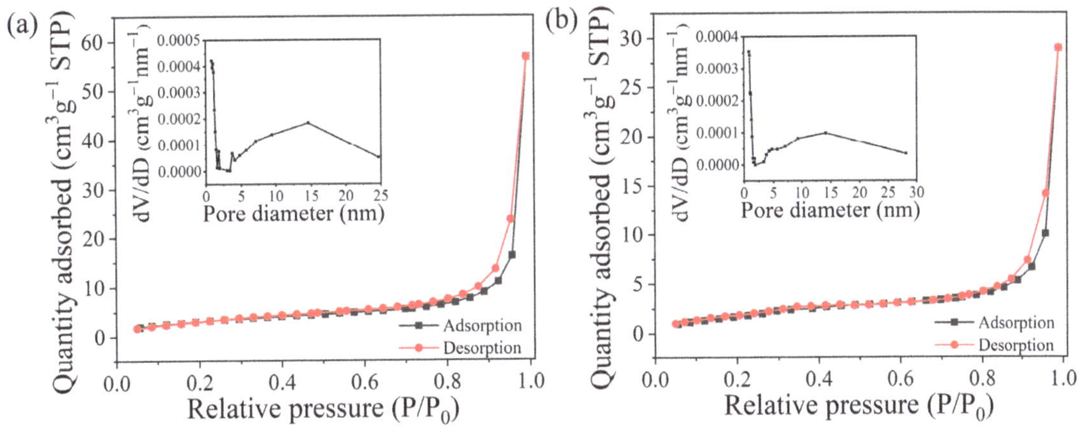

Figure 4. Nitrogen adsorption/desorption isotherms of (**a**) Ag@SnO_2; (**b**) pure SnO_2 nanosheets. The inset image is the pore size distribution.

The element components and valences of the Ag@SnO_2 nanosheets were examined by X-ray photoelectron spectroscopy (XPS). Figure 6a exhibits the survey spectrum of the product, demonstrating the existence of Sn, O, and Ag. As shown in Figure 6b, there were two peaks at ca. 486.1 eV and 494.5 eV in the Sn 3d spectrum, which could be ascribed to Sn $3d_{5/2}$ and Sn $3d_{3/2}$ [39], respectively. This result suggests the formation of Sn^{4+} in SnO_2. Figure 6c exhibits the Ag 3d spectrum, where two peaks at 367.1 eV and 373.1 eV are ascribed to Ag $3d_{5/2}$ and Ag $3d_{3/2}$, respectively [22,40]. From the O 1s spectrum in

Figure 6d, three peaks at ca. 529.8 eV, 530.6 eV, and 531.9 eV were ascribed to lattice oxygen ($O_{L.}$ 53.4%), vacancy oxygen ($O_{V.}$ 27.8%), and chemisorbed oxygen ($O_{C.}$ 18.8%), respectively. Abundant chemisorbed oxygen formed in the composites, which improved the sensing performance. Figure S5 shows the C 1s spectrum of cross-linked Ag@SnO_2 nanosheets, which may come from the adsorbed CTAB on the sample. Furthermore, the weight percentage of Ag among the Ag@SnO_2 composites was calculated to be 9.7 wt % through the XPS results.

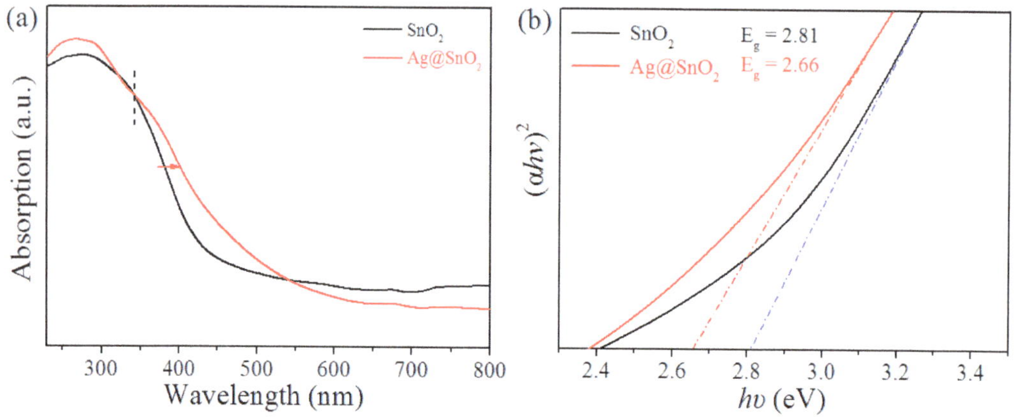

Figure 5. (**a**) UV-vis absorption spectra of pure cross-linked SnO_2 and cross-linked Ag@SnO_2 nanosheets. (**b**) Bandgap determinations using $(\alpha h\nu)^2$ vs. hv of cross-linked SnO_2 nanosheets and Ag@SnO_2 nanosheets.

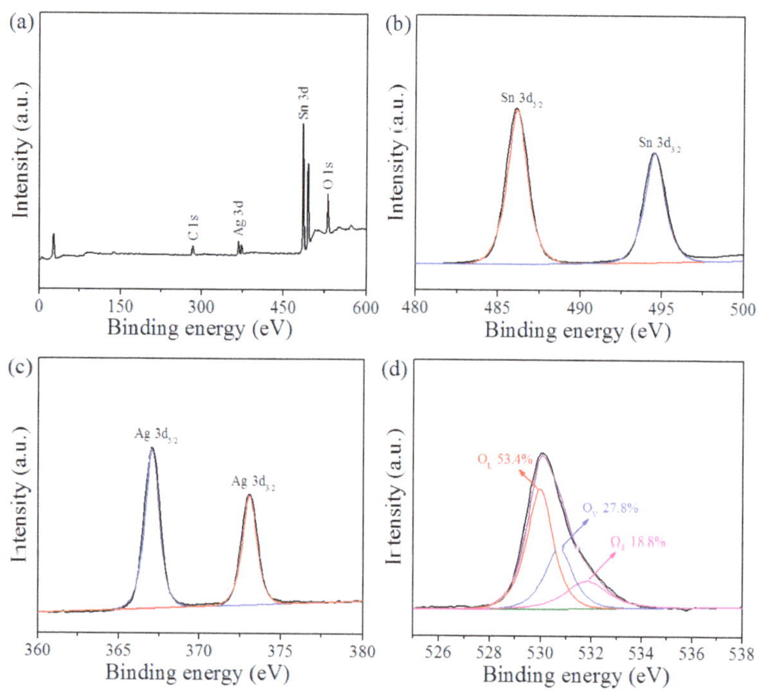

Figure 6. XPS spectra of cross-linked Ag@SnO_2 nanosheets of (**a**) survey spectrum; (**b**) Sn 3d; (**c**) Ag 3d; (**d**) O 1s spectra.

3.2. Sensing Performance of Ag@SnO_2 Nanosheet Sensors

The working temperatures are always considered an important factor of a gas sensor for wide applications. At different working temperatures, the responses of cross-linked Ag@SnO_2 and pure cross-linked SnO_2 nanosheet sensors exposed to 100 ppm formaldehyde vapor were recorded, as shown in Figure 7a. In the temperature range of 80–200 °C, the responses of the composites increased as the working temperature was increased. As for reductive gases, the response was defined as R_a/R_g, where R_a is sensor resistance in dry air, and R_g is sensor resistance in dry air containing test gas. At 140 °C, the response reached the maximum value of 101.4, with a decrease at higher temperatures. As a result, 140 °C is the optimal working temperature of the Ag@SnO_2 sensor. For the pure SnO_2 sensor, the maximum response was 45.5 at temperatures up to 180 °C. Thus, 180 °C is considered the optimal working temperature of the SnO_2 sensor, which is higher than that of the Ag@SnO_2 sensor. Due to the introduction of Ag NPs in the Ag@SnO_2, the strong catalytic effect enables formaldehyde molecules to be oxidized at a lower temperature. As shown in Figure 7b, formaldehyde, ethanol, isopropanol, acetone, methanol, toluene, ammonia, and benzene vapors are detected based on the Ag@SnO_2 and pure SnO_2 nanosheets sensors at 140 °C. The responses of the Ag@SnO_2 sensor to these seven vapors were 101.4, 66.8, 60.5, 42.8, 17.2, 12.2, 7.5, and 5.1, respectively. The corresponding responses of the pure SnO_2 sensor were 45.5, 42.2, 35.2, 29.3, 15.7, 9.8, 6.5, and 4.8, respectively. The results suggest that the sensing performance of the cross-linked Ag@SnO_2 nanosheet sensor is superior to that of the pure cross-linked SnO_2 nanosheet sensor. The high response of formaldehyde is mainly because different gases have different dissociation energies. The dissociation energy of formaldehyde (364 kJ mol^{-1}) is lower than that of ethanol (436 kJ mol^{-1}), methanol (439 kJ mol^{-1}), benzene (431 kJ mol^{-1}), acetone (393 kJ mol^{-1}), and toluene (368 kJ mol^{-1}). The low dissociation energy of formaldehyde is conducive to the reaction with the absorbed oxygen on the sensing material, thus showing strong response and high selectivity.

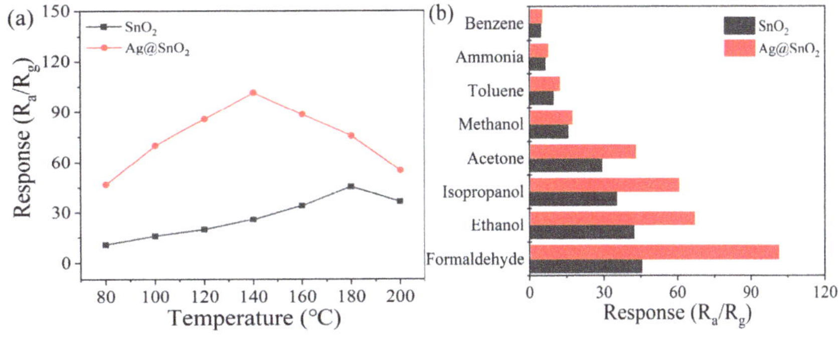

Figure 7. (**a**) Sensor responses of Ag@SnO_2 sensor and SnO_2 sensor to 100 ppm formaldehyde at various working temperatures; (**b**) response curves of two sensors to eight kinds of vapors (100 ppm) at 140 °C.

The gas sensors fabricated by the cross-linked Ag@SnO_2 and pure cross-linked SnO_2 nanosheets were applied to detect some VOCs at an optimal working temperature of 140 °C. Figure 8a exhibits the real-time response curve of the Ag@SnO_2 sensor and the pure SnO_2 sensor after exposure to formaldehyde vapor at different concentrations (1–400 ppm). The output current increased rapidly when the gas sensor was exposed to the formaldehyde vapor and maintained a stable condition. When the formaldehyde vapor broke away from the gas sensor, the output current quickly returned to the original value and maintained a stable current. Thus, the Ag@SnO_2 sensor possesses a distinct gas-sensing response and recovery. In Figure 8, the Ag@SnO_2 sensor exhibited an overshoot at a high concentration, which is the sensor response just after exposure to the target gas was highest, followed by a signal decay up to the end of the pulse. Particularly, the overshoot is drastic in ethanol and isopropanol detection. This was mainly caused by uneven gas diffusion concentration after injecting the target gas. The local concentration of the target gas was high at the beginning of injection, and then diffused to the surrounding area. The Ag@SnO_2 sensor exhibits a fast strong response due to the catalytic effect of the Ag nanoparticles, resulting in an overshoot at high concentrations and signal decay, which cannot be obviously observed in the slow, weak response of the pure SnO_2 sensor. From the response curve of the sensor (insets), the response increased gradually as the formaldehyde concentration was increased. The calculated responses at a very low (1 ppm) and high concentrations (100 ppm) were 6.4 and 101.4, respectively. Figure 8b–d shows the real-time response curves of the Ag@SnO_2 sensor for ethanol, isopropanol, and acetone vapors, respectively. The inserts present the respective response curves. Exposure to a fixed concentration of vapors (100 ppm) revealed responses to ethanol, isopropanol, and acetone vapors of 66.8, 60.5, and 42.8, respectively. Furthermore, for the pure cross-linked SnO_2 nanosheet sensor, the corresponding responses for formaldehyde, ethanol, isopropanol, and acetone vapors were 45.5, 42.2, 35.2, and 29.3. In addition, after exposure to 100 ppm formaldehyde, the response and recovery times of Ag@SnO_2 sensor were 21 s and 23 s, respectively, as shown in Figure S6. The response and recovery times of the SnO_2 nanosheet sensor were 28 s and 26 s, respectively, which are longer than those of the Ag@SnO_2 sensor. Compared with the SnO_2 nanoparticles [13], SnO_2 nanoflowers [20], flower-like SnO_2 microspheres [41], Pt-SnO_2 nanospheres [14], Ag-SnO_2 nanoparticles [34], hollow PdO/ZnO/SnO_2 nanospheres [42], and flower-like Bi-SnO_2 nanostructures [43], the cross-linked Ag@SnO_2 nanosheets showed exceptional sensing performance, including a higher response, quicker response, and short recovery times, as shown in Table S1. Two additional sensors were fabricated with the cross-linked Ag@SnO_2 nanosheets under the same process. Similar gas-sensing properties were obtained, which suggests the reproducibility of the sensors.

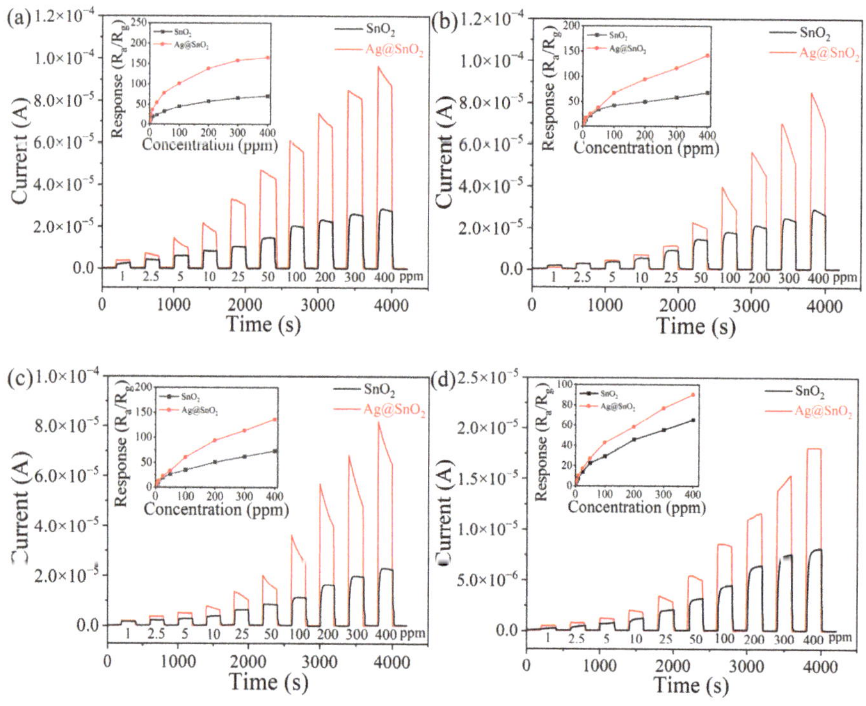

Figure 8. Real-time response curves of sensor devices upon exposure to different concentrations of (**a**) formaldehyde; (**b**) ethanol; (**c**) isopropanol; (**d**) acetone at 140 °C. The insets show the corresponding sensor response curves.

The repeatability of the Ag@SnO_2 sensor was assessed by conducting real-time response tests to 100 ppm formaldehyde ten times at the optimal working temperature, as shown in Figure 9a. The output current increased sharply when the sensor was exposed to the formaldehyde vapor. The output current quickly returned to the initial state when the formaldehyde vapor was away from the sensor. Over ten cycles, the sensor maintained a similar current to the first, indicating reasonable repeatability of the Ag@SnO_2 sensor for formaldehyde vapor detection. Figure S7 shows the long-term stability of the Ag@SnO_2 sensor towards 100 ppm formaldehyde vapor at the working temperature of 140 °C. The response of the sensor remained relatively stable within 14 days, and the sensor performance remained unchanged throughout the long-term testing process, indicating its stability. Figure 9b shows the response curves for different formaldehyde concentrations. The linear regression coefficients for Ag@SnO_2 and SnO_2 were 0.9642 and 0.9884, respectively.

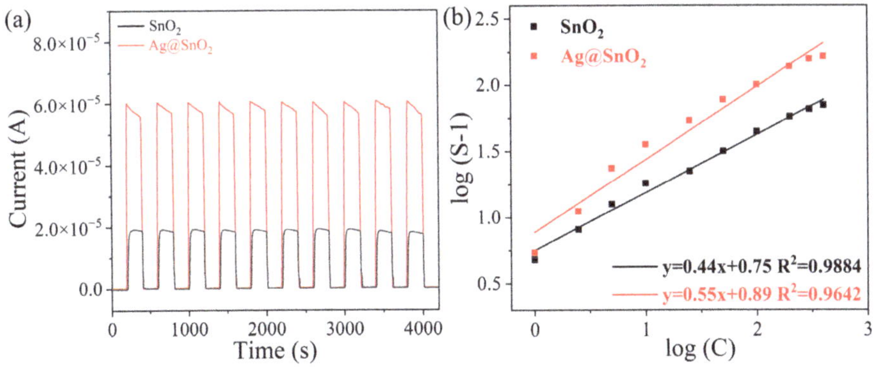

Figure 9. (**a**) Current change curves of Ag@SnO_2 sensor towards 100 ppm formaldehyde vapor for 10 tests at 140 °C; (**b**) the linear fitting curve of cross-linked Ag@SnO_2 and SnO_2 nanosheet sensors towards formaldehyde in the concentration range of 1–400 ppm.

3.3. Gas-Sensing Mechanism

The sensor based on SnO_2 abides by a gas-sensing mechanism with a typical *n*-type semiconductor material. Figure 10 presents a schematic illustration of the sensing mechanism of the Ag@SnO_2 nanosheet sensor. The internal resistance of the sensing material will change due to large differences in resistance when the sensors are exposed to air and reducing gases [44]. When the SnO_2 sensor is exposed to air, oxygen molecules will be adsorbed on the SnO_2 surface. At the working temperature, oxygen molecules transform to oxygen species (O_2^- and O^-) with strong oxidation by capturing electrons from the conduction band (CB) of SnO_2. An electron-depleted layer (EDL) will form because of the loss of electrons in the CB of SnO_2, which leads to a rapid increase in sensor resistance [45,46]. When the sensor is exposed to formaldehyde vapor, the oxygen species on the material surface will quickly react with formaldehyde vapor to produce CO_x and $H_2O(g)$. At the same time, the electrons captured by the oxygen anions are released back to the CB of the SnO_2, narrowing the EDL thickness. This contributes to a rapid decrease in sensor resistance. The reaction process is shown in the following formulae [47].

$$O_{2(gas)} \rightarrow O_{2(ads)}, \tag{4}$$

$$O_{2(gas)} + e^- \rightarrow O^{2-}{}_{(ads)}, \tag{5}$$

$$O^{2-}{}_{(ads)} + e^- \rightarrow 2O^-{}_{(ads)}, \tag{6}$$

$$CH_2O_{(ads)} + 2O^-{}_{(ads)} \rightarrow CO_{2(g)} + H_2O_{(g)} + 2e^-. \tag{7}$$

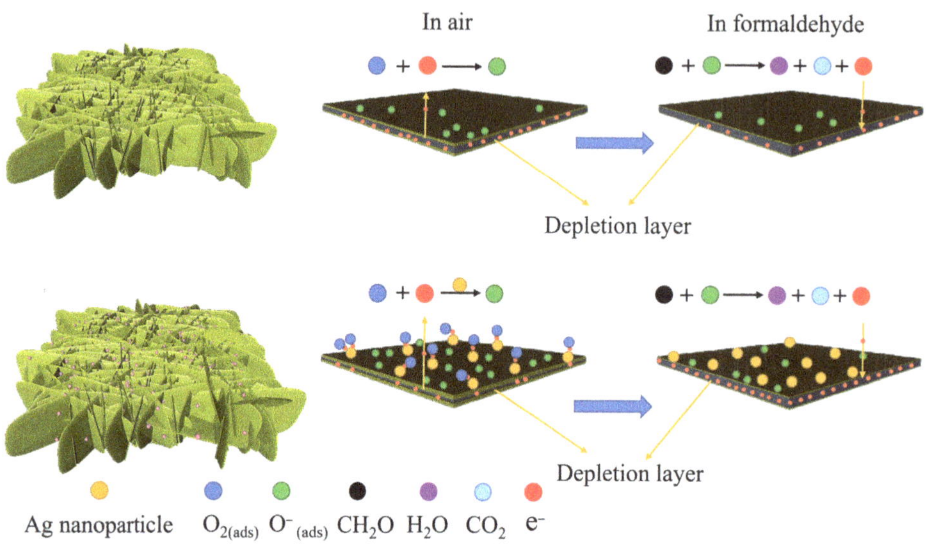

Figure 10. Schematic diagram of the sensing mechanism of cross-linked Ag@SnO_2 nanosheet sensor.

The mechanism for the enhanced sensing performance of the Ag@SnO_2 sensor can be described from the following aspects. After Ag nanoparticles are attached to the SnO_2 surface, the spillover and catalytic effects will show [34]. Ag NPs will increase the chemisorbed oxygen on the SnO_2 surface, which provides more reactive sites for redox reactions. Moreover, the barrier variation of the SnO_2-based sensors is greatly influenced by the amount and type of oxygen adsorbed on the material surface. The Ag@SnO_2 composites have more chemisorbed oxygen than the pure SnO_2, which is due to the spillover effect between Ag nanoparticles and the material surface, accelerating the decomposition rate of the oxygen. The faster oxygen decomposition rate provides more oxygen species on the material surface. Hence, the reaction rate between oxygen species and formaldehyde molecules is accelerated, and the resistance transition rate inside the sensor is also increased.

Furthermore, the work function of Ag (5.1 eV) is higher than that of SnO_2 (4.9 eV). When Ag NPs are modified on the surface of the SnO_2, electrons flow from the surface of the SnO_2 to the surface of the Ag NPs, and the Schottky junction is formed between them [48,49]. When the Ag@SnO_2 sensor is exposed to air, the electrons accumulating on the surface of the Ag nanoparticles react quickly with oxygen, which promotes the ionization of absorbed oxygen. In addition, the amount of oxygen species converging on the surface of the Ag NPs will overflow to the material surface, resulting in an increasing number of oxygen species on the surface [50]. Finally, the electron depletion layer of Ag@SnO_2 is thicker than that of SnO_2, which further increases the resistance of sensing material in air. When the Ag@SnO_2-based sensor is exposed to formaldehyde gas, the large number of oxygen species adsorbed on the sensing material quickly react with formaldehyde molecules with the help of Ag catalysis. As the reaction progresses, many free electrons return, decreasing the thickness of the electron depletion layer and increasing the current of the Ag@SnO_2 sensor.

4. Conclusions

Numerous cross-linked SnO_2 nanosheets decorated with Ag NPs were prepared using a hydrothermal method, followed by a heat treatment and liquid reduction process. Many Ag NPs were dispersed uniformly over the surface of the cross-linked SnO_2 nanosheets with a thickness of ca. 10 nm. In contrast to the pure cross-linked SnO_2 nanosheet sensor, the Ag-decorated SnO_2 nanosheet sensor showed an enhanced sensing performance for VOC detection. For the 100 ppm formaldehyde vapor, the Ag@SnO_2 sensors exhibited a stronger response than the pure SnO_2 sensor at a low optimal working temperature. The exceptional sensing behaviors of the Ag@SnO_2 sensors were attributed to the favorable cross-linked nanosheet structure, large specific surface area, and the synergistic effect of Ag NPs and cross-linked SnO_2 nanosheets.

Supplementary Materials: The following supporting information can be downloaded at: https://www.mdpi.com/article/10.3390/chemosensors11020116/s1, Figure S1: Experimental setup; Figure S2: Photograph of the sensor; Figure S3: High-magnification SEM image of cross-linked SnO_2 nanosheets; Figure S4: High-magnification SEM image of cross-linked Ag@SnO_2 nanosheets; Figure S5: C 1s spectrum of cross-linked Ag@SnO_2 nanosheets; Figure S6: Response and recovery characteristics of cross-linked Ag@SnO_2 nanosheets and pure cross-linked SnO_2 nanosheet sensors to 100 ppm formaldehyde vapor; Figure S7: Long-term stability of the Ag@SnO_2 sensor towards 100 ppm formaldehyde vapor at the working temperature of 140 °C; Table S1: Responses of SnO_2-based gas sensors to different concentrations of formaldehyde vapor.

Author Contributions: Conceptualization, H.W.; methodology, H.W., X.D. and H.R.; validation, H.R.; investigation, H.W. and X.D.; resources, Y.S. and S.W.J.; writing—original draft preparation, H.W.; writing—review and editing, Y.S., J.H. and S.W.J.; supervision, Y.S.; funding acquisition, S.W.J. All authors have read and agreed to the published version of the manuscript.

Funding: This research was funded by the National Research Foundation of Korea (NRF-2019R1A5A 8080290).

Institutional Review Board Statement: Not applicable.

Informed Consent Statement: Not applicable.

Data Availability Statement: The data presented in this study are available on request from the corresponding author.

Conflicts of Interest: The authors declare no conflict of interest.

References

1. Hui, C.; Sun, L.; Li, G.; Zou, X. Well-tuned surface oxygen chemistry of cation off-stoichiometric spinel oxides for highly selective and sensitive formaldehyde detection. *Chem. Mater.* **2018**, *30*, 2018–2027.
2. Wang, D.; Wan, K.; Zhang, M.; Li, H.; Wang, P.; Wang, X.; Yang, J. Constructing hierarchical SnO_2 nanofiber/nanosheets for efficient formaldehyde detection. *Sens. Actuators B Chem.* **2019**, *283*, 714–723. [CrossRef]
3. Zhang, D.; Cao, Y.; Yang, Z.; Wu, J. Nanoheterostructure construction and DFT study of Ni-doped In_2O_3 nanocubes/WS_2 hexagon nanosheets for formaldehyde sensing at room temperature. *ACS Appl. Mater. Interfaces* **2020**, *12*, 11979–11989. [CrossRef]
4. Li, N.; Fan, Y.; Shi, Y.; Xiang, Q.; Wang, X.; Xu, J. A low temperature formaldehyde gas sensor based on hierarchical SnO/SnO_2 nano-flowers assembled from ultrathin nanosheets: Synthesis, sensing performance and mechanism. *Sens. Actuators B Chem.* **2019**, *294*, 106–115. [CrossRef]
5. Ji, H.; Zeng, W.; Li, Y. Gas sensing mechanisms of metal oxide semiconductors: A focus review. *Nanoscale* **2019**, *11*, 22664–22684. [CrossRef]
6. Huang, B.; Zeng, W.; Li, Y. Synthesis of ZIF-8 coating on ZnO nanorods for enhanced gas-sensing performance. *Chemosensors* **2022**, *10*, 297. [CrossRef]
7. Deng, Z.; Zhang, Y.; Xu, D.; Zi, B.; Zeng, J.; Lu, Q.; Xiong, K.; Zhang, J.; Zhao, J.; Liu, Q. Ultrasensitive formaldehyde sensor based on SnO_2 with rich adsorbed oxygen derived from a metal organic framework. *ACS Sens.* **2022**, *7*, 2577–2588. [CrossRef]
8. Meng, F.; Qi, T.; Zhang, J.; Zhu, H.; Yuan, Z.; Liu, C.; Qin, W.; Ding, M. MoS_2-templated porous hollow MoO_3 microspheres for highly selective ammonia sensing via a Lewis acid-base interaction. *IEEE Trans. Ind. Electron.* **2022**, *69*, 960–970. [CrossRef]
9. Li, Q.; Zeng, W.; Zhou, Q.; Wang, Z. Highly sensitive ethanol sensing using NiO hollow spheres synthesized via hydrothermal method. *Chemosensors* **2022**, *10*, 341. [CrossRef]
10. Ou, Y.; Zhu, G.; Liu, P.; Jia, Y.; Zhu, L.; Nie, J.; Zhang, S.; Zhang, W.; Gao, J.; Lu, H.; et al. Anchoring platinum clusters onto oxygen vacancy-modified In_2O_3 for ultraefficient, low-temperature, highly sensitive, and stable detection of formaldehyde. *ACS Sens.* **2022**, *7*, 1201–1212. [CrossRef] [PubMed]
11. Liu, H.; Liu, B.; Li, P.; Kang, W.; Zhang, Y. High sensitivity and anti-humidity gas sensor for nitrogen dioxide based on Ce/SnO_2 nanomaterials. *Sensor Actuat. A-Phys.* **2022**, *344*, 113717.
12. Xu, R.; Zhang, L.; Li, M.; Yin, Y.; Yin, J.; Zhu, M.; Chen, J.; Wang, Y.; Bie, L. Ultrathin SnO_2 nanosheets with dominant high-energy {001} facets for low temperature formaldehyde gas sensor. *Sens. Actuators B Chem.* **2019**, *289*, 186–194. [CrossRef]
13. Yang, T.; Zhu, M.; Gu, K.; Zhai, C.; Zhao, Q.; Yang, X.; Zhang, M. Facile synthesis of SnO_2 nanoparticles for improved formaldehyde detection. *New J. Chem.* **2018**, *16*, 3185–14014.

14. Tang, Y.; Han, Z.; Qi, Y.; Yang, Z.; Han, H.; Jiang, Y.; Zhang, X.; Wu, L.; Wang, Z.; Liu, J.; et al. Enhanced ppb-level formaldehyde sensing performance over Pt deposited SnO_2 nanospheres. *J. Alloy. Compd.* **2022**, *899*, 163230. [CrossRef]
15. Tonezzer, M. Selective gas sensor based on one single SnO_2 nanowire. *Sens. Actuators B Chem.* **2019**, *288*, 53–59. [CrossRef]
16. Sharma, A.P.; Dhakal, P.; Pradhan, D.K.; Behera, M.K.; Xiao, B.; Bahoura, M. Fabrication and characterization of SnO_2 nanorods for room temperature gas sensors. *Aip Adv.* **2018**, *8*, 095219. [CrossRef]
17. Su, P.; Li, W.; Zhang, J.; Xie, X. Chemiresistive gas sensor based on electrospun hollow SnO_2 nanotubes for detecting NO at the ppb level. *Vacuum* **2022**, *199*, 110961. [CrossRef]
18. Xiang, C.; Chen, T.; Zhao, Y.; Sun, J.; Jiang, K.; Li, Y.; Zhu, X.; Zhang, X.; Zhang, N.; Guo, R. Facile hydrothermal synthesis of SnO_2 nanoflowers for low-Concentration formaldehyde detection. *Nanomaterials* **2022**, *12*, 2133. [CrossRef]
19. Li, Y.; Chen, N.; Deng, D.; Xing, X.; Xiao, X.; Wang, Y. Formaldehyde detection: SnO_2 microspheres for formaldehyde gas sensor with high sensitivity, fast response/recovery and good selectivity. *Sens. Actuators B Chem.* **2017**, *238*, 264–273. [CrossRef]
20. Hu, J.; Wang, H.; Chen, M.; Zhang, Y.; Zhao, X.; Zhang, D.; Lu, Q.; Zhang, J.; Liu, Q. Constructing hierarchical SnO_2 nanoflowers for enhanced formaldehyde sensing performances. *Mater. Lett.* **2020**, *263*, 126843. [CrossRef]
21. Zhang, J.; Zhang, B.; Yao, S.; Li, H.; Chen, C.; Bala, H.; Zhang, Z. Improved triethylamine sensing properties of fish-scale-like porous SnO_2 nanosheets by decorating with Ag nanoparticles. *J. Materiomics* **2022**, *8*, 518–525. [CrossRef]
22. Yu, H.; Yang, T.; Zhao, R.; Xiao, B.; Li, Z.; Zhang, M. Fast formaldehyde gas sensing response properties of ultrathin SnO_2 nanosheets. *RSC Adv.* **2015**, *126*, 104574–104581. [CrossRef]
23. Han, J.; Zheng, Y.; Fan, M. Qualitative and quantitative recognition method of drug-producing chemicals based on SnO_2 gas Sensor with dynamic measurement and PCA weak separation. *Sens. Actuators B Chem.* **2021**, *348*, 130698.
24. Zhao, R.; Zhang, X.; Peng, S.; Hong, P.; Zou, T.; Wang, Z.; Xing, X.; Yang, Y.; Wang, Y. Shaddock peels as bio-templates synthesis of Cd-doped SnO_2 nanofibers: A high performance formaldehyde sensing material. *J. Alloys Compd.* **2020**, *813*, 152170. [CrossRef]
25. Li, X.; Liu, W.; Huang, B.; Liu, H.; Li, X. Layered SnSe2 microflakes and $SnSe_2/SnO_2$ heterojunctions for low-temperature chemiresistive-type gas sensing. *J. Mater. Chem. C* **2020**, *8*, 15804–15815. [CrossRef]
26. Meng, D.; Liu, D.; Wang, G.; Shen, Y.; San, X.; Li, M.; Meng, F. Low-temperature formaldehyde gas sensors based on NiO-SnO_2 heterojunction microflowers assembled by thin porous nanosheets. *Sens. Actuators B Chem.* **2018**, *273*, 418–428. [CrossRef]
27. Kou, X.; Meng, F.; Chen, K.; Wang, T.; Sun, P.; Liu, F.; Yan, X.; Sun, Y.; Liu, F.; Shimanoe, K.; et al. High-performance acetone gas sensor based on Ru-doped SnO_2 nanofibers. *Sens. Actuators B Chem.* **2020**, *320*, 128292. [CrossRef]
28. Wang, F.; Hu, K.; Liu, H.; Zhao, Q.; Wang, K.; Zhang, Y. Low temperature and fast response hydrogen gas sensor with Pd coated SnO_2 nanofiber rods. *Int. J. Hydrogen Energ.* **2020**, *45*, 7234–7242. [CrossRef]
29. Meng, F.; Zheng, H.; Chang, Y.; Zhao, Y.; Li, M.; Wang, C.; Sun, Y.; Liu, J. One-step synthesis of $Au/SnO_2/RGO$ nanocomposites and their VOC sensing properties. *IEEE Trans. Nanotechnol.* **2018**, *172*, 212–219. [CrossRef]
30. Duy, N.V.; Thai, N.X.; Ngoc, T.M.; Le, D.T.T.; Hung, C.M.; Nguyen, H.; Tonezzer, M.; Hieu, N.V.; Hoa, N.D. Design and fabrication of effective gradient temperature sensor array based on bilayer SnO_2/Pt for gas classification. *Sens. Actuators B Chem.* **2022**, *351*, 130979.
31. Yuan, X.; Sun, L.; Wang, J.; Liao, D.; Sun, J. Metal organic frameworks derived SnO_2 microsphere doped Ag for monitoring low concentration ethanol. *Mater. Sci. Semicond. Proc.* **2021**, *136*, 106110. [CrossRef]
32. Li, G.; Cheng, Z.; Xiang, Q.; Yan, L.; Wang, X.; Xu, J. Bimetal PdAu decorated SnO_2 nanosheets based gas sensor with temperature-dependent dual selectivity for detecting formaldehyde and acetone. *Sens. Actuators B Chem.* **2019**, *283*, 590–601. [CrossRef]
33. Peng, S.; Hong, P.; Li, Y.; Xing, X.; Yang, Y.; Wang, Z.; Zou, T.; Wang, Y. Pt decorated SnO_2 nanoparticles for high response CO gas sensor under the low operating temperature. *J. Mater. Sci. Mater. Electron.* **2019**, *30*, 3921–3932. [CrossRef]
34. Liu, D.; Pan, J.; Tang, J.; Liu, W.; Bai, S.; Luo, R. Ag decorated SnO_2 nanoparticles to enhance formaldehyde sensing properties. *J. Phys. Chem. Solids* **2019**, *124*, 36–43. [CrossRef]
35. Ovsianytskyi, O.; Nam, Y.; Tsymbalenko, O.; Lan, P.; Moon, M.; Lee, K. Highly sensitive chemiresistive H_2S gas sensor based on graphene decorated with Ag nanoparticles and charged impurities. *Sens. Actuators B Chem.* **2018**, *257*, 278–285. [CrossRef]
36. Ma, X.; Gao, R.; Zhang, T.; Sun, X.; Li, T.; Gao, S.; Zhang, X.; Xu, Y.; Cheng, X.; Huo, L. Mesoporous SnO_2 nanospheres sensor for fast detection of HCHO and its application in safety detection of aquatic products. *Sens. Actuators B Chem.* **2023**, *374*, 132844. [CrossRef]
37. Luque, P.A.; Garrafa-Galvez, H.E.; Nava, O.; Olivas, A.; Martínez-Rosas, M.E.; Vilchis-Nestor, A.R.; Villegas-Fuentes, A.; Chinchillas-Chinchillas, M.J. Efficient sunlight and UV photocatalytic degradation of methyl orange, methylene blue and rhodamine B, using Citrus×paradisi synthesized SnO_2 semiconductor nanoparticles. *Ceram. Int.* **2021**, *47*, 23861–23874. [CrossRef]
38. Suriya, P.; Prabhu, M.; Kumar, E.; Jagannathan, K. Effect of Ag doping on structural, optical, complex impedance and photovoltaic properties of SnO_2 nanoparticles prepared by co-precipitation method for dye sensitized solar cell application. *Optik* **2022**, *260*, 168971. [CrossRef]
39. Bhangare, B.; Ramgir, N.S.; Jagtap, S.; Debnathb, A.K.; Muthe, K.P.; Terashima, C.; Aswal, D.K.; Gosavi, S.W.; Fujishima, A. XPS and Kelvin probe studies of SnO_2/RGO nanohybrids based NO_2 sensors. *Appl. Surf. Sci.* **2019**, *487*, 918–929. [CrossRef]
40. Shi, C.; Wang, X.; Wang, Y.; Wang, T.; Li, H.; Yi, G.; Sun, G.; Zhang, Z. Synthesis, characterization, and gas-sensing properties of macroporous Ag/SnO_2 composite by a template method. *Mater. Sci. Semicond. Process.* **2022**, *138*, 106256. [CrossRef]

41. Re, P.; Qi, L.; You, K.; Shi, Q. Hydrothermal synthesis of hierarchical SnO_2 nanostructures for improved formaldehyde gas sensing. *Nanomaterials* **2022**, *12*, 228.
42. Jiao, S.; Xue, W.; Zhang, C.; Li, F.; Meng, B.; Zhan, Z. Improving the formaldehyde gas sensing performance of the ZnO/SnO_2 nanoparticles by PdO decoration. *J. Mater. Sci.-Mater. Electron.* **2020**, *31*, 684–692.
43. Zhu, K.M.; Ma, S.Y. Preparations of Bi-doped SnO_2 hierarchical flower-shaped nanostructures with highly sensitive HCHO sensing properties. *Mater. Lett.* **2019**, *607*, 357–366. [CrossRef]
44. Kim, S.; Singh, G.; Oh, M.; Lee, K. An analysis of a highly sensitive and selective hydrogen gas sensor based on a 3D Cu-doped SnO_2 sensing material by efficient electronic sensor interface. *ACS Sens.* **2021**, *6*, 4145–4155. [CrossRef] [PubMed]
45. Manikandan, V.; Petrila, L.; Vigneselvan, S.; Mane, R.S.; Vasile, B.; Dharmavarapu, R.; Lundgaard, S.; Juodkazis, S.; Chandrasekaran, J. A reliable chemiresistive sensor of nickel-doped tin oxide (Ni-SnO_2) for sensing carbon dioxide gas and humidity. *RSC Adv.* **2020**, *10*, 3796. [CrossRef]
46. Chen, Y.; Cao, Y. Ultrasensitive and low detection limit of acetone gas sensor based on ZnO/SnO_2 thick films. *RSC Adv.* **2020**, *10*, 35958. [CrossRef]
47. Zhou, S.; Wang, H.; Hu, J.; Lv, T.; Rong, Q.; Zhang, Y.; Zi, B.; Chen, M.; Zhang, D.; Wei, J.; et al. Formaldehyde gas sensor with extremely high response employing cobalt-doped SnO_2 ultrafine nanoparticles. *Nanoscale Adv.* **2022**, *4*, 824. [CrossRef]
48. Xu, H.; Li, J.; Fu, Y.; Li, P.; Luo, W.; Tian, Y. Ag/Ag_2S Nanoparticle-induced sensitization of recovered sulfurdoped SnO_2 nanoparticles for SO_2 detection. *ACS Appl. Nano Mater.* **2020**, 3, 8075–8087. [CrossRef]
49. Li, Z.; Li, H.; Wu, Z.; Wang, M.; Luo, J.; Torun, H.; Hu, P.; Yang, C.; Grundmann, M.; Liu, X.; et al. Advances in designs and mechanisms of semiconducting metal oxide nanostructures for high-precision gas sensors operated at room temperature. *Mater. Horiz.* **2019**, *6*, 470–506. [CrossRef]
50. Tomer, V.K.; Duhan, S. Ordered mesoporous Ag-doped TiO_2/SnO_2 nanocomposite based highly sensitive and selective VOC sensors. *J. Mater. Chem. A* **2016**, *4*, 1033. [CrossRef]

chemosensors

Article

Study on Denoising Method of Photoionization Detector Based on Wavelet Packet Transform

Zengyuan Liu [1,2], Xiujuan Feng [1,2,*], Chengliang Dong [1] and Mingzhi Jiao [3]

1 School of Mines, China University of Mining and Technology, Xuzhou 221116, China
2 Rare Earth Research Institute, China University of Mining and Technology, Xuzhou 221116, China
3 National and Local Joint Engineering Laboratory of Internet Application Technology on Mine, China University of Mining and Technology, Xuzhou 221116, China
* Correspondence: xjfeng@cumt.edu.cn

Abstract: Aiming at the task of noise suppression caused by the photoionization detector (PID) monitoring signal of volatile organic compounds (VOCs) due to local non-uniformity of the photocathode surface of PID in the ionization chamber, this paper proposes an analytical method of a PID signal with the adaptive weight of the small wave package decomposition node. The PID signal is transmitted to the upper machine software through the single-chip microcontroller. The appropriate wavelet packet decomposition level is determined according to the time frequency characteristics of the original signal of the PID, and the optimal wavelet packet base is selected through the polynomial fitting of the signal quality evaluation index. By comparing the quality of signals processed by the traditional wavelet packet denoising method and the denoising method presented in this paper, the superiority of the proposed method in the denoising signals of PID was verified. This method can eliminate the noise generated by local non-uniformity on the photocathode surface of the PID ionization chamber in a high humidity environment, which lays a foundation for the accurate monitoring of VOCs in a high humidity environment.

Keywords: wavelet packet transform; evaluation of quality; optimal wavelet packet base; signal denoising; VOCs

Citation: Liu, Z.; Feng, X.; Dong, C.; Jiao, M. Study on Denoising Method of Photoionization Detector Based on Wavelet Packet Transform. *Chemosensors* **2023**, *11*, 146. https://doi.org/10.3390/chemosensors11020146

Academic Editors: Fanli Meng, Zhenyu Yuan and Dan Meng

Received: 25 December 2022
Revised: 12 February 2023
Accepted: 13 February 2023
Published: 16 February 2023

1. Introduction

Volatile organic compounds (VOCs) usually refer to organic compounds with a boiling point between 50 °C and 260 °C under normal pressure [1]. VOCs affect the environmental quality and are carcinogenic and mutagenic to the human body [2–4]. At present, the monitoring of VOCs mainly includes photoionization detection (PID), Fourier transform infrared spectrometry (FTIR), and gas chromatography-mass spectrometry (GC-MS), flame ionization detection (FID), metal oxide semiconductor sensors (MOS), differential optical absorption spectroscopy (DOAS), and other methods. The maintenance cost of FTIR optical equipment is high and it requires a large maintenance amount. The detection cycle of the GC-MS samples is relatively long. In addition, different detectors have different amounts of sensitivity, selectivity, accuracy, and equipment maintenance. When FID monitors VOCs, oxygen, moisture, and organics containing nitrogen, oxygen, or halogen atoms in the flue gas can interfere with the test; VOCs detected by DOAS are of limited variety. At present, it can mainly monitor benzene, toluene, and other benzene series [5]. Currently, portable VOC monitoring methods are widely used including PID and MOS. Fanli Meng et al. [6–9] used $Au/SnO_2/RGO$ nanocomposites to prepare MOS sensors, which can monitor 1–1000 ppm of ethanol. MOS sensors made of reduced GO/carbon monoxide nanocomposites can monitor 1 ppb of xylene. The response time of the sensor was shortened by $LaCoO_3$ modified ZnO, and the optimum temperature of four alcohol homologous gases monitored by a ZnO sensor was studied. However, MOS gas sensors have poor selectivity.

PID is one of the most widely used portable VOC gas detection technologies. The measurement accuracy of the PID is used through the proper denoising method, which has always been employed by scholars both domestic and abroad. Rezende [10] used the digital moving average filtering method to decrease the noise of the PID sensor data to obtain a lower limit of toluene monitoring. Agbroko et al. [11] used first-order low-pass filtering to reduce the noise of the PID sensor, and obtained a sensor signal with a high signal-to-noise ratio for the analysis of the total content of VOCs and the content of each component. Je C et al. reported that in a multi-channel VOC monitoring system, the average value filtering was used to lower the noise of the PID sensing data, and the optimal average value sampling number was found through experiments. Liu Chang [12] compared the advantages and disadvantages of VOC monitoring based on PID. The effects of the limited filtering method, median filtering method, arithmetic average filtering method, recursive average filtering method, median average filtering method, the median value average filtering method, and the median average filtering method have been discussed in detail. Finally, the median average filtering method is applied for the optimal sensor data, and the data of the VOC monitoring sensor with a high signal-to-noise ratio can be obtained. At present, the noise of the PID signal is mostly filtered by median filtering and the low-pass filtering method [13]. The two filtering means separate the useful signal and the noise signal in the time domain by Fourier transform in the frequency domain, but the separation can be completed only when the useful signal and the noise signal are not mixed together, and the noise of the signal is often distributed in the whole frequency domain, so when removing the noise, the useful signal is often removed. Considering that the noise of the signal is often distributed in the entire frequency domain, useful signals are often removed simultaneously while removing noise [14–17]. Therefore, at present, the method of the denoising of PID sensors used for VOC monitoring at home and abroad takes effect at the cost of sacrificing signals. The wavelet analysis method can analyze the time frequency of the signal, overcoming the shortcomings of traditional filtering method that works only for frequency domain analysis [18–20]. Based on the analysis of traditional wavelet packet transform, this article proposes a method of decomposing the energy-adaptable weight of the small wave bag decomposition node. This method overcomes the problem of the traditional wavelets when removing the noise of the signal. Therefore, the accuracy of the VOC quantitative analysis is improved. This article verifies the superiority of the method of removing noise in small traditional wavelets through the signal quality evaluation results.

2. VOC Monitoring Device

The composition of the VOC monitoring device is as follows (Figure 1): single-chip module, PID module, A/D conversion module, serial port transmission module, and upper machine module. The PID enables VOCs to generate a current under high UV lamps. The current in the PID will generate a voltage value through a fixed resistor. After the analog voltage is converted by an A/D converter, it is displayed on a single-chip micro tube in a digital voltage form. The single-chip control serial port to send the data to the upper machine software was developed by C#. The time domain waveform generated by the sensor can be displayed and stored in the computer. The frequency domain signal stored by the upper machine software can be imported into MATLAB to perform denoising work on the light ionized sensor signal.

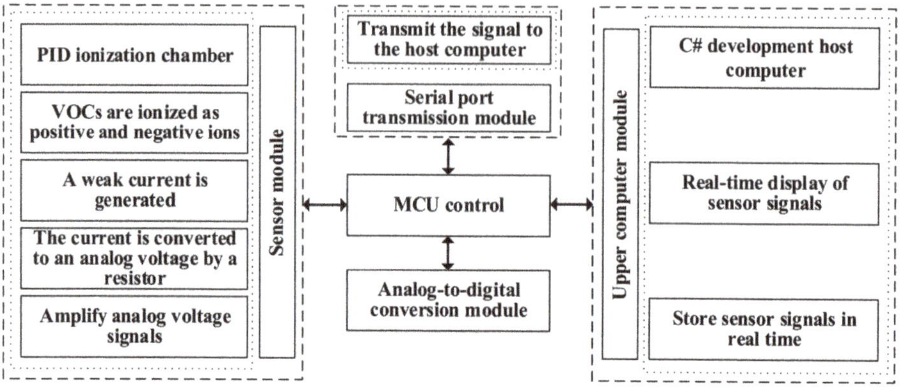

Figure 1. Composition of the VOC monitoring device.

2.1. PID Module

Photoionization detection (PID) is the simplest technology to detect VOCs [21–23]. Its working principle (Figure 2) [24–26] is that a UV lamp acts as the light source, and the detected gas is passed into the ionization chamber, which is irradiated by UV light and decomposed into detected signals—electrons and ions. Under the action of an electric field, electrons and ions move to the positive and negative electrodes, respectively, to form the current, and the current is output after amplification. After the resistance of the fixed resistance value, the voltage value is output to realize the quantitative test of its gas concentration. The principle of a photoionization detector is shown in Figure 2.

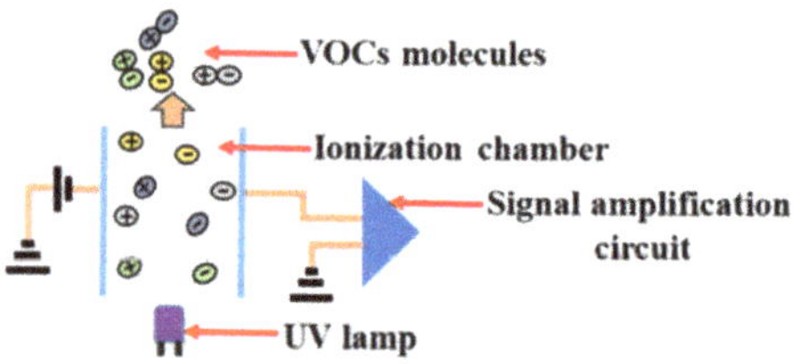

Figure 2. Working principle diagram of the PID of VOCs.

In this device, a Shenyang Magnesium Technology Light ion gas sensor is used for VOC monitoring. The sensor has an ultra-small UV lamp of 10.6 eV (i.e., manufacturer: Shenyang Magnesium Technology, type: krypton gas lamp, window: magnesium fluoride, power: 0.5 W) that can monitor VOCs such as benzine, ketones, and aldehydes with an ionization potential of less than 10.6 eV. When the VOC concentration is 0, the PID voltage value is 0.043 V, the response time of PID is less than 3 s, and a VOC gas of 200 ppb–2000 ppm can be detected [27]. It also has different response values for different VOCs. The working temperature range is −20–60 °C. The sensor has three pins, which are power supply voltage pins, ground pins, and signal transmission pins, respectively. The XPT2046 modulus conversion chip is controlled by the single chip microcomputer in the signal transmission pins.

2.2. Single-Machine Serial Port Transmission Module

In the VOC monitoring device, the data acquisition and control unit of the light ionized sensor uses a STC89C52RC single-chip microcomputer to complete the content of the unit [28,29]. The STC89C52RC single-chip microcomputer is an 8-bit micro-controller developed by Hongjing Company, Wuhu, China, with an operating frequency of 12 MHz, a built-in 512 B memory and an 8 kB storage space. This device uses Keil Uvision 5 to record the C language into the single-chip machine.

The STC89C52 single-chip microcomputer does not have an A/D conversion function, so the module conversion module uses the XPT2046 chip controlled by a single-chip microcomputer. The XPT2046 contains a 12-bit resolution 125 kHz conversion rate to approach the A/D converter one by one to meet the real-time monitoring of the PID sensor to monitor VOCs. The PID signal output pins are connected to the AIN3 pin of the single-chip microcomputer with the DuPont line, and the PID and XPT2046 chip are together.

The serial port is a very widely used communication interface as it is cheap, easy to use, and the communication lines are simple, which can realize the interoperability of the single-chip microcomputer and the upper machine. The STC89C52RC single-chip microcontroller comes with UART (Universal Asynchronous Receiver Transmitter). The sensor data are passed to the upper machine by the serial port.

2.3. Design of the Upper Machine Module

The development of the upper-bit machine in the serial communication system used by a single-chip microcomputer for volatile organic matter monitoring was implemented in the development environment of Visual Studio 2015 [30]. The main function of the upper machine is to receive the signal value of the PID sensor through the microcontroller serial port and draw the waveform of the signal to generate the signal. First, we added serial ports, then added the GROUPBOX control to the window design, next, we added ComboBox1 to the selection of the port slogan in the GROUPBOX control range, added Combobox2 for the choice of baud rate, and then we finally added two radio button controls to select the receiving serial port transmission data. The pattern, the number, and character mode were chosen, three button controls were added to control the open port, closed port, and waveform display, then textbox control was added to display the signal value of the sensor.

2.4. VOCs Online Detection Device and Method

Since the response value of PID is easily affected by environmental factors such as temperature and humidity, the experimental device (Figure 3) was built in a fume hood room with stable environmental conditions. The indoor temperature of the ventilation cabinet was 25 degrees Celsius and the air pressure was 1.01×10^5 Pa. The standard gas used in this article was derived from Jining Xieli Specialty Co. Ltd., Jining, China. A standard gas bottle capacity is 8 L, the concentration of the standard gas is 2000 ppm benzene gas, and the gas is filled with nitrogen. A quality flowmeter was used to control the amount of the standard gas to calculate the content of the standard gas. The unit of the quality flowmeter is in L/min. Through Equation (1), the concentration of the standard gas can be calculated and its unit can be converted to mg/L to facilitate the calculation of the gas content in the gas cylinder. Among them, C is the mass concentration (mg/m^3) of benzene gas, X is the volume concentration 2000 ppm, and M is the molar mass of benzene.

$$X = C \times 24.5/M \quad (1)$$

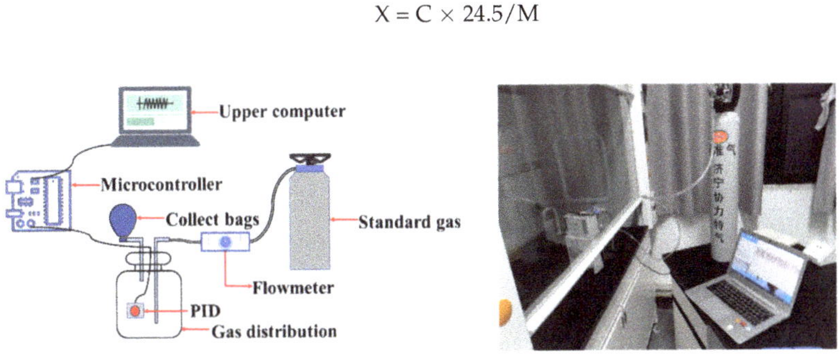

Figure 3. Experimental device of the PID module denoising for the online detection of VOCs.

The gas from the standard gas cylinder is introduced into the collecting cylinder with a capacity of 3 L. A certain concentration of benzene gas can be prepared in the collecting cylinder by controlling the flow rate and time of gas in the standard cylinder. The VOC monitoring module is shown in Figure 3.

The standard gas flow rate was 0.05 L/min, and the time was 10 s. The phenyl gas concentration in the gas cylinder was 5.57 ppm. The noise-containing PID response signal under this concentration is shown in Figure 4.

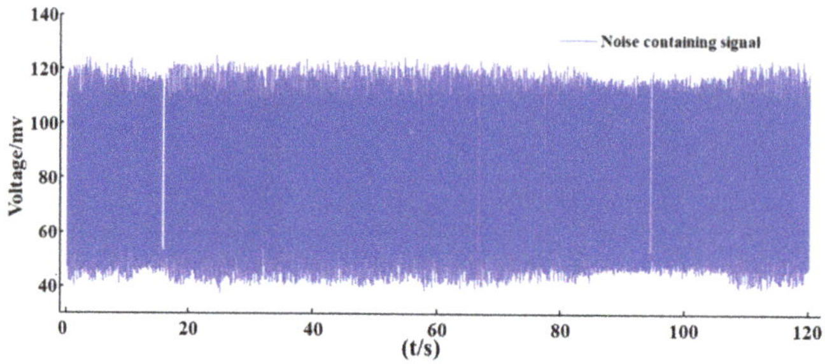

Figure 4. PID original signal.

3. PID Sensor Signal Denoising Method Based on Wavelet Analysis

3.1. Selection of a Wavelet Decomposition Level

Given the orthogonal scale function $\varphi(x)$ and its corresponding small wave function $\psi(x)$, there is [31–34]:

$$\begin{cases} \varphi(x) = \sqrt{2}\sum\limits_{k\in Z} q_k\varphi(2x-k) \\ \psi(x) = \sqrt{2}\sum\limits_{k\in Z} g_k\varphi(2x-k) \end{cases} \tag{2}$$

where q_k and g_k are the conjugation filters; q_k is the low-pass filter coefficient; g_k is the high-pass filter coefficient.

Set $\mu_0 = \varphi(x)$, $\mu_1 - \psi(x)$, at this time:

$$\begin{cases} \mu_{2n}(x) = \sqrt{2}\sum\limits_{k\in Z} q_k\mu_n(2x-k) \\ \mu_{2n+1}(x) = \sqrt{2}\sum\limits_{k\in Z} g_k\mu_n(2x-k) \end{cases} \tag{3}$$

Among them, n is the n-wave sequence when the decomposition scale is j.

Define V_j^n as a closure space for the function $\mu_n(x)$, V_j^{2n} is the closure space of $\mu_{2n}(x)$, V_j^{2n+1} is the closure space of $\mu_{2n+1}(x)$, then V_j^n can be decomposed as:

$$V_j^n = V_j^{2n} + V_j^{2n+1} \tag{4}$$

where V_j^{2n} and V_j^{2n+1} meet the conditions of $V_j^{2n} \perp V_j^{2n+1}$.

If the small wave packet coefficients of the signal on V_j^{2n} and V_j^{2n+1} are p_j^{2n} and p_j^{2n+1}, respectively, then the small wave bag decomposition can be obtained by $\{p_k^{j+1,2n}\}$ and $\{p_k^{j+1,\,2n+1}\}$, which is expressed as [35]:

$$\begin{cases} p_k^{j+1,2n} = \sum\limits_l h_{2l-k}p_l^{j,n} \\ p_k^{j+1,2n+1} = \sum\limits_l g_{2l-k}p_l^{j,n} \end{cases} \tag{5}$$

Among them, $p_k^{j,n}$ indicates that the decomposition coefficient of the k wavebase on the n fork tree with a decomposition scale is j and $p_k^{j+1,2n}$ and $p_k^{j+1,2n+1}$ are $p_k^{j,n}$ of two bit [36].

When the PID signal is ready, the decomposition level of the wavelet packet is first determined, where the decomposition level is too low. The noise and useful information of the signal cannot be effectively separated, and the signal noise cannot be effectively removed. In contrast, if the selected decomposition level is too high, a large amount of useful information of the signal will also be removed to cause the new distortion while removing the noise. Therefore, choosing the appropriate wavelet breakdown level is an important step for sensors to generate the signal noise. This article conducted a 4-layer small wave package decomposition experiment on the signal generated by the light ionization sensor. The number of decomposition layers is shown in Figure 5.

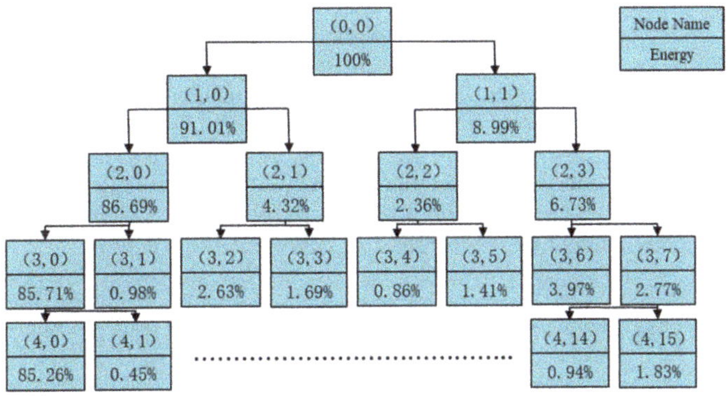

Figure 5. Wavelet packet decomposition node and energy diagram.

According to the Parseval theorem, the total energy of a signal in its time domain is equal to the total energy in the frequency domain. For small waves, just separate the signals and low-frequency components and changes in the form of signals, but the total energy before and after decomposition always remains equal. If a signal X (T) is decomposed to the lower layer, the energy of each sub-band signal can be calculated according to Equation (6) [37–39].

$$E_{ij} = \int |S_{i,j}(t)|^2 dt = \sum_{k=1}^{n} |x_{i,j}(k)|^2 \tag{6}$$

The distribution of the corresponding energy distribution corresponding to the decomposition trees and decomposition of each node is shown in Figure 5.

The signal-to-noise ratio (SNR) and mean-square error (MSE) of the signal of different wavelets were calculated to decompose the levels to determine the best wavelet breakdown level. SNR represents the ratio of the effective signal energy to noise signal energy. The larger the value of the SNR, the lower the noise effect. The calculation method of the SNR is as shown in Equation (7). MSE represents the degree of similarity between the original signal and the signal after noise. The MSE calculation formula is shown in Equation (8). In this paper, SNR and MSE were used to evaluate the PID signal quality.

$$\mathrm{SNR} = 10\lg(\mathrm{s}/\mathrm{n}) \tag{7}$$

$$\mathrm{MSE} = \frac{1}{n}\sum_{i=1}^{n}(Y_i - \hat{Y}_i)^2 \tag{8}$$

where s and n in SNR represents the effective power of signals and noise, respectively; Y_i in MSE is the original signal; $\hat{Y}_i$ is the signal after noise.

The values of the MSE and SNR of the PID signal after two, three, and four layers of the signal are shown in Table 1.

Table 1. The MSE and SNR of different decomposition levels under the signal.

Evaluation Indicator	2-Layer Decomposition	3-Layer Decomposition	4-Layer Decomposition
SNR	33.0418	36.6291	6.0453
MSE	3.1490	1.3867	6.3679

As shown in Table 1, when the wavelet packet decomposition level was three layers, the SNR value of PID signal was the largest, the MSE was the smallest, and the PID signal processing effect was the best. Therefore, this paper carried out the 3-layer wavelet packet decomposition of the PID signal.

The PID signal collected by the upper machine was decomposed by a three-layer wavelet, the signal was divided into eight frequency bands from low frequency to high frequency, as shown in Figure 6.

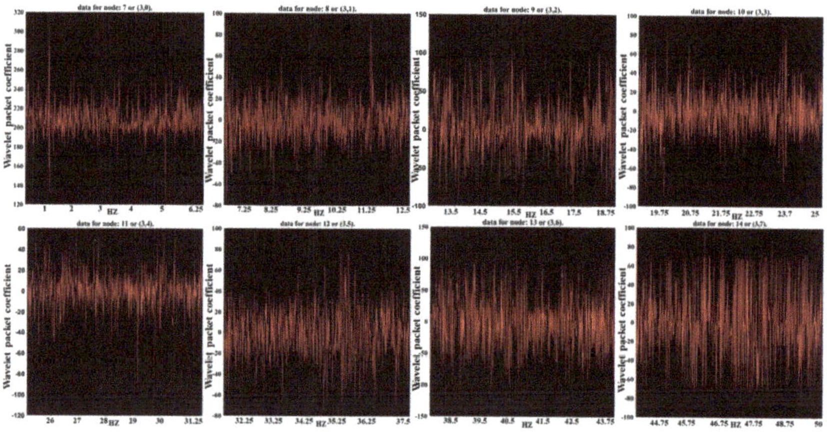

Figure 6. Biorl.3 wavelet packet 3−layer decomposition of 5.57 ppm benzene gas PID with the noise signal.

3.2. Selection of the Wavelet Packet Base

When the wavelet packet changes the PID signal noise, whether the selection of the wavelet base is appropriate will have an important impact on the signal quality after the PID signal processing. After determining the decomposition of the small wave bag as three layers of decomposition, this work used different wavelet packets to perform the experimental analysis of the PID signal denoising, and determined the optimal wavelet packet base based on the results of the signal quality evaluation. The Daubechies (db) wavelet packet base and Symlets (sym) wavelet packet base have an N-order. The Daubechies wavelet packet has good regularity, that is, the smooth error introduced by the wavelet packet as a sparse basis is not easy to detect, which makes the signal reconstruction process smooth. The characteristics of the db wavelet packet are that with the increase in order (sequence N), the larger the order of vanishing moment, the higher the vanishing moment, the better the smoothness, the stronger the localization ability of the frequency domain, the better the division effect of the frequency band, but it will weaken the time domain tight support, and the calculation amount is greatly increased, and the real-time becomes worse. sym is an improvement in the db function. The sym wavelet has better symmetry, which can reduce the phase distortion of signal analysis and reconstruction to a certain extent. The wavelet packet base SNR and MSE of different levels of db and sym were calculated.

By calculating the quality of the 12-order wavelet packet processing signal, the db and sym wavelet packet base were treated with PID. The best order of the PID signal and the optimal wavelet were obtained. The values of the MSE and SNR of the wavelet packet base between db and sym are shown in Table 2.

Table 2. The MSE and SNR of the db and sym wavelet packet base at different levels.

Wavelet Packet Base	MSE	SNR	Wavelet Packet Base	MSE	SNR
db2	1.3541	36.7320	sym2	1.3541	36.7320
db3	1.3213	36.8389	sym3	1.3213	36.8389
db4	1.3892	36.6203	sym4	1.3995	36.5880
db5	1.3719	36.6747	sym5	1.4136	36.5441
db6	1.3722	36.6734	sym6	1.3415	36.7721
db7	1.4312	36.4900	sym7	1.3853	36.6322
db8	1.4004	36.5848	sym8	1.4062	36.5668
db9	1.3803	36.6479	sym9	1.4184	36.5291
db10	1.3886	36.6219	sym10	1.4133	36.5446
db11	1.4221	36.5177	sym11	1.3713	36.6761
db12	1.3832	36.6383	sym12	1.4075	36.5627
db13	1.4323	36.4864	sym13	1.4272	36.5022

Fourier curve fitting was performed on the values of MSE corresponding to the wavelet packet bases of different orders of db, and the Fourier fitting formula is shown as follows.

$$f(a) = a_0 + a_1\cos(aw) + b_1\sin(aw) + a_2\cos(2aw) + b_2\sin(2aw) + a_3\cos(3aw) + b_3\sin(3aw) + a_4\cos(4aw) + b_4\sin(4aw) + a_5\cos(5aw) + b_5\sin(5aw) \quad (9)$$

The formula's corresponding parameters are shown in Table 3.

Table 3. The db series wavelet packet base corresponded to the parameter value of the MSE fitting function.

$a_0 = 1.4$	$a_1 = 0.0148$	$b_1 = -0.02251$	$a_2 = 0.02968$
$b_2 = -0.02415$	$a_3 = 0.02019$	$b_3 = -0.01056$	$a_4 = 0.03571$
$b_4 = -0.01956$	$a_5 = 0.01173$	$b_5 = -0.0007733$	$w = 0.471$

The corresponding fitting curve is shown in Figure 7.

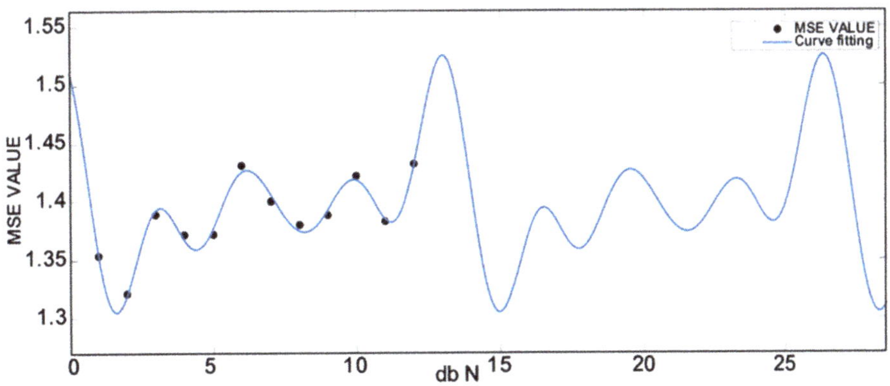

Figure 7. The MSE fitting curve of the db wavelet packet base.

This article evaluated the accuracy of the fitting curve by calculating its decision coefficient. The calculation method is shown in Equation (10).

$$R^2 = 1 - \frac{SS_{\text{res}}}{SS_{\text{tot}}} = 1 - \frac{\Sigma(y_i - \hat{y}_i)^2}{\Sigma(y_i - \overline{y})^2} \tag{10}$$

The R^2 of this polynomial fitting was 0.9826 degrees to meet the minimal value requirements of the MSE of the db wavelet packet base. The minimal value of the MSE of the db wavelet packet base $f(a)min$ was 1.3213

Fourier curve fitting was performed on the SNR values corresponding to the wavelet packet bases of different orders of db, and the Fourier fitting formula was obtained as follows.

$$\begin{aligned} f(b) = {} & a_0 + a_1\cos(bw) + b_1\sin(bw) + a_2\cos(2bw) + b_2\sin(2bw) + a_3\cos(3bw) + \\ & b_3\sin(3bw) + a_4\cos(4bw) + b_4\sin(4bw) + a_5\cos(5bw) + b_5\sin(5bw) \end{aligned} \tag{11}$$

The formula's corresponding parameters are shown in Table 4.

Table 4. The db series wavelet packet base corresponded to the parameter values of the SNR fitting function.

$a_0 = 36.59$	$a_1 = 0.04017$	$b_1 = 0.06998$	$a_2 = -0.08918$
$b_2 = 0.0732$	$a_3 = -0.06034$	$b_3 = 0.03041$	$a_4 = -0.1132$
$b_4 = 0.05443$	$a_5 = -0.03743$	$b_5 = -0.0002794$	$w = 0.4731$

The corresponding fitting curve is shown in Figure 8.

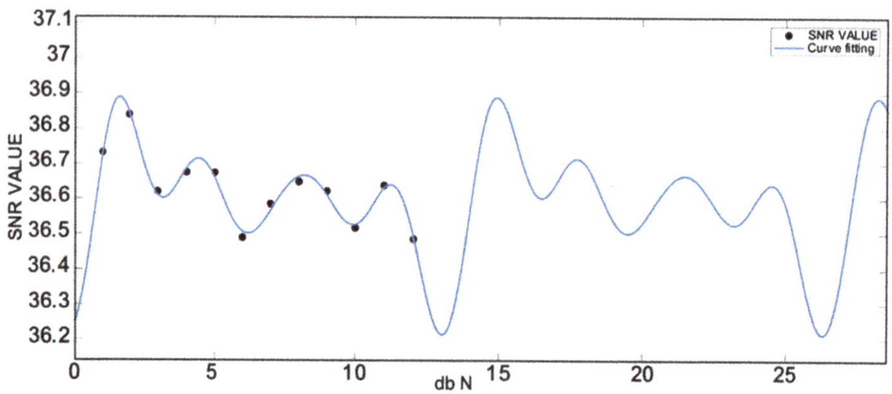

Figure 8. The SNR fitting curve of the db wavelet packet base.

The R^2 of this polynomial fit was 0.9844, and the fitting accuracy met the requirements of the maximum value of the SNR of the db wavelet packet base. The db wavelet packet base SNR value $f(b)max$ was 36.8389.

Fourier curve fitting was performed on the MSE values corresponding to the wavelet packet bases of different orders of sym, and the Fourier fitting formula was obtained as follows.

$$\begin{aligned} f(x) = {} & a_0 + a_1\cos(xw) + b_1\sin(xw) + a_2\cos(2xw) + b_2\sin(2xw) + a_3\cos(3xw) + \\ & b_3\sin(3xw) + a_4\cos(4xw) + b_4\sin(4xw) + a_5\cos(5xw) + b_5\sin(5xw) \end{aligned} \tag{12}$$

The corresponding parameters of the fitting formula are shown in Table 5.

Table 5. The sym series wavelet packet base corresponded to the parameter value of the MSE fitting function.

$a_0 = 1.389$	$a_1 = -0.003555$	$b_1 = -0.02018$	$a_2 = 0.0003253$
$b_2 = -0.005518$	$a_3 = 0.02447$	$b_3 = -0.02445$	$a_4 = 0.01532$
$b_4 = 0.003341$	$a_5 = -0.01182$	$b_5 = 0.0009305$	$w = 0.5068$

The corresponding fitting curve is shown in Figure 9.

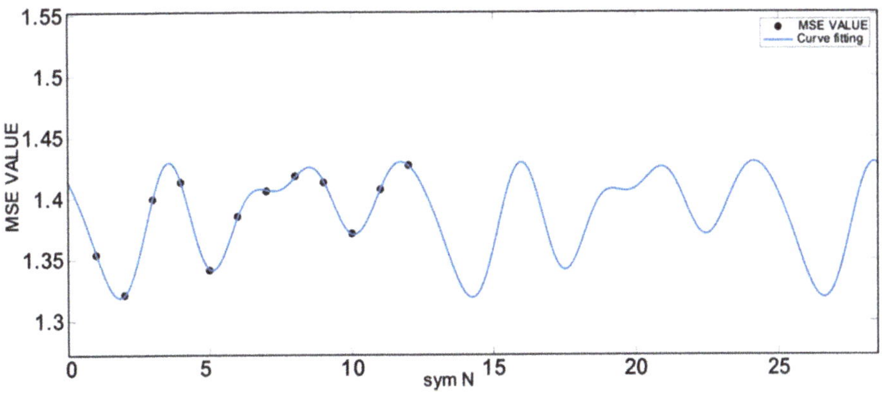

Figure 9. The sym wavelet packet base MSE fitting curve.

The R^2 of the polynomial fit was 0.9996, and the fitting accuracy met the extremely small value of the MSE of the sym wavelet packet base. The minor value of the MSE of the sym wavelet packet base $f(c)min$ was 1.3213.

Fourier curve fitting was performed on the SNR values corresponding to the wavelet packet bases of different orders of sym, and the Fourier fitting formula was obtained as follows.

The corresponding parameters of the fitting formula are shown in Table 6.

$$f(x) = a_0 + a_1\cos(xw) + b_1\sin(xw) + a_2\cos(2xw) + b_2\sin(2xw) + a_3\cos(3xw) + b_3\sin(3xw) + a_4\cos(4xw) + b_4\sin(4xw) + a_5\cos(5xw) + b_5\sin(5xw) \quad (13)$$

Table 6. The sym series wavelet packet base corresponded to the parameter values of the SNR fit function.

$a_0 = 36.62$	$a_1 = 0.01159$	$b_1 = 0.06443$	$a_2 = -0.0009322$
$b_2 = 0.0179$	$a_3 = -0.07799$	$b_3 = 0.07653$	$a_4 = -0.04938$
$b_4 = -0.01174$	$a_5 = 0.03721$	$b_5 = -0.002947$	$w = 0.5074$

The corresponding fitting curve is shown in Figure 10.

The R^2 of the polynomial fit was 0.9996, and the fitting accuracy met the requirements of the great value of the SNR of the sym wavelet packet base. The maximum value of the SNR of the sym wavelets $f(x)max$ was 36.8389.

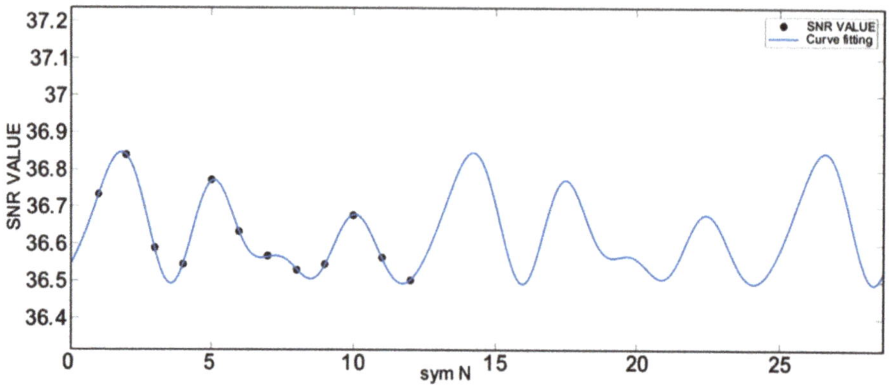

Figure 10. The SNR fit curve of the sym wavelet packet base.

The values of the MSE and SNR of Haar (haar), Dmeyer (dmey), BiorSplines (bior), ReverseBior (rbio), Coiflets (coif), and the Fejér–Korovkin (fk) finite order wavelet packet base are shown in Table 7.

Table 7. The MSE and SNR of the haar, dmey, bior, rbio, coif, fk wavelet packet base.

Wavelet Packet Base	MSE	SNR	Wavelet Packet Base	MSE	SNR
haar	1.3867	36.6291	rbio 1.3	1.4672	36.3840
dmey	1.4553	36.4168	rbio 1.5	1.6119	35.9748
bior 1.3	1.3166	36.8543	rbio 2.2	1.4232	36.5151
bior 1.5	1.4407	36.4621	rbio 2.4	1.3220	36.8363
bior 2.2	1.6239	35.9424	rbio 2.6	1.3933	36.6075
bior 2.4	1.3496	36.7465	rbio 2.8	1.3866	36.6285
bior 2.6	1.3824	36.6414	rbio 3.1	2.2907	34.4489
bior 2.8	1.3646	36.6978	rbio 3.3	1.4753	36.3598
bior 3.1	6.4503	29.9554	rbio 3.5	1.4547	36.4198
bior 3.3	1.9288	35.1958	rbio 3.7	1.3718	36.6752
bior 3.5	1.5646	36.1035	rbio 3.9	1.3718	36.6750
bior 3.7	1.4093	36.5579	fk 4	1.3307	36.8085
coif 1	1.4168	36.5348	fk 6	1.3724	36.6733
coif 2	1.3994	36.5879	fk 8	1.3713	36.6767
coif 3	1.4208	36.5218	fk 14	1.3773	36.6571
coif 4	1.4008	36.5836	fk 18	1.4258	36.5062
coif 5	1.4443	36.4499	fk 22	1.4485	36.4374

From Tables 2 and 7, fit Equations (9), (11)–(13), the signal can be decomposed by three layers, the maximum SNR removal noise was 36.8543, and the minimum MSE was 1.3166, corresponding to the small wavebase bior 1.3 wavelet packet base, so this article selected the bior 1.3 wavelet packet base to remove the signal of the PID.

3.3. *Self-Adaptive Weight Threshold Denoise Method Based on a Wavelet Decomposition Node Energy*

The wavelet decomposition was divided by the frequency band of the signal, and the sub-generation component of the low-frequency signal was obtained, so the characteristic information of the original signal also existed in the signal of each child. The characteristics of the signal can be analyzed through the energy of the wavelet decomposition node. After the signal is decomposed by a wavelet, the energy distribution characteristics in each frequency band can be used as an important basis for the signal distribution features.

The determination of the threshold and threshold function is the key to the problem of threshold. Traditional threshold-free noise is divided into two types: soft threshold function and hard threshold function. The wavelets are changed to adjust the wavelet coefficient sequence θ + γ. Among them, θ is the small wave packet coefficient of the signal, and γ is the wavelet coefficient of the wavelet.

Soft threshold: η is a threshold to destroy the son, that is, the threshold function; λ is the threshold.

$$\eta = \begin{cases} \operatorname{sgn}(\theta+\gamma)\,(|\theta+\gamma|-\lambda), & |\theta+\gamma| \geq \lambda \\ 0, & |\theta+\gamma| < \lambda \end{cases} \tag{14}$$

Hard threshold:

$$\eta = \begin{cases} \theta+\gamma, & |\theta+\gamma| \geq \lambda \\ 0, & |\theta+\gamma| < \lambda \end{cases} \tag{15}$$

Since the low-frequency part of the signal's small waves contain most of the energy of the signal, as shown in the energy chart of the small wave decomposition node in Figure 5, we kept all of the low-frequency part of the signal. For the weight of the non-low-frequency part according to the weight of its signal energy, it will give corresponding weights to the corresponding weight, so the value retains its energy to remove the PID signal noise. The weight of each node's energy is as shown in Equation (14). Among them, $E_{i,j}$ can be calculated by Equation (2), S is the weight of the energy of the wavelet decomposition node.

$$S = \frac{E_{i,j}}{Etotal} \tag{16}$$

Based on the threshold of the tide decomposition node energy self-adaptive weight, the noise removal method is:

$$\eta = \begin{cases} \operatorname{sgn}(S\theta+\gamma)\,(|S\theta+\gamma|-\lambda), & |\theta+\gamma| \geq \lambda \\ 0, & |\theta+\gamma| < \lambda \end{cases} \tag{17}$$

According to the weight value of the wavelet coefficient, the preservation ratio of the energy of the non-low-frequency node is determined, so wavelet coefficient reconstruction was performed to remove the noise of the signal. After the signal noise removal, the SNR of the signal was 36.8543, and the MSE was 1.3166; the traditional wavelet package noise SNR was 33.8054, and the MSE was 2.6828, added indicators than the use of traditional wavelets, as shown in Figure 11.

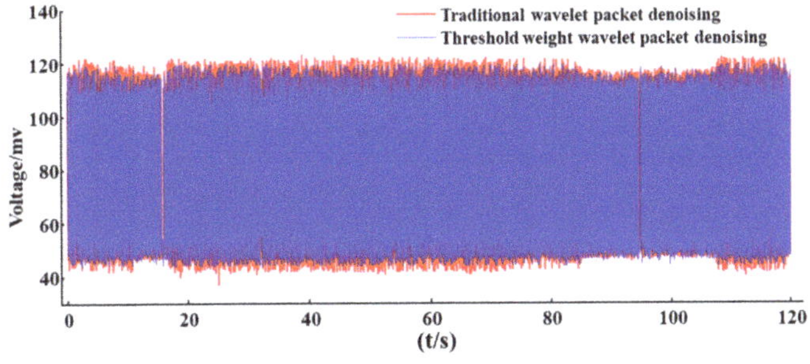

Figure 11. Comparison of the denoising effect between the traditional wavelet packet and threshold weight wavelet packet.

4. Conclusions

This study controlled the sensor through a single-chip microcomputer to achieve the sensor signal collected by the single-chip microcomputer and passed the sensor signal to the computer through the serial port. The optimal scale of wavelet decomposition was determined by the signal quality evaluation method, and the extreme values of MSE and SNR in the db and sym wavelet packet basis function with N-order wavelet packet base function were determined by the Fourier curve fitting method. The SNR and MSE values of the signal processed by the finite order wavelet base were calculated. The bior 1.3 wavelet packet base was determined to be the optimal wavelet packet base for PID signal denoising. Aiming at the problem that the high frequency signal is not processed after wavelet packet decomposition, which leads to the loss of more details after signal denoising, the PID signal denoising method based on the adaptive weight of the node energy of the wavelet packet decomposition was proposed, and the signal quality after the traditional wavelet packet denoising and the signal processing method in this paper were compared. It was verified that the proposed method has the advantage of improving the high quality of the PID signal.

Author Contributions: Conceptualization, X.F.; methodology, X.F. and Z.L.; software, Z.L. and C.D.; validation, Z.L., C.D. and M.J.; formal analysis, Z.L.; investigation, Z.L., X.F., C.D. and M.J.; resources, X.F.; data curation, Z.L. and C.D.; writing—original draft preparation, Z.L.; writing—review and editing, X.F.; visualization, Z.L. and C.D.; supervision, Z.L., X.F., C.D. and M.J.; project administration, X.F.; funding acquisition, X.F. All authors have read and agreed to the published version of the manuscript.

Funding: This research was funded by [Shandong Province Key R&D Program] grant number [2021CXGC011206] And [The Graduate Innovation Program of China University of Mining and Technology] grant number [2022WLJCRCZL008] And [The Postraduate Reserach&Practice Innovation Program of Jiangsu Province] grant number [SJCX22_1144].

Institutional Review Board Statement: Exclude this statement.

Informed Consent Statement: Exclude this statement.

Data Availability Statement: Exclude this statement.

Conflicts of Interest: The authors declare no conflict of interest.

References

1. Kesselmeier, J.; Staudt, M. Biogenic volatile organic compounds (VOC): An overview on emission, physiology and ecology. *J. Atmos. Chem.* **1999**, *33*, 23–88. [CrossRef]
2. Kamal, M.S.; Razzak, S.A.; Hossain, M.M. Catalytic oxidation of volatile organic compounds (VOCs)—A review. *Atmos. Environ.* **2016**, *140*, 117–134. [CrossRef]
3. Zhang, Q.; Xia, B.; Yang, Z.; Chen, J. Research on the definition and characterization of volatile organic compounds. *Pollut. Control Technol.* **2014**, *27*, 3–7.
4. Zhang, X.; Xue, Z.; Sun, X.; Chai, H. Control status and countermeasures of atmospheric volatile organic compounds in China. *Environ. Sci. Manag.* **2014**, *39*, 16–19.
5. Wang, Q.; Zhou, G.; Zhong, Q.; Zhao, J.; Yang, K. Status and needs research for on-line monitoring of VOCs emissions from stationary sources. *Huan Jing Ke Xue = Huanjing Kexue* **2013**, *34*, 4764–4770.
6. Meng, F.; Zheng, H.; Chang, Y.; Zhao, Y.; Li, M.; Wang, C.; Sun, Y.; Liu, J. One-step synthesis of Au/SnO_2/RGO nanocomposites and their VOC sensing properties. *IEEE Trans. Nanotechnol.* **2018**, *172*, 212–219. [CrossRef]
7. Meng, F.; Li, X.; Yuan, Z.; Lei, Y.; Qi, T.; Li, J. Ppb-Level Xylene Gas Sensors based on Co3O4 Nanoparticles coated Reduced Graphene Oxide(rGO) Nanosheets Operating at Low Temperature. *IEEE Trans. Instrum. Meas.* **2021**, *70*, 9511510. [CrossRef]
8. Qin, W.; Yuan, Z.; Gao, H.; Zhang, R.; Meng, F. Perovskite-structured $LaCoO_3$ modified ZnO gas sensor and investigation on its gas sensing mechanism by first principle. *Sens. Actuators B Chem.* **2021**, *341*, 130015. [CrossRef]
9. Meng, F.; Shi, X.; Yuan, Z.; Ji, H.; Qin, W.; Shen, Y.; Xing, C. Detection of Four Alcohol Homologue Gases by ZnO Gas Sensor in Dynamic Interval Temperature Modulation Mode. *Sens. Actuators B Chem.* **2022**, *350*, 130867. [CrossRef]
10. Rezende, G.C. Microfabricated Modular Photoionization Detector for Volatile Organic Compounds. Doctoral Dissertation, University of Limerick, Limerick, Ireland, 2019.
11. Agbroko, S.O.; Covington, J. A novel, low-cost, portable PID sensor for the detection of volatile organic compounds. *Sens. Actuators B Chem.* **2018**, *275*, 10–15. [CrossRef]
12. Liu, C. Development of On-Line VOC Detector. Master's Thesis, Harbin Institute of Technology, Harbin, China, 2016.
13. Bílek, J.; Maršolek, P.; Bílek, O.; Buček, P. Field Test of Mini Photoionization Detector-Based Sensors—Monitoring of Volatile Organic Pollutants in Ambient Air. *Environments* **2022**, *9*, 49. [CrossRef]
14. Je, C.; Stone, R.; Oberg, S.G. Development and application of a multi-channel monitoring system for near real-time VOC measurement in a hazardous waste management facility. *Sci. Total Environ.* **2007**, *382*, 364–374. [CrossRef] [PubMed]
15. Mahmud, M.M.; Seok, C.; Wu, X.; Şennik, E.; Biliroğlu, A.Ö.; Adelegan, O.J.; Kim, I.; Jur, J.S.; Yamaner, F.Y.; Oralkan, Ö. A Low-Power Wearable E-Nose System Based on a Capacitive Micromachined Ultrasonic Transducer (CMUT) Array for Indoor VOC Monitoring. *IEEE Sens. J.* **2021**, *21*, 19684–19696. [CrossRef]

16. Hagens, L.A.; Verschueren, A.R.; Lammers, A.; Heijnen, N.F.; Smit, M.R.; Nijsen, T.M.; Geven, I.; Schultz, M.J.; Bergmans, D.C.; Schnabel, R.M.; et al. Development and validation of a point-of-care breath test for octane detection. *Analyst* **2021**, *146*, 4605–4614. [CrossRef] [PubMed]
17. Praveenchandar, J.; Vetrithangam, D.; Kaliappan, S.; Karthick, M.; Pegada, N.K.; Patil, P.P.; Rao, S.G.; Umar, S. IoT-Based Harmful Toxic Gases Monitoring and Fault Detection on the Sensor Dataset Using Deep Learning Techniques. *Sci. Program.* **2022**, *2022*, 7516328. [CrossRef]
18. Alyasseri, Z.A.A.; Khader, A.T.; Al-Betar, M.A.; Abasi, A.K.; Makhadmeh, S.N. EEG signals denoising using optimal wavelet transform hybridized with efficient metaheuristic methods. *IEEE Access* **2019**, *8*, 10584–10605. [CrossRef]
19. Zhang, Y.; Wang, R.; Li, S.; Qi, S. Temperature sensor denoising algorithm based on curve fitting and compound kalman filtering. *Sensors* **2020**, *20*, 1959. [CrossRef]
20. Hall, L.T.; Maple, J.L.; Agzarian, J.; Abbott, D. Sensor system for heart sound biomonitor. *Microelectron. J.* **2000**, *31*, 583–592. [CrossRef]
21. Rezende, G.C.; Le Calvé, S.; Brandner, J.J.; Newport, D. Micro photoionization detectors. *Sens. Actuators B Chem.* **2019**, *287*, 86–94. [CrossRef]
22. Langhorst, M.L. Photoionization detector sensitivity of organic compounds. *J. Chromatogr. Sci.* **1981**, *19*, 98–103. [CrossRef]
23. Ševčik, J.; Krýsl, S. A photoionization detector. *Chromatographia* **1973**, *6*, 375–380. [CrossRef]
24. Li, H. VOCs Detection Based on Photoionization Technology. Master's Thesis, Chongqing University of Posts and Telecommunications, Chongqing, China, 2019.
25. Zhou, Q.; Zhang, S.; Zhang, X.; Zhou, W.; Li, J.; Zhao, Y. Optimization of ionization chamber structure and experimental study of photoionization detector. *Chin. J. Transducer Technol.* **2018**, *31*, 323–328.
26. Liu, R.; Hu, H. Design of Photoionization sensor for VOC Gas Detection. *Instrum. Tech. Sens.* **2020**, 1–5.
27. Wang, J. Research on the Technology of VOC Gas Concentration Detection System Based on PID Principle. Master's Thesis, North University of China, Taiyuan, China, 2021.
28. Luo, G.; Liang, X.; Li, H.; Lu, C.; Pan, L.; Zhang, S. Intelligent Kitchen Based on STC89C52RC Micro Control Unit. In *Cyber Security Intelligence and Analytics, Proceedings of the 2020 International Conference on Cyber Security Intelligence and Analytics (CSIA 2020), Haikou, China, 28–29 February 2020*; Springer: Cham, Switzerland, 2020; pp. 348–353.
29. Cong, Y.; Ding, L. Design of Intelligent Express Robot Control System Based on 51 Single Chip Microcomputer. *J. Sens. Technol. Appl.* **2019**, *7*, 150–157.
30. Putri, S.F.; Rahayu, I.F.D. Perancangan Dan Implementasi Sistem Informasi Penjualan Dan Pembelian Pada Koperasi Syari'ah Muslim Mandiri Sejahtera (Kammis) Menggunakan Visual Studio 2015 Dan MySQL. *J. TEDC* **2020**, *14*, 183–194.
31. Cody, M.A. The wavelet packet transform: Extending the wavelet transform. *Dr. Dobb's J.* **1994**, *19*, 44–46.
32. Himeur, Y.; Alsalemi, A.; Bensaali, F.; Amira, A. Robust event-based non-intrusive appliance recognition using multi-scale wavelet packet tree and ensemble bagging tree. *Appl. Energy* **2020**, *267*, 114877. [CrossRef]
33. Altabey, W.A. The Damage Identification in Laminate Composite Plate under Fatigue Load Using Wavelet Packet Energy Curvature Difference Transform. *Compos. Part C Open Access* **2022**, *9*, 100304. [CrossRef]
34. Wang, K.; Su, G.; Liu, L.; Wang, S. Wavelet packet analysis for speaker-independent emotion recognition. *Neurocomputing* **2020**, *398*, 257–264. [CrossRef]
35. Hindi, A.; Dwairi, M.O.; Alqadi, Z. Analysis of Digital Signals using Wavelet Packet Tree. *Int. J. Comput. Sci. Mob. Comput.* **2020**, *9*, 96–103.
36. Huang, D.; Cui, S.; Li, X. Wavelet packet analysis of blasting vibration signal of mountain tunnel. *Soil Dyn. Earthq. Eng.* **2019**, *117*, 72–80. [CrossRef]
37. Ekici, S.; Yildirim, S.; Poyraz, M. Energy and entropy-based feature extraction for locating fault on transmission lines by using neural network and wavelet packet decomposition. *Expert Syst. Appl.* **2008**, *34*, 2937–2944. [CrossRef]
38. Sun, Y.; Cao, Y.; Li, P. Fault diagnosis for train plug door using weighted fractional wavelet packet decomposition energy entropy. *Accid. Anal. Prev.* **2022**, *166*, 106549. [CrossRef]
39. Ling, T.H.; Li, X.B.; Dai, T.G.; Peng, Z.B. Features of energy distribution for blast vibration signals based on wavelet packet decomposition. *J. Cent. South Univ. Technol.* **2005**, *12*, 135–140. [CrossRef]

chemosensors

MDPI

Article

Nickel-Doped ZnO Porous Sea Urchin Nanostructures with Various Amounts of Oxygen Defects for Volatile Organic Compound Detection

Haibo Ren [1,*], Huaipeng Weng [1], Xumeng Dong [1], Jiarui Huang [2,*] and Sang Woo Joo [3,*]

1 School of Materials Science and Engineering, Anhui Polytechnic University, Wuhu 241000, China
2 Key Laboratory of Functional Molecular Solids of the Ministry of Education, Anhui Laboratory of Molecule-Based Materials, College of Chemistry and Materials Science, Anhui Normal University, Wuhu 241002, China
3 School of Mechanical Engineering, Yeungnam University, Gyeongsan 712749, Gyeoungbuk, Republic of Korea
* Correspondence: renhaibo@ahpu.edu.cn (H.R.); jrhuang@mail.ahnu.edu.cn (J.H.); swjoo@yu.ac.kr (S.W.J.)

Abstract: Porous sea urchin-like nickel-doped ZnO with various nickel contents and high specific surface area were synthesized using a solution method followed by calcination. The nickel-doped ZnO products consisted of numerous porous nanoleaves. The Ni content in these products ranged from 5% to 20%. The Ni dopants in the ZnO lattice were verified by X-ray diffraction and X-ray photoelectron spectroscopy. The sensors based on nickel-doped ZnO sea urchins showed superior sensing performance for some volatile organic compounds (VOCs). ZnO sea urchins with 10% nickel doping exhibited the best gas-sensing performance, including a low working temperature, short response/recovery time, and high sensor response. In particular, the 10% Ni-doped ZnO sea urchin sensor exhibited a response of 84.4 with response/recovery times of 17/20 s towards 100 ppm formaldehyde vapor. These superior sensing behaviors were attributed mainly to a suitable Ni content with high content of oxygen defects, small nanocrystals, and a porous hierarchical structure with a high specific surface area.

Keywords: zinc oxide; nickel doping; zinc hydroxide carbonate; nanoleaf; gas sensor; volatile organic compounds

Citation: Ren, H.; Weng, H.; Dong, X.; Huang, J.; Joo, S.W. Nickel-Doped ZnO Porous Sea Urchin Nanostructures with Various Amounts of Oxygen Defects for Volatile Organic Compound Detection. *Chemosensors* **2023**, *11*, 223. https://doi.org/10.3390/chemosensors11040223

Academic Editor: Boris Lakard

Received: 11 February 2023
Revised: 25 March 2023
Accepted: 2 April 2023
Published: 4 April 2023

1. Introduction

Volatile organic compounds (VOCs) are often emitted from decoration and building materials. As the gaseous toxins are widespread in both indoors and outdoors settings, prolonged exposure to VOCs causes harm to the eyes and damages the nervous system, causing a series of diseases [1]. Moreover, VOCs at low concentrations can easily cause serious respiratory problems. For example, formaldehyde at very low concentrations (0.75 ppm) can cause cancer [2]. Therefore, convenient gas sensors with a low detection limit, high response, and low energy consumption are needed urgently. A variety of metal oxide semiconductors that function as excellent sensing materials, such as SnO_2 [3], ZnO [4], In_2O_3 [5], Fe_2O_3 [6], NiO [7], and V_2O_5 [8], were developed for detecting volatile harmful gases. Among the metal oxide semiconductors, ZnO, a typical n-type semiconductor, possesses a direct bandgap of 3.37 eV. ZnO has drawn considerable attention due to its physical/chemical stability, easy preparation, abundance, high electron mobility, and excellent electric conductivity [9,10]. It is currently being used for detecting various flammable, volatile, and harmful gases [11–14]. On the other hand, for a pure ZnO sensing material, the high detection limit, high working temperature, and low response greatly limit the widespread applications of ZnO sensors. Therefore, it is urgent to develop new ZnO-based sensors with relatively low working temperature, high sensitivity, and good selectivity.

Some strategies were adopted to overcome the above problems. The morphology of ZnO partially affects the sensitivity, working temperature, and response rate of gas sensing behaviors. Especially, a 3D porous structure with a high surface area could provide numerous surface active sites for accepting more gas molecules and holding surface chemical reactions. Another strategy for improving gas-sensing behaviors is associated with the surface decoration of noble metals and doping ZnO with different metal elements. The commonly adopted noble metals Pt [15], Pd [16], Ag [17], and Au [18] are adopted for modification of the ZnO surface. They can greatly optimize the sensing behaviors of the gas sensor by improving the catalytic activity of the sensing materials. Despite the obvious improvement in the sensitivity to VOCs, the high cost of noble metals has limited their wider applications as gas-sensing materials. Therefore, doping with different elements (e.g., Fe [19], Co [20], Ni [21], Cu [22], and Al [23]) of low cost and wide availability is considered as one of the practical strategies to enhance the sensing performance of ZnO. For example, Mo et al. prepared a mesoporous Co-doped ZnO hierarchical structure by the calcination of its precursor [20]. The 5% Co-doped ZnO hierarchical structure sensor exhibited a sensitivity of 54 towards 50 ppm ethanol at 180 °C. El Khalidi et al. synthesized various Ni-doped ZnO films by spray pyrolysis [21]. Among them, the 2% Ni-doped ZnO film sensor showed the best gas-sensing performance, with the highest sensitivity (90) for 100 ppm acetone at 450 °C. The response and recovery times were 50 s and 90 s, respectively. Of the various metal dopants, Ni-doped ZnO has exceptional features. Ni could induce a spillover–sensitization effect [21]. Ni^{2+}-doped ZnO produces more donor defects and can increase the level of adsorbed gas molecules. In addition, it can facilitate the ionization of gas molecules and the dissociation of targeted gases from the surface of the sensing materials, leading to higher sensitivity and rapid response/recovery behaviors in ZnO sensors. The empty 3d orbital in Ni^{2+} allows the production of more oxygen vacancies when Zn^{2+} is replaced with Ni^{2+}. Thus, the transportation of charges accelerates and the number of adsorbed oxygen anions increases significantly [24].

So far, some strategies have been adopted for preparing the doped ZnO with various nanostructures, including co-precipitation [24], spray pyrolysis [25], the hydrothermal method [26], the sol–gel method [27], high-temperature calcination [28], RF sputtering [29], and the electrospinning method [30]. Kamble et al. fabricated various Ni-doped content nanorod-like ZnO nanostructure sensors by a co-precipitation method [24]. The 2% Ni-ZnO nanorod sensor presented a response of 356 at 200 °C. The detected gas was NO_2 gas with 100 ppm. Modaberi et al. prepared a Ni-doped content nanorod-like ZnO nanostructure by calcining the precursor [28]. The 8% Ni-doped nanorod-like ZnO sensor exhibited the highest response towards H_2S (100 ppm) gas at 200 °C. The response value reached 45.3. Among the synthetic methods of Ni-doped ZnO, the cooperation of an aqueous solution method and calcination treatment has attracted much attention because of its simple operation, low cost, and easy synthesis. Thus far, a high content of Ni-doped porous sea urchin-like ZnO nanostructures assembled from numerous nanosheets has rarely been achieved using an aqueous solution method and a subsequent heat treatment. The effect of ZnO doped with a high concentration of Ni on the sensitivity of gas sensors also needs to be investigated.

In this study, porous sea urchin-like ZnO nanostructures with various Ni doping contents were prepared using an aqueous solution method and a subsequent heat treatment. The sensing performance of the Ni-doped ZnO sea urchins towards VOCs was also studied. Compared to other VOCs, the Ni-doped ZnO sea urchin sensors exhibited a strong response to formaldehyde vapor. The Ni doping and unique porous sea urchin-like structure promote an outstanding sensing property. The mechanism of the gas-sensing behavior to formaldehyde vapor is also discussed.

2. Experimental Details

2.1. Synthesis of Ni-Doped ZnO Nanostructures

Porous Ni-doped ZnO nanostructures were synthesized as follows. Briefly, 0.0211 g of $NiCl_2{\cdot}6H_2O$ and 0.5021 g of $Zn(NO_3)_2{\cdot}6H_2O$ were dissolved in 35 mL of deionized water; the atomic ratio of Ni and Zn was 0.5:9.5. A saturated NH_4HCO_3 solution was then prepared. The pH was adjusted to 6.5 by introducing CO_2 gas for approximately 3 h. Subsequently, 15 mL of the saturated NH_4HCO_3 solution was poured into the previously prepared 35 mL solution under stirring, and then was placed at a temperature atmosphere of 6 °C for 8 h. After that, the precursor was collected after washing and dried at 50 °C for 15 h. The 5% Ni-doped ZnO nanostructure was obtained by high-temperature calcination at 350 °C for 1 h. The 10% Ni-doped ZnO sample, the 15% Ni-doped ZnO sample, and the 20% Ni-doped ZnO sample were synthesized using a similar procedure to that of the 5% Ni-doped ZnO sample, with the amount of $NiCl_2{\cdot}6H_2O$ changed from 0.0211 g to 0.0446 g, 0.0708 g, and 0.0997 g, respectively. The synthesis of the pure ZnO sample was similar to that of the 5% Ni-doped ZnO sample without adding $NiCl_2{\cdot}6H_2O$.

2.2. Material Characterization

The products were characterized by X-ray diffraction (XRD, Shimadzu XRD-6000, Shimadzu, Kyoto, Japan) using high-intensity Cu Kα radiation with a wavelength of 1.54178 Å, field-emission scanning electron microscopy (FESEM, Hitachi S-4800, Hitachi, Kyoto, Japan, operated at 5 kV), high-resolution transmission electron microscopy (HRTEM, JEOL-2010 TEM, JEOL, Kyoto, Japan) with an acceleration voltage of 200 kV, thermogravimetric analysis (TGA, NETZSCH STA449C, NETZSCH, Selb, Germany), and N_2 adsorption–desorption isotherm measurements (Nova 2000E, Quantachrome, Boynton Beach, FL, USA). The pore size distribution was determined from the desorption branch of the isotherm using the Barrett–Joyner–Halenda (BJH) method. The TGA was carried out in air with a set heating rate (10 °C min^{-1}) in the temperature scope (room temperature to 600 °C). X-ray photoelectron spectroscopy (XPS, ESCALAB 250, ESCALAB, Ciudad de México, México) was also performed. The elemental distribution of the products was determined by energy dispersive spectroscopy (FESEM, Hitachi S-4800, operated at 15 kV).

2.3. Fabrication and Testing of the Gas Sensor

Figure S1 in the Supplementary Material presents the experimental device for gas-sensing measurements. The fabrication process of the gas sensors was as follows: 0.1 g of porous sea urchin-like 5% Ni-ZnO, 10% Ni-ZnO, 15% Ni-ZnO, and 20% Ni-ZnO composites, and porous sea urchin-like ZnO were dispersed in a 0.5 mL ethanol solution to obtain a uniform suspension. The above suspension was smeared uniformly on the outer surface of the tube-shaped substrate made from aluminum, where a pair of Au electrodes was tightly printed on the bottom of the aluminum-made substrate. The as-prepared sensors were dried in air at 60 °C for 3 h, followed by high-temperature calcination of 280 °C for 2 h. Subsequently, the working temperature for the gas sensors can be achieved by placing a Ni-Cr heating coil into the ceramic tube, as shown in Figure S2a. The sensor was kept at 280 °C for two days in air to improve its long-term stability. A stationary-state gas distribution method was used to test the gas response. The sensor response measurement was performed on an electrochemical workstation (CHI-660E, Shanghai Chenhua Instruments Co., Ltd., Shanghai, China) using chronoamperometry at 0.7 V. Figure S2b presents a diagram of the testing principle of the gas-sensing measurement system. Detecting gases, such as ethanol vapor (headspace vapor of ethanol), formaldehyde vapor (headspace vapor of a 38.5 wt.% formaldehyde solution), and NH_3/N_2 mixtures (1 vol%), were injected into a test chamber (1000 mL) and mixed with air. To detect 1 ppm of ethanol vapor, formaldehyde vapor, and NH_3, 0.017 mL headspace vapor of ethanol, 1.0 mL headspace vapor of a 38.5 wt.% formaldehyde solution, and 0.1 mL NH_3/N_2 mixtures were injected into the test chamber, respectively. The response of the sensor was defined as R_a/R_g (reducing gases), where R_a is the resistance of the sensor in dry air and in dry air mixed with the test gases. In the measurement system, the response was also calculated using the following equation: I_a/I_g, where I_g and I_a are the output currents in the test gas and air, respectively. The response or recovery time is expressed as the time required for the sensor output to reach 90% saturation after applying or switching off the gas in a step function.

3. Results and Discussion

3.1. Structural and Morphological Characterization

Figures 1a and S3a show the characteristic diffraction peaks of the precursors. All XRD peaks of the samples corresponded to the crystal planes of the monoclinic $Zn_5(CO_3)_2(OH)_6$ (JCPDS no. 19-1458). After heat treatment in air, the XRD patterns of all Ni-doped ZnO products displayed in Figures 1a and S3a could be assigned to hexagonal ZnO (JCPDS no. 36-1451). The XRD pattern of the 20% Ni-doped ZnO sample showed a very weak peak at approximately 43.3°, which was ascribed to the crystal plane (200) of bunsenite NiO (JCPDS no. 47-1049). No other obvious peaks related to nickel compounds were observed in the XRD patterns of the ZnO products doped with different nickel contents, indicating that nickel ions systematically enter the lattice without destroying the original crystal structure of ZnO. The mean crystallite size of ZnO crystallites could be calculated using the Scherer formula:

$$D = \frac{0.9\lambda}{\beta \cdot \cos\theta} \quad (1)$$

where λ, β, and θ are the wavelength of the X-ray beam, width at half maximum, and diffraction angle, respectively. The mean crystallite size of the 0%, 5%, 10%, 15%, and 20% Ni-ZnO sea urchins were 18.2 nm, 15.1 nm, 12.5 nm, 15.3 nm, and 14.0 nm, respectively. The lattice constants of the samples listed in Table S1 are comparable with standard data. a (a = b) and c were extracted from the peaks (002) and (100). The calculated lattice parameters decreased with the increase of Ni content, which refers to the substitution of Zn^{2+} with Ni^{2+} with the ionic radius of nickel (0.055 nm) slightly smaller than that of zinc (0.060 nm). However, the calculated lattice parameters increased after 10% Ni, which may be due to the formation of a NiO phase decreasing the amount of Ni dopants.

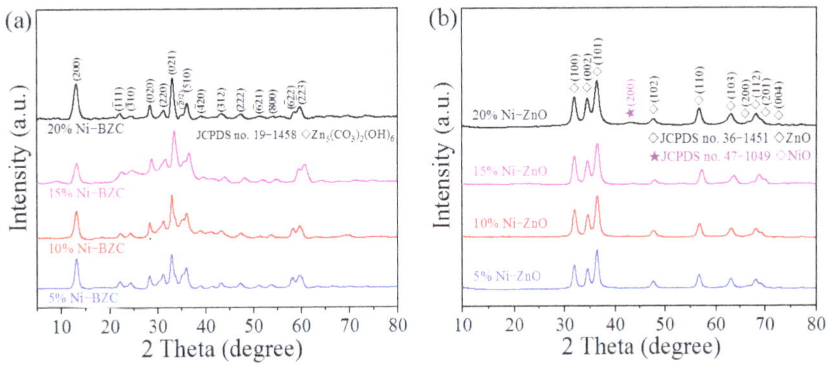

Figure 1. XRD patterns of (**a**) Ni-doped $Zn_5(CO_3)_2(OH)_6$ precursors and (**b**) Ni-doped ZnO products.

Furthermore, Figure S3b shows the EDS analysis of the 10% Ni-ZnO product. Only Zn, Ni, and O were observed, suggesting that the final product was Ni-doped ZnO.

Figures 2 and S4 present SEM images of the sea urchin-like Ni-doped $Zn_5(CO_3)_2(OH)_6$ precursor before and after calcination. Observing the SEM images shown in Figure 2a,b, the precursors exhibited a sea urchin-like structure with a mean diameter of 4–5 μm. The unique structure comprised numerous nanoleaves with a mean thickness of ~12 nm. These nanoleaves with smooth surfaces grew perpendicularly outward to form the sea urchin-like structure. After calcination in air, the 10% Ni-ZnO sample maintained the sea urchin-like morphology, as shown in Figure 2c,d. With increasing Ni content in the Ni-ZnO sea urchins, the mean diameter of the sea urchin-like Ni-ZnO becomes larger. In Figure S4a,c,e, we can find that 5%, 15%, and 20% Ni-doped $Zn_5(CO_3)_2(OH)_6$ precursors all showed similar structures: a sea urchin-like morphology. Numerous ultrathin nanoleaves lead to the unique structure in a self-assembling manner. The sea urchin-like morphology was retained after the calcination of the products, as shown in Figure S4b,d,f. The EDS mapping images of 10% Ni-ZnO sea urchins are exhibited in Figure S5. Ni, Zn, and O were dispersed uniformly in nano-sea urchins. The morphologies of the precursors (basic zinc carbonate) and the pure ZnO products are exhibited in FESEM images (Figure S6). Obviously observed, the pure $Zn_5(CO_3)_2(OH)_6$ precursor also displays a sea urchin-like morphology, as shown in Figure S6a,b. This favorable structure is constructed by the self-assemble of abundant nanoleaves. The diameter of the pure ZnO precursors averages around 8.6 μm. FESEM images of the pure ZnO products are shown in Figure S6c,d. After calcination (450 °C, 2 h), the post-calcination sea urchin morphology is maintained without any damage to the morphology, which can be observed in the FESEM images (Figure S6c,d) of the pure ZnO products.

Figure 2. Representative SEM images of (**a**,**b**) the sea urchin-like 10% Ni-doped $Zn_5(CO_3)_2(OH)_6$ precursor and (**c**,**d**) the corresponding porous sea urchin-like Ni-doped ZnO.

The detailed structure of 10% Ni-doped ZnO sea urchins was studied further by transition electron microscopy (TEM). Figure 3a–c presents the TEM and HRTEM images of the sample. The sample had a sea urchin morphology composed of numerous ultra-small nanoleaves. There were many irregular nanopores on the nanoleaves (Figure 3b). These pores of various sizes were produced by the decomposition of the $Zn_5(CO_3)_2(OH)_6$ precursor to CO_2 gas and gaseous H_2O, which can improve the sensing performance. The interplanar distances were 0.281, 0.261, and 0.163 nm, which correspond to the hexagonal phase of ZnO planes (100), (002), and (110), respectively (Figure 3c) [31]. Figure 3d shows the selected area electron diffraction (SAED) pattern of the sample, which can be assigned to the (100), (101), (110), (103), (112), and (201) planes of hexagonal ZnO [32]. The TEM results were consistent with the structural and morphological characteristics of the 10% Ni-ZnO sample obtained from XRD (Figure 1b) and SEM (Figure 2).

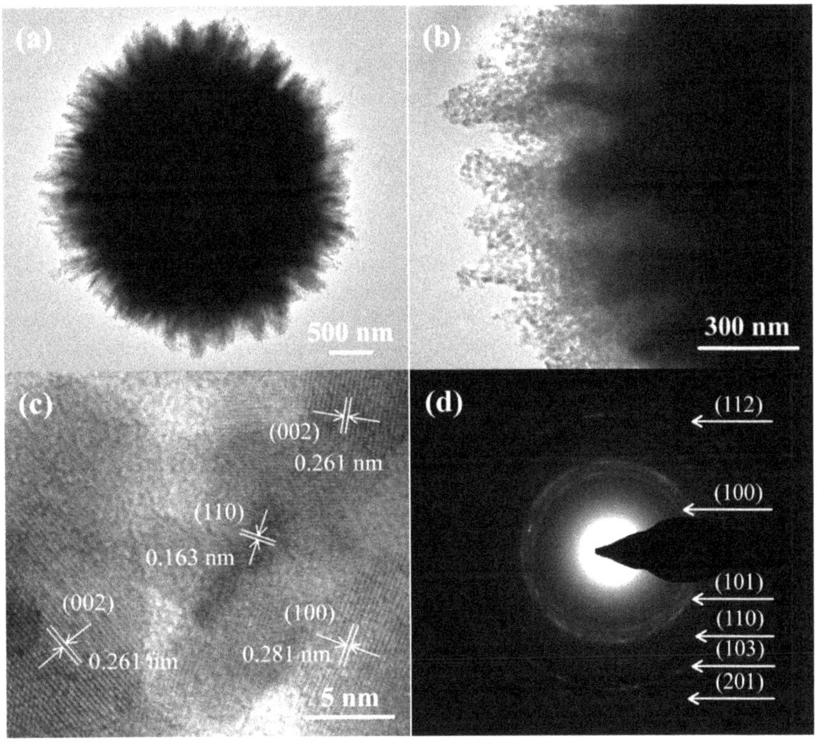

Figure 3. (**a**) TEM image, (**b**) high-magnification TEM image, (**c**) HRTEM image, and (**d**) SAED pattern of 10% Ni-doped ZnO sea urchins.

TGA was performed to study the thermal decomposition process of Ni-doped $Zn_5(CO_3)_2(OH)_6$ precursors (Figure S7). Before 180 °C, a slight mass loss occurred because of absorbed water. At temperatures from 180 °C to 400 °C, a mass loss of 27.9% occurred, which is consistent with the theoretical value of 28.8%. This is primarily because the Ni-doped $Zn_5(CO_3)_2(OH)_6$ precursors decomposed to Ni-doped ZnO, CO_2 gas, and gaseous H_2O. A large volume of CO_2 gas and gaseous H_2O were released from the inner structure of the precursors, resulting in a large number of mesopores in the final products after calcination. These mesopores will significantly enhance the specific surface area of Ni-doped ZnO sea urchins and provide more available sites for catalytic redox reactions. The decomposition chemical equation can be described as follows:

$$Zn_5(CO_3)_2(OH)_6 = 5ZnO + 3H_2O \text{ (mass 13.8\%)} + 2CO_2 \text{ (mass 15.0\%)} \quad (2)$$

Figure S8 shows the FTIR spectra of 10% Ni-ZnO sea urchins and the corresponding precursor. The main characteristic peaks at 472, 837, 1416, and 1510 cm^{-1} (Figure S8a) were attributed to the CO_2^{3-} bending vibrations in the precursor. The peak at ca. 3391 cm^{-1} can be assigned to the –OH group in the precursor. In the FTIR spectrum of 10% Ni-ZnO sea urchins (Figure S8b), a strong characteristic peak at 432 cm^{-1} can be observed clearly because of the vibration of Ni–O and Zn–O in the Ni-doped ZnO [33,34]. Because some peaks referring to Ni–O in NiO overlapped with those of Zn–O in ZnO, there is only a characteristic peak ranging from 430 cm^{-1} to 440 cm^{-1} in the FTIR spectrum. The characteristic peaks associated with the CO_2^{3-} bending vibrations have disappeared due to the thorough decomposition of carbonate in the precursor to CO_2 and H_2O. In addition, two weak peaks at 1386 and 1638 cm^{-1} were associated with the O–H bending vibration of the H_2O molecule absorbed on Ni-doped ZnO sea urchins [35,36]. The broadest peak was found at approximately 3448 cm^{-1}. This peak is assigned to the stretching vibration of the O–H bond from the absorbed H_2O molecule [33].

The N_2 adsorption–desorption isotherms of the samples are shown in Figures S9 and S10. A plot of the pore size distribution suggests that the overwhelming majority of pore sizes are distributed from around 1.2 to 29.7 nm. The specific surface areas of the 0%, 5%, 10%, 15%, and 20% Ni-ZnO sea urchins were 30.3 $m^2\ g^{-1}$, 53.4 $m^2\ g^{-1}$, 116.5 $m^2\ g^{-1}$, 78.3 $m^2\ g^{-1}$, and 15.8 $m^2\ g^{-1}$, respectively. The 10% Ni-ZnO sea urchins exhibited the highest specific surface area, which is mainly due to their microstructure with different sizes and the effect of Ni dopants decreasing the crystal size of ZnO. As for 15% and 20% Ni-ZnO sea urchins, the NiO phase observed in the XRD patterns would reduce the amount of Ni dopant, resulting in a low specific surface area. The high specific surface area likely produces more metal vacancies and oxygen species on the surface and inside the porous structure, generating more unsaturated active sites [37].

Figure 4a shows the UV-Vis spectra of the porous Ni-ZnO and pure ZnO sea urchins. There is a distinct absorption peak at ca. 360 nm. In addition, the absorption intensity of Ni-doped ZnO sea urchins was higher than that of pure ZnO sea urchins within the wavelength range of 300–800 nm. Moreover, the Tauc equation of $\alpha h\upsilon = C(h\upsilon - E_g)^{1/2}$ was applied to estimate the bandgap of pure ZnO and Ni-ZnO sea urchins. As shown in Figure 4b, the calculated bandgap of the 5%, 10%, 15%, and 20% Ni-ZnO sea urchins were 2.91, 2.89, 2.92, and 2.93 eV, respectively, which are smaller than that of pure ZnO sea urchins (3.01 eV). This suggests that introducing the Ni dopant in ZnO leads to less energy for the electronic transition than the pure ZnO sea urchins. Compared with the pure ZnO sample, the decrease in the bandgap of Ni-doped ZnO samples is due to the formation of energy levels of the $^5D_4$ state of Ni^{2+} under the $^3F_4$ state of Zn^{2+}, i.e., immediately decreasing the conduction band of the ZnO sample [24].

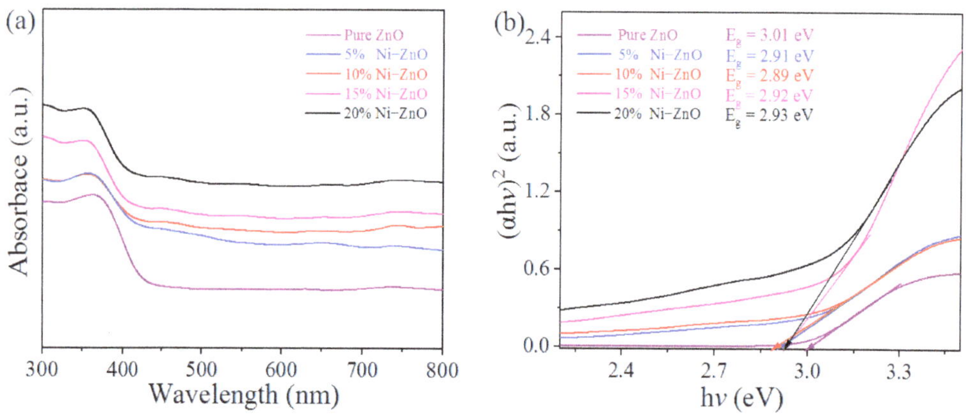

Figure 4. (**a**) UV-vis absorption spectra of the porous pure sea urchin-like ZnO and the porous urchin-like Ni-ZnO. (**b**) Bandgap determination from $(\alpha h\upsilon)^2$ vs. $h\upsilon$ of the porous pure sea urchin-like ZnO and the porous sea urchin-like Ni-ZnO.

Figure 5a shows the X-ray photoelectron spectroscopy (XPS) full-scale spectrum of 10% Ni-ZnO sea urchins. All peaks were assigned to the four elements of Zn, Ni, O, and C [13]. The Zn 2p high-resolution XPS is exhibited in Figure 5b. The binding energies at approximately 1044.5 and 1021.7 eV correspond to Zn $2p_{1/2}$ and Zn $2p_{3/2}$, respectively [38]. Figure 5c exhibits the Ni 2p high-resolution XPS. The peaks at ca. 855.2 and 860.8 eV belong to Ni $2p_{3/2}$, and the distinct peaks with binding energies of around 872.6 and 878.5 eV were attributed to the Ni $2p_{1/2}$ [39,40]. This indicates the presence of Ni^{2+} in the Ni-ZnO sea urchins. As displayed in Figure 5d, the O 1s peak was asymmetric, and it can be divided into three peaks. One peak ($O_{lat.}$) at 530.3 eV was assigned to lattice oxygen in wurtzite

ZnO, and the peak ($O_{def.}$) at 530.9 eV corresponds to the defect sites with low-oxygen coordination [41]. The $O_{def.}/O_{lat.}$ value can indicate the oxygen defect concentration in Ni-ZnO sea urchins. Figure S11 presents the O 1s high-resolution XPS spectra of the 5%, 10%, 15%, and 20% Ni-ZnO sea urchins and pure ZnO sea urchins. The $O_{def.}/O_{lat.}$ values of the 5%, 10%, 15%, and 20% Ni-ZnO, and of the pure ZnO sea urchins were 0.34, 0.97, 0.32, 0.47, and 0.31, respectively, as listed in Table S2. The last peak ($O_{ads.}$) at 532.5 eV was due to the physi/chemisorbed H_2O. Based on the above analysis results, the 10% Ni-ZnO sea urchins showed the highest oxygen defect concentration among these ZnO-based sea urchin samples, which is due to their high specific surface area and the effect of Ni dopants decreasing the crystal size of ZnO and introducing oxygen-related defects. As for 15% and 20% Ni-ZnO sea urchins, the NiO phase observed in the XRD patterns was verified by the O 1s spectra shown in Figure S11c,d.

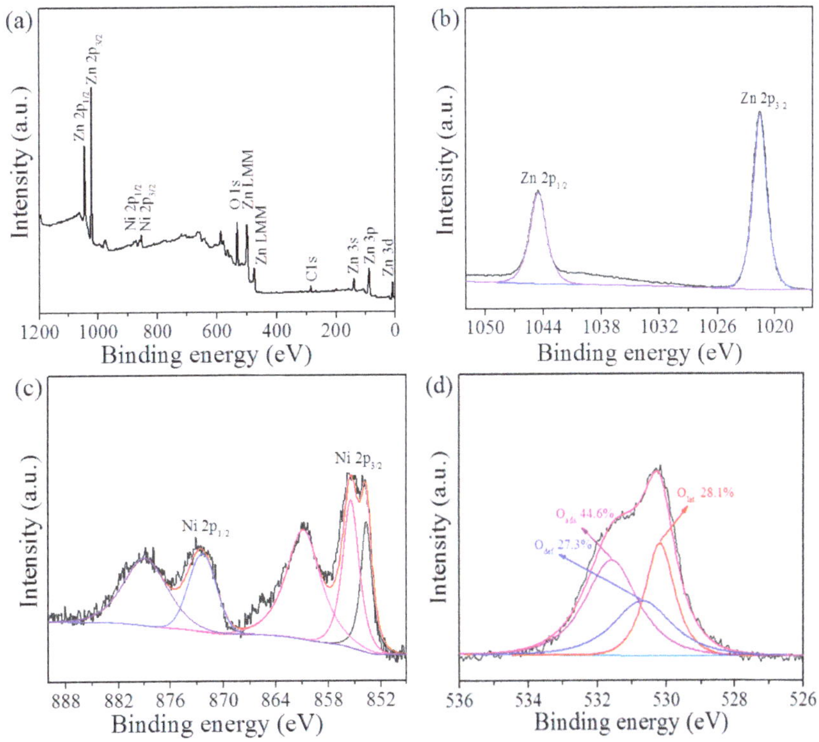

Figure 5. XPS spectra of porous urchin-like Ni-ZnO composites: (**a**) survey spectrum, (**b**) Zn 2p, (**c**) Ni 2p, and (**d**) O 1s spectra.

3.2. Gas-Sensing Performance

Figure 6a shows the changes in resistance (R_a) of the ZnO and Ni-doped ZnO sea urchin sensors. The working temperatures started from 140 °C to 260 °C. With regard to all the prepared sensors, the R_a value of the sensors decreases as the operating temperature continues to rise. The temperature dependences of the resistance of the obtained samples followed a linear function. The more abundant electrons were forced into a conductance band of the sensing materials activated by the sufficient energy provided by the high temperatures [42]. So, as is well known, the gas sensors belong to the type of surface-controlled one based on the curves changes of resistance temperatures. Under the same working temperature, the R_a sequence of the prepared sensors was present as follows: 20% Ni-ZnO > 15% Ni-ZnO > 10% Ni-ZnO > 5% Ni-ZnO > pure ZnO. This was mainly due to the effect of Ni dopants which can introduce oxygen-related defects [43]. As the

Ni-doped content of the ZnO increases, the crystal size of ZnO decreases and more oxygen molecules are adsorbed on the ultra-small ZnO nanoleaves, forming oxygen ions. In air, these increasing oxygen ions could capture more electrons from ZnO, leading to the higher resistance of the Ni-ZnO sensing film. Furthermore, when the amount of Ni dopants is high, the appearance of NiO and the formation of NiO–ZnO p–n junctions would also increase the resistance of the Ni-ZnO sensing film.

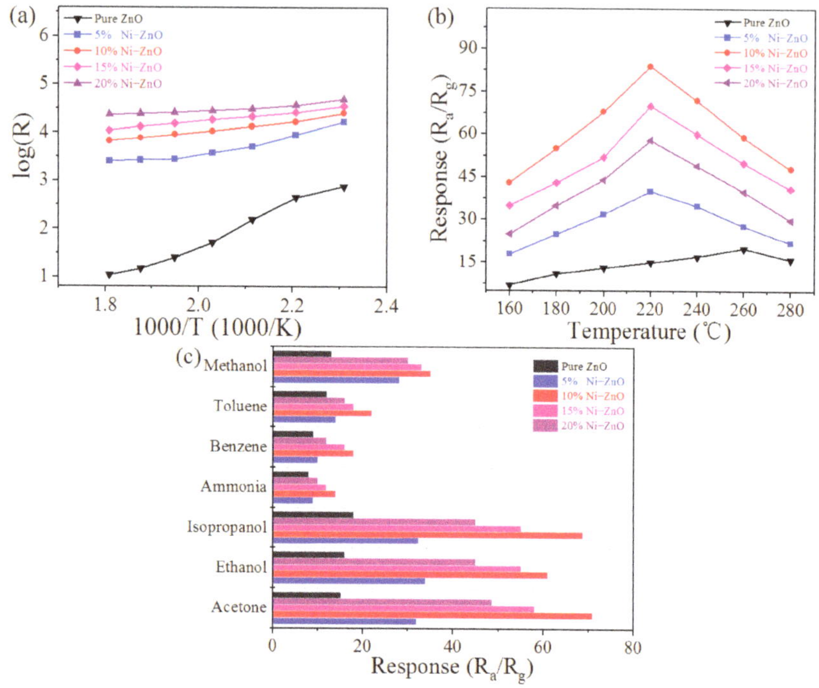

Figure 6. (**a**) Resistance of the ZnO-based sensors in air at various working temperatures. (**b**) Responses of ZnO-based sensors upon exposure to 100 ppm formaldehyde vapor at different working temperatures. (**c**) Responses of ZnO-based sensors upon exposure to different gases (100 ppm) at their optimal temperatures.

In Figure 6b, the response of the sensors first increased with increasing working temperature. When the temperature was over 220 °C, however, the response decreased with increasing working temperature. Thus, for the Ni-ZnO sensors, the optimal working temperature can be thought to be 220 °C. This phenomenon conforms to the adsorption/desorption kinetics and thermodynamics on sensing material surface. The gas molecules and oxygen species cannot react fully because of the lack of sufficient energy. When excessive energy was applied to the sensors, the gas molecules adsorbed on the material surface were easily desorbed, resulting in fewer effective surface-absorbed oxygen molecules and targeted molecules. As to the pure ZnO sensor, the optimal working temperature was 260 °C, which is much higher than that of the Ni-ZnO sensors. This demonstrated the decrease in working temperature due to the introduction of Ni dopants. In addition, the response of the Ni-ZnO sea urchin sensor towards 100 ppm formaldehyde vapor was much higher than that of the pure ZnO sea urchin sensor. This was likely attributed to the high concentration of oxygen defects, as verified by XPS. The sample possessed uniform sea urchin-like morphology and larger specific surface area and pore size, likely promoting electron transfer and providing more effective active sites for the redox reaction on the sensing material.

The selectivity of the sensors was studied, as shown in Figure 6c. Among them, the 10% Ni-ZnO sea urchin sensor showed the best sensing performance. The responses for

100 ppm formaldehyde, acetone, ethanol, isopropanol, ammonia, benzene, toluene, and methanol vapors were 84.4, 71.1, 61.8, 68.8, 14.3, 18.2, 22.5, and 35.3, respectively. Therefore, the 10% Ni-ZnO sensor exhibited good sensitivity towards formaldehyde, acetone, ethanol, and isopropanol vapors. Of the sensors, the 10% Ni-ZnO sea urchin sensor exhibited the strongest response to these vapors. Moreover, a good gas-sensing performance was also exhibited by the other sensors for various VOCs. The content of Ni in the ZnO sea urchins has an important effect on the sensors. The presence of oxygen defects enhanced the overall ionic potential of the composites. Thus, the large amount of adsorbed oxygen could strongly capture the material surface. Formaldehyde with a stronger Bronsted acid could react with more adsorbed oxygen (O_2^- and O^-) [44].

Figure 7 presents the response curves of different sensors based on pure ZnO, 5%, 10%, 15%, and 20% Ni-ZnO porous sea urchin-like structures exposed to various concentrations of formaldehyde, acetone, ethanol, and isopropanol at an operating temperature of 220 °C. For the 5% Ni-ZnO sensor, the curve of real-time response to formaldehyde gas at different concentrations was observed. The curves in the insets refer to the relationship between the concentration and the response when the sensors are exposed to formaldehyde vapor at different concentrations. In detail, even when exposed to very low concentrations (1 ppm), the response to formaldehyde vapor was close to ca. 6.2. The response was 40.2 when the concentration was increased to 100 ppm. The increasing concentration of detected gases improves the response to formaldehyde, acetone, ethanol, and isopropanol, as shown in the insets. The response to acetone, ethanol, and isopropanol (100 ppm) was 33.6, 34.2, and 32.4, respectively. For the 10% Ni-ZnO sea urchin sensor, the response values to 100 ppm formaldehyde, acetone, ethanol, and isopropanol were 84.4, 71.1, 61.8, and 68.8, respectively. In addition, the 15% Ni-ZnO sea urchin sensor, the 20% Ni-ZnO sea urchin sensor, and the pure ZnO sea urchin sensor showed response values to formaldehyde of 70.4, 62.2, and 20.5, respectively, to acetone of 58.1, 48.6, and 15.2, respectively, to ethanol of 55.2, 45.1, and 16.9, respectively, and to isopropanol of 55.9, 45.7, and 18.5, respectively. Furthermore, for 100 ppm formaldehyde vapor, the response and recovery times were also investigated, as exhibited in Figure S12. The response times (~22 s, ~17 s, ~19 s, ~20 s, and ~22 s) and the recovery times (~30 s, ~20 s, ~22 s, ~24 s, and ~32 s) were obtained for 5%, 10%, 15%, 20% Ni-ZnO, and pure ZnO sea urchin sensors, respectively. Compared to the other sensors, the 10% Ni-ZnO sensor with the largest specific surface area exhibited superior gas-sensing behaviors upon exposure to formaldehyde, acetone, ethanol, and isopropanol vapors. The largest specific surface area could hold more passages and "surface accessibility" to accelerate the transfer of the target molecule. This leads to an easy gas molecule (detected gases or oxygen) adsorption/dissociation from the sensing materials attributed to the production of more active sites. Moreover, the excellent sensing performance was attributed to the smaller size of porous nanoleaves and the smaller size of the pores distributed over the surface of the numerous nanoleaves. In general, the specific surface area, Ni-doping content, and porous sea urchin-like structure jointly affect the sensing properties. Compared to ZnO-based sensors, the 10% Ni-ZnO sensors presented better sensing behaviors to formaldehyde vapor than 3 mol% NiO/ZnO microflowers [45], CdO-ZnO nanorices [46], 3 wt.% Ag-In_2O_3/ZnO nanocomposites [47], SnO_2-ZnO/PdO nanoparticles [48], ZnO/SnO_2 hollow nanospheres [49], Co_3O_4/ZnO hollow spheres [50], and 1% Er-ZnO nanowires [51], as listed in Table S3.

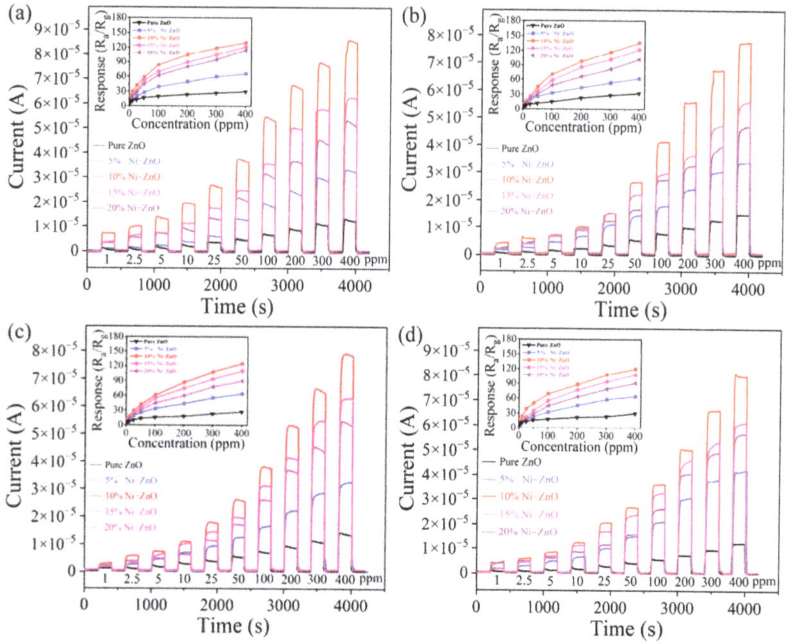

Figure 7. Real-time response curves of the sensor devices upon exposure to different concentrations of (**a**) formaldehyde, (**b**) acetone, (**c**) ethanol, and (**d**) isopropanol at a working temperature of 220 °C. The insets show the corresponding sensor response curves.

Figure 8a shows ten reversible cycles of the sea urchin-shaped Ni-ZnO sensor upon exposure to 100 ppm formaldehyde vapor. When the sensor comes in contact with formaldehyde vapor, the current quickly rises to a high level. The current decreases sharply to the initial level when the sensor is off formaldehyde vapor. The response currents of the sensors were stable and repeatable. Therefore, all Ni-ZnO sea urchin sensors showed excellent cycling stability and repeatability for detecting VOCs. Furthermore, Figure 8b shows the good linear relationship of the responses of the prepared sensors to the various concentrations of formaldehyde. Linear regression coefficients (R^2) of the 5%, 10%, 15%, and 20% Ni-ZnO sea urchin sensors, and of the pure ZnO sea urchin sensor, were 0.978, 0.991, 0.977, 0.978, and 0.931, respectively. Clearly, the 10% Ni-ZnO sea urchin sensor shows a better linear relationship between the response and formaldehyde concentration.

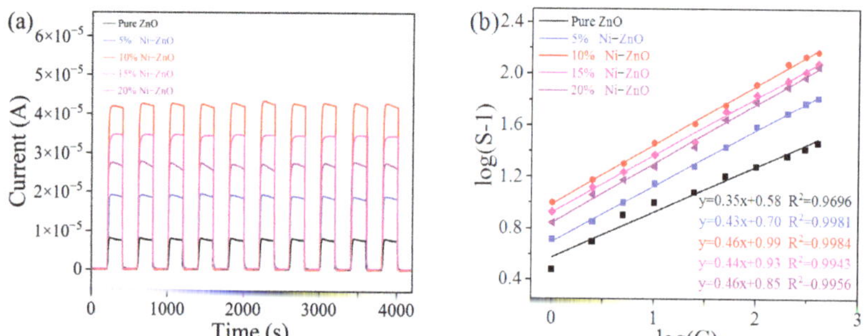

Figure 8. (**a**) Current change curves of the 10% Ni-ZnO sensor upon exposure to 100 ppm formaldehyde vapor for 10 tests at the optimal working temperature (220 °C). (**b**) Linear relationship of the response to different concentrations of formaldehyde.

3.3. Gas-Sensing Mechanism

The sensing mechanism of the ZnO-based sensor was attributed to the surface adsorbed oxygen species and the redox reactions between the target gases and the oxygen species, which leads to variations in the sensor resistance. For the pure ZnO sensor, when the sensor is exposed to the air, oxygen molecules will adsorb on ZnO surface, generating thin electron depletion layers. Electrons in the ZnO conduction band are drawn by oxygen molecules to form oxygen species with strong oxidizing properties (O_2^- and O^-). Each target gas is dissociated/ionized on the ZnO surface and reacts with ionosorbed oxygen elements (O_2^-, O^-, and O_2), forming H_2O (g) and CO_2 (g). The enhancement of the conductivity will be produced due to a thinner electron depletion layer as the number of adsorbed oxygen decreases. Consequently, the electrons return into the ZnO conduction band when surface catalytic reactions occur between oxygen anions and HCHO (see Equation (3)). Compared to pure ZnO sea urchins, the sensing performance of Ni-ZnO sea urchin sensors was improved greatly, which can be attributed to the increase of the charge transfer by doping, producing electronic sensitization that makes ZnO more sensitive to VOC vapors. The Ni-doped ZnO nanocrystals provide abundant active sites for formaldehyde molecule adsorption which, combined with defects of Ni-doped ZnO nanocrystals, can generate the intermediates and react with absorbed oxygen on Ni-doped ZnO (see Equations (4) and (5)). Therefore, Ni dopants can lead to the dissociation and ionization of VOCs, making ZnO more sensitive and selective. The addition of Ni increased the VOC adsorption amount on the surface and accelerated its ionization [51,52]. This process decreases the working temperature of the sensor due to water formation.

$$HCHO_{(ads)} + 2O^-(O_2^-) \rightarrow CO_2 + H_2O + 2e^- \quad (3)$$

$$HCHO_{(ads)} + h^+ + O^-(O_2^-) \rightarrow HCOOH \quad (4)$$

$$HCHO_{(ads)} + 2h^+ + 2O^-(O_2^-) \rightarrow CO_2 + H_2O \quad (5)$$

Moreover, the sensing property of the ZnO-based sensor depends greatly on its average grain diameter (D) [52]. When $D \leq 2L$ (L = thickness of the depletion layer), the grain can be fully depleted, and its conductance is grain-controlled [18]. The diameter of the ZnO crystallites (10% Ni-ZnO) averaged around 12.5 nm, which is the same as the ZnO Debye length (L_D) [18], indicating that 10% Ni-ZnO are almost fully depleted. The nanosize effect implies that many surface zinc atoms can play an important part in surface catalytic reactions. Furthermore, the largest specific surface area (116.5 $m^2\ g^{-1}$) could hold more passages and "surface accessibility" to accelerate the transfer of the target molecule. This leads to an easy gas molecule (detected gases or oxygen) adsorption/dissociation from the sensing materials, attributed to the production of more active sites. On the other hand, due to the difference of lattice spacing, some defects and vacancies could be created near the region of n-ZnO/p-NiO heterocontacts, which will lead to more active sites for gas adsorption and surface reaction compared with pure ZnO. Therefore, the Ni-ZnO sea urchins as sensing materials can improve sensitivity and shorten response/recovery times significantly.

4. Conclusions

The Ni-ZnO sea urchins consisting of numerous porous nanoleaves were prepared by an aqueous solution method and a subsequent heat treatment. The Ni-ZnO sea urchins with various amounts of Ni dopants possessed small nanocrystals and a high specific surface area. Used as sensing material, they exhibited a strong response, fine stability, and short response–recovery times for detecting formaldehyde vapor. Furthermore, among these ZnO-based gas sensors, the 10% Ni-ZnO sea urchin sensor exhibited the highest gas-sensing performance towards some VOC vapors, such as formaldehyde, acetone, ethanol, and isopropanol. The high specific surface area, stable 3D porous structure, and moderate content of Ni dopants with a high content of oxygen defects contribute to the surface catalytic reaction, gas absorption–desorption, and effective electron transfer in the gas-sensing process.

Supplementary Materials: The following supporting information can be downloaded at: https://www.mdpi.com/article/10.3390/chemosensors11040223/s1. Figure S1: Experimental setup; Figure S2: (a) Photograph of the sensor and (b) diagram of the testing principle of the gas-sensing measurement system (V_h: heating voltage); Figure S3: XRD patterns of (a) $Zn_5(CO_3)_2(OH)_6$ precursor and calcined ZnO product, and (b) EDS analysis of 10% Ni-ZnO product; Figure S4: SEM images of (a) 5% Ni-$Zn_5(CO_3)_2(OH)_6$ precursor and (b) calcined 5% Ni-ZnO product. SEM images of (c) 15% Ni-$Zn_5(CO_3)_2(OH)_6$ precursor and (d) calcined 15% Ni-ZnO product. SEM images of (e) 20% Ni-$Zn_5(CO_3)_2(OH)_6$ precursor and (f) calcined 20% Ni-ZnO product; Figure S5: (a) SEM image and corresponding EDS mapping images of the sea urchin-like 10% Ni-ZnO product: (b) Ni, (c) Zn, and (d) O; Figure S6: SEM images of (a,b) the $Zn_5(CO_3)_2(OH)_6$ precursor and (c,d) the calcined ZnO product; Figure S7: TGA curves of the sea urchin-like Ni-doped $Zn_5(CO_3)_2(OH)_6$ precursor; Figure S8: FTIR spectra of (a) the sea urchin-like 10% Ni-$Zn_5(CO_3)_2(OH)_6$ precursor and (b) sea urchin-like 10% Ni-ZnO; Figure S9: Nitrogen adsorption and desorption isotherms of (a) 5% Ni-ZnO, (b) 10% Ni-ZnO, (c) 15% Ni-ZnO, and (d) 20% Ni-ZnO sea urchins. The insets show the corresponding pore size distributions TGA curves of sea urchin-like Ni-doped $Zn_5(CO_3)_2(OH)_6$ precursor; Figure S10: Nitrogen adsorption–desorption isotherms of pure urchin-like ZnO. The inset is the corresponding pore size distribution; Figure S11: O 1s high-resolution XPS spectrum of (a) 5%, (b) 10%, (c) 15%, and (d) 20% Ni-ZnO sea urchins, and (e) pure ZnO sea urchins; Figure S12: Dynamic response-recovery curves of porous urchin-like Ni-ZnO sensors and the porous urchin-like ZnO sensor towards 100 ppm formaldehyde; Table S1: Structural parameters of Ni-doped ZnO sea urchins according to the Joint Committee on Powder Diffraction Standards (JCPDS no. 36-1451 (a = 3.250 Å and c = 5.207 Å)); Table S2: Fitting results of the O 1s XPS spectra of Ni-ZnO and pure ZnO sea urchins; Table S3: Responses of various ZnO-based sensing materials to different concentrations of formaldehyde vapor.

Author Contributions: Conceptualization, H.W.; methodology, H.W., X.D. and H.R.; validation, H.R.; investigation, H.W. and X.D.; resources, J.H. and S.W.J.; writing—original draft preparation, H.W.; writing—review and editing, J.H. and S.W.J.; supervision, H.R.; funding acquisition, S.W.J. All authors have read and agreed to the published version of the manuscript.

Funding: This research was funded by the National Research Foundation of Korea (NRF-2019R1A 5A8080290), Breeding Project of Anhui Polytechnic University (KZ42022076) and Open Research Found of Anhui Key Laboratory of High-performance Non-ferrous Metal Materials (YSJS-2023-07).

Institutional Review Board Statement: Not applicable.

Informed Consent Statement: Not applicable.

Data Availability Statement: The data presented in this study are available on request from the corresponding author.

Conflicts of Interest: The authors declare no conflict of interest.

References

1. Shooshtari, M.; Salehi, A. An electronic nose based on carbon nanotube-titanium dioxide hybrid nanostructures for detection and discrimination of volatile organic compounds. *Sens. Actuators B Chem.* **2018**, *357*, 131418. [CrossRef]
2. Broza, Y.; Vishinkin, R.; Barash, O.; Nakhleh, M.; Haick, H. Synergy between nanomaterials and volatile organic compounds for non-invasive medical evaluation. *Chem. Soc. Rev.* **2018**, *47*, 4781–4859. [CrossRef]
3. Kim, K.; Choi, P.; Itoh, T.; Masuda, Y. Catalyst-free highly sensitive SnO_2 nanosheet gas sensors for parts per billion-level detection of acetone. *ACS Appl. Mater. Interfaces* **2020**, *12*, 51637–51644. [CrossRef]
4. Huang, B.; Zeng, W.; Li, Y. Synthesis of ZIF-8 coating on ZnO nanorods for enhanced gas-sensing performance. *Chemosensors* **2022**, *10*, 297. [CrossRef]
5. Zhang, S.; Lin, Z.; Song, P.; Sun, J.; Wang, Q. MOF-derived In_2O_3 nanotubes/Cr_2O_3 nanoparticles composites for superior ethanol gas-sensing performance at room temperature. *Ceram. Int.* **2022**, *48*, 28334–28342. [CrossRef]

6. Wang, M.; Wang, Y.; Li, X.; Ge, C.; Hussain, S.; Liu, G.; Qiao, G. WO_3 porous nanosheet arrays with enhanced low temperature NO_2 gas sensing performance. *Sens. Actuators B Chem.* **2020**, *316*, 128050. [CrossRef]
7. Li, Q.; Zeng, W.; Zhou, Q.; Wang, Z. Highly sensitive ethanol sensing using NiO hollow spheres synthesized via hydrothermal method. *Chemosensors* **2022**, *10*, 341. [CrossRef]
8. Vijayakumar, Y.; Nagaraju, P.; Sreekanth, T.; Rushidhar, U.; Reddy, P. Effect of precursor volume on chemically sprayed V_2O_5 thin films for acetaldehyde detection. *Superlattices Microstruct.* **2021**, *153*, 106870. [CrossRef]
9. Beitollahi, H.; Tajik, S.; Nejad, F.; Safaei, M. Recent advances in ZnO nanostructure-based electrochemical sensors and biosensors. *J. Mater. Chem. B* **2020**, *8*, 5826–5844. [CrossRef]
10. Liu, B.H.; Wang, S.; Yuan, Z.; Duan, Z.H.; Zhao, Q.N.; Zhang, Y.J.; Su, Y.J.; Jiang, Y.D.; Xie, G.Z.; Tai, H.L. Novel chitosan/ZnO bilayer film with enhanced humidity-tolerant property: Endowing triboelectric nanogenerator with acetone analysis capability. *Nano Energy* **2020**, *78*, 105256. [CrossRef]
11. Cui, X.S.; Lu, Z.R.; Wang, Z.C.; Zeng, W.; Zhou, Q. Highly sensitive SF_6 decomposition byproducts sensing platform based on CuO/ZnO heterojunction nanofibers. *Chemosensors* **2023**, *11*, 58. [CrossRef]
12. Dilova, T.; Atanasova, G.; Dikovska, A.; Nedyalkov, N. The effect of light irradiation on the gas-sensing properties of nanocomposites based on ZnO and Ag nanoparticles. *Appl. Surf. Sci.* **2020**, *505*, 144625. [CrossRef]
13. Si, W.; Du, W.; Wang, F.; Wu, L.; Liu, J.; Liu, W.; Cui, P.; Zhang, X. One-pot hydrothermal synthesis of nano-sheet assembled NiO/ZnO microspheres for efficient sulfur dioxide detection. *Ceram. Int.* **2020**, *46*, 7279–7287. [CrossRef]
14. Dang, T.; Son, N.; Lanh, N.; Phuoc, P.; Viet, N.; Thong, L.; Hung, C.; Duy, N.; Hoa, N.; Hieu, N. Extraordinary H_2S gas sensing performance of ZnO/rGO external and internal heterojunctions. *J. Alloy. Compd.* **2021**, *879*, 160457. [CrossRef]
15. Yuan, Z.; Feng, Z.; Kong, L.; Zhan, J.; Ma, X. Simple synthesis of porous ZnO nanoplates hyper-doped with low concentration of Pt for efficient acetone sensing. *J. Alloy. Compd.* **2021**, *865*, 158890. [CrossRef]
16. Cao, P.; Yang, Z.; Navale, S.; Han, S.; Liu, X.; Liu, W.; Lu, Y.; Stadler, F.; Zhu, D. Ethanol sensing behavior of Pd-nanoparticles decorated ZnO-nanorod based chemiresistive gas sensors. *Sens. Actuators B Chem.* **2019**, *298*, 126850. [CrossRef]
17. Ahemad, M.; Le, T.; Kim, D.; Yu, Y. Bimetallic AgAualloy@ZnO core-shell nanoparticles for ultra-high detection of ethanol: Potential impact of alloy composition on sensing performance. *Sens. Actuators B Chem.* **2022**, *359*, 131595. [CrossRef]
18. Drmosh, Q.; Wajih, Y.; Alade, I.; Mohamedkhair, A.; Qamar, M.; Hakeem, A.; Yamani, Z. Engineering the depletion layer of Au-modified ZnO/Ag core-shell films for high-performance acetone gas sensing. *Sens. Actuators B Chem.* **2021**, *338*, 129851. [CrossRef]
19. Guo, W.; Zhao, B.; Zhou, Q.; He, Y.; Wang, Z.; Radacsi, N. Fe-doped ZnO/reduced graphene oxide nanocomposite with synergic enhanced gas sensing performance for the effective detection of formaldehyde. *ACS Omega* **2019**, *4*, 10252–10262. [CrossRef]
20. Mo, Y.; Shi, F.; Qin, S.; Tang, P.; Feng, Y.; Zhao, Y.; Li, D. Facile fabrication of mesoporous hierarchical Co-doped ZnO for highly sensitive ethanol detection. *Ind. Eng. Chem. Res.* **2019**, *58*, 8061–8071. [CrossRef]
21. Elkhalidi, Z.; Hartiti, B.; Siadat, M.; Comini, E.; Arachchige, H.; Fadili, S.; Thevenin, P. Acetone sensor based on Ni doped ZnO nanostructues: Growth and sensing capability. *J. Mater. Sci. Mater. Electron.* **2019**, *30*, 7681–7690. [CrossRef]
22. Rahman, M. Efficient formaldehyde sensor development based on Cu-codoped ZnO nanomaterial by an electrochemical approach. *Sens. Actuators B Chem.* **2020**, *305*, 127541. [CrossRef]
23. Namgung, G.; Ta, Q.; Yang, W.; Noh, J. Diffusion-driven Al-doping of ZnO nanorods and stretchable gas sensors made of doped ZnO nanorods/Ag nanowires bilayers. *ACS Appl. Mater. Interfaces* **2019**, *11*, 1411–1419. [CrossRef]
24. Kamble, V.; Navale, Y.; Patil, V.; Desai, N.; Salunkhe, S. Enhanced NO_2 gas sensing performance of Ni-doped ZnO nanostructures. *J. Mater. Sci. Mater. Electron.* **2021**, *32*, 2219–2233. [CrossRef]
25. Badawi, A.; Althobaiti, M.; Ali, E.; Alharthi, S.; Alharbi, A. A comparative study of the structural and optical properties of transition metals (M = Fe, Co, Mn, Ni) doped ZnO films deposited by spray-pyrolysis technique for optoelectronic applications. *Opt. Mater.* **2022**, *124*, 112055. [CrossRef]
26. Jiang, B.; Yang, S. Nickel-doped ZnO nanowalls with enhanced electron transport ability for electrochemical water splitting. *Nanomaterials* **2021**, *11*, 1980. [CrossRef]
27. Minhas, H.; Kumar, D.; Kumar, A. Preparation, characterization and electromagnetic interference shielding effect of Ni-doped ZnO thin films. *Mater. Res. Express* **2019**, *6*, 105049. [CrossRef]
28. Modaberi, M.; Rooydell, R.; Brahma, S.; Akande, A.; Mwakikunga, B.; Liu, C. Enhanced response and selectivity of H_2S sensing through controlled Ni doping into ZnO nanorods by using single metal organic precursors. *Sens. Actuators B Chem.* **2018**, *273*, 1278–1290. [CrossRef]
29. Bhati, V.; Ranwa, S.; Fanetti, M.; Valant, M.; Kumar, M. Efficient hydrogen sensor based on Ni-doped ZnO nanostructures by RF sputtering. *Sens. Actuators B Chem.* **2018**, *255*, 588–597. [CrossRef]
30. Wu, J.; Li, T.; Meng, G.; Xiang, Y.; Hai, J.; Wang, B. Carbon nanofiber supported Ni-ZnO catalyst for efficient and selective hydrogenation of pyrolysis gasoline. *Catal. Sci. Technol.* **2021**, *11*, 4216–4225. [CrossRef]
31. Zhang, S.; Li, Y.; Sun, G.; Zhang, B.; Wang, Y.; Cao, J.; Zhang, Z. Synthesis of NiO-decorated ZnO porous nanosheets with improved CH_4 sensing performance. *Appl. Surf. Sci.* **2019**, *497*, 143811. [CrossRef]
32. Zhang, H.; Chen, W.; Li, Y.; Song, Z.; Zeng, W.; Tang, S.; Wang, S.; Zhou, D. Hierarchical heterostructures of nanosheet-assembled NiO-modified ZnO microflowers for high performance acetylene detection. *Ceram. Int.* **2020**, *46*, 3574–3581. [CrossRef]

33. Lei, C.S.; Pi, M.; Cheng, B.; Jiang, C.J.; Qin, J.Q. Fabrication of hierarchical porous ZnO/NiO hollow microspheres for adsorptive removal of Congo red. *Appl. Surf. Sci.* **2018**, *435*, 1002–1010. [CrossRef]
34. Cun, T.; Dong, C.; Huang, Q. Ionothermal precipitation of highly dispersive ZnO nanoparticles with improved photocatalytic performance. *Appl. Surf. Sci.* **2016**, *384*, 73–82. [CrossRef]
35. Niu, H.; Zhou, D.; Yang, X.; Li, X.; Wang, Q.; Qu, F. Towards three-dimensional hierarchical ZnO nanofiber@Ni$(OH)_2$ nanoflake core-shell heterostructures for high-performance asymmetric supercapacitors. *J. Mater. Chem. A* **2015**, *3*, 18413–18421. [CrossRef]
36. Starukh, G.; Rozovik, O.; Oranska, O. Organo/Zn-Al LDH nanocomposites for cationic dye removal from aqueous media. *Nanoscale Res. Lett.* **2016**, *11*, 228–237. [CrossRef]
37. Mao, Y.; Cheng, Y.; Wang, J.; Yang, H.; Li, M.; Chen, J.; Chao, M.; Tong, Y.; Liang, E. Amorphous NiO electrocatalyst overcoated ZnO nanorod photoanodes for enhanced photoelectrochemical performance. *New J. Chem.* **2016**, *4*, 107. [CrossRef]
38. Su, C.; Zhang, L.; Han, Y.; Ren, C.; Li, B.; Wang, T.; Zeng, M.; Su, Y.; Hu, N.; Zhou, Z.; et al. Glucose-assisted synthesis of hierarchical NiO-ZnO heterostructure with enhanced glycol gas sensing performance. *Sens. Actuators B Chem.* **2021**, *329*, 129167. [CrossRef]
39. Ouyang, Y.; Xia, X.; Ye, H.; Wang, L.; Jiao, X.; Lei, W.; Hao, Q. Three-dimensional hierarchical structure ZnO@C@NiO on carbon cloth for asymmetric supercapacitor with enhanced cycle stability. *ACS Appl. Mater. Inter.* **2018**, *10*, 3549–3561. [CrossRef] [PubMed]
40. Bhatia, P.; Nath, M. Green synthesis of p-NiO/n-ZnO nanocomposites: Excellent adsorbent for removal of congo red and efficient catalyst for reduction of 4-nitrophenol present in wastewater. *J. Water Process Eng.* **2020**, *33*, 101017. [CrossRef]
41. Zhou, Q.; Zeng, W.; Chen, W.; Xu, L.; Kumar, R.; Umar, A. High sensitive and low-concentration sulfur dioxide (SO_2) gas sensor application of heterostructure NiO-ZnO nanodisks. *Sens. Actuators B Chem.* **2019**, *298*, 126870. [CrossRef]
42. Wei, S.; Wang, S.; Zhang, Y.; Zhou, M. Different morphologies of ZnO and their ethanol sensing property. *Sens. Actuators B Chem.* **2014**, *192*, 480–487. [CrossRef]
43. Yuan, Z.; Li, J.; Meng, F. High response n-propanol sensor based on co-modified ZnO nanorods. *J. Alloy. Compd.* **2022**, *910*, 164971. [CrossRef]
44. Gao, L.; Fu, H.; Zhu, J.; Wang, J.; Chen, Y.; Liu, H. Synthesis of SnO_2 nanoparticles for formaldehyde detection with high sensitivity and good selectivity. *J. Mater. Res.* **2020**, *35*, 2208–2217. [CrossRef]
45. San, X.; Li, M.; Liu, D.; Wang, G.; Shen, Y.; Meng, D.; Meng, F. A facile one-step hydrothermal synthesis of NiO/ZnO heterojunction microflowers for the enhanced formaldehyde sensing properties. *J. Alloy. Compd.* **2018**, *739*, 260–269. [CrossRef]
46. Umar, A.; Ibrahim, A.; Kumar, R.; Algadi, H.; Albargi, H.; Alsairi, M.; Alhmami, M.; Zeng, W.; Ahmed, F.; Akbar, S. CdO-ZnO nanorices for enhanced and selective formaldehyde gas sensing applications. *Environ. Res.* **2021**, *200*, 111377. [CrossRef]
47. Dong, C.; Xu, L.; Hang, B.; Deng, S.; Xiao, X.; Wang, Y. Nonaqueous synthesis of Ag-functionalized In_2O_3/ZnO nanocomposites for highly sensitive formaldehyde sensor. *Sens. Actuators B Chem.* **2016**, *224*, 193–200. [CrossRef]
48. Jiao, S.; Xue, W.; Zhang, C.; Li, F.; Meng, B.; Zhan, Z. Improving the formaldehyde gas sensing performance of the ZnO/SnO_2 nanoparticles by PdO decoration. *J. Mater. Sci. Mater. Electron.* **2020**, *31*, 684–692.
49. Sun, Y.; Yang, H.; Zhao, Z.; Suematsu, K.; Li, P.; Yu, Z.; Zhang, W.; Hu, J. Fabrication of ZnO quantum dots@SnO_2 hollow nanospheres hybrid hierarchical structures for effectively detecting formaldehyde. *Sens. Actuators B Chem.* **2020**, *318*, 128222. [CrossRef]
50. Bai, S.; Guo, J.; Shu, X.; Xiang, X.; Luo, R.; Li, D.; Chen, A.; Liu, C. Surface functionalization of Co_3O_4 hollow spheres with ZnO nanoparticles for modulating sensing properties of formaldehyde. *Sens. Actuators B Chem.* **2017**, *245*, 359–368. [CrossRef]
51. Zhao, S.; Shen, Y.; Li, A.; Chen, Y.; Gao, S.; Liu, W.; Wei, D. Effects of rare earth elements doping on gas sensing properties of ZnO nanowires. *Ceram. Int.* **2021**, *47*, 24218–24226. [CrossRef]
52. Han, M.; Kim, H.; Lee, H.; Park, J.; Lee, H. Effects of porosity and particle size on the gas sensing properties of SnO_2 films. *Appl. Surf. Sci.* **2019**, *481*, 133–137. [CrossRef]

chemosensors

Article

Ppb-Level NO_2 Sensor with High Selectivity Fabricated by Flower-like Au-Loaded In_2O_3

Ji Zhang [1], Fangfang Zhang [1], Xu Li [2,*] and Qingji Wang [1,*]

1 State Key Laboratory of Marine Resource Utilization in South China Sea, College of Information and Communication Engineering, Hainan University, Haikou 570228, China
2 School of Chemical Engineering & Light Industry, Guangdong University of Technology, Guangzhou 510006, China
* Correspondence: lixu@gdut.edu.cn (X.L.); wangqingji@hainanu.edu.cn (Q.W.)

Abstract: With increasingly serious environmental problems caused by the improvement in people's living standards, the number of cars has increased sharply in recent years, which directly leads to the continuous increase in the concentration of NO_2 in the air. NO_2 is a common toxic and irritant gas, which is harmful to both the human body and the environment. Therefore, this research focuses on NO_2 detection and is committed to developing high-performance, low detection limit NO_2 sensors. In this study, flower-like Au-loaded In_2O_3 was successfully fabricated using the hydrothermal method and the wet impregnation method. The morphological features and chemical compositions of the as-prepared samples were characterized using SEM, TEM, XRD and XPS. A variety of sensors were fabricated and the gas-sensing properties of sensors were investigated. The results indicate that the sensor based on 0.5 mol% Au/In_2O_3 shows a response value of 1624 to 1 ppm NO_2 at 100 °C, which is 14 times that based on pure In_2O_3. Meanwhile, the detection limit of the sensor based on 0.5 mol% Au/In_2O_3 for NO_2 is 10 ppb, and the response value is 10.4. In addition, the sensor based on 0.5 mol% Au/In_2O_3 also has high selectivity to NO_2 among CO, CO_2, H_2, CH_4, NH_3, SO_2 and H_2S. Finally, the sensitization mechanism of Au/In_2O_3 was discussed, and the reasons for improving the performance of the sensor were analyzed. The above results and analysis demonstrate that the gas-sensing attributes of the sensor based on 0.5 mol% Au/In_2O_3 to NO_2 improved remarkably; at the same time, it has been proved that the composite material has extensive potential in practical applications.

Keywords: NO_2 sensor; *flower-like*; Au-loaded In_2O_3; low detection limit; high selectivity

Citation: Zhang, J.; Zhang, F.; Li, X.; Wang, Q. Ppb-Level NO_2 Sensor with High Selectivity Fabricated by Flower-like Au-Loaded In_2O_3. *Chemosensors* **2023**, *11*, 289. https://doi.org/10.3390/chemosensors11050289

Academic Editor: Boris Lakard

Received: 28 March 2023
Revised: 1 May 2023
Accepted: 8 May 2023
Published: 12 May 2023

1. Introduction

With the rapid development of science and technology, the process of urbanization and industrialization has also accelerated. While bringing convenience to people's lives, the environmental pollution caused cannot be underestimated. On the one hand, the increase in car ownership directly leads to an increase in the content of NO_2 in the atmosphere; on the other hand, the exhaust gas emitted by factories is also one of the main sources of NO_2 [1–3]. As is known to all, NO_2 is a kind of common toxic and tangy gas, which is mainly generated from the emission of automobile exhaust and industrial waste gas. NO_2 is the main culprit of environmental problems such as acid rain and smog, which cause atmospheric pollution that affects the ecological balance of the planet [4–6]. Moreover, as long as the human body inhales 1 ppm of NO_2, it can cause lung disease and breathing difficulties, which seriously affect physical health [7]. Concurrently, the World Health Organization specifies that the standard value of NO_2, which is harmful to human health, is 40 $\mu g/m^3$ (~21.25 ppb) [8]. Therefore, how to detect NO_2 quickly and effectively has become one of the urgent problems to be solved. At present, commonly used gas detection methods in the market include mass spectrometry, chromatography and so on. Compared with these large-scale detection instruments, semiconductor gas sensors stand out in the field of gas detection because of their advantages of low cost, high performance, good portability and easy integration [9–11]. However, with the improvement in people's environmental awareness, the technical requirements for gas sensors are higher than before, such as high response values, prominent selectivity and low detection limits. Thus, understanding how to develop NO_2 sensors with excellent sensing properties has attracted extensive attraction among researchers.

Up until now, among various semiconductor gas sensors, indium oxide (In_2O_3) is considered to be the most promising gas-sensing material due to its wide bandgap and high conductivity [12–15]. In recent years, it has been confirmed that In_2O_3 of different morphologies has great gas-sensing performances in NO_2 gas sensors, such as rod-shaped, sheet-shaped and flower-shaped [16]. According to reports, Shen et al. developed a NO_2 sensor with In_2O_3 nanorods using hydrothermal process, which has a response of 20.9 to 1 ppm NO_2 and a detection limit of 100 ppb (1.4) [17]. Yang et al. prepared a NO_2 sensor with In_2O_3 nanosheets via a hydrothermal process, and the sensor based on In_2O_3 nanosheets had a response value of 5.31 to 1 ppm NO_2 and a detection limit of 100 ppb (1.69) [7]. Zhou et al. fabricated a NO_2 sensor based on a In_2O_3 nanoflower using the hydrothermal method, which demonstrated a detection limit of 1 ppm NO_2 and a response of 2.1 [18]. In view of the above reports, it is not difficult to find that there are some defects regarding detection limits and response values because of the inherent properties of pure In_2O_3. In order to enhance the gas-sensing properties of the NO_2 sensor based on In_2O_3, people have attempted to load noble metal [19], construct heterojunctions [20] and compound conducting polymers [21], which have become research focuses in promoting gas-sensing properties. In the above properties-enhanced techniques, noble metal is not only used to provide high catalytic and high electroconductibility [22], but also to enhance adsorption ability for target gas on the surface of oxide, thus accelerating the electron transfer process between oxide and target gas [23–25]. In view of the above advantages, noble metal loading has proved to be one of the effective ways to improve sensor performance. Therefore, noble metal loading is used as the experimental method in this work to improve the performance of In_2O_3-based NO_2 sensors. In addition, the relevant literature and reports regarding In_2O_3-based sensors used to improve NO_2 performance through different strategies are summarized and listed, as shown in Table 1.

Table 1. The gas-sensing attributes of NO_2 sensor based on various In_2O_3.

Materials	Temp. (°C)	Conc. (ppm)	Response	DL (ppm)	Ref.
In_2O_3 microspheres	80	0.5	737.8	0.05	[26]
In_2O_3 microtubes	92	10	193	0.05	[27]
In_2O_3 nanowires	RT	5	740	0.01	[28]
In_2S_3/In_2O_3 nanoflower	160	10	251	2	[16]
Rb-doped flower-like In_2O_3	75	5	1502	0.1	[29]
In_2O_3 nanoparticles on GO	225	40	78	10	[30]
Au-loaded mesoporous In_2O_3	65	0.5	472.4	0.01	[24]
Flower-like 0.5 mol% Au/In_2O_3	100	1	1624	0.01	this work

Response: R_g/R_a; DL: detection limit.

In this work, various molar ratios of Au-loaded In_2O_3 (0 mol%, 0.3 mol%, 0.5 mol%, 1 mol%) were successfully fabricated via the hydrothermal method and the wet impregnation method. The morphological structure and chemical composition of samples prepared were characterized using XRD, SEM, TEM and XPS. Meanwhile, the as-prepared samples of different ratios were made into sensors, and gas-sensing performances were evaluated using a static test system. The results indicate that the sensors based on 0.5 mol% Au/In_2O_3 possess high response, excellent selectivity and a low detection limit toward NO_2. In addition, the gas-sensing mechanism was discussed by analyzing characterization and test results, which explained the reasons for improving the performance of sensors based on Au/In_2O_3. As such, it is demonstrated that this material has potential applications for NO_2 sensors and provides a reliable gas-sensitive material for NO_2 detection.

2. Experimental Section

2.1. Synthesis of Peach-Pit In_2O_3

In this experiment, all chemicals were purchased from Aladdin Reagent. All reagents were of analytical grade and used without further purification. Pure In_2O_3 was synthesized by a hydrothermal method. In a typical process, $InCl_3{\cdot}4H_2O$ (0.5 mmol) was dissolved in 15 mL of deionized water and then 15 mL of glycerol was successively added into the solution during gentle stirring until a homogenous solution was formed. Subsequently, trisodium citrate dihydrate $Na_3C_6H_5O_7{\cdot}2H_2O$ (1.75 mmol) was added into the above solution with vigorous stirring. When the mixed solution was stirred until homogeneity was reached, 250 μL of NaOH (0.1 M) was added into the solution and stirred to obtain a precursor solution. Finally, the precursor solution was transferred into a 50 mL PTFE-lined autoclave for hydrothermal process and heated at 190 °C for 16 h. After the autoclave was naturally cooled to room temperature, the samples were harvested by centrifugation with deionized water and absolute alcohol several times. Then, the samples were dried at 80 °C for 12 h and the white powders were collected. After that, the powders were transferred into a muffle furnace and annealed at 400 °C for 2 h with a heating rate of 2 °C/min to obtain peach-pit In_2O_3.

2.2. Synthesis of Flower-like Au-Loaded In_2O_3

Flower-like Au-loaded In_2O_3 was fabricated via the wet impregnation process. A total of 100 mg of the as-prepared peach-pit In_2O_3 was added into 10 mL of ethanol and processed by ultrasonic treatment for 20 min to obtain a uniformly dispersed solution. Subsequently, a suitable amount of $HAuCl_4{\cdot}3H_2O$ was added into the solution and stirred in a water bath at 40 °C until ethanol volatilized completely. After that, the precursor was collected and transferred into an Al_2O_3 boat. Then, the precursor was annealed at 300 °C for 2 h (2 °C/min) in a muffle furnace to obtain flower-like Au-loaded In_2O_3. Under the same experimental process, a series of Au-loaded In_2O_3 was successfully fabricated, and the molar ratios between Au and In_2O_3 were 0.3%, 0.5% and 1%.

2.3. Characterization of Samples

X-ray diffraction (XRD, Rigaku MiniFlex 600 X with Cu Kα1 radiation λ = 1.5406 Å) was used to obtain the crystal structure at 40 kV, 15 mA. A scanning electron microscope (PHENOM SCIENTIFIC ProX G5, Phenom, Rotterdam, Netherlands) was used to characterize the morphology and structure of the as-prepared samples. The detailed morphology of the as-prepared samples was characterized by transmission electron microscopy (FEI Tecnai G2 F30, FEI, Hillsboro, TX, USA). The chemical compositions of the as-prepared samples were obtained via X-ray photoelectron spectroscopy (Thermo escalab 250Xi, Thermo Fisher Scientific, Waltham, MA, USA).

2.4. Fabrication and Measurement of Gas Sensors

In this work, an Al_2O_3 ceramic tube structure (length: 4 mm, internal diameter: 0.8 mm, external diameter: 1.2 mm) was used. There was a pair of Au ring-shaped electrodes at each end of the Al_2O_3 ceramic tube with two Pt wires on each electrode as pins. The sensor was fabricated as follows: The as-prepared samples were mixed with deionized water to form a paste, which was then coated evenly onto the surface of the Al_2O_3 ceramic tube to form a sensing layer, and the tube was dried by infrared lamp for 15 min. Finally, the sensing device was annealed at 300 °C for 1 h with a heating rate of 2 °C/min to enhance the stability of the sensor. A sensor was used to insert a nickel chromium alloy wire into an aluminum oxide ceramic tube as a heater and weld it to a hexagonal base to obtain a sensor. The prepared device was aged for 24 h in a standard test environment for subsequent gas-sensitivity testing.

The gas-sensing performances were evaluated by a static test system (evaluation condition: 50% RH, 25 °C), where the heating current of the sensor was provided by a DC-regulated power supply, the resistance was recorded by multimeter and the data were registered by computers. The response value of the NO_2 sensor is defined as R_g/R_a, where R_a is the resistance value of the sensor after stabilization after exposure to NO_2 and R_g is the resistance value of the sensor after stabilization in pure air. In addition, the response and recovery time of a sensor is defined as the time required for the resistance value of the sensor to reach 90% of the total resistance value change during the adsorption and desorption process.

3. Results and Discussion

3.1. Characterization of Material Structure

To investigate the crystal phase and purity of Au-loaded In_2O_3, the XRD of the as-prepared samples is shown in Figure 1. From the XRD images, the diffraction peaks of the as-prepared samples at 2θ angles of 22.37°, 30.99°, 32.61°, 45.61°, 50.25°, 57.20° and 58.19° can be observed, in which the diffraction peaks at the corners are consistent with the refractive indices of the crystal faces (012), (104), (110), (024), (116), (214) and (300), respectively. This corresponds to hexagonal In_2O_3 (JCPDS 22-0366). There are no diffraction peaks of other impurities observed in the XRD of Au-loaded In_2O_3, which can verify that the Au-loaded In_2O_3 composite had a certain high purity. The diffraction peaks of (006) and (113) crystal planes at angles of 37.2° and 37.7° gradually become wider peaks with the increase in Au content, which may be attributed to the (111) plane diffraction peak of Au NPs at an angle of 38.2° [31]. In addition, the other Au NPs diffraction peaks were not observed, probably due to the low content of Au [32]. The XRD results prove that the Au-loaded In_2O_3 samples were prepared successfully.

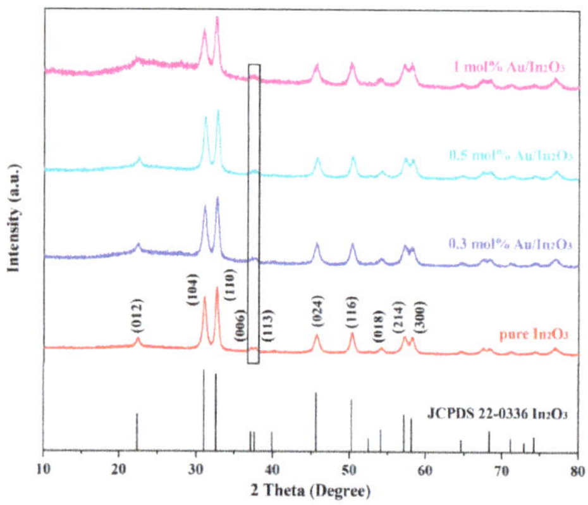

Figure 1. XRD spectra of the as-prepared samples.

Morphology and structure are two of the important factors affecting gas-sensing properties. The morphology and structure of peach-pit In_2O_3 and 0.5 mol% Au/In_2O_3 were characterized by SEM, as shown in Figure 2. It can be seen from Figure 2a,b that the diameter of peach-pit In_2O_3 was about 400 nm. Meanwhile, peach-pit In_2O_3 was identified to be pure phase because any other morphologies were not found in SEM. The morphological features of 0.5 mol% Au/In_2O_3 is shown in Figure 2c,d. Obviously, the morphology of 0.5 mol% Au/In_2O_3 became flower-like and its diameter was about 650 nm. Each flower was closely interwoven with nanosheets with a thickness of about 25 nm, constituting a flower-like uniform in size. At the same time, flower-like 0.5mol% Au/In_2O_3 was well dispersed without aggregation, and nanosheets are regularly stacked to form a flower-like structure. All the flowers bloomed, and the ultra-thin nanosheets were very loose, which provided rich space for gas diffusion. Through SEM images, it was found that the morphology and structure of Au/In_2O_3 changed, which can prove that gold-loaded Au/In_2O_3 was synthesized successfully.

Figure 2. SEM images of peach-pit In_2O_3 (**a**,**b**) and 0.5 mol% Au/In_2O_3 (**c**,**d**).

Morphology details and crystal structures were characterized by TEM and HRTEM, as exhibited in Figure 3a–d. TEM images of 0.5 mol% Au/In_2O_3 are presented in Figure 3a,b, which show that the flower-like structure of 0.5 mol% Au/In_2O_3 has independent dispersion. HRTEM images of 0.5 mol% Au/In_2O_3 clearly show crystal lattice stripes in Figure 3c,d, which can prove that 0.5 mol% Au/In_2O_3 has high crystallinity. In addition, the lattice spacing of 0.28 nm and 0.27 nm was obtained from HRTEM images, which match with In_2O_3 (104) and (110) crystal planes, respectively. The lattice spacing of 0.14 nm corresponds to (111) planes of Au nanoparticles [33]. Additionally, In, Au and O elements are regularly dispersed from the element mapping images of 0.5 mol% Au/In_2O_3, as shown in Figure 3e–h. The elemental mapping of 0.5 mol% Au/In_2O_3 can further testify that Au-loaded In_2O_3 is developed successfully.

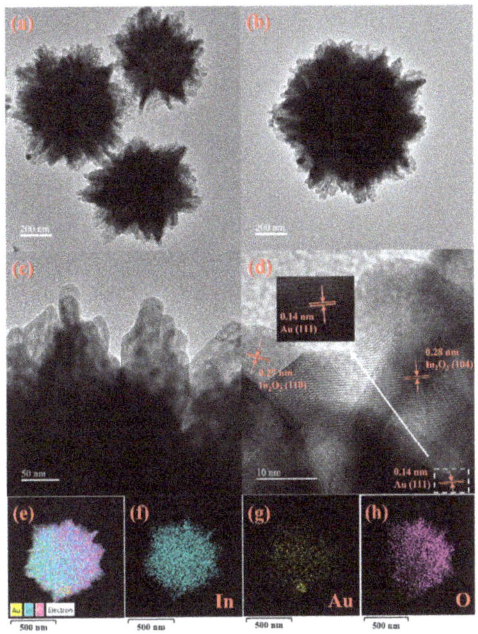

Figure 3. TEM and HRTEM (**a**–**d**) elemental mapping (**e**–**h**) of 0.5 mol% Au/In_2O_3.

The chemical compositions of In_2O_3 and 0.5 mol% Au/In_2O_3 were characterized by XPS, as shown in Figure 4a,b. The In 3d spectra of pure In_2O_3 and 0.5 mol% Au/In_2O_3 both presented two peaks at 444.0 eV, 451.6 eV and 444.1 eV, 451.7 eV, respectively, each of which is fitted to the In-O bond of In $3d_{5/2}$ and In $3d_{3/2}$ [34]. The O 1s spectra of In_2O_3 and 0.5 mol% Au/In_2O_3 are divided into three peaks by fitting process, as is exhibited in Figure 4b. The fitting peaks emerged at 529.5 eV, 531.2 eV, 532.1 eV and 529.6 eV, 531.2 eV 532.1 eV in pure In_2O_3 and 0.5 mol% Au/In_2O_3, which are assigned to lattice oxygen (OL), oxygen vacancy (OV) and chemical-adsorbed oxygen (OC), respectively [35]. Moreover, the content of oxygen species in pure In_2O_3 and 0.5 mol% Au/In_2O_3 are compared in Figure 4d, where the content of OL, OV and OC in pure In_2O_3 and Au/In-0.5 are 65.87%, 21.56%, 12.57% and 51.8%, 30.5%, 17.7%, respectively. Compared to pure In_2O_3, the contents of OV and OC in 0.5 mol% Au/In_2O_3 increased, which may be responsible for the improvement in gas-sensing properties. The Au 4f spectra of 0.5 mol% Au/In_2O_3 are shown in Figure 4c, where the peaks at 83.3 eV and 87.1 eV separately conform to Au $4f_{7/2}$ and Au $4f_{5/2}$ [36].

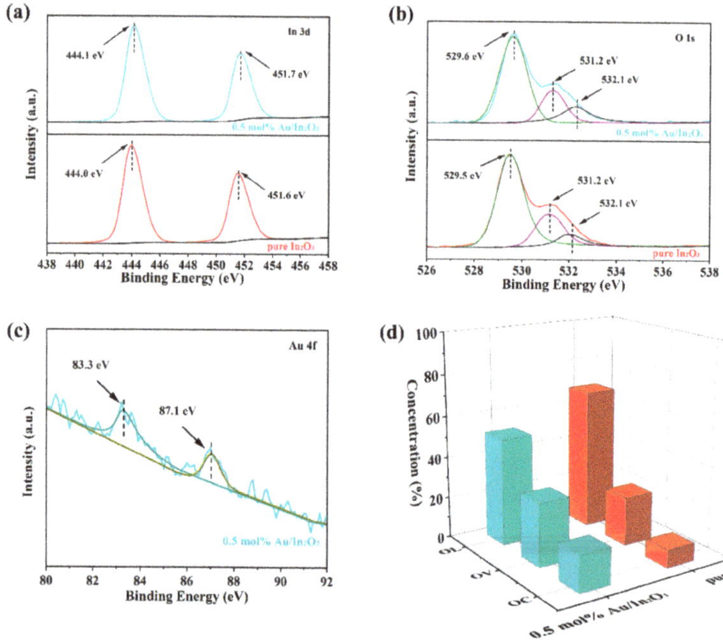

Figure 4. XPS spectrums of samples. The In 3d and O 1s of samples (**a–d**); the Au 4f of 0.5 mol% Au/In_2O_3 (**c**); oxygen species content of pure In_2O_3 and 0.5 mol% Au/In_2O_3 (**d**).

3.2. Gas-Sensing Properties

The operating temperature is the primary consideration for sensors, which is appraised by the response values of different sensors to 1 ppm NO_2 at different temperatures, as displayed in Figure 5a. It is apparent that the responses values of different gas sensors to 1 ppm NO_2 reach their maximum at 100 °C. The response values of sensors based on 0.5 mol% Au/In_2O_3 and pure In_2O_3 are1624 and 117 to 1 ppm NO_2 at 100 °C, respectively, and the response value of sensors based on 0.5 mol% Au/In_2O_3 is 14 times that of those based on pure In_2O_3. Therefore, 100 °C is regarded as the optimum operating temperature in this paper. The air resistance (R_a) of different sensors at different temperatures is exhibited in Figure 5b. Undoubtedly, R_a decreased as the temperature increased. Additionally, we found that the resistance of sensors based on different content of Au-loaded In_2O_3 was lower than that of pure In_2O_3 [25]. The transient curves of different sensors to 1 ppm NO_2 are exhibited in Figure 5c–f. The resistances of different sensors ascended in a NO_2

atmosphere, which conforms to the gas-sensing characteristics based on In_2O_3 sensors. Meanwhile, it is apparent that in NO_2, resistance changed drastically in the sensor based on 0.5 mol% Au/In_2O_3 compared to the other sensors, and that the gas-sensing attributes significantly improved.

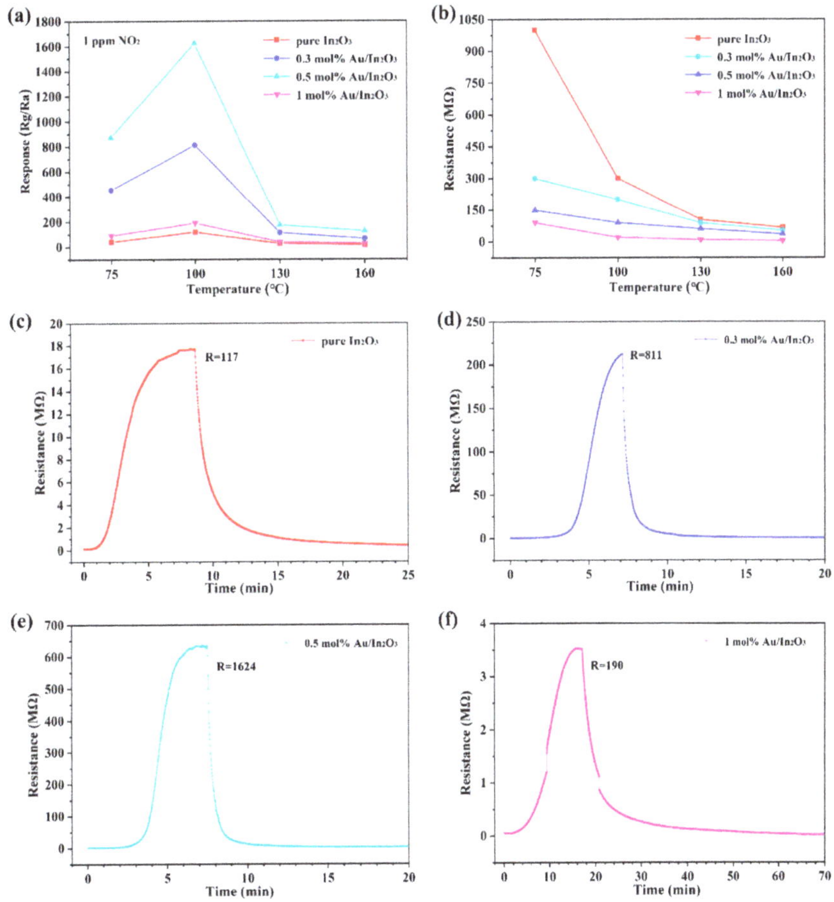

Figure 5. Response values and resistance values of different sensors at 75 °C to 160 °C (**a**,**b**); transient curve of different sensors at 100 °C (**c**–**f**).

The stability of the sensor represents the adaptability of the device to different concentrations of gas, while also reflecting the detection concentration range. Dynamic curves are used to further compare the gas-sensing properties of sensors based on 0.5 mol% Au/In_2O_3 and pure In_2O_3, as shown in Figure 6a,b. At different NO_2 concentrations, the gas-sensing properties of the sensor based on 0.5 mol% Au/In_2O_3 were promoted remarkably, in comparison with the sensor based on pure In_2O_3. Additionally, the response value greatly improved while the response recovery time did. Simultaneously, the detection limit of the sensor based on 0.5 mol% Au/In_2O_3 was boosted significantly. The detection limit of the sensor based on pure In_2O_3 was 50 ppb NO_2 and the response value was only 10. The detection limit of the sensor based on 0.5 mol% Au/In_2O_3 was 10 ppb NO_2 and had a response value of 10.4. Moreover, the response values of the sensor based on 0.5 mol% Au/In_2O_3 were positively correlated with increasing concentrations of NO_2, as displayed in Figure 6c. The transient response of the sensor based on 0.5 mol% Au/In_2O_3 to10 ppb NO_2 is shown in Figure 6d, where the response and recovery time is 390s and 270s, respectively.

To sum up, although low concentrations of NO_2, the sensor based on 0.5 mol% Au/In_2O_3 shows excellent gas-sensing properties. The above test and evaluation results prove that 0.5 mol% Au/In_2O_3 has good stability and a low detection limit, and that the detection range for NO_2 is very wide.

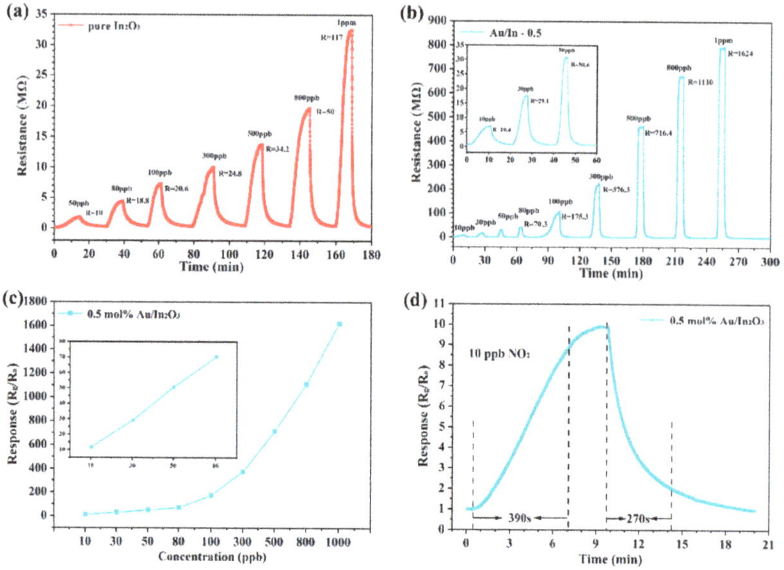

Figure 6. The dynamic response curve of sensors based on pure In_2O_3 and 0.5 mol% Au/In_2O_3 at 100 °C (**a**,**b**); response value of sensor based on 0.5 mol% Au/In_2O_3 (**c**); sensor based on 0.5 mol% Au/In_2O_3 vs. 10 ppb NO_2 transient curve (**d**).

Reproducibility and long-term stability jointly determine whether a sensor can be applied to practical indicators; these two performance parameters directly affect the service life and maintenance frequency of the sensor, which are very important in practical applications. The reproducibility of the sensor based on 0.5 mol% Au/In_2O_3 was evaluated in 500 ppb NO_2 five times, as is represented in Figure 7a. In the five-cycle experiment, the sensor based on 0.5 mol% Au/In_2O_3 was continuously and alternately exposed to 500 ppb NO_2 and air, which showed the settled response and recovery properties. Figure 7b signals the response values of the sensor based on 0.5 mol% Au/In_2O_3 to 500 ppb NO_2 in the five-cycle experiment, and it can be noticed that the response values did not change evidently throughout the five tests; meanwhile, the response characteristics were basically consistent. According to the above results, the sensor based on 0.5 mol% Au/In_2O_3 was provided with great reproducibility. The long-term stability of the sensor based on 0.5 mol% Au/In_2O_3 is exhibited in Figure 7c. The response values of the sensor based on 0.5 mol% Au/In_2O_3 at 500 ppb NO_2 presented no downward trend in 2 weeks and the variation was within 5%, which can prove that the sensor based on 0.5 mol% Au/In_2O_3 has good stability. Furthermore, the selectivity of the sensor based on 0.5 mol% Au/In_2O_3 is shown in Figure 7d, where the response values of the sensor based on 0.5 mol% Au/In_2O_3 to CO, CO_2, H_2, CH_4, NH_3 of 500 ppm, as well as SO_2 and H_2S of 10 ppm, were lower than that of 10 ppb NO_2. It is worth noting that the response value of the sensor based on 0.5 mol% Au/In_2O_3 of 10 ppb NO_2 was 6–10 times of that other interference gases ($R_{10\ ppb\ NO2}/R_{500\ ppm\ NH3}$ = 10, $R_{10\ ppb\ NO2}/R_{10\ ppm\ H2S}$ = 6), which shows that the sensor based on 0.5 mol% Au/In_2O_3 has bodacious selectivity to NO_2. In view of the test and evaluation results reproducibility, long-term stability and selectivity, it is shown that the sensor based on 0.5 mol% Au/In_2O_3 has certain practical application value. This article provides a new type of gas-sensing material for NO_2 sensors.

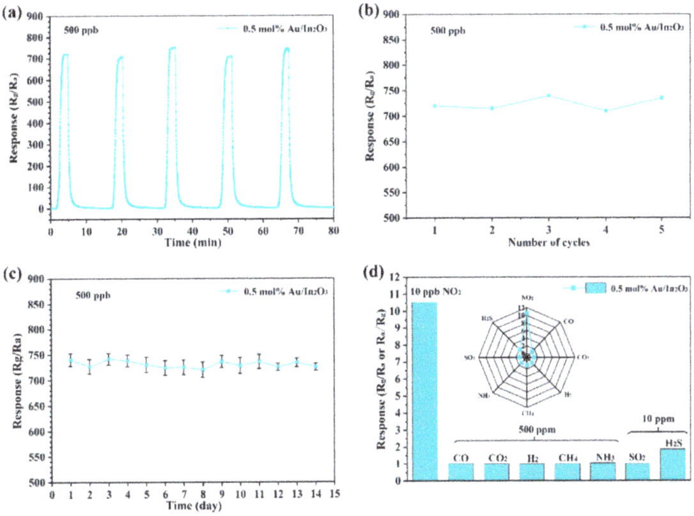

Figure 7. Five-cycle transient curve and its corresponding response values of sensor based on 0.5 mol% Au/In_2O_3 at 100 °C (**a**,**b**); long-term stability and selectivity (**c**,**d**).

3.3. Gas-Sensing Mechanism

The sensing mechanism of the sensor based on In_2O_3 is a process in which the target gas reacts with the adsorbed oxygen on the In_2O_3 surface to change the resistance [19]. In_2O_3 is a typical n-type semiconductor oxide, and when the sensor based on In_2O_3 is in the air, oxygen molecules combine with electrons in In_2O_3 to form active oxygen (O_2^-), as shown in Equations (1) and (2) [37]. When the sensor based on In_2O_3 makes contact with NO_2, the reaction is as follows: NO_2 not only reacts with the active oxygen on the surface of In_2O_3, but also occupies the electrons in the conduction band of In_2O_3. In short, NO_2 is paired with O_2^- and e^-, and the reaction is shown in Equations (3) and (4) [38,39]. When the sensor based on In_2O_3 is separated from NO_2, it returns to the initial state, as shown in Equation (5) [29].

$$O_2\ (gas) \rightarrow O_2\ (ads) \tag{1}$$

$$O_2\ (ads) + e^- \rightarrow O_2^-\ (ads) \tag{2}$$

$$NO_2\ (gas) + e^- \rightarrow NO_2^-\ (ads) \tag{3}$$

$$NO_2\ (gas) + O_2^-\ (ads) + 2e^- \rightarrow NO_2^-\ (ads) + 2O^-\ (ads) \tag{4}$$

$$NO_2^-\ (ads) + 2O^-\ (ads) \rightarrow NO_2\ (gas) + O_2\ (gas) + 3e^- \tag{5}$$

The reaction process of flower-like 0.5 mol% Au/In_2O_3 after contact with NO_2 is shown in Figure 8. The reasons for the remarkable improvement in gas-sensing performances based on 0.5 mol% Au/In_2O_3 can be attributed to the following three points: The first reason is that they can benefit from the specific morphology and structure. According to SEM and TEM images, the morphology of 0.5 mol% Au/In_2O_3 becomes a flower-like and is stacked by nanosheets. Moreover, flower-like 0.5 mol% Au/In_2O_3 has better dispersion than peach-pit In_2O_3, and there is no obvious aggregation of flower-like morphology and structure. The size of the small flowers is relatively uniform and well dispersed. In addition, abundant gaps are clearly displayed between adjacent nanosheets, which facilitates gas diffusion and enables sufficient reaction between the target gas and the sensing material. Therefore, dispersive and uniform morphology may provide a lot of space for the diffusion of gas molecules [39,40], which may promote gas-sensing performances; The second reason is due to the high catalytic activity of the noble metal Au. When In_2O_3-

based sensors come into contact with NO_2, they undergo an oxidation reaction on the surface of In_2O_3. Due to the inherent characteristics of In_2O_3 with insufficient surface activity, the reaction is not intense enough, resulting in poor gas-sensing performance. Owing to the high catalytic activity of Au, the activation energy of the chemical adsorption reaction of gas molecules can be reduced, and the chemical reaction rate can be increased. Accordingly, the gas-sensing performances of the sensor is enhanced [37]. The third reason can be attributed to the fact that 0.5 mol% Au/In_2O_3 has more OV and OC than pure In_2O_3. Based on the XPS characterization results, the content of OV and OC in 0.5 mol% Au/In_2O_3 is significantly higher than that in pure In_2O_3. The increase in OC content may make In_2O_3 have more reactive oxygen [38,41],which makes the oxidation reaction on the surface of In_2O_3 easier. As such, this may be another reason for improving gas-sensing performances. In consideration of the above three reasons, the performance of Au/In_2O_3-based sensors have been significantly improved, providing reliable gas-sensitive material for NO_2 detection.

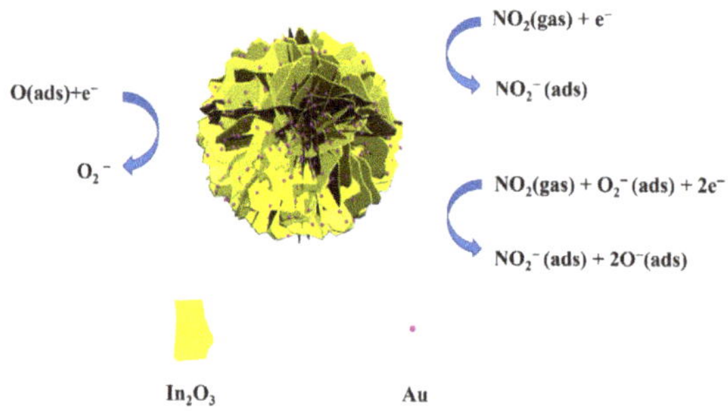

Figure 8. Diagram of gas-sensing mechanism of sensor based on 0.5 mol% Au/In_2O_3 to NO_2.

4. Conclusions

In summary, peach-pit In_2O_3 and flower-like 0.5 mol% Au/In_2O_3 were successfully prepared by the hydrothermal method and wet impregnation. The sensors based on different Au-loaded In_2O_3 were prepared for the evaluation of gas-sensing performances. Among those sensors, sensors based on 0.5 mol% Au/In_2O_3 exhibited excellent gas-sensitive properties. The specific performance is as follows: The response value of sensors based on 0.5 mol% Au/In_2O_3 is 1624 to 1 ppm NO_2 at 100 °C, which is 14 times higher than pure In_2O_3. Furthermore, the response value is 10.4 of sensors based on 0.5 mol% Au/In_2O_3 to 10 ppb NO_2, which can prove that this sensor has a low detection limit. The detection limit of the sensor based on pure In_2O_3 is 50 ppb, and the response value is only 10. In the meantime, the sensor based on 0.5 mol% Au/In_2O_3 also has good selectivity ($R_{10\ ppb\ NO2}/R_{500\ ppm\ NH3}$ = 10, $R_{10\ ppb\ NO2}/R_{10\ ppm\ H2S}$ = 6) and reliable repeatability. The sensitization mechanism is discussed through XRD, SEM, TEM, and XPS characterization analysis, and the improvement in the sensor based on 0.5 mol% Au/In_2O_3 performance can be attributed to the unique flower-like morphology, high catalytic activity of Au and the increase in oxygen species. The above reasons play a positive role in the diffusion and adsorption of NO_2 molecules for the sensor based on 0.5 mol% Au/In_2O_3. In conclusion, this work provides a new type of gas-sensing composite via a simple experimental method for NO_2 sensors and proves that this composite has potential application value.

Author Contributions: J.Z. and F.Z.: Writing—original draft; X.L.: Writing—review; Q.W.: Conceptualization and Supervision. All authors have read and agreed to the published version of the manuscript.

Funding: This work was funded by the National Nature Science Foundation of China (61803172), Hainan Provincial Natural Science Foundation of China (621RC509) and the Start-up Research Foundation of Hainan University (KYQD(ZR)1910).

Institutional Review Board Statement: Not applicable.

Informed Consent Statement: Not applicable.

Data Availability Statement: The data can be obtained from the corresponding author.

Acknowledgments: This work was funded by the National Nature Science Foundation of China (61803172), Hainan Provincial Natural Science Foundation of China (621RC509) and the Start-up Research Foundation of Hainan University (KYQD(ZR)1910).

Conflicts of Interest: The authors declare no conflict of interest.

References

1. Lin, H.; Wang, J.; Xu, S.; Zhang, Q.; Cheng, Y.; Han, D.; Wang, H.; Zhuo, K. Au-WO_3 Nanowire-Based Electrodes for NO_2 Sensing. *ACS Appl. Nano Mater.* **2022**, *5*, 14311–14319. [CrossRef]
2. Wang, Q.J.; Kou, X.Y.; Liu, C.; Zhao, L.J.; Lin, T.T.; Liu, F.M.; Yang, X.L.; Lin, J.; Lu, G.Y. Hydrothermal synthesis of hierarchical CoO/SnO_2 nanostructures for ethanol gas sensor. *J. Colloid Interface Sci.* **2018**, *513*, 760–766. [CrossRef] [PubMed]
3. Gu, D.; Liu, W.; Wang, J.; Yu, J.; Zhang, J.W.; Huang, B.Y.; Rumyantseva, M.N.; Li, X.G. Au Functionalized SnS_2 Nanosheets Based Chemiresistive NO_2 Sensors. *Chemosensors* **2022**, *10*, 165. [CrossRef]
4. Ahmadi, A.; Abbaspour, M.; Arjmandi, R.; Abedi, Z. Air Quality Risk Index (AQRI) and its application for a megacity. *Int. J. Environ. Sci. Technol.* **2015**, *12*, 3773–3780. [CrossRef]
5. Bai, M.J.; Chen, M.; Li, X.; Wang, Q.J. One-step CVD growth of ZnO nanorod/SnO_2 film heterojunction for NO_2 gas sensor. *Sens. Actuator B Chem.* **2022**, *373*, 132738. [CrossRef]
6. Bonyani, M.; Zebarjad, S.M.; Janghorban, K.; Kim, J.Y.; Kim, H.W.; Kim, S.S. Au-Decorated Polyaniline-ZnO Electrospun Composite Nanofiber Gas Sensors with Enhanced Response to NO_2 Gas. *Chemosensors* **2022**, *10*, 388. [CrossRef]
7. Yang, W.; Chen, H.; Lu, J. Assembly of stacked In_2O_3 nanosheets for detecting trace NO_2 with ultrahigh selectivity and promoted recovery. *Appl. Surf. Sci.* **2021**, *539*, 148217. [CrossRef]
8. Chowdhury, S.; Haines, A.; Klingmuller, K.; Kumar, V.; Pozzer, A.; Venkataraman, C.; Witt, C.; Lelieveld, J. Global and national assessment of the incidence of asthma in children and adolescents from major sources of ambient NO_2. *Environ. Res. Lett.* **2021**, *16*, 11. [CrossRef]
9. Wang, Q.; Wang, C.; Sun, H.; Sun, P.; Wang, Y.; Lin, J.; Lu, G. Microwave assisted synthesis of hierarchical Pd/SnO_2 nanostructures for CO gas sensor. *Sens. Actuators B Chem.* **2016**, *222*, 257–263. [CrossRef]
10. Wang, Q.J.; Bao, L.W.; Cao, Z.Q.; Li, C.Y.; Li, X.; Liu, F.M.; Sun, P.; Lu, G.Y. Microwave-assisted hydrothermal synthesis of Pt/SnO_2 gas sensor for CO detection. *Chin. Chem. Lett.* **2020**, *31*, 2029–2032. [CrossRef]
11. Meng, F.L.; Shi, X.; Yuan, Z.Y.; Ji, H.Y.; Qin, W.B.; Shen, Y.B.; Xing, C.Y. Detection of four alcohol homologue gases by ZnO gas sensor in dynamic interval temperature modulation mode. *Sens. Actuator B Chem.* **2022**, *350*, 130867. [CrossRef]
12. Tuerdi, A.; Yan, P.; He, F.G.; Abdukayum, A. Enhanced photocatalytic activity of a flower-like $In_2O_3/ZnGa_2O_4$:Cr heterojunction composite with long persisting luminescence. *RSC Adv.* **2022**, *12*, 34874–34881. [CrossRef] [PubMed]
13. Nasriddinov, A.; Tokarev, S.; Fedorova, O.; Bozhev, I.; Rumyantseva, M. In_2O_3 Based Hybrid Materials: Interplay between Microstructure, Photoelectrical and Light Activated NO_2 Sensor Properties. *Chemosensors* **2022**, *10*, 135. [CrossRef]
14. Ueda, T.; Boehme, I.; Hyodo, T.; Shimizu, Y.; Weimar, U.; Barsan, N. Enhanced NO_2-Sensing Properties of Au-Loaded Porous In_2O_3 Gas Sensors at Low Operating Temperatures. *Chemosensors* **2020**, *8*, 72. [CrossRef]
15. Meng, F.L.; Wang, H.; Yuan, Z.Y.; Zhang, R.Z.; Li, J. Ppb-Level Triethylamine Gas Sensors Based on Palladium Nanoparticles Modified Flower-Like In_2O_3 Grown on rGO Nanosheets Operating at Low Temperature. *IEEE Trans. Instrum. Meas.* **2022**, *71*, 1–9. [CrossRef]
16. Wang, B.-R.; Liu, L.-Y.; Guo, G.-C.; Bai, Y.-J.; Tu, J.-C.; Wang, R.-Z. Interface enhancement effect of hierarchical In_2S_3/In_2O_3 nanoflower heterostructures on NO_2 gas sensitivity. *Appl. Surf. Sci.* **2022**, *584*, 152669. [CrossRef]
17. Shen, Y.; Zhong, X.; Zhang, J.; Li, T.; Zhao, S.; Cui, B.; Wei, D.; Zhang, Y.; Wei, K. In-situ growth of mesoporous In_2O_3 nanorod arrays on a porous ceramic substrate for ppb-level NO_2 detection at room temperature. *Appl. Surf. Sci.* **2019**, *498*, 143873. [CrossRef]
18. Zhou, P.; Shen, Y.; Lu, W.; Zhao, S.; Li, T.; Zhong, X.; Cui, B.; Wei, D.; Zhang, Y. Highly selective NO_2 chemiresistive gas sensor based on hierarchical In_2O_3 microflowers grown on clinoptilolite substrates. *J. Alloys Compd.* **2020**, *828*, 154395. [CrossRef]
19. Liu, Y.; Li, S.; Xiao, S.; Du, K. In_2O_3 microtubes decorated with Ag nanoparticles for NO_2 gas detection at room temperature. *Vacuum* **2022**, *202*, 111197. [CrossRef]
20. Yuan, Z.-Y.; Yang, F.; Zhu, H.-M.; Meng, F.-L.; Ibrahim, M. High-response n-butanol gas sensor based on ZnO/In_2O_3 heterostructure. *Rare Met.* **2022**, *42*, 198–209. [CrossRef]
21. Han, C.; Li, X.; Liu, J.; Dong, H.; Cheng, W.; Liu, Y.; Xin, J.; Li, X.; Shao, C.; Liu, Y. $In_2O_3/g\text{-}C_3N_4$/Au ternary heterojunction-integrated surface plasmonic and charge-separated effects for room-temperature ultrasensitive NO_2 detection. *Sens. Actuators B Chem.* **2022**, *371*, 132448. [CrossRef]
22. Fan, Y.-Y.; Tu, H.-L.; Pang, Y.; Wei, F.; Zhao, H.-B.; Yang, Y.; Ren, T.-L. Au-decorated porous structure graphene with enhanced sensing performance for low-concentration NO_2 detection. *Rare Met.* **2020**, *39*, 651–658. [CrossRef]

23. Ding, M.D.; Xie, N.; Wang, C.; Kou, X.Y.; Zhang, H.; Guo, L.L.; Sun, Y.F.; Chuai, X.H.; Gao, Y.; Liu, F.M.; et al. Enhanced NO_2 gas sensing properties by Ag-doped hollow urchin-like In_2O_3 hierarchical nanostructures. *Sens. Actuator B Chem.* **2017**, *252*, 418–427. [CrossRef]
24. Li, S.; Cheng, M.; Liu, G.; Zhao, L.; Zhang, B.; Gao, Y.; Lu, H.; Wang, H.; Zhao, J.; Liu, F.; et al. High-response and low-temperature nitrogen dioxide gas sensor based on gold-loaded mesoporous indium trioxide. *J. Colloid Interface Sci.* **2018**, *524*, 368–378. [CrossRef]
25. Ueda, T.; Boehme, I.; Hyodo, T.; Shimizu, Y.; Weimar, U.; Barsan, N. Effects of Gas Adsorption Properties of an Au-Loaded Porous In_2O_3 Sensor on NO_2-Sensing Properties. *ACS Sens.* **2021**, *6*, 4019–4028. [CrossRef] [PubMed]
26. Hu, X.L.; Zhou, X.; Wang, B.; Sun, P.; Li, X.W.; Wang, C.; Liu, J.Y.; Lu, G.Y. Facile synthesis of hollow In_2O_3 microspheres and their gas sensing performances. *Rsc Adv.* **2015**, *5*, 4609–4614. [CrossRef]
27. Wang, N.; Ye, J.-X.; Sun, J.-B.; Zhang, X.-F.; Deng, Z.-P.; Xu, Y.-M.; Huo, L.-H.; Gao, S. Rapid and accurate detection of highly toxic NO_2 gas based on catkins biomass-derived porous In_2O_3 microtubes at low temperature. *Sens. Actuator B Chem.* **2022**, *361*, 131692. [CrossRef]
28. Zhang, B.; Bao, N.; Wang, T.; Xu, Y.; Dong, Y.; Ni, Y.; Yu, P.; Wei, Q.; Wang, J.; Guo, L.; et al. High-performance room temperature NO_2 gas sensor based on visible light irradiated In_2O_3 nanowires. *J. Alloys Compd.* **2021**, *867*, 159076. [CrossRef]
29. Wang, Y.; Yao, L.; Xu, L.; Wu, W.; Lin, W.; Zheng, C.; Feng, Y.; Gao, X. Enhanced NO_2 gas sensing properties based on Rb-doped hierarchical flower-like In_2O_3 microspheres at low temperature. *Sens. Actuator B Chem.* **2021**, *332*, 129497. [CrossRef]
30. Shah, S.; Han, S.; Hussain, S.; Liu, G.; Shi, T.; Shaheen, A.; Xu, Z.; Wang, M.; Qiao, G. NO_2 gas sensing responses of In_2O_3 nanoparticles decorated on GO nanosheets. *Ceram. Int.* **2022**, *48*, 12291–12298. [CrossRef]
31. Li, Z.; Zhang, Y.; Zhang, H.; Yi, J.X. MOF-derived Au-loaded Co_3O_4 porous hollow nanocages for acetone detection. *Sens. Actuator B Chem.* **2021**, *344*, 9. [CrossRef]
32. Lv, X.H.; Lan, H.; Guo, J.; Guo, M.X.; Yan, Y. Synthesis of Au-loaded $AgInS_2$ nanoparticles with highly enhanced visible light photocatalytic performances. *J. Mater. Sci.-Mater. Electron.* **2020**, *31*, 22284–22296. [CrossRef]
33. Ponnuvelu, D.V.; Dhakshinamoorthy, J.; Prasad, A.K.; Dhara, S.; Kamruddin, M.; Pullithadathil, B. Geometrically Controlled Au-Decorated ZnO Heterojunction Nanostructures for NO_2 Detection. *ACS Appl. Nano Mater.* **2020**, *3*, 5898–5909. [CrossRef]
34. Yu, Q.; Jin, R.; Zhao, L.; Wang, T.; Liu, F.; Yan, X.; Wang, C.; Sun, P.; Lu, G. MOF-Derived Mesoporous and Hierarchical Hollow-Structured In_2O_3-NiO Composites for Enhanced Triethylamine Sensing. *ACS Sens.* **2021**, *6*, 3451–3461. [CrossRef]
35. Liu, Y.; Liu, J.; Pan, Q.; Pan, K.; Zhang, G. Metal-organic framework (MOF) derived In_2O_3 and g-C_3N_4 composite for superior NOx gas-sensing performance at room temperature. *Sens. Actuator B Chem.* **2022**, *352*, 131001. [CrossRef]
36. Yang, G.; Ding, H.; Chen, D.M.; Feng, J.J.; Hao, Q.; Zhu, Y.F. Construction of urchin-like $ZnIn_2S_4$-Au-TiO_2 heterostructure with enhanced activity for photocatalytic hydrogen evolution. *Appl. Catal. B Environ.* **2018**, *234*, 260–267. [CrossRef]
37. Zhu, Q.; Gu, D.; Liu, Z.; Huang, B.; Li, X. Au-modified 3D SnS_2 nano-flowers for low-temperature NO_2 sensors. *Sens. Actuator B Chem.* **2021**, *349*, 130775. [CrossRef]
38. Kou, X.Y.; Wang, C.; Ding, M.D.; Feng, C.H.; Li, X.; Ma, J.; Zhang, H.; Sun, Y.F.; Lu, G.Y. Synthesis of Co-doped SnO_2 nanofibers and their enhanced gas-sensing properties. *Sens. Actuator B Chem.* **2016**, *236*, 425–432. [CrossRef]
39. Han, D.; Zhai, L.; Gu, F.; Wang, Z. Highly sensitive NO_2 gas sensor of ppb-level detection based on In_2O_3 nanobricks at low temperature. *Sens. Actuator B Chem.* **2018**, *262*, 655–663. [CrossRef]
40. Fan, F.Y.; Tang, P.G.; Wang, Y.Y.; Feng, Y.J.; Chen, A.F.; Luo, R.X.; Li, D.Q. Facile synthesis and gas sensing properties of tubular hierarchical ZnO self-assembled by porous nanosheets. *Sens. Actuator B Chem.* **2015**, *215*, 231–240. [CrossRef]
41. Wang, Z.H.; Hou, C.L.; De, Q.M.; Gu, F.B.; Han, D.M. One-Step Synthesis of Co-Doped In_2O_3 Nanorods for High Response of Formaldehyde Sensor at Low Temperature. *ACS Sens.* **2018**, *3*, 468–475. [CrossRef] [PubMed]

chemosensors

Article

Engineering Band Structure of SnO_2 Nanoparticles via Coupling with g-C_3N_4 Nanosheet for the Detection of Ethanolamine

Jiuyu Li [1], Kerui Xie [1], Yating Wang [1], Ruihua Zhao [1], Yangyang Shang [1] and Jianping Du [1,2,3,*]

1 College of Chemical and Engineering Technology, Taiyuan University of Technology, 79 Yingze West Street, Taiyuan 030024, China
2 College of Chemistry, Taiyuan University of Technology, 79 Yingze West Street, Taiyuan 030024, China
3 Key Laboratory of Gas Energy Efficient and Clean Utilization, Taiyuan University of Technology, Taiyuan 030024, China
* Correspondence: dujp518@163.com

Abstract: Volatile organoamines are important industrial raw materials and chemicals. Long-term exposure to amines could be harmful to human health and even cause serious pollution. In this study, SnO_2 decorated g-C_3N_4 material was fabricated and used as a sensor material for the detection of ethanolamine (EA). The structures, morphology, surface chemical states, and band structure were characterized, and gas sensing was studied. The results showed that SnO_2 nanoparticles were dispersed on g-C_3N_4, and band structure was dependent on g-C_3N_4 doping. Notably, the interface heterojunction was conducive to electron transferring and O_2 molecule adsorption; the formed reactive oxygen species enhanced the reaction between oxygen and EA, thus leading to high sensitivity to EA. This composite exhibited a high response that was 2.6 times higher than that of pure SnO_2, and the detection limit reached 294 ppb. A g-C_3N_4/SnO_2-based sensor displayed a high selectivity to EA with a fast response time (1 s) and recovery time (20 s) at low operating temperatures. In particular, this sensor exhibited a linear relationship between the response and concentration, which is required for quantitative analysis.

Keywords: g-C_3N_4/SnO_2; heterojunction; band structure; ethanolamine; sensors

Citation: Li, J.; Xie, K.; Wang, Y.; Zhao, R.; Shang, Y.; Du, J. Engineering Band Structure of SnO_2 Nanoparticles via Coupling with g-C_3N_4 Nanosheet for the Detection of Ethanolamine. *Chemosensors* **2023**, *11*, 296. https://doi.org/10.3390/chemosensors11050296

Academic Editor: Simonetta Capone

Received: 28 March 2023
Revised: 26 April 2023
Accepted: 10 May 2023
Published: 16 May 2023

1. Introduction

Volatile organic amine compounds (VOACs) have a wide range of sources [1]. Natural sources mainly include animal excrement, animal and plant remains, and microbial products; non-natural sources mainly include chemical production, industrial waste discharge, and domestic waste. The harm inflicted on living organisms and the ecological environment by VOACs should not be underestimated. The excessive presence of VOACs will cause biological poisoning and environmental deterioration [2]. In China, the emissions from non-natural sources are much higher than those from natural sources [3]. It is not enough to rely on the self-regulation of the ecosystem. Therefore, it is necessary to deal with these harmful compounds artificially. Strengthening the monitoring of VOACs such as triethylamine (TEA) and ethanolamine (EA) is an important means to control their emissions and reduce their harm, so suitable detection methods and detection materials are particularly needed.

VOCs are usually detected via spectrometry, chromatography, etc., which have the advantages of low error rates and strong stability. However, their operation is complicated; they usually need to be used in combination with other instruments, and these factors restrict their portable applications. Therefore, it is necessary to develop sensing material to quickly detect VOCs on a large scale. In this regard, semiconductor metal oxide materials, due to their simple preparation and low cost, are recognized as materials with applications in chemical sensors [4–8]. Our previous research reported strategies for the application of ZnO and SnO_2 via non-metal and metal doping, and we found that different, modified oxides can respond to different amines, such as S-ZnO for EA [9] and Cr-SnO_2 for TEA [10]. Another case is CuO/SnO_2: the heterojunction between the interface of CuO and SnO_2 cause this composite to exhibit a high response to TEA [11]. For comparison, SnO_2 is also a common sensing material [12–15]. For example, Zhang et al. prepared nanostructural SnO_2 for ethanol detection and improved its performance [16]. Ma et al. prepared mesoporous SnO_2 for the fast detection of formaldehyde [17]. Din et al. developed NiO/SnO_2 heterojunction material to detect ethanol, and it exhibited a high performance [18]. Dai et al. prepared rod-like Au-modified ZnO nanoflowers for EA detection [19]. However, doped and oxide-modified SnO_2 is only efficient in the identification of TEA and other VOCs. In order to improve its response to EA, we chose to disperse SnO_2 on a matrix and construct a heterojunction by introducing non-metallic materials to this material.

Graphitic carbon nitride (g-C_3N_4), which possesses a two-dimensional structure and is composed of carbon and nitrogen atoms, is a promising sensitization material [20]. It is an n-type semiconductor material with many advantages, such as a high specific surface area, excellent catalytic properties, remarkable two-dimensional material properties, and a gas adsorption ability [21], so it has attracted the attention of many researchers. g-C_3N_4 has been used in photocatalysis [22,23], energy storage [24], gas sensing [25], and optoelectronic devices [26]. Because of its distinctive two-dimensional structure, it can be used to improve the gas sensitivity of semiconductor metal oxides for applications in gas sensors.

g-C_3N_4 can form heterogeneous structures with semiconductor metal oxides, which can not only cause metal oxide nanoparticles to have good dispersibility but also optimize their gas sensitivity. Cao et al. [27] synthesized a SnO_2 compound material with two-dimensional g-C_3N_4 nanosheets via a hydrothermal route, and the compound demonstrated a high sensitivity to ethanol gas. The sensitivity to 500 ppm ethanol gas was 240 at 300 °C. Lu et al. [28] prepared a 2D/2D ZnO/g-C_3N_4 heterojunction material, and the results showed excellent NO_2 sensing performance and a response to 7 ppm NO_2 that reached 44.8. Niu et al. prepared mesoporous Co_3O_4 nanowires modified with g-C_3N_4 nanosheets, and this composite displayed a high level of response to toluene gas [29]. The high surface area and great electronic structure of two-dimensional g-C_3N_4 are beneficial to improving the gas sensitivity of semiconductors. These composites have enhanced response rates and selectivity compared to pure metal oxides.

In this work, SnO_2-decorated g-C_3N_4 materials were synthesized and characterized, and the band structure was regulated by changing g-C_3N_4. Gas-sensing performance was studied based on the response to volatile organic compounds.

2. Experimental

2.1. Chemicals and Reagents

All reagents were analytical grade and were not further purified before use. Urea was purchased from Tianjin Cameo Chemical Reagent Co., Ltd., Tianjin, China. Stannous chloride was purchased from Tianjin Guangfu Technology Development Co., Ltd., Tianjin, China. Thiourea was purchased from Tianjin Kaitong Chemical Reagent Co. Ltd., Tianjin, China. EA was purchased from Tianjin FuchenChemical Reagent Co. Ltd., Tianjin, China.

2.2. Preparation of g-C_3N_4

Graphitic carbon nitride (g-C_3N_4) was directly synthesized by pyrolysis urea in a muffle furnace. First, 20 g urea was put into an alumina crucible with a cover, then kept at 550 °C for 4 h (heating rate v = 10 °C/min). The light yellow g-C_3N_4 sample was collected. A certain amount of light yellow g-C_3N_4 powder was ultrasonically treated in deionized water for 60 min and then dried.

2.3. Synthesis of SnO_2 and g-C_3N_4/SnO_2 Materials

The procedure of material synthesis is shown in Scheme 1. In a typical synthesis of g-C_3N_4/SnO_2-5, 16 mmol stannous chloride and 16 mmol thiourea were dissolved in 120 mL deionized water and stirred for 50 h, resulting in a solution labeled A. The theoretical content of 5 wt% g-C_3N_4 was stirred and mixed with 30 mL A suspension for 30 min and then kept at 140 °C for 8 h. The g-C_3N_4/SnO_2-5 sample was obtained via centrifugation with deionized water and dried, followed by heat treatment for 2 h at 550 °C in air. According to the same procedure, pure SnO_2 and SnO_2 hybrid synthesized by changing g-C_3N_4 amount, which were labeled as g-C_3N_4/SnO_2-x (x is the theoretical mass fraction of g-C_3N_4 to SnO_2, i.e., 2.5% and 10%).

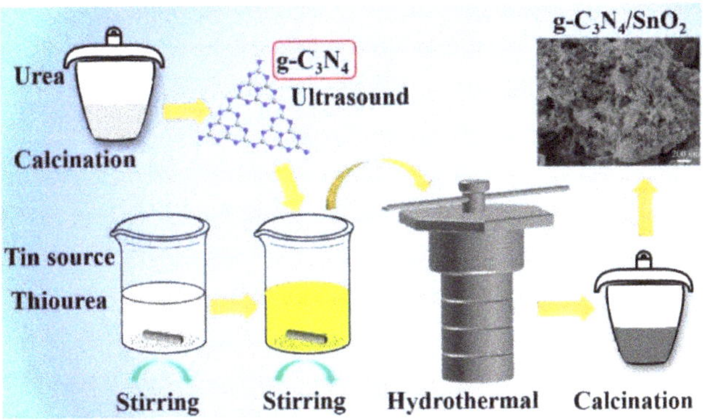

Scheme 1. Synthetic procedure of g-C_3N_4 decorated SnO_2 materials.

2.4. Characterization of g-C_3N_4/SnO_2 Materials

The crystal structures of as-synthesized g-C_3N_4/SnO_2 materials were analyzed via XRD (Shimadzu, Tokyo, Japan, XRD-6000, with high-intensity Cu Kα radiation, λ = 0.15418 nm). The morphology and microstructure were observed via SEM (Hitachi SU8010, Hitachi, Japan) and TEM (JEOL, JEM-2100F, Tokyo, Japan) at the accelerating voltage of 5 kV and 200 kV, respectively. The specific surface areas of the materials were recorded via the Bronner–Emmett–Teller (BET) method, and the pore structures were analyzed using N_2 adsorption–desorption technique (MicromeriticsTristar II 3020, Atlanta, GA, USA).

2.5. Fabrication of Sensors and VOCs Detection

The gas sensors for the tests were fabricated as follows. Briefly, the SnO_2-based materials are dispersed by mixing the as-synthesized SnO_2 or g-C_3N_4/SnO_2 powder with ethanol. A few drops of dispersion were coated thinly and evenly onto the surface of alumina ceramic tube with Au electrodes, Pt wires, and Ni-Cr wire as heater (Scheme 2a). Ceramic tubes coated with SnO_2-based materials were dried at 80 °C for 1h and then heated at 300 °C for 2 h. Finally, the element should be welded on the pedestal and then inserted in the test circuit (Scheme 2b).

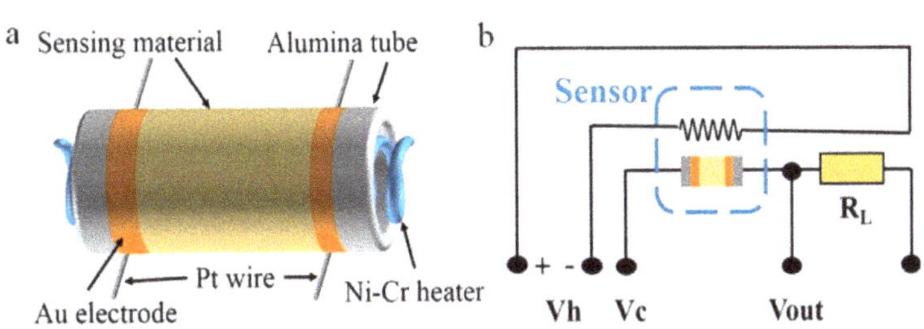

Scheme 2. The construction of gas sensor (**a**) and electric circuit of test device (**b**).

The gas sensing properties were measured using WS-30A static test system (Winsen Electronics Co. Ltd., Zhengzhou, China). The sensitivity (S) of sensor is defined as R_a/R_g, where R_a and R_g denote the resistance in air and target gas, respectively. The environment's humidity is about 35%.

3. Results and Discussion

Figure 1a illustrates the synthetic route of SnO_2 nanoparticles anchored on the g-C_3N_4 (g-C_3N_4/SnO_2) by simple hydrolysis at room temperature, hydrothermal, and heat treatment. Thiourea was used as the accelerant and stabilizer, and the yellow precursor was obtained by hydrolysis, dehydration, and partial oxidation of stannum ion. This is a slow process in which only a small amount of Sn^{2+} is oxidized to tin dioxide. In the stirring process, thioureas (CH_4N_2S), which produce amino ($-NH_2$), imino($-NH$), and mercaptan ($-SH$), react with HCl generated via hydrolysis of tin tetrachloride ($SnCl_4$), which accelerates the continuous hydrolysis process. Moreover, the coupling of Sn^{4+} with sulfhydryl groups plays a stabilizing role, contributing to the formation of stable nanoscale particles. During the hydrothermal process, g-C_3N_4 is agglomerated on the surface, forming a sheet structure of g-C_3N_4 coated with tin dioxide precursor nanoparticles. Finally, the g-C_3N_4 composite coated with SnO_2 nanoparticles was prepared via drying and calcination.

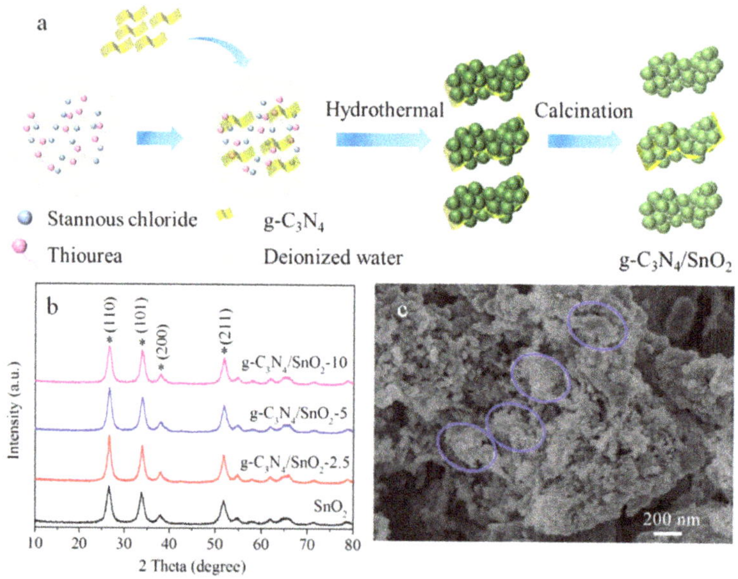

Figure 1. (**a**) Schematic diagram of the formation process; (**b**) XRD patterns of as-synthesized g-C_3N_4/SnO_2 (* denotes SnO_2 characteristic peaks); (**c**) SEM image of g-C_3N_4/SnO_2-5 material.

The phase structure and crystal structure were characterized via XRD. Figure S1 shows (100) and (002) planes corresponding to g-C_3N_4 (JCPDS, No. 87-1526) at 13.0° and 27.6°, respectively. Figure 1b shows the XRD patterns of the pure SnO_2 nanoparticles and g-C_3N_4/SnO_2 composites. The positions of 26.6°, 33.9°, 37.9°, and 51.8°, where the diffraction peaks are located, corresponding to (110), (101), (200), and (211) planto of the tetragonal rutile structure SnO_2 (JCPDS, No. 41-1445), respectively. However, no other peaks of impurity, including g-C_3N_4 diffraction peaks, were detected in the XRD patterns, which may be due to the trace amounts of g-C_3N_4 in the g-C_3N_4/SnO_2 composite materials, which did not reach the detection limit of the XRD. It is also possible that the diffraction peak of g-C_3N_4 near 27.6° and that of SnO_2 at 26.6°overlaps. There is no obvious diffraction peak shift in the diffraction pattern, indicating that g-C_3N_4 and SnO_2 do not exist independently, which means that n-n heterojunctions form on the contact interface between the two materials. Figure 1c shows the morphology of the g-C_3N_4/SnO_2-5 composite. SnO_2 nanoparticles are dispersed on the nanosheets of g-C_3N_4 and form a supported particle group (shown with a blue oval). The SEM images of g-C_3N_4, SnO_2, and other g-C_3N_4/SnO_2 show layer, nanoparticles aggregation, and nanosheets-supported particle group (Figure S2).

In order to analyze the chemical composition and state of g-C_3N_4/SnO_2-5 material, XPS characterization was performed. Figure S3 shows the full XPS spectrum of g-C_3N_4/SnO_2-5 material, in which the peaks are related to C, N, O, and Sn. It indicates the successful synthesis of g-C_3N_4/SnO_2 composite material, which indirectly proves the existence of heterojunction between g-C_3N_4 and SnO_2.

Figure 2a shows the high-resolution spectrum of C 1s, and the binding energies of 284.5, 285.1, 286.2, and 286.9 eV belong to the C-C bond of sp^2, C-N, C-NH_2, and C=NH in the triazine ring, respectively [30]. Figure 2b shows the spectrum of N 1s, and the binding energies of 399.7, 401.7, 402.00 eV and 402.65 eV belong to sp^2 hybrid nitrogen (C-N=C), tertiary nitrogen (N-$(C)_3$), N-O and C-N-H structure, respectively [31]. It indicates the existence of g-C_3N_4 in the g-C_3N_4/SnO_2-5 composite. Figure 2c is the spectrum of O 1s, and the peaks appear at 530.95 and 531.8 eV. The peak at 530.95 eV is attributed to the lattice oxygen of SnO_2, and the peak at 531.8 eV is attributed to defective oxygen. In Figure 2d, the peaks of 495.5 eV and 487.15 eV correspond to Sn $3d_{3/2}$ and Sn $3d_{5/2}$, indicating that Sn exists in the form of Sn^{4+}. The results show that the material is composed of g-C_3N_4 and SnO_2.

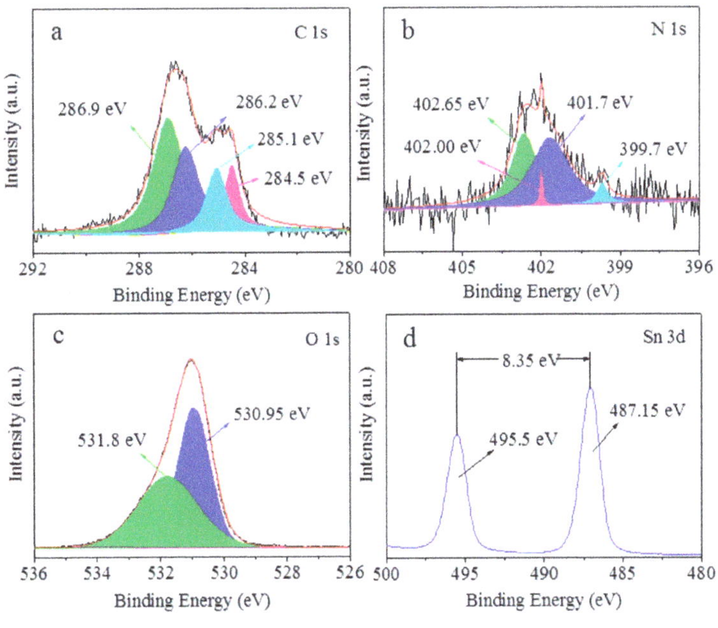

Figure 2. XPS spectra of the C 1s (**a**), N 1s (**b**), O 1s (**c**), and Sn 3d (**d**) of as-synthesized g-C_3N_4/SnO_2-5 nanomaterial.

Figure 3a shows that SnO_2 and g-C_3N_4/SnO_2-5 materials possess characteristics of type IV isotherms, and the specific surface areas of SnO_2 and g-C_3N_4/SnO_2-5 are 24.5 and 25.7 m^2 g^{-1}, respectively. When the relative pressure P/P_0 is between 0.4 and 1.0, the adsorption capacity rises rapidly, showing a H3-type hysteresis loop, indicating that a narrow-slit aggregate structure is formed between g-C_3N_4 and SnO_2 nanoparticles, and there are abundant mesoporous structures. Figure 3b shows the pore size distribution of SnO_2 and g-C_3N_4/SnO_2-5. It can be seen from the figure that both materials have smaller pore size distributions, and the average pore sizes of SnO_2 and g-C_3N_4/SnO_2-5 are 8.2 nm and 10.8 nm, respectively. The increased specific surface area is beneficial to improve the gas sensing property.

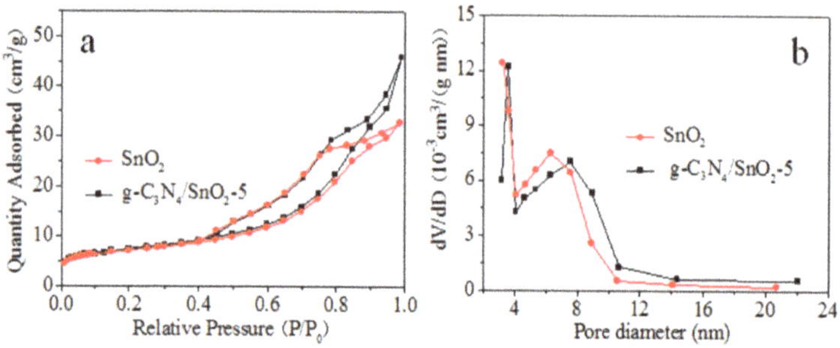

Figure 3. N_2 adsorption–desorption isotherms (**a**) and pore size distribution (**b**) of as-synthesized g-C_3N_4/SnO_2−5.

The gas-sensing properties of these materials in the atmosphere were investigated. The working temperature has a great influence on the reaction kinetics of gas molecules and oxygen on the surface of the material, and it also has an effect on the gas-sensing behavior of the material. The optimal working temperature of SnO_2 and g-C_3N_4/SnO_2 materials for detecting ethanolamine (EA) was explored. Figure 4a shows the response of the sensor to 100 ppm EA gas in the test range of 200–280 °C. The optimal operating temperature is 240 °C for the four samples obtained by changing the g-C_3N_4 content. The response to EA increases with increasing temperature when the temperature is less than 240 °C. The possible reason for this is that when the optimal working temperature is lower than 240 °C, the activity of adsorbed oxygen on the surface of SnO_2-based materials gradually increases with the increase in temperature, and the sensitivity to EA shows an upward trend. When the optimal working temperature is higher than 240 °C, the oxygen molecules, which are adsorbed on the surface of SnO_2-based materials, are rapidly desorbed before reaction with EA molecules as temperature increases, resulting in a decrease in the sensitivity to EA. It can also be seen that the sensitivity to EA increases first and then decreases with the increase in g-C_3N_4 content in SnO_2-based materials at the optimum working temperature. The g-C_3N_4/SnO_2-5 shows the best gas sensitivity. The responses of pure SnO_2, g-C_3N_4/SnO_2-2.5, g-C_3N_4/SnO_2-5, and g-C_3N_4/SnO_2-10 to 100 ppm EA are 16.5, 18.7, 23.3, and 22, respectively. The optimal amount of g-C_3N_4 doped in SnO_2 is conducive to better dispersion of the composite. The formation of heterojunctions and the increase in defects, which are between the 2D sheet g-C_3N_4 and SnO_2 nanoparticles, provide more active sites for gas adsorption and reaction, and active sites improve the gas sensitivity. However, when the content of g-C_3N_4 in the composite exceeds a certain threshold (5 wt%), g-C_3N_4 nanosheets can be connected to form a micro-bridge on the surface. The micro-bridge may result in the reduction of the resistance of the g-C_3N_4/SnO_2 composite, finally leading to a decrease in the performance of gas sensing.

Selectivity is also one of the important parameters to measure the gas-sensing performance of materials. At the optimum operating temperature of 240 °C, the sensing properties of SnO_2-based materials with different g-C_3N_4 contents to 100 ppm of volatile organic compounds are tested. The results indicate good selectivity to EA (Figure 4b), which may be due to the fact that the bond energy of the C-N bond in the EA molecule is 307 kJ mol^{-1}, indicating EA has strong reducibility compared with other volatile organic compounds. Therefore, as-prepared material exhibits better selectivity to EA. In addition, selectivity is dependent on the type of oxides and structure of composites. This is the reason that ZnO/g-C_3N_4 has no response to EA and only to DMA [21].

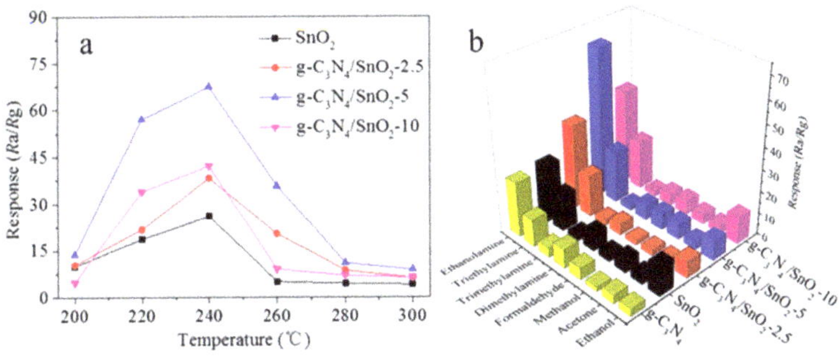

Figure 4. (**a**) The response to 100 ppm EA at different temperatures; (**b**) selectivity to EA on exposure to various volatile gases.

Figure 5a shows the transient response–recovery characteristics of SnO_2 and $g\text{-}C_3N_4/SnO_2$ to EA in the concentration range of 1 to 100 ppm at 240 °C. The results show that materials have a fast response and recovery ability for EA detection in a wide concentration range, and the response increases with increasing concentration. Notably, these materials exhibit an obvious response to 1–10 ppm of low-concentration EA (Figure 5b). Figure 6a displays a good linear relationship between response and concentration of EA. The detection limit of the $g\text{-}C_3N_4/SnO_2$-5 sensor to EA is 294 ppb. To explore the stability of the $g\text{-}C_3N_4/SnO_2$-5 sensor, the responses to low-concentration EA were recorded within nine days (Figure 6b), and the response value to 100 ppm EA at interval of 2 days reaches around 50 after 5 days, indicating $g\text{-}C_3N_4/SnO_2$-5 sensor has better stability to EA.

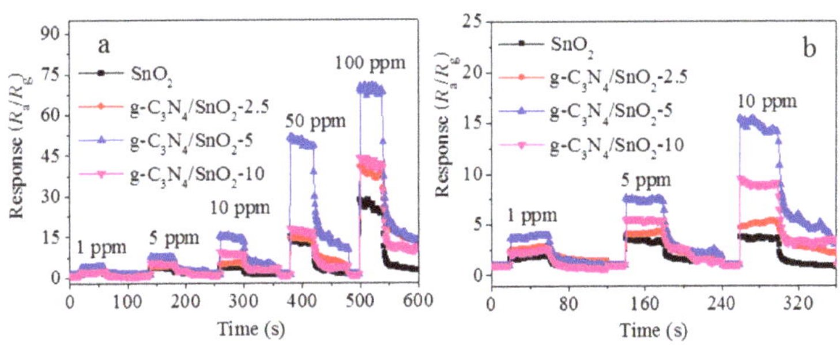

Figure 5. Transient response–recovery profiles of composite material at 240 °C after exposure to (**a**) different concentration EA and (**b**) low concentration (1–5 ppm) EA.

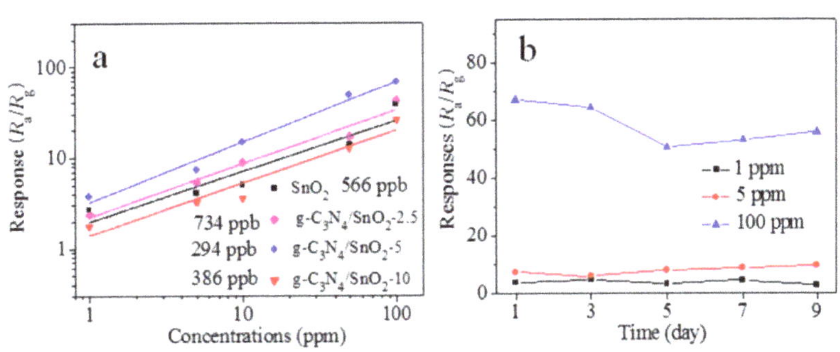

Figure 6. (**a**) Linear relationships between response and concentration of EA; (**b**) stability of material within nine days.

The effect of band structure changes on sensing performance has been analyzed by calculating the band gap energy of as-synthesized g-C_3N_4/SnO_2 from UV-vis results (Figure S4 and Figure 7a,c), and the linear extrapolation formula is referred to as our previous work [21]. The band gap energy is 2.92 and 2.86 eV for g-C_3N_4 and SnO_2, and 2.81 eV for g-C_3N_4/SnO_2-5 (Figure 7b,d). The results show that less energy can drive electrons to transfer from the valence to the conduction band, and the increased free electrons are helpful in improving the response to EA.

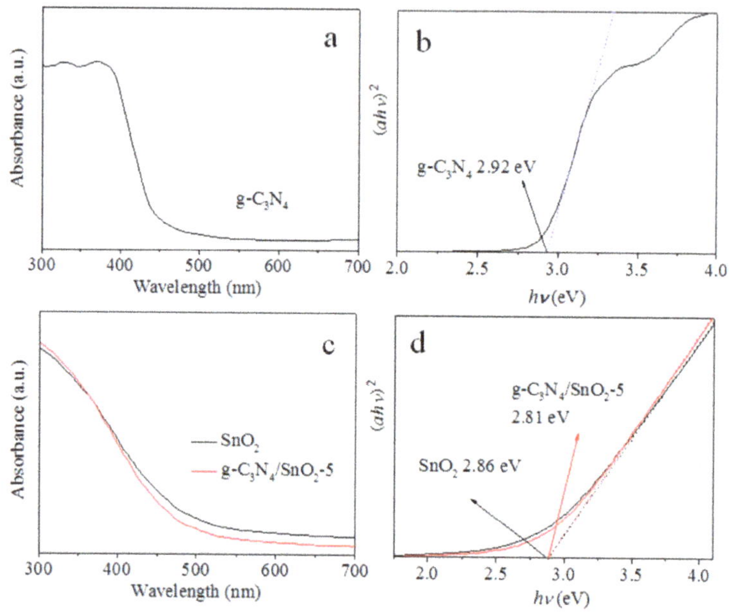

Figure 7. UV-vis DRS (**a**,**c**) and the band gap energy (**b**,**d**) of g-C_3N_4, SnO_2, and g-C_3N_4/SnO_2-5 materials (The dashed line is tangent extension line).

The sensing mechanism of n-type semiconductor materials conforms to the space charge layer model. When the sensor is moved from air to VOCs gases, the electron depletion layer formed on the surface of g-C_3N_4/SnO_2-5 changes, which can explain the gas sensitivity of g-C_3N_4/SnO_2-5 [30]. Therefore, the effect of different environmental conditions on the depletion layer width of g-C_3N_4/SnO_2-5 material is studied to explain the resistance variation of g-C_3N_4/SnO_2-5 sensor in air and EA. Figure 8 illustrates the formation of the electron depletion layer of the g-C_3N_4/SnO_2-5 sensor in air and EA gas. When g-C_3N_4/SnO_2-5 sensing material is in the air, oxygen molecules are adsorbed on the surface of g-C_3N_4/SnO_2-5 sensing material to form surface adsorbed oxygen (Figure 8), which will capture free electrons from the conduction band of g-C_3N_4/SnO_2-5 sensing material to form surface adsorbed oxygen ions ($O^{\alpha-}$). There is a strong interaction between gas and surface [32]. A wide electron depletion layer is generated on the surface of g-C_3N_4/SnO_2-5, which causes a decrease in the carrier concentration and an increase in the resistance of the gas sensor. The formulas are as follows:

$$O_2\ (gas) = O_2\ (ads) \tag{1}$$

$$O_2\ (ads) + \alpha e^- = O^{\alpha-}\ (ads) \tag{2}$$

When the gas sensor is exposed to EA gas, it will react with chemisorbed oxygen (Figure 8). The reaction is represented as follows [23]:

$$N(CH_2CH_3)_3\ (gas) + O^{\alpha-}\ (ads) = CO_2 + H_2O + N_2 + \alpha e^- \tag{3}$$

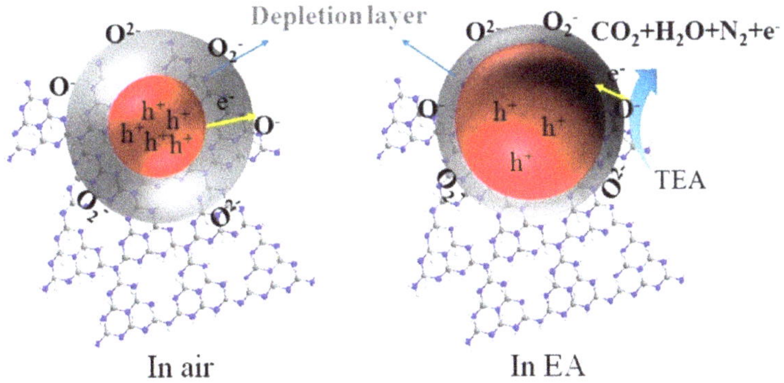

Figure 8. Schematic diagram of EA gas sensing mechanism of $g-C_3N_4/SnO_2$ material.

The captured electrons are released back into the conduction band of g-C_3N_4/SnO_2-5, and the thickness of the electron depletion layer of g-C_3N_4/SnO_2-5 material decreases, which results in an increase in the carrier concentration and decrease in the resistance of the g-C_3N_4/SnO_2-5 sensor. When g-C_3N_4/SnO_2-5 is exposed to air again, the chemisorption oxygen is produced, and the electron depletion layer becomes wide. Compared with pure SnO_2, g-C_3N_4/SnO_2-5 exhibits superior gas sensitivity to EA. The possible reason is that g-C_3N_4 acts as a matrix to disperse SnO_2 nanoparticles and prevent aggregation of SnO_2 nanoparticles, which facilitates the adsorption and diffusion of oxygen molecules and EA molecules, and thus enhances the reaction of EA molecules with adsorbed oxygen ions, which are dependent on types of nano or microparticles and structure of sensing materials. In addition, the n-n heterojunction formed between the g-C_3N_4 and SnO_2 interface. When EA molecules cross through this interface, the electrical properties change at the heterojunction. The electrons transfer from the conduction band of g-C_3N_4 to the conduction band of SnO_2, and the electrons and holes separate and finally a high potential barrier is created. The heterostructure between g-C_3N_4 and SnO_2 may inhibit the recombination of electron–hole pairs and promote the rapid transfer of electrons from EA to the surface of g-C_3N_4/SnO_2. As a result, the electrical conductivity of the heterojunction increases, which results in a high response.

To obtain theoretical evidence of energy band structure and conduction, model construction and calculation are conducted, and the details are put in Supporting Information according to similar methods [30,33,34]. Figure 9a–c show the structure models of g-C_3N_4, SnO_2, and g-C_3N_4/SnO_2 materials, and the calculated energy band structures are shown in Figure 9d–f. The band gaps of the pure g-C_3N_4 and SnO_2 are marked in Figure 9d–f as 1.26 and 1.00 eV, respectively. It can also be seen that the top of the valence band (VBT) and the bottom of the conduction band (CBB) of g-C_3N_4 are centered at the Γ point and the A point, respectively (Figure 9d), while the VBT and CBB of SnO_2 are located at the S and X point (Figure 9e), which imply that g-C_3N_4 is an indirect band gap semiconductor and SnO_2 was a direct band gap semiconductor. [30] For the g-C_3N_4/SnO_2 material, it is clear that the energy band curves are more fine and close after the heterojunction is formed between SnO_2 and g-C_3N_4 (Figure 9f), leading to the electrons being reconstructed, and the band gap is closed to 0.0 eV (Figure 9f), indicating that the conductivity of g-C_3N_4/SnO_2 material has been improved significantly.

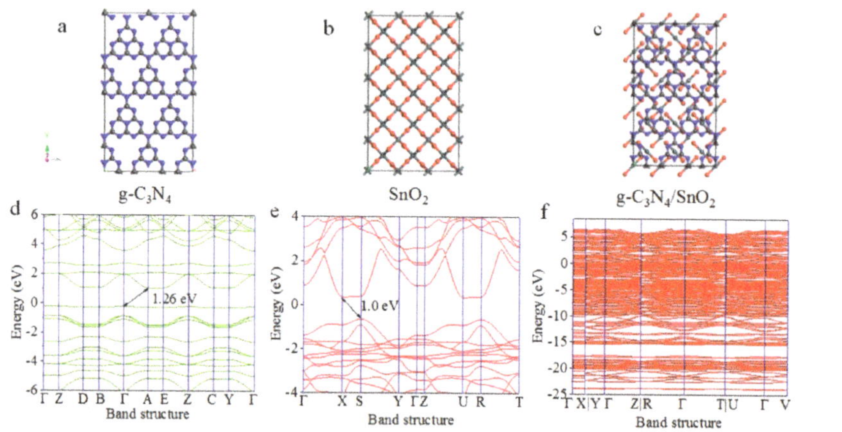

Figure 9. The crystal structure of (**a**) g−C_3N_4, (**b**) SnO_2, and (**c**) g−C_3N_4/SnO_2. The calculated energy band structure of (**d**) g−C_3N_4, (**e**) SnO_2, and (**f**) g−C_3N_4/SnO_2.

To further explore the change in conductivity of g-C_3N_4/SnO_2 material, the total density of states (TDOS) and partial density of states (PDOS) of g-C_3N_4, SnO_2, and g-C_3N_4/SnO_2 were calculated, and the results are summarized in Figure 10. The main contribution of the valence band edge is C2p and N2p orbitals, and the conduction band edge is C2p and N2p orbitals for pure g-C_3N_4 (Figure 10a). The top of the SnO_2 valence band is Sn5s, 5p orbitals, and the bottom of the conduction band is Sn5p,4d and O2p orbitals (Figure 10b). For comparison, the DOS peaks shift to the left, and the peak at −15 eV increases. Notably, a new peak occurs near the Fermi level after SnO_2 modification (Figure 10c,d), showing that there is a strong orbital hybridization between SnO_2 and g-C_3N_4, and there is significant electron exchange between them.

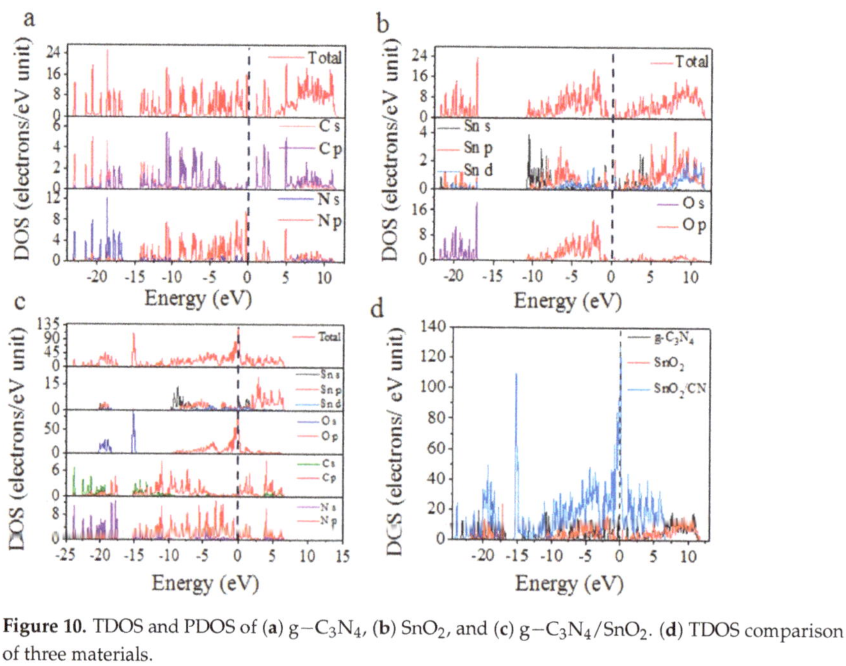

Figure 10. TDOS and PDOS of (**a**) g−C_3N_4, (**b**) SnO_2, and (**c**) g−C_3N_4/SnO_2. (**d**) TDOS comparison of three materials.

4. Conclusions

g-C_3N_4/SnO_2 material was synthesized via a hydrothermal and pyrolysis route. Structural and morphology characterizations show that g-C_3N_4 acts as a matrix of SnO_2 to prevent the particles from agglomeration. As-obtained g-C_3N_4/SnO_2 exhibited higher response and selectivity to EA compared with pure SnO_2. The response to 100 ppm EA reaches ca. 68 at 240 °C, and it is superior to that of SnO_2. Enhanced sensing performance is attributed to the morphology, energy band, and structure of composites consisting of nanoparticles anchored on g-C_3N_4 layers, and the heterojunctions formed between SnO_2 and g-C_3N_4 interface resulting in fast electron transfer and increasing interstitial lattice defect sites, which can adsorb more O_2 molecules and provide more reactive oxygen species to react with EA molecules. This research provides a strategy to prepare new composite material for the effective detection of EA and other related organoamines released from tobacco and chemical industries.

Supplementary Materials: The following supporting information can be downloaded at: https://www.mdpi.com/article/10.3390/chemosensors11050296/s1, Figure S1: XRD patterns of g-C_3N_4; Figure S2: SEM images of as-synthesized materials (a) g-C_3N_4,(b) SnO_2, (c) g-C_3N_4/SnO_2-2.5, and (d) g-C_3N_4/SnO_2-10; Figure S3: XPS spectra of the full range spectrum of g-C_3N_4/SnO_2-5 material; Figure S4: (a) UV–Vis absorption spectra of g-C_3N_4, SnO_2, and g-C_3N_4/SnO_2 and (b) bandgap energy of g-C_3N_4, SnO_2, and g-C_3N_4/SnO_2.

Author Contributions: J.L.: Experimental work, data curation, and writing—original draft. K.X.: Data analysis and draft revision. Y.W.: Data analysis and draft revision. R.Z.: Supervision, conceptualization, and writing—review and editing. Y.S.: data curation and draft review and revision. J.D.: Experimental design, supervision, conceptualization, and writing—revision and editing. All authors have read and agreed to the published version of the manuscript.

Funding: This research is funded by the National Natural Science Foundation of China (No. 51572185), the Key R&D program of Shanxi Province (International Cooperation, No. 201903D421079), and the National Natural Science Foundation of Shanxi (No. 20210302123173).

Data Availability Statement: Not applicable.

Conflicts of Interest: The authors declare that they have no known competing financial interests.

References

1. Lv, D.; Lu, S.; Tan, X.; Shao, M.; Wang, L. Source profiles, emission factors and associated contributions to secondary pollution of volatile organic compounds (VOCs) emitted from a local petroleum refinery in Shandong. *Environ. Pollut.* **2021**, *3*, 116589. [CrossRef] [PubMed]
2. Hung, F.H.; Tsang, K.F.; Wu, C.K.; Liu, Y.; Wan, W.H. An Adaptive Indoor Air Quality Control Scheme for Minimizing Volatile Organic Compounds Density. *IEEE Access* **2020**, *8*, 22357–22365. [CrossRef]
3. Qin, J.; Wang, X.; Yang, Y.; Qin, Y.; Shi, S.; Xu, P.; Chen, R.; Zhou, X.; Tan, J.; Wang, X. Source apportionment of VOCs in a typical medium-sized city in North China Plain and implications on control policy. *J. Environ. Sci.* **2021**, *107*, 26–37. [CrossRef] [PubMed]
4. Yuan, Z.; Han, E.; Meng, F.; Zuo, K. Detection and identification of volatile organic compounds based on temperature-modulated ZnO sensors. *IEEE Trans. Instrum. Meas.* **2020**, *69*, 4533–4544. [CrossRef]
5. Wang, X.; Chen, F.; Yang, M.; Guo, L.; Xie, N.; Kou, X.; Song, Y.; Wang, Q.; Sun, Y.; Lu, G. Dispersed WO_3 nanoparticles with porous nanostructure for ultrafast toluene sensing. *Sens. Actuators B Chem.* **2019**, *289*, 195–206. [CrossRef]
6. Meng, F.; Zheng, H.; Chang, Y.; Zhao, Y.; Li, M.; Wang, C.; Sun, Y.; Liu, J. One-stepsynthesis of Au/SnO_2/RGO nanocomposites and their VOC sensing properties. *IEEE Trans. Nanotechnol.* **2018**, *17*, 212–219. [CrossRef]
7. Zhang, P.; Zhang, L.; Xu, H.; Xing, Y.; Chen, J.; Bie, L. Ultrathin CeO_2 nanosheets as bifunctional sensing materials for humidity and formaldehyde detection. *Rare Met.* **2021**, *40*, 1614–1621. [CrossRef]
8. Bhat, P.; Kumar, S.N. Evaluation of IDE-based flexible thin film ZnO sensor for VOC sensing in a custom designed gas chamber at room temperature. *J. Mater. Sci.-Mater. Electron.* **2022**, *33*, 1529–1541. [CrossRef]
9. Zhang, H.; Shi, W.; Gao, N.; Zhao, R.; Ahmed, M.; Li, J.; Du, J.; Asefa, T. Highly sensitive and selective gas-phase ethanolamine sensor by doping sulfur into nanostructured ZnO. *Sens. Acutators B Chem.* **2019**, *296*, 126633. [CrossRef]
10. Zhang, K.; Xie, K.; Ahmed, M.; Chai, Z.; Zhao, R.; Li, J.; Du, J. Cr-doped SnO_2 microrod adhering nanoparticles for enhanced triethylamine sensing performance. *Mater. Lett.* **2022**, *312*, 131684. [CrossRef]
11. Shang, Y.; Shi, W.; Zhao, R.; Ahmed, M.; Li, J.; Du, J. Simple self-assembly of 3D laminated CuO/SnO_2 hybrid for the detection of triethylamine. *Chin. Chem. Lett.* **2020**, *31*, 2055–2058. [CrossRef]

12. Du, J.; Xie, Y.; Yao, H.; Zhao, R.; Li, J. Size-dependent gas sensing and selectivity of SnO_2 quantum dots toward volatile compounds. *Appl. Surf. Sci.* **2015**, *346*, 256–262. [CrossRef]
13. Acharyya, S.; Jana, B.; Nag, S.; Saha, G.; Guha, P.K. Single resistive sensor for selective detection of multiple VOCs employing SnO_2 hollow spheres and machine learning algorithm: A proof of concept. *Sens. Acutators B Chem.* **2020**, *321*, 128484. [CrossRef]
14. Meng, F.; Ji, H.; Yuan, Z.; Chen, Y.; Gao, H. Dynamic Measurement and Recognition Methods of SnO_2 Sensor to VOCs under Zigzag-rectangular Wave Temperature Modulation. *IEEE Sens. J.* **2021**, *21*, 10915–10922. [CrossRef]
15. Du, J.; He, S.; Zhao, R.; Chen, S.; Guo, T.; Wang, H. Facile self-assembly of SnO_2 nanospheres for volatile amines gas sensing. *Mater. Lett.* **2017**, *186*, 318–321. [CrossRef]
16. Zhang, R.; Xu, Z.; Zhou, T.; Fei, T.; Wang, R.; Zhang, T. Improvement of gas sensing performance for tin dioxide sensor through construction of nanostructures. *J. Colloid Interface Sci.* **2019**, *557*, 673–682. [CrossRef]
17. Ma, X.; Gao, R.; Zhang, T.; Sun, X.; Li, T.; Gao, S.; Zhang, X.; Xu, Y.; Cheng, X.; Huo, L. Mesoporous SnO_2 nanospheres sensor for fast detection of HCHO and its application in safety detection of aquatic products. *Sens. Actuators B Chem.* **2022**, *374*, 132844. [CrossRef]
18. Din, S.U.; Ul Haq, M.; Sajid, M.; Khatoon, R.; Chen, X.; Li, L.; Zhang, M.; Zhu, L. Development of high-performance sensor based on NiO/SnO_2 heterostructures to study sensing properties towards various reducing gases. *Nanotechnology* **2020**, *31*, 395502. [CrossRef]
19. Dai, H.; Ding, J.; Chen, H.; Fu, H. Improvement of ethanolamine sensing performance based on Au-modified ZnO rod-like nanoflowers. *Mater. Lett.* **2023**, *340*, 134183. [CrossRef]
20. Zou, Z.; Zhao, Z.; Zhang, Z.; Tian, W.; Yang, C.; Jin, X.; Zhang, K. Room-temperature optoelectronic gas sensor based on core-shellg-$C_33N_4@WO_3$ heterocomposites for efficient ammonia detection. *Anal. Chem.* **2023**, *95*, 2110–2118. [CrossRef]
21. Bai, K.; Cui, Z.; Li, E.; Ding, Y.; Zheng, J.; Liu, C.; Zheng, Y. Adsorption of gas molecules on group III atoms adsorbed g-C_3N_4: A first-principles study. *Vacuum* **2020**, *175*, 109293. [CrossRef]
22. Chen, M.; Guo, C.; Hou, S.; Lv, J.; Xu, J. A novel Z-scheme AgBr/P-g-C_3N_4 heterojunction photocatalyst: Excellent photocatalytic performance and photocatalytic mechanism for ephedrine degradation. *Appl. Catal. B-Environ.* **2020**, *266*, 118614. [CrossRef]
23. Li, L.; Huang, Z.; Li, Z.; Li, H.; Wang, A. Defect-rich porous g-C_3N_4 nanosheets photocatalyst with enhanced photocatalytic activity. *J. Mater. Sci.-Mater. Electron.* **2021**, *32*, 6465–6474. [CrossRef]
24. Liu, P.; Zhang, Z.; Hao, R.; Huang, Y.; Liu, K. Ultra-highly stable zinc metal anode via 3D-printed g-C_3N_4 modulating interface for long life energy storage systems. *Chem. Eng. J.* **2020**, *403*, 126425. [CrossRef]
25. Lin, Z. Preparation of ZnO/layered g-C_3N_4 and their n-butanol gas sensing properties. *Mater. Sci.* **2020**, *10*, 278–286.
26. Yuan, B.; Wang, Y.; Elnaggar, A.Y.; Azab, I.H.E.; Huang, M.; Mahmoud, M.H.H.; El-Bahy, S.M.; Guo, M. Physical vapor deposition of graphitic carbon nitride (g-C_3N_4) films on biomass substrate: Optoelectronic performance evaluation and life cycle assessment. *Adv. Compos. Hybrid. Mater.* **2022**, *5*, 813–822. [CrossRef]
27. Cao, J.; Qin, C.; Wang, Y.; Zhang, B.; Gong, Y.; Zhang, H.; Sun, G.; Bala, H.; Zhang, Z. Calcination method synthesis of SnO_2/g-C_3N_4 composites for a high-performance ethanol gas sensing application. *Nanomaterials* **2017**, *7*, 98. [CrossRef]
28. Wang, H.; Bai, J.; Dai, M.; Liu, K.; Liu, Y.; Zhou, L.; Liu, F.; Li, F.; Gao, Y.; Yan, X.; et al. Visible light activated excellent NO_2 sensing based on 2D/2D ZnO/g-C_3N_4 heterojunction composites. *Sens. Actuators B Chem.* **2020**, *304*, 127287. [CrossRef]
29. Niu, J.; Wang, L.; Xu, J.; Jin, H.; Hong, B.; Jin, D.; Peng, X.; Ge, H.; Wang, X. Mesoporous Co_3O_4 nanowires decorated with g-C_3N_4 nanosheets for high performance toluene gas sensors based on p-n heterojunction. *Mater. Chem. Phys.* **2023**, *293*, 126980. [CrossRef]
30. Xie, K.; Wang, Y.; Zhang, K.; Zhao, R.; Chai, Z.; Du, J.; Li, J. Controllable band structure of ZnO/g-C_3N_4 aggregation to enhance gas sensing for the dimethylamine detection. *Sens. Actuators Rep.* **2022**, *4*, 100084. [CrossRef]
31. Ni, J.; Wang, Y.; Liang, H.; Kang, Y.; Liu, B.; Zhao, R.; Wang, Y.; Shuai, X.; Shang, Y.; Guo, T.; et al. Facile template-free preparation of hierarchically porous graphitic carbon nitrides as high-performance photocatalyst for degradation of methyl violet dye. *ChemistrySelect* **2021**, *6*, 7130–7135. [CrossRef]
32. Stefan, K.; Chris, B. Atomistic Descriptions of gas–surface interactions on tin dioxide. *Chemosensors* **2021**, *9*, 270.
33. Wang, V.; Xu, N.; Liu, J.C. VASPKIT: A user-friendly interface facilitating high-throughput computing and analysis using VASP code. *Computer Phys. Commun.* **2021**, *267*, 108033. [CrossRef]
34. Liangruksa, M.; Sukpoonprom, P.; Junkaew, A.; Photaram, W.; Siriwong, C. Gas sensing properties of palladium-modified zinc oxide nanofilms: A DFT study. *Appl. Surf. Sci.* **2021**, *544*, 148868. [CrossRef]

MDPI AG
Grosspeteranlage 5
4052 Basel
Switzerland
Tel.: +41 61 683 77 34

Chemosensors Editorial Office
E-mail: chemosensors@mdpi.com
www.mdpi.com/journal/chemosensors

www.ingramcontent.com/pod-product-compliance
Lightning Source LLC
Chambersburg PA
CBHW061843210926
51850CB00006B/961
* 9 7 8 3 7 2 5 8 2 2 4 5 4 *